MINDS ON SCIENCE

Middle and Secondary School Methods

JACK HASSARD

Georgia State University

HarperCollinsPublishers

This book is dedicated to
C. Stuart Brewster
for your encouragement and support

Executive Editor: Christopher Jennison
Full-Service Manager: Michael Weinstein
Production Coordinator: Cindy Funkhouser
Project Coordinator, Text and Cover Design:
  Caliber/Phoenix Color Corp.
Cover Photo: Lou Jones/The Image Bank, Inc.
Production Manager: Priscilla Taguer
Compositor: Caliber/Phoenix Color Corp.
Printer and Binder: Malloy Lithographers
Cover Printer: The Lehigh Press, Inc.

MINDS ON SCIENCE

Library of Congress Cataloging-in-Publication Data

Hassard, Jack.
   Minds on science : middle and secondary school
methods / Jack Hassard.
       p. cm.
  Includes bibliographical references and index.
  ISBN 0-06-500019-6
  1. Science—Study and teaching (Secondary) I. Title.
Q181.H372 1992
507.1'2—dc20   91–43797
                                  CIP

95 94 93   9 8 7 6 5 4 3 2

BRIEF CONTENTS

CONTENTS

CHAPTER

3

THE GOALS AND HISTORY OF SCIENCE EDUCATION

CHAPTER

4

THE SCIENCE CURRICULUM: MIDDLE SCHOOL PATTERNS AND PROGRAMS

CHAPTER

5

THE SCIENCE CURRICULUM: HIGH SCHOOL PATTERNS AND PROGRAMS

CHAPTER
6

SCIENCE, TECHNOLOGY, AND SOCIETY IN THE CLASSROOM

CHAPTER
7

MODELS OF SCIENCE TEACHING

CHAPTER
8

STRATEGIES FOSTERING THINKING IN THE SCIENCE CLASSROOM

CHAPTER
9

DESIGNING AND ASSESSING SCIENCE UNITS AND COURSES OF STUDY

CHAPTER
10

FACILITATING LEARNING IN THE SCIENCE CLASSROOM

CHAPTER

11

SCIENCE
FOR ALL

Inquiry Activities

PREFACE

Helping students understand ideas about the world of science through personal experience is one of the fundamental goals of science teaching in today's schools. Helping you learn how to do this is the central purpose of *Minds on Science*.

The way in which teachers approach the practice of science teaching is rooted in core beliefs about the purpose and role of science in society, and about how students come to learn science. Science is understood when students have the opportunity to engage in active inquiry, to collaborate with peers to discuss their ideas, to use their tools of learning—in short to be immersed in an experiential learning environment.

The preparation of teachers who are able to approach science in this way should be based on science teacher education that is experiential, inquiry oriented, and reflective. *Minds on Science* has been written and designed to foster these ideas.

Minds on Science is a science methods book designed for preservice and in-service middle school and high school science teachers. Science education for the 1990s will be characterized by a period of innovation and change. At present there are several large-scale projects (Project 2061 of the American Association for the Advancement of Science and the Scope, Sequence and Coordination Project of the National Science Teachers Association, for example) aimed at reforming science education, with a special effort directed at middle and high school science. *Minds on Science* is a methods book designed to help prepare a new cadre of science teachers who can cope with the demands of science teaching as well as be prepared to be the leaders in science education in the twenty-first century.

Science teaching can be perceived as a performing art. Although you will be introduced to and asked to learn a variety of teaching skills, the craft of teaching is a wholistic one, in which the teacher helps students "invent" and "construct" ideas. This is accomplished when students connect hands-on science experiences with reflective or minds-on science experiences. The facilitation of this method is the heart of the art of teaching.

Minds on Science is organized around a number of major unifying themes of teaching clustered in three areas: the nature of science, the nature of teaching, and the nature of learning. These include themes such as how students learn, the goals of science education, curriculum, models of teaching, strategies of teaching, planning and evaluation, and the facilitation of learning. A theme that runs throughout the entire book and is used as the underlying theme in the concluding chapter of the book is that of "science for all."

Minds on Science provides the science teacher with a wealth of teaching strategies based not only on contemporary research but on the basis of the wisdom of practice. Readers of *Minds on Science* will be engaged in inquiry activities (pedagogical activities designed to explore science teaching) and will participate in a variety of learning experiences described in the Science Teaching Gazette.

Each chapter is divided into two sections: the first focuses on the content of a unifying theme; the second part is a newspaperlike feature called the Science Teaching Gazette, containing a variety of strategies to help to learn about science teaching.

A number of special features showcase important ways to learn about the art of science teaching. In the first part of each of the chapters, I have designed inquiry activities that are integrated into the content of the text, enabling you to reflect on the important concepts of science teaching in the minds-on-science tradition. There are 28 of these engaging activities. They integrate Piagetian and Vygotskian principles of learning by engaging you with new ideas, and then providing you with an opportunity to reflect on these ideas in peer learning teams. You might take a look at the list of inquiry activities on page xii to get the flavor of these experiences.

The Science Teaching Gazette sections contain several interesting learning tools. These include case studies, which encourage reflective thinking on a number of science teaching problems, excerpts from the literature of science and science teaching, interviews with practicing science teachers, and a research matters column, which gives practical tips based on research in science teaching. Also included are cooperative learning and individual activities under the headings Problems and Extensions, and Think Pieces. Finally, each Science Teaching Gazette contains an annotated collection of science teaching resources.

One feature I think you will find extremely powerful is the Science Teachers Talk column in each Science Teaching Gazette. These are craft-talk interviews with several practicing science teachers. These "wisdom-of-practice" sections enable us to learn about important aspects of science teaching by reading the thoughtful and compelling ideas from the minds of outstanding science teachers.

The work of the science education research and development community is acknowledged throughout the book. One feature of the book, however, is a special section in the Science Teaching Gazette entitled Research Matters that describes practical actions science teachers can take based on the results of research.

Science education is entering into another period of reform, at the national as well as at the global level. These reform efforts are not only documented in the book; they are integrated into the philosophy of science teaching that is part of the whole book. I have also tried to go beyond the boarders of the United States when thinking about and describing science education. You will find discussions of the nature of science education in a number of countries including China, Japan, and Nigeria. As you begin your career as a science teacher in the 1990s, I hope that you will see yourself connected to a community of science teachers around the world.

A book of this kind is impossible to write without the help of many people. I wish to acknowledge and thank a number of people who helped me write this one.

First, I couldn't have done this without my wife, Mary-Alice, who gave me support, love, and encouragement. Dr. Dvorah Naveh worked with me throughout the project. She researched and wrote the annotated references for each chapter and helped develop the appendices. She also reviewed and provided invaluable feedback on all parts of the book. I also wish to thank Cathy Warfield, Gwinnett County Schools, and Barbara Wilson, Lisa Walker, and Lolita Farrell from Georgia State University for assisting me during various stages.

Several science teachers were willing to complete lengthy questionnaires that I used to create the Science Teachers Talk sections of the book. I want to thank Ginny Almeder, Bob Miller, Jerry Pelletier, John Ricciardi, Dale Rosene, and Mary Wilde. I also want to thank Eliot Wigginton for granting me permission to use his teacher interview questionnaire. I also wish to express thanks to Robert Cross and Ronald Price for helping prepare the sections on

Australia and China, and Babatunde Abayoni for help with the section on Nigeria.

I was fortunate to have several outstanding colleagues review the entire manuscript. I want to thank Professor Betsy Balzano of the State University of New York at Brockport, Professor Emmet Wright of Kansas State University, and Professor Bob Yager of the University of Iowa.

I could not have developed this book without the research and writing of many scientists, science education researchers, and science teachers. I have tried to reference their works to enable you to go directly to the original ideas for more in-depth study.

During the time that I developed this book, I had the privilege (and continue to) to direct the Global Thinking Project, a telecomputing and global ecological curriculum project that has brought together science educators from the United States and the Russian Republic. These science educators have been, and continue to be, an inspiration to me, and gave me great moral support throughout the writing of this book.

I want to thank my colleagues at Georgia State University, especially my colleagues in the Department of Curriculum and Instruction for their support over the more than 20 years that I have been associated with them.

Christopher Jennison, Executive Editor of Harper-Collins, provided the publishing environment and support that I value enormously. He believed in my initial proposal and gave me the encouragement to continue the project to the end. Cindy Funkhouser of HarperCollins directed the production of the book and made the whole process very smooth and manageable. David Banner and Richard Benz of Caliber helped prepare the manuscript for publication and performed numerous eleventh-hour duties flawlessly.

1

MINDS ON SCIENCE: A RECONNAISSANCE

"The most important discovery made by scientists was science itself," said Jacob Bronowski, a mathematician, philosopher of science, and teacher. What about science teaching? Is there a comparable discovery made by science educators about science teaching? Perhaps the discovery approach to learning itself is one candidate. Or perhaps the discovery that students don't learn science through direct instruction; rather they construct their own knowledge from formal and informal experiences on their own. Or perhaps that all students are capable of learning.

There are many candidates for important "discoveries" that have been made by science educators. If you are interested in finding out about the world of science teaching and fascinating discoveries made by science teachers and researchers about learning, curriculum, and instruction, this book is for you.

We'll start our exploration of science teaching with a reconnaissance of the field. Just as a scout goes out ahead and looks around to get a view of the scene, so it is with this chapter. You'll look ahead by examining and comparing some of your ideas about science teaching with those of other science teachers. Before you know it, you'll be teaching and participating in science lessons prepared by and taught to your peers. Then you'll investigate some conceptions about the nature and philosophy of science, and relate this to approaches to science teaching such as inquiry and cooperative learning.

We'll then introduce you to some students via brief vignettes designed to capture the intensity, personal dimensions, and holistic character of the students you'll teach. Finally the chapter will close with some insights and wisdom about teaching by hearing from a high school teacher who has been involved with secondary school students for over 20 years.

PREVIEW QUESTIONS

- What are your current views about science teaching? How do these compare with the views of professional science teachers?
- What are some major conceptual ideas about science teaching?
- Why do you want to be a science teacher?
- What do science teachers like most about teaching?
- What are some of the important characteristics of science?
- Is inquiry teaching a valid method in the secondary science classroom? Are there other valid approaches? What fosters inquiry?
- Do scientists and students represent two cultures? If so, how can these cultures be bridged?
- Who are the students we teach? What are they like?
- What characterizes an effective science teacher?

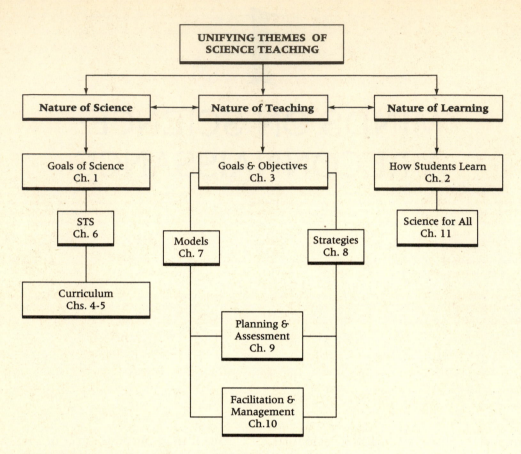

FIGURE 1.1   Unifying Themes of Science Teaching

UNIFYING THEMES
OF SCIENCE TEACHING

What knowledge do you need to become a science teacher? To help you in this quest the content in the book has been organized around a set of unifying themes of science teaching that taken as a whole consititute a partial answer to the question, What does a person need to know to be a science teacher? Some science educators refer to unifying themes as conceptual themes. A conceptual theme is a big idea—a unifying notion—an organizational structure. The conceptual themes that been have chosen are well documented in the literature; they make sense to the practicing science teacher.

The themes are clustered into three organizers: the nature of science, the nature of teaching, and the nature of learning and the learner (Figure 1.1).

The conceptual themes should be recognizable to you, and furthermore, you come to this course with knowledge and ideas about these conceptual themes. I suggest that the place to begin your study of science teaching is to evaluate your present views about teaching, learning, students, and the curriculum—in short, what do you bring to this course in terms of knowledge, attitudes, and your appreciation of science teaching.

INQUIRY ACTIVITY   1.1

Exploring Your Ideas About Science Teaching

Later in this book you will discover that secondary students come to science class with existing ideas about the science content that you will teach. Novices of a field of study, such as students in your future sci-

ence classes, possess initial conceptions of the field, say earth science or physics. Many of these ideas or initial conceptions are actually misconceptions or naive ideas. Nevertheless, these misconceptions represent a good place to begin instruction. Thus, this activity is designed to help you think about and *explore your existing ideas or frameworks regarding science teaching*. It's not a pretest but rather an opportunity to discuss your initial ideas about science teaching from a holistic and problem-oriented vantage point.

Materials

Index cards
Information in Table 1.1

Procedure

1. Read each of the situations given for the conceptual themes listed in Table 1.1.
2. Write the themes on individual index cards. Shuffle the cards, and place them face down on a table around which four to six students are sitting.
3. Select one person to start. The person selects a card from the top of the deck of index cards to identify the unifying theme. Read the problem situation associated with the theme aloud to your group. Use the questions listed in the third column to guide your exploration of the theme. To explore the theme, you can

   a. Give your initial point of view and share it with the group. You can ask other group members if they agree or disagree.
   b. Ask each group member to write a brief statement, and then read them aloud to the group.
   c. If the problem situation merits it, role play the situation with other members of your group. The person drawing the card and selecting this method asks for volunteers and directs the role-playing scene. The enactment should take no more than two or three minutes.

Minds on Strategies

1. How do your initial ideas compare with other students in your class?
2. What is a framework? How do frameworks develop? How can they change?
3. In what ways do you think your initial ideas or frameworks regarding science teaching reflect the most recent research and practice of science teaching?

TABLE   1.1

EXPLORING YOUR INITIAL
CONCEPTIONS OF SCIENCE TEACHING

Unifying Theme of Science Teaching	Problem Situation	Assessing Your Initial Conception
Nature of Science	Carl Sagan says that "science is a way of thinking, much more than it is a body of knowledge."	What is your view of science? Do you agree or disagree with Sagan? What are the implications of Sagan's definition for science teaching?
Learning	You overhear a science teacher explaining to her eighth grade earth science class that intelligence is incremental, not fixed. She believes that this will encourage students to try harder, especially when learning new and difficult ideas and concepts.	What is your view of intelligence? Do you think teaching students about human intelligence might help them learn science?
Goals	According to a report on science teaching written by a prestigious group, the main goal of science teaching is to produce a scientifically literate society.	Do you agree with this? Are there other goals that are worthy and should be an integral part of science teaching?
Curriculum	The title of a keynote address at a major conference on science teaching is "The Science Curriculum: A Nonchanging Phenomenon!"	What is the science curriculum, anyway? Is it nonchanging, or has the curriculum changed over the years?

(Table continued on p. 4)

(Table continued from p. 3)

Unifying Theme of Science Teaching	Problem Situation	Assessng Your Initial Conception
Science, Technology, Society	A science teacher announces at a departmental meeting that she is going to include the following topics in her survey biology class: ethics and animal rights, birth control methods, abortion, and AIDS counseling. One teacher objects saying, "These are too controversial, we'll have half the parents in here."	What do you think? Should topics like these be part of the science curriculum? Why?
Models	A first-year teacher used a nontraditional teaching model during the first week of school. It was a small group activity with hands-on materials. Students used meter sticks and were asked to measure various heights and lengths with the sticks. Students were confused. How could they measure something bigger than the meter stick? Another pair of students was carving symbols and words in the meter stick, and another group couldn't decide whether the smallest marks were centimeters or millimeters.	Are nontraditional models of teaching prone to problems and the unexpected? Should they be avoided by first-year teachers until they get their feet wet?
Strategies	The most common strategy used in high school science teaching is lecture and discussion. Many science teachers claim that this is an inadequate strategy for most students and suggest other strategies.	What do you think? Isn't lecture an efficient way to teach science? Are there other strategies that might reach more students? What are they?
Planning	At a conference between a student science teacher and her college supervisor, the student expresses anger that the students didn't enjoy the lesson that she had spent three hours planning. She just can't believe they were rude during a lesson she planned so hard for.	How important is planning for lessons? Does this student teacher have expectations that are too high? How would you react in such a situation?
Assessing	A teacher announces that he is going to let students work in small teams on three quizzes each term. The students will turn in one paper, and each will receive the group's grade.	Do you think this is a good idea? Why? Would you employ such an assessment plan in your class?
Management and Facilitation	A fellow student returns from observing high school science classes with two maps drawn of the classrooms visited (see Figure 1.2).	What can you infer about each teacher's view of classroom management? How do their views of facilitating learning compare?
Science for All	A number of schools around the country with large numbers of at-risk students have adopted an approach called integrative learning. This holistic approach appears to be successful with students who are disinterested in school, and normally end up dropping out.	How would you teach the at-risk student, the student who has had a continous record of failure in school, and clearly is prone to drop out of school? Can all students learn science?

FIGURE 1.2 Diagram of two classrooms

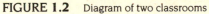

SCIENCE TEACHING: YOUR CAREER CHOICE

For a variety of reasons, you've chosen to be a secondary science teacher. In studies to find out why people choose a career in science teaching, interest in the subject matter of the field is rated as the most important.[1] Is this true for you? Other reasons why people choose a career in science teaching include such factors as

- They feel their abilities are well suited to teaching
- They like the opportunities to work with young people
- Teaching contributes to the betterment of society

You may also have made a choice between middle school/junior high and high school level teaching.[2] Perhaps you are interested in working with early adolescent students in either a middle school or a junior high school. Or you've decided that you want to work with older students and have geared your preparation to high school science teaching. In either case, being a science teacher will require you to blend knowledge of science, pedagogy, and how students learn. How can this be done so that students learn and develop an appreciation for science, and you perceive science as a rewarding career? What will this entail?

As you begin your study of science teaching, keep in mind that these conceptual themes will be helpful organizational ideas for you; nonetheless, you should also acknowledge that having a theoretical base for these notions will not ensure your success in the classroom. Teaching requires an integration of theory and practice. You will therefore find in this book a number of practical, laboratory-oriented activities designed to help you translate some of the theoretical ideas into practice. There is a good chance that the course you are taking will also involve some practical work in a middle school or junior high or a high school. These

[1]Linda Darling-Hammond, Lisa Hudson, and Sheila Nataraj Kirby. *Redesigning Teacher Education: Opening the Door for New Recruits to Science and Mathematics Teaching* (Santa Monica, California: The RAND Corporation, 1990) p. 56.

[2]It should be noted at this point that there are clear differences between middle schools and junior high schools in terms of philosophy, curriculum organizations, and instruction. These differences, and differences between middle school or junior high and high schools, will be discussed later in the book.

opportunities during your teacher preparation experience are important as you develop your own professional outlook on science teaching. To gain more insight into your career choice, let's hear from some practicing science teachers, and what they have to say about the rewards of science teaching.

WISDOM OF PRACTICE

In preparation for this book, I interviewed (a copy of the interview questions is in the Science Teaching Gazette in this chapter) a number of practicing middle and high school science teachers because I wanted to include their ideas—their wisdom of practice—as we explored science teaching. I wanted these teachers to report to you how they deal with the main concepts and ideas of science teaching. In this chapter I will introduce you to these teachers, all of whom are real practicing science teachers. The teachers are

> Ginny Almeder, a biology teacher from Georgia
> Bob Miller, a biology teacher from Texas
> Jerry Pelletier, a junior high science teacher from California
> John Ricciardi, a physics and astronomy teacher from Nevada
> Dale Rosene, a middle school science and computer teacher from Michigan
> Mary Wilde, a middle school science teacher from Georgia

In subsequent chapters these wisdom of practice interviews will be found in the Science Teaching Gazette under the section entitled "Science Teachers Talk." The comments made by the teachers are brief but candid and are here to give you some insights into teaching from a practitioner's point of view.

Many teachers report that science teaching can be a very rewarding career. What do science teachers like most about teaching? Surely this will give us some insight into the profession of science teaching, and help you formulate your own goals and strategies for making your choice of science teaching a successful and positive one.

Science Teachers Talk: What Do You Like Most About Teaching?

John Ricciardi. What I like most about science is that I can be myself, which is being part of a body of teenagers. Their spirit, ambience, and energy can become the self that is me and who I am becomes naturally part of them. For me, teaching science is becoming myself by becoming one with all that "sciencing" is in my students.

Ginny Almeder. Science is my way of questioning the universe, a pursuit we appear compelled to follow by our human nature. Teaching high school provides me with an opportunity to share my love of science with young people. Students are generally enthusiastic and open-minded about their world. It is a good time to introduce them to the joys of science. I appreciate having the opportunity to help young people realize their potential, especially in the area of science. It is gratifying to observe students improving their skills, becoming more questioning, and developing a healthy self-concept.

Jerry Pelletier. I am fascinated by science. It encompasses a myriad of subjects and experiences and is an ever changing and developing field. Some ideas have remained unchanged for hundreds of years, while others have changed many times through the centuries. I find that my excitement for the subject of science can easily be transmitted to the students. I enjoy observing students interacting while trying to understand and solve scientific concepts. Science lends itself to the inductive method of teaching. Students are constantly questioning themselves and their observations. In essence science is fun for students as well as myself.

Mary Wilde. What I like most about teaching science is the variety of ways and techniques one can use to teach a particular concept. You can prevent yourself from becoming "burned out" because there are always new demonstrations, activities, and experiments to incorporate in your curriculum that can explain old concepts. It is very exciting to be part of the new discoveries, new theories, and new conceptual ideas that take place in the scientific world. What is even more "thrilling" is the sharing of these new theories and discoveries with our young people. The teaching of new scientific principles, or old scientific principles in new ways, stimulates a curiosity and creative desire within the student. Thus, for me, science is a very successful tool to help the student develop creative skills, thought process skills, and problem-solving skills while learning factual content and conceptual theories that explain how this world "ticks." Science is the *why* and *how*, and isn't that what everyone wants to know?

These teachers enjoy interacting with students and believe that science can be fun for students and can provide an opportunity to introduce students to the joys of science. Let's begin our study of science teaching by visiting a high school classroom where science content, pedagogy, students, and teachers meet: the science-teaching interface.

ON THE NATURE
OF SCIENCE TEACHING

"What are these?"
"Where did they come from?"
"How old are they?"
"Where did you get them?"
"Are they all the same?"
"What are they used for?"

Questions asked by the the teacher? No! These are questions asked by students in a ninth-grade teacher's physical science class at Southside High School in the Atlanta Public Schools. The teacher, one of thousands of new science teachers in the United States,[3] began the first day of school with a very brief activity. He gave each student a fossil crinoid stem, placing one in each student's hand and telling them they could not look at the object until he said they could. The students were instructed to explore the object without looking, and to write down observations of the object, and to make a small drawing of it as well. Still without looking, the teacher asked the students to call out some of the observations (hard, breaks easily, gritty, grainy, cylindrical, about 2 centimeters in diameter, grooves along the side, a hole in the center). Then the teacher provided each student with modeling clay and asked each student to make a replica of the crinoid (still without looking).

Finally the teacher asked the students to guess what they thought it was (rock, bone, dog biscuit, pottery) and then to look at the object. Without telling the students very much, he asked the students if they had any questions about the object. Their curiosity led to several questions and then to a discussion of these

400 million year old fossils from Silurian rock beds of north Georgia.[4] The next day, the teacher divided his students into groups and assigned a different task to each group. Later in the lesson a student from each group reported their results to the class.[5]

The teacher began his class by actively engaging his students with natural materials, having them work in groups, and encouraging them to use their observation skills and creative abilities to solve problems and participate in interesting tasks.

However, this lesson and the way his students feel about science is in stark contrast with what is known about science teaching in the United States and other countries around the world. In general, students see science class as dull, no fun, and a place where they do not wish to be. Students do not like the typical or traditional science classroom.[6] Although studies about science teaching reveal that there are many factors that seem to make science teaching more interesting and result in high achievement (these will be discussed in the next chapter), one factor that seems to be very important is engagement. What this refers to is the active involvement of students in the learning process. Students were engaged in handling, operating, or practicing on or with physical objects as part of the lesson.[7]

[4]For the next day's lesson, the teacher divided the students into groups. Each group was involved in a different activity and shared the results with the whole class. See Inquiry Activity 1.2 and try these with a group of students.
[5]See Inquiry Activity 1.2 for a description of the tasks.
[6]David Holdzkom and Pamela B. Lutz, *Research Within Reach: Science Education* (Charleston, W.Va.: Research and Development Interpretation Service, Appalachia Education Laboratory, Inc., 1986).
[7]K. C. Wise and J. R. Okey, "A Meta-analysis of the Effects of Various Science Teaching Strategies on Achievement," *Journal of Research in Science Teaching* 20 (1983): pp. 419–435.

[3]N. Carey, B. Mittman, and L. Darling-Hammond, *Recruiting Mathematics and Science Teachers Through Nontraditional Programs: A Survey* (Santa Monica, Calif.: The RAND Corporation, 1988).

FIGURE 1.3   Students in this class are actively engaged with natural materials, working in collaborative groups, and are encouraged to use scientific thinking skills and creative abilities to solve problems. This is in stark contrast to the typical science lesson. © 1987 Hazel Hankin.

Perhaps the one metaphor about science teaching that has become a password for good science teaching is that science teaching should be *hands-on*. In recent years, however, this metaphor has been enriched and expanded with the use of the phrase "minds on science." These metaphors seems like a simple and logical step in the teaching process, but the evidence from science education research studies is quite the contrary:[8]

1. The predominant method of teaching in science is recitation (discussion), with the teacher in control. We will call this the delivery mode of teaching, contrasting it with the engagement mode

2. The secondary science curriculum is usually organized with the textbook at the core, and the main goal of the teacher is to cover (or deliver) all the content in the book

3. The science demonstration ranks second as the most frequently observed science "activity." Two out of five classes perform demonstrations once a week. And please note that in most demonstrations, students are typically passive observers

4. Student reports and projects are used about only once a month or more in half the classes

5. Because of the anxiety to cover the text, the use of inquiry techniques is discouraged and is rarely observed. Instead activities are generally workbook exercises in following directions and verifying information given by the textbook or teacher

A great controversy exists in the field of science education surrounding the issue of engagement versus delivery. Which model is more effective? Which is more efficient? How do students react to these models? Which model helps most students understand science? Which model do you prefer?

Let's explore the differences between the delivery and engagement models of science teaching in order

[8]See Holdzkom and Lutz, *Research Within Reach*, pp. 43–44 for a discussion of current science teaching practices.

to develop a better notion of the nature of science teaching. To do this you will plan and teach two lessons using a microteaching format. Microteaching is a scaled-down version of teaching in which you present a short lesson (usually 5–15 minutes in length) to a small group (5–7) of peers or secondary students, videotape the episode, and evaluate the lesson, and then make recommendations for possible changes. You will find more details on microteaching in Chapter 9.

INQUIRY ACTIVITY   1.2

The Crinoid Stem and the Nature of Science Teaching

To explore the nature of science teaching, you are going to plan at least two microteaching lessons based on the following ideas and carry them out with a group of peers or a group of secondary students. One of the microteaching lessons should be selected from the list entitled "Engagement Mode," and the other microteaching lesson should be selected from the list entitled "Delivery Mode."

Materials

Collection of fossils of the same species, metric rulers, crayons or marking pens, newsprint, bell caps, string and glue, and other materials and equipment to teach the microteaching lesson.

Procedure

1. Divide into groups and select a task from either the engagement or the delivery mode of teaching.

Your group is to prepare a ten-minute microteaching lesson based on the task you selected. You can teach the lesson to either a peer group or a group of secondary students. You may want to videotape the lesson so that you can replay it.

2. When groups are finished, one member should present each group's results to the whole class.

Minds on Strategies

1. Evaluate the lesson by comparing the engagement mode to the delivery mode of instruction by considering the following questions: Was there evidence of curiosity on the part of the students during the lesson? Did the students show their creativity? Did they ask questions? Was there an aesthetic dimension in the lesson? Which lesson model did the students (learners) prefer? Which lesson did the teachers prefer?
2. Which approach do you think is more motivational? Why?

TABLE   1.2

ENGAGEMENT VERSUS DELIVERY MODES OF TEACHING

Engagement Mode Tasks

Task 1. Part a. You are a group of scientists. Make as long a list of observations of the crinoid as your group can. When your group has completed the list, ask the instructor for the second part of your activity.

Note: be sure to write your list on a large sheet of chart paper; you can use more than words!

Part b. Classify each of the observations your group made according to the human sense used for each observation, e.g., F = feel, touch; T = taste; S = smell; E = sight, eyes; H = hearing, sound; O = other senses.

Delivery Mode Tasks

Task 1. Lecture and carry out a discussion on the physical characteristics of the fossil. Be sure to include observations that require the use of the five senses.

(Table continued on p. 10)

(Table continued from p. 9)

Engagement Mode Tasks

Task 2. You are a group of mathematicians. Measure the diameters (in centimeters) of at least 20 crinoids. (You will have to visit other groups to get a total of 20 measurements. Send out four of your group to measure 5 crinoids each while the remaining ones measure your crinoids.) Make a population graph of the crinoids you measured. Set up the graph like the one shown in Figure 1.4

Note 1: Draw your graph on a piece of chart paper; make it large and colorful.

Note 2: Seek out another group that did this task. Compare your graphs. Are your populations different? How do you know?

Task 3. You are a group composed of historians, anthropologists, and geologists. Use your imaginative side and draw a complete picture of what your team thinks the crinoid looks like. (See Figure 1.5.) You are looking at only a piece of the animal. How do you think it looks as a complete creature? Does it have a head? Does it have feet? How does it move? Special note: When you draw your creature, put the creature in the context of an environment. Ask your group; Where does this creature live? Does it live alone? Or are there others about? What does it eat for food? How does it get its food? Who are its predators?

Delivery Mode Tasks

Task 2. Lecture and discuss the population characteristics of the fossil. Focus attention on one characteristic, namely diameter (if you use crinoids; choose another characteristic for other fossils). Explain the terms *fossil, population graph,* and *diameter* to the students.

Task 3. Introduce the students to the concept of environment. Use the fossil crinoid as the species to study. Use diagrams and pictures so that the students will be able to describe the ecological characteristics of the crinoids' environment.

(Table continued on p. 11)

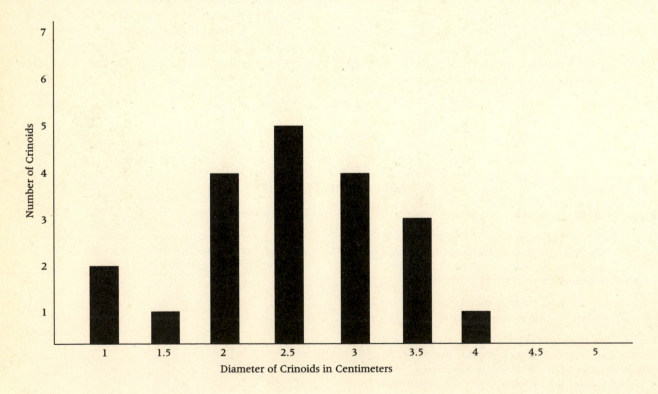

FIGURE 1.4   Crinoid Population Graph

(Table continued from p. 10)

Engagement Mode Tasks

Task 4. You are a group of writers. Poets! Your task is to prepare several poems about the crinoid that your team will read to a group of fellow teachers. Write several poems, called *syntus*, using the following formula:

Line 1: Single word or concept such as fossil, crinoid, age, or time
Line 2: An observation of line 1
Line 3: An inference about line 1
Line 4: A feeling about line 1
Line 5: A synonym of line 1

Note 1: Brainstorm observations, inferences, and feelings about the crinoid. Try to think about being the crinoid, living when it lived (400 million years ago). Use the results of your brainstorming to create your syntus.

Note 2: Write your final products on sheets of chart paper. Make them colorful and easy to read from a distance.

Task 5. You are a group of artists. Your task is to make pendants using the crinoids, bell caps, gold or silver chain, and glue. After your group has made pendants, show other groups how to do the same.

Delivery Mode Tasks

Task 4. Give a brief lecture on the fossils so that students will be able to describe what fossils are, how they are formed, what they tell us about the earth, and what they are used for. The students should be able to write a brief essay on fossils as a result of your presentation.

Task 5. Deliver a lecture on the artistic and practical aspects of fossils. How are fossils used in arts and crafts? What people in the community would have a use for fossils?

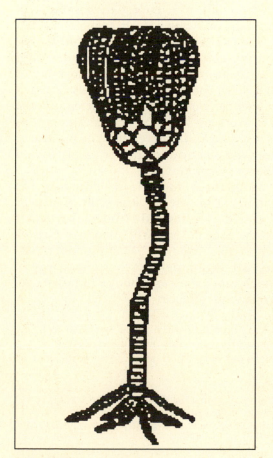

FIGURE 1.5   A fossil crinoid

THE NATURE
OF SCIENCE

You are entering the science teaching profession at a time when many science educators, scientists, and the general public are calling for new directions in science education. Because of the growing impact of science and technology on societal and individual affairs, people from many sectors of society have expressed the desire that science education be reappraised and that new direction be charted. Paul DeHart Hurd suggests that the science curriculum of the future be based on interrelationships between human beings, natural phenomena, advancements in science and technology, and the quality of life.[9] He suggests that science teachers examine closely the nature of science, especially the multidimensional changes in science, technology, and society. He and other educators criticize the content of the present science curricula as being remote from human needs and social benefits, reflecting the concern that science is alien and separate from individual and public interests. To make science understandable and useful to people, it is essential that the nature of science be communicated to students in the science curriculum.

What Is Science?

One scientist who had an effect not only on the scientific community but on the nonscience community as well was Richard Feynman, a theoretical physicist and popularizer of science. In his book *Surely You're Joking, Mr. Feynman* he said, "Before I was born, my father told my mother, 'If it's a boy, he's going to be a scientist.'" Not only did he become a scientist, he also was a winner of the Nobel Prize.[10] Feynman saw science as an attempt to understand the world. To him

understanding the world was analogous to understanding the rules of a game, like chess:

> We can imagine that this complicated array of moving things which constitutes "the world" is something like a great chess game being played by the gods, and we are observers of the game. We do not know what the rules of the game are; all we are allowed to do is to *watch* the playing. Of course, if we watch long enough, we may eventually catch on to a few of the rules. *The rules of the game* are what we mean by *fundamental physics*. Even if we know every rule, however...what we really can explain in terms of those rules is very limited, because almost all situations are so enormously complicated that we cannot follow the plays of the game using the rules, much less tell what is going to happen next. We must, therefore, limit ourselves to the more basic question of the rules of the game. If we know the rules, we consider that we "understand" the world.[11]

Another scientist, a chemist named Michael Polanyi, explored the nature of science and said that there was a "republic of science," a community of independent men and women freely cooperating, collaborating, and exchanging ideas and information. This community cuts across national borders and brought scientists from all over the globe together as a cooperative global community. Polanyi also claimed that to be a scientist, one had to be inducted into the profession by working with a master as an apprentice. Interestingly, he also believed that the practice of science was not a science but rather an art passed from one scientist to another.[12]

If you look up the word *science* in a dictionary, the usual definition is "knowledge, especially of facts or principles, gained by systematic study; a particular branch of knowledge dealing with a body of facts or truths systematically arranged."[13] Yet prominent scientists, like Carl Sagan, define science as a way of thinking much more than as a body of knowledge. Sagan says this about science:

> Its goal is to find out how the world works, to seek what regularities there may be, to penetrate the connections of things—from sub-nuclear particles

[9]Paul Dehart Hurd, "New Directions in Science Education," in LaMoine L. Motz and Gerry M. Madrazo, *Third Sourcebook for Science Supervisors* (Washington, D.C.: National Science Teachers Association, 1988), pp. 3–7.

[10]Feynman was a member of the Rogers Commission, which investigated and reported on the cause of the space shuttle Challenger explosion. In the sequel to *Surely You're Joking, Mr. Feynman*, entitled *What Do You Care What Other People Think?* Feynman writes about his experiences as a member of the commission. It would be an understatement to say that his actions and thoughts were not always welcomed by the commission chair, but through the eyes of this scientist we witness the inner workings of the commission and get an insight into the confusion and misjudgment that characterized the management of NASA leading up to the disaster. All Americans observed this scientist using his knowledge to reveal the cause of the disaster when Feynman dropped a ring of rubber into a glass of cold water and pulled it out, misshapen.

[11]Richard P. Feynman et al., *The Feynman Lectures on Physics*, v. I (Menlo Park: Addison-Wesley, 1963), p. 2–1.

[12]Michael Polanyi, *The Republic of Science* (NY: Roosevelt University, 1962), p. 5.

[13]The International Webster New Encyclopaedia Dictionary (Chicago: The English Language Institute of America, 1975).

which may be the constituents of all matter, to living organisms, the human social community, and thence to the cosmos as a whole. Our perceptions may be distorted by training and prejudice or merely because of the limitations of our sense organs which of course perceive directly but a small fraction of phenomena of the world....Science is based on experiment, on a willingness to challenge old dogma, on an openness to see the universe as it really is. Accordingly science sometimes requires courage—at the very least the courage to question the conventional wisdom.[14]

Exploring the nature of science a little further, and relating it to the views of Feynman, Polyani, and Sagan, we might consider the following relationships:

Science and Courage

One human quality that is important in science is courage. If we put this in terms of one's willingness, as Sagan said, to question conventional wisdom, we are led to an important notion: questioning all things is a fundamental value underlying thinking in science. For example, Nicolaus Copernicus, the sixteenth-century Polish scientist, questioned the conventional wisdom of the Ptolemaic earth-centered universe. His questioning of an old idea led to a new one: that the sun was the center of the solar system and the planets revolved around the sun rather than the earth. About a hundred years after the publication of Copernicus's book, Galileo Galelei narrowly escaped the rack of the Holy Inquisition by recanting his support for the Copernican concept of the universe.

Questioning well-established ideas or proposing a radically different hypothesis to explain data is a courageous act. Quite often people who propose such ideas are shunned, considered crazy, or rejected by the "establishment." For example, in 1920 Alfred Wegener, a German meteorologist, proposed that the continents were not stationary masses but moving platforms of rock that had drifted apart over millions of years of geologic time. At the time his idea was considered farfetched and crazy. Physicists and geologists pointed out that there were no forces within the earth to move billions of kilograms of rock. Fifty years later, most geologists supported the theory of plate tectonics, that the earth's crust is composed of large plates that move about, colliding, spreading apart, and sliding past each other.

A more recent example of courage is the case involving Dr. Frances Oldham Kelsey. In a book entitled

[14]Carl Sagan, *Broca's Brain: Reflections on the Romance of Science* (New York: Random House, 1979), p. 13.

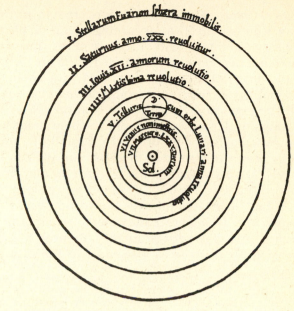

a. Copernican sun-centered universe

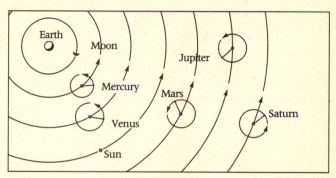

b. Ptolemaic earth-centered universe

FIGURE 1.6 Copernicus's idea of a sun-centered universe questioned the conventional wisdom of geocenteredness. His questioning of a firmly held conception is an example of *courage* in the practice of science.

Earth 225 Million years ago

Earth Today

FIGURE 1.7 Alfred Wegener envisioned the continents as belonging at a time in the geologic past to a single super land mass called Pangea. Most scientists thought his idea was absurd.

Women of Courage, Frances Kelsey is referred to as "the doctor who said no."[15] After earning a Ph.D. in pharmacology (an infant field of science at the time she earned her degree) and then a medical degree from the University of Chicago, she moved with her husband and two children to Washington and took a job with the Food and Drug Administration (FDA) in Washington. Her job was to evaluate applications for licenses to market new drugs. In the fall of 1960, shortly after she arrived at the FDA, the William S. Merrell Company applied for a license to market a new drug called Kevadon. Kevadon was a sleeping pill. It had been used all over the world, was very effective in relieving pregnant women from morning sickness, and was very profitable. Frances Kelsey showed great courage as a scientist in the case of Kevadon:

> While Merrell's application was being reviewed by Kelsey at the FDA, they were distributing two thousand kilograms of the drug. At the time this was a legal practice as long as the drug company labeled the drug "experimental." Merrell, in their advertising and marketing materials, informed their salespeople that they had firmly established the safety, dosage and usefulness of the drug by both foreign and U.S. laboratory clinical studies. They had not.
>
> At the FDA, Merrell's application was being reviewed. Dr. Kelsey and her research team were not satisfied with the information Merrell provided as part of their application. For example, the drug when administered to animals showed no sign of toxicity but did not make the animals sleepy. The drug was being distributed to humans as a sleeping pill. Two days before the 60 day approval period was up, Dr. Kelsey told the Merrell Company that their application was not approved and that they would have to submit further information.
>
> This initial rejection (November 10, 1960) of Merrell's application to distribute the drug was followed by a series of episodes between Dr. Kelsey and the Merrell Company. There were attempts by the Merrell Company to go over Kelsey's head and in so doing try to embarrass her in front of her superiors. This did not work. Merrell even supplied research reports supposedly documenting the safety of the drug. Upon investigation it was discovered that the researcher's name that appeared on the report did not even write it. And Merrell threatened Kelsey with a law suit saying that one of her letters to them was libelous. Through all this Kelsey stood firm and boldly held her ground. It culminated with the banning of the drug in December 1961 when thalidomide had been traced to an outbreak of deformities in newborn babies by the thousands in Europe. Then in 1963, the American public was stunned when they read stories and saw the horrible pictures in their newspapers that

one gallant woman doctor had stood between them and a repetition of this disaster in the United States. On August 7, 1963, President Kennedy presented Dr. Kelsey with the Distinguished Federal Civilian Service Medal. Kennedy applauded Dr. Kelsey's work saying she had defended the hopes that all of us have for our children. The courageous behavior of Frances Kelsey also lead to an increase in the FDA's staff and a change in the laws regulating the distribution and sale of drugs to humans.[16]

Science, Problem Solving and the Human Mind

Thinking in science is often associated with creativity and problem solving. These are important aspects of science and should be an essential goal of the science curriculum. In his book *How Creative Are You?*, Eugene Raudsepp identifies a long list of qualities that are characteristic of people who think creatively: they are innovative, risk takers, and mold breakers; they are willing to ask questions, fearless adventurers, unpredictable, persistent, and highly motivated; they are able to think in images and to toy with ideas, and to tolerate ambiguity and anticipate productive periods. The social implications of creative thinking are that we live in an ever changing world.

Many popularizers of science and creative thinking believe that all people are creative and are able to deal with change. Science courses have traditionally focused only on helping students learn scientific facts and concepts and then stopped. Rarely are students encouraged to tackle real problems, thereby putting to use the facts and concepts they have learned. But as educators like Hurd warn, the future science curriculum should present problems to solve that are desirable to students, that is, ones in which students have a stake in the solution, such as nutrition, chemical safety, space exploration, human ecosystems, drugs, population growth, ecocrises, and the quality of life.

To solve problems, to deal with situations creatively, requires the use of imagination. Historians of scientific discovery often point out that imagery and imagination have played important roles in intellectual discoveries and breakthroughs. This fact is nicely conveyed in the title of a book by June Goodfield, *An Imagined World: A Story of Scientific Discovery*. The book describes the drama of scientific discovery and sheds light on the role of creativity and imagination in this endeavor.[17] The world of imagery is safe harbor for thoughts and images, and for the mind's participation

[15]Margaret Truman, *Women of Courage* (New York: Bantam, 1976).

[16]Truman, *Women of Courage*, pp. 178–196. Used with permission of William Morrow and Company.
[17]See June Goodfield, *An Imagined World: A Story of Scientific Discovery* (New York: Harper & Row, 1981).

in problem solving. For example, Einstein's famous thought experiments and his images led him to many of his concepts of space and time. Indeed, he used imagery to experience what he thought it would be like to ride on a beam of light.

Jacob Bronowski believed that imagination was one of the important qualities of the mind. In *A Sense of the Future*, he said this about imagination, the human mind, and science:

> All great scientists have used their imagination freely, and let it ride them to outrageous conclusions without crying "Halt." Albert Einstein fiddled with imaginary experiments from boyhood, and was wonderfully ignorant of the facts that they were supposed to bear on. When he wrote the first of his beautiful papers on the random movement of atoms, he did not know that the Brownian motion which it predicted could be seen in any laboratory. He was sixteen when he invented the paradox that he resolved ten years later, in 1905, in the theory of relativity, and it bulked much larger in his mind than the experiment of Albert Michelson and Edward Morley which had upset every other physicist since 1881. All his life Einstein loved to make up teasing puzzles like Galileo's, about falling lifts and the detection of gravity; and they carry the nub of the problems of general relativity on which he was working.[18]

Science and Human Values

When society acknowledges the importance of qualities of the mind such as independence in thinking, originality, freedom to think, or dissidence it is elevating them to social values. And as social values they are given special protection through laws governing society's behavior. Since science is an activity of men and women, certain values must guide their work. Bronowski claims that because of this science is not value free, and that the work of science is based on a search for truth. In his book *A Sense of the Future*, Bronowski discusses the human values that are indeed the values that guide science:

> If truth is to be found, and if it is to be verified in action, what other conditions are necessary, and what other values grow of themselves from this?
>
> First, of course, comes independence, in observation and thence in thought. The mark of independence is originality, and one of its expressions is dissent. Dissent in turn is the mark of freedom. That is, originality and independence are private needs of the truthful man, and dissent and freedom are public means to protect them. This is why society ought to offer the safeguard of free thought, free speech, free inquiry, and tolerance; for these are needs which

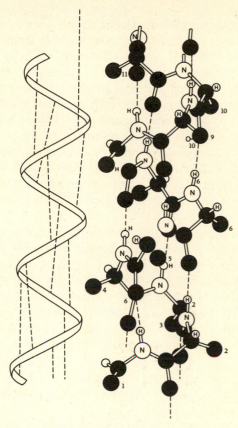

FIGURE 1.8   DNA: Midas's Gold of 1950s science

follow logically when men are committed to explore the truth. They have, of course, never been granted, and none of the values which I have advanced have been prized in a dogmatic society.[19]

Sometimes the values that motivate scientists result in behavior that wouldn't hold up to Bronowski's ideas. For instance, in the 1950s the race was on to discover the structure of the DNA molecule. Horace Freeland Judson in his book *The Eighth Day of Creation* said, "DNA, you know, is Midas's gold. Everyone who touches it goes mad."[20] In this case we ask; What part does ambition, achievement, and success play in the practice of science? How does a scientist's gender affect relationships? Are women scientists left behind their male counterparts? Is it possible for a scientist to literally "go mad" in the pursuit of what may be an astonishing discovery? Are scientists sometimes motivated by blind ambition? The story that follows will enable you to think about these questions.

The setting for this tale is in England about 100 years after Charles Darwin and Alfred Russell Wallace codiscovered a theory of evolution.

[18]Jacob Bronowski, *A Sense of the Future: Essays in Natural Philosophy* (Cambridge, Mass.: MIT Press, 1977), p. 28.

[19]Bronowski, *A Sense of the Future*; p. 209. Used with permission of the MIT Press.

[20]Horace Freeland Judson, *The Eighth Day of Creation* (New York: Simon & Schuster, 1979), p. 70.

In the 1950s a race was on to be the first to discover and unlock the secret that would reveal the basis for life. That secret was locked away in the structure of the DNA molecule—the substance of life. Two persons emerge at first, in this story: James B. Watson, a 24 year old American born scientist fresh out of graduate school with a new Ph.D., and Francis Crick, a 38 year old graduate student at Cambridge University, England, still working on his Ph.D.

Watson and Crick teamed up and decided to go all out to discover the structure of DNA. Their reward, if they could make the discovery before the famous American chemist Linus Pauling, would be the Nobel Prize.

The process of discovering the structure of the DNA molecule was multifaceted. A driving force in the discovery was their motivation to discover and report their findings before Pauling did. Pauling, 6000 miles away in Pasadena, California, was working diligently on the DNA problem as well. Watson and Crick, but especially Watson, were worried that news would break from Pasadena. Watson wrote that "no further news emerged from Pasadena before Christmas. Our spirits slowly went up, for if Pauling had found a really exciting answer, the secret could not be kept for long."[21]

The process also involved a collaboration with Maurice Wilkins and Rosalind Franklin, both of whom were researchers at King's College, England. Wilkins was trained in physics but became interested in the structure of the DNA and had been pursuing its structure for years.

At the time that Watson and Crick entered the DNA search, Wilkins was the only researcher in England giving serious attention to the DNA problem. However, there was Rosalind Franklin. She was trained in the study of crystals and how they are arranged. She used X-rays to study the structure of crystals and she was probably one of the most competent researchers in this field at the time.

Wilkins thought of her as his assistant. Rosalind Franklin thought of herself not as Wilkins assistant but as a bona fide researcher pursuing the DNA structure as her main line of research. She was in fact hired to work in the same laboratory, but as the head of a research group, a position equal to that of Wilkins.[22]

Vivian Gornick writes that the relationship that Watson and Crick had "was its own double helix: all attracting opposites and catalytic joinings. These two

ate, drank, slept, and breathed DNA."[23] Rosalind Franklin did not have this kind of relationship with anyone. "If she had someone to talk to, chances are she would have gotten to DNA first, it was all there in her notes and photographs, she just didn't know what to make of what she had."[24]

Gornick, in the introduction of her book *Women in Science*, raises questions about the work of women in science: What was it like to be a woman scientist? What if a woman working in science feels it is not so accessible to her? What if a woman in science feels she must prove herself many times more often than a man does; that her work is more often challenged and less often supported?[25]

Was Rosalind Franklin not allowed in on the discussions between Watson, Crick, and Wilkins because she was a woman? James Watson's book *The Double Helix* provides some insight into this:

> Clearly Rosy had to go or be put in her place. The former was obviously preferable because, given belligerent moods, it would be difficult for Maurice (Wilkins) to maintain a dominant position that would allow him to think unhindered about DNA....Unfortunately, Maurice could not see any decent way to give Rosy the boot. To which, she had been given to think that she had a position for several years. Also, there was no denying she had a good brain. If she could only keep her emotions under control, there would be a good chance she could really help him....The real problem, then, was Rosy. The thought could not be avoided that the best home for a feminist was in another person's lab.[26]

Anne Sayre, author of *Rosalind Franklin and DNA*, finds Watson's description of Rosalind quite different from her own view. In her biography of Franklin, Sayre questions the accuracy of some of Watson's facts:

> A question arose concerning the accuracy of some of Watson's facts, simply because he presented in *The Double Helix* a character named "Rosy" who represented, but did not really coincide with, a woman named Rosalind Franklin....The technique used to change Rosalind Franklin into "Rosy" was subtle, but really not unfamiliar. Part of it, at the simplest level, was the device of the nickname itself, one that was never used by any friend of Rosalind's, and certainly not to her face....For we are presented with a picture

[21]James Watson, *The Double Helix* (New York: Atheneum, 1968), pp. 17–18.

[22]Anne Sayre, *Rosalind Franklin and DNA* (New York: W.W. Norton, 1975), p. 17.

[23]Vivian Gornick, *Women in Science* (New York: Simon & Schuster, 1990), p. 12.

[24]Vivian Gornick, *Women in Science* (New York: Simon & Schuster, 1990), p. 12.

[25]Gornick, *Women in Science*, p. 13.

[26]Watson, *The Double Helix*, pp. 17–18. Used with permission.

of a deplorable situation. The progress of science is being impeded, and by what? Why, by a woman, to begin with, one labeled as subordinate, meant—or even destined—to occupy that inferior position in which presumably all women belong, even those with good brains....But perhaps the progress of science is also being impeded somewhat by a man as well, one too inhibited by decency to be properly ruthless with female upstarts, and so to get on with the job.[27]

Rosalind Franklin's work on the DNA problem was brilliant. Had she lived until 1962 (she died of cancer in 1958 at the age of 37), she no doubt would have shared the Nobel Prize awarded to Watson, Crick, and Wilkins.

Since the time of these events, the nature of science has been influenced by the increased participation of women in the field of science. However, the participation of women and minorities in science has been discouraged by the nature of school science, and the negative effect school science has had in attracting women to careers in science. In Chapter 11 we will explore this in more depth.

Since the 1970s there has been a movement in the field of science and science education that has supported an approach to science teaching based on women's studies, methods, and theories to attract women to courses in science in middle school and high school, and encourage women to choose fields in mathematics, science, and engineering.[28]

Science and Democracy

When science is examined as an enterprise that involves the values of independence, freedom, the right to dissent, and tolerance, it is clear that as a social activity, science cannot flourish in an authoritarian climate. Some philosophers of science such as Bronowski claim that science cannot be practiced in authoritarian regimes. In a democratic environment old ideas can be challenged and rigorously criticized, albeit with some difficulty because of the human desire to hold on to old ideas, especially by the original proposers. Yet it is the essence of scientific think-

ing to propose alternative ideas, and then to test these alternative ideas against existing concepts. As is pointed out in the American Association for the Advancement of Science report *Science for All Americans*, "Indeed, challenges to new ideas are the legitimate business of science in building valid knowledge."[29] The principles upon which democracy is built are the very concepts that describe the scientific enterprise. Earlier it was pointed out that Polanyi felt science was organized as a republic of science in which independent people freely cooperated to explore and solve problems about the natural world. The values of a democratic society are the values that undergird Polanyi's concept of a republic of science.

The Scientific Enterprise and Teaching

The concepts that have been presented about the nature of science have implications for science teaching. There should be a consistency between discussions about the nature of science and the nature of teaching science. If we are trying to convey to students not only scientific facts and information, but the process of science, we are obliged to establish environments in classrooms that presume the same values that guide the practice of science. Questions that we can raise about our classrooms in this regard are as follows: To what extent are students given the opportunity to challenge ideas? Are activities planned in which there are alternative methods, answers, and solutions? Are students encouraged to identify and then try to solve problems relevent to themselves? Is it acceptable for students to disagree with ideas and propose new ones? Do the problems that students work on have any consequence in their lives now?

Science is defined as much by what is done and how it is done as it is by the results of science. To understand science as a way of thinking and doing as much as "bodies of knowledge" requires that science teaching emphasize the thought processes and activities of scientists. Thus we are led to explore one of the fundamental thought processes in science, namely, inquiry.

[27]Sayre, *Rosalind Franklin and DNA*, pp. 18–19. Used with permission.
[28]There is a large body of literature on women in science. See for example Sue V. Rosser, *Female-Friendly Science* (New York: Pergamon, 1990). Also refer to the readings in Chapter 11.

[29]*Science for All Americans* (Washington, D.C.: American Association for the Advancement of Science, 1989).

INQUIRY ACTIVITY   **1.3**

Surveying Students' Views of Science

Knowing what your students think of science can play an important role in influencing your day-to-day lesson plans. This activity is designed to help you detect and describe secondary students' view of science. Three methods are described in this activity.

Materials

Drawing paper, pencils and crayons, copies of science survey instrument.

Procedure

Choose one of the methods and survey 10 to 15 students on their view of science. After you have surveyed the students, arrange a time in which you can discuss the results of the survey with the students. (Note: this activity can also be carried out with a group of your teacher-training peers.)

Minds on Strategies

1. Summarize the results of your method of investigation by analyzing and drawing conclusions from the drawings, perhaps by creating a poster of the drawings, or tabulating the results of the survey.
2. Compare the method you used with the other two methods. How do students view science? Do they have a positive view of science? What is their image of a scientist? How do students compare science and technology? What effect does science have on society? Society on science?

Method 1: The Essay.   Have the students write an essay explaining what they think science is, and how scientists do their work. To help the students you might give them one of several sentence starters as a vehicle to begin. Some examples:

> Science is…
> Scientists believe that…
> The purpose of science is…

Method 2: The Drawing.   Have students make a drawing of a scientist. Ask the students to show the scientist at work. You might also have the students write a brief statement explaining their drawing.

Method 3: The Questionnaire.   Survey the viewpoints that middle and high school students hold on the following items, which are based on an instrument developed to survey Canadian high school students and which have been modified for use here.[30] Have a group of students respond to the items. Asterisk a different item on each student's questionnaire. Ask the student to write out reasons for his or her choice for that item. The purpose of each item is given in parentheses, which you do not need to include when you distribute the questionnaire to the students. The purpose of each item will be important when you analyze the results, however.

[30]Based on R. W. Fleming, "High-School Graduates' Beliefs about Science-Technology-Society. II. The Interaction Among Science, Technology and Society," *Science Education* 71(2): pp 163–186. Used with permission.

Opinions About Science

Instructions

Please check whether you agree or disagree with the following statements. If you cannot agree or disagree check "can't tell" for the statement. For the item with an asterisk, please write the reasons for your choice.

1. In the United States, science and technology have little to do with each other. (relationship between science and technology)

    _____ agree   _____ disagree   _____ can't tell

2. In the United States, technology gets ideas from science and science gets new processes and instruments from technology.

           ____ agree   ____ disagree   ____ can't tell

3. To improve the quality of living in the United States, it would be better to invest money in technological research rather than scientific research. (science, technology, and quality of life)

           ____ agree   ____ disagree   ____ can't tell

4. Although advances in science and technology may improve living conditions in the United States and around the world, science and technology offer little help in resolving such social problems as poverty, crime, unemployment, overpopulation, and the threat of nuclear war. (science, technology and social problems)

           ____ agree   ____ disagree   ____ can't tell

5. Scientists and engineers should be given the authority to decide what types of energy the United States will use in the future (e.g., nuclear, hydro, solar, or coal) because scientists and engineers are the people who know the facts best. (technocratic and democratic decision-making postures)

           ____ agree   ____ disagree   ____ can't tell

6. The U.S. government should give scientists research money only if the scientists can show that their research will improve the quality of living in the United States today. (mission-oriented perspective)

           ____ agree   ____ disagree   ____ can't tell

7. The U.S. government should give scientists research money to explore the unknowns of nature and the universe. (the basic science perspective)

           ____ agree   ____ disagree   ____ can't tell

8. Communities or government agencies should not tell scientists what problems to investigate because scientists themselves are the best judges of what needs to be investigated. (role of government and communities in the choice of research problems)

           ____ agree   ____ disagree   ____ can't tell

9. Science would advance more efficiently in the United States if it were more closely controlled by our government. (government control of research)

           ____ agree   ____ disagree   ____ can't tell

10. Science would advance more efficiently in the United States if it were independent of government influence.

           ____ agree   ____ disagree   ____ can't tell

11. The political climate of the United States has little effect on U.S. scientists because they are pretty much isolated from U.S. society. (effect of political climate on scientists)

           ____ agree   ____ disagree   ____ can't tell

SCIENCE TEACHING AND INQUIRY

Imagine science classrooms in which

- The teacher pushes a steel needle through a balloon and the balloon does not burst. The teacher asks the students to find out why the balloon didn't burst
- Students are dropping objects into jars containing liquids with different densities and recording the time it takes each object to reach the bottom of the jar. They are trying to find out about viscosity
- Students are using probes connected to a microcomputer to measure the heart rates of students before and after doing five minutes of exercise. They are investigating the effect of exercise on pulse rate
- Students are reading newspaper articles on the topic "toxic waste dumps" to form opinions about a proposed dump being established in their community

In each case students are actively involved in measuring, recording data, and proposing alternative ideas to solve problems, find meaning, and acquire information. In these situations students were involved in the process of inquiry. The greatest challenge to those who advocate inquiry teaching is the threat to the traditional and dominant role of the teacher in secondary education. We are going to discuss inquiry teaching first because of its relationship to the essence of science, but also because of the philosophical implications of siding with an inquiry approach. By taking a stand in favor of inquiry teaching, the teacher is saying, "I believe that students are capable of learning how to learn; they have within their repertoire the abilities, as well as the motivation, to question and to seek knowledge; they are persons and therefore learners in their own right, not incomplete adults." The philosophy of inquiry implies that the teacher views the learner as a thinking, acting, responsible person.

Characteristics of Inquiry

Inquiry is a term used in science teaching that refers to a way of questioning, seeking knowledge or information, or finding out about phenomena. Many science educators have advocated that science teaching should emphasize inquiry. Wayne Welch, a science

educator at the University of Minnesota, argues the techniques needed for effective science teaching are the same as those used for effective scientific investigation.[31] Thus, the methods used by scientists should be an integral part of the methods used in science classrooms. We might think of the method of scientific investigation as the inquiry process. Welch identifies five characteristics of the inquiry process:

Observation: Science begins with the observation of matter or phenomena. It is the starting place for inquiry. However, as Welch points out, asking the right questions is a crucial aspect of the process of observation.

Measurement: Quantitative description of objects and phenomena is an accepted practice of science and is desirable because of the value placed in science on precision and accurate description.

Experimentation: Experiments are designed to test questions and ideas and as such are the cornerstone of science. Experiments involve questions, observations, and measurements.

Communication: Communicating results to the scientific community and the public is an obligation of the scientist and is an essential part of the inquiry process. The values of independent thinking and truthfulness in reporting the results of observations and measurements are essential in this regard. As was pointed out earlier in the section on the nature of science, the "republic of science" is dependent on the communication of all its members. Generally, this is done by articles published in journals and discussions at professional meetings and seminars.

Mental Processes: Welch describes several thinking processes that are integral to scientific inquiry: inductive reasoning, formulating hypotheses and theories, and deductive reasoning, as well as analogy, extrapolation, synthesis, and evaluation. The mental processes of scientific inquiry may also include other processes such as the use of imagination and intuition.[32]

[31]Wayne Welch, "A Science-based Approach to Science Learning," in David Holdzkom and Pamela B. Lutz, *Research Within Reach: Science Education* (Washington, D.C.: National Science Teachers Association, 1985) pp. 161–170.

[32]Welch, "A Science-based Approach to Science Learning," pp. 161–170.

Inquiry teaching is a method of instruction, yet not the only method that secondary science teachers employ. However, because of the philosophical orientation of this book toward an inquiry approach to teaching, we will explore it first but also highlight three other methods (direct or interactive teaching, cooperative learning, and conceptual change teaching) that contemporary science teachers use in their classrooms.

Inquiry in the Science Classroom

Secondary science classrooms should involve students in a wide range of inquiry activities. The description of "scientific inquiry" is a general description of the inquiry model of teaching. The inquiry model of teaching presented in this book includes guided and unguided inductive inquiry, deductive inquiry, and problem solving.[33] Students engaged in a variety of inquiry activities will be able to apply the general model of inquiry to a wide range of problems. Thus, the biology teacher who takes the students outside and asks them to determine where the greatest number of wild flowers grow in a field is engaging the students in guided inquiry. The students would be encouraged to make observations and measurements of the flowers and the field, perhaps create a map of the field, and then draw conclusions based on these observations. In an earth science class, a teacher has been using inductive inquiry to help students learn about how rocks are formed and now wants the students to devise their own projects and phenomena to study about rocks. Inductive inquiry is a teacher-centered form of instruction.

On the other hand, unguided inductive inquiry is student-centered inquiry, in that the student will select the phenomena and the method of investigation, not the teacher. However, this does not mean that the teacher is not involved. The teacher may gather the class together for a brainstorming session to discuss potential phenomena to explore and study, based on the class's work to date. Small teams of students are then organized. The teams discuss the list of topics and phenomena generated in the brainstorming session, and then proceed to devise a project of their own.

In both forms of inductive inquiry, students are engaged in learning about concepts and phenomena by using observations, measurements, and data to develop conclusions. We might say the student is moving from specific cases to the general concept. In *deductive inquiry* the student starts with the big idea, conclusion, or general concept and moves to specific cases. In classroom situations, a physics teacher, for instance, may want the class to test the principle that light is refracted when it passes from one medium to another. The students perform a laboratory exercise in which they make observations of light as it is passed through water, glycerine, and glass. The lab is designed to help students confirm the concept. Many of the laboratory activities that are embedded in secondary science textbooks are deductive inquiry exercises. Is deductive inquiry teacher-centered or student-centered? Why do you think so?

Learning how to solve problems is another form of inquiry teaching. Challenging problems such as these can be investigated by secondary students: How did life originate on the earth? What will the consequences be if the earth's average temperature continues to rise? How can AIDS be prevented? What is the effect of diet and exercise on the circulatory system? What solid waste products are the most environmentally hazardous? What resources are most critically in short supply? Posing problems such as these brings real-world problems into the science classroom and furthers students' appreciation for the process of inquiry. Teachers who use problem solving are providing a perspective for students in which they will propose solutions to problems and make recommendations toward what should be done to change, improve, correct, prevent, or better the situation. Involving students in solving problems that are important to the culture and themselves is an important goal of science teaching. Paul DeHart Hurd comments that "a problem-oriented societal context for science courses provides the framework essential for the development of such intellectual skills as problem solving, decision making, and the synthesis of knowledge."[34]

Environments That Foster Inquiry

The classroom environment has psychological, sociological, philosophical and physical dimensions affected by the curriculum, the students, the teachers, the school, the community, and the nation. Yet in much of the research investigating classroom environments, the teacher's role is often seen as a powerful determinant of the classroom climate. In his book *Teaching*

[33]See Donald C. Orlich, "Science Inquiry and the Commonplace," *Science and Children*, March, 1989, pp. 22–24; and Donald Orlich et al., *Teaching Strategies: A Guide to Better Instruction*, 2nd ed. (Lexington, Mass.: D.C. Heath, 1985), especially chap. 8.

[34]Hurd, "New Directions in Science Education," p. 6.

Science as Inquiry, Steven Rakow points out that behaviors and attitudes of the teacher play an essential role in inquiry teaching, and he identifies the following as characteristic of successful inquiry teachers:

1. They model scientific attitudes
2. They are creative
3. They are flexible
4. They use effective questioning strategies
5. They are concerned both with thinking skills and with science content

Yet the overriding characteristic of the environments that foster inquiry is the attitude of the teacher toward the nature of students and the nature of science knowledge. Departing from the traditional role as primary givers of information, the science teacher that adopts the inquiry philosophy is more of a facilitator of learning and a manager of the learning environment. The student is placed in the center of the inquiry teacher's approach to teaching, thereby fostering the student's self-concept and development. These teachers bring to the classroom an assortment of approaches designed to meet the needs of the array of students that fill their classrooms. Although inquiry centralizes these teachers' philosophy, they look to other methods of teaching.

LIFE BEYOND INQUIRY

There are many approaches to help students understand science besides inquiry. There is more than one way to learn; there is more than one way to teach.

Students will come to your classroom with different learning styles, and more important, you will develop a teaching style that should not only be congruent with contemporary research on teaching but equally based on your personality, experience, values, and goals. We will explore a spectrum of approaches in Chapters 7 and 8 to help you develop a repertoire of methods and strategies. Here, very briefly, are a few of these methods.

Direct or Interactive Teaching

Think for a moment about the roles and interactions between students and teachers in most secondary science classrooms. You probably can envision the teacher working directly with the whole class, perhaps presenting a brief lecture, and then engaging the students by asking questions. The students might also be observed doing seat work, sometimes on their own, at other times with a partner or in a small group. Homework is assigned near the end of the class period, and students might get a head start on the assignment before class ends. Various models, sometimes refered to as direct or interactive instruction, have emerged over the past few years based on the relationship between observing teacher behavior and relating these behaviors to student learning. A large number of studies have supported a general pattern of key instructional behaviors, as shown in Table 1.3.

Cooperative Learning

Cooperative learning is an approach to teaching in which groups of students work together to solve problems and complete learning assignments.[35] Cooperative learning is a deliberate attempt to influence the culture of the classroom by encouraging cooperative actions among students.[36] Cooperative learning is a strategy easily integrated with an inquiry approach to teaching. Furthermore, science teachers have typically had students work in at least pairs, if not larger small groups, during lab. Cooperative learning strategies have been shown to be effective in enhancing problem solving and high-level thinking goals. We will explore a variety of cooperative learning models in Chapter 7 that are easily put into practice.

Conceptual Change Teaching

A growing number of science education practitioners and researchers have developed an approach to science teaching that focuses on the problem of conceptual change. According to these science educators, students come to the science class with naive conceptions or misconceptions about science concepts and phenomena. Further, these science educators suggest that concepts students hold are constructed; they are

[35]David Johnson and Roger Johnson, *Circles of Learning,* (Alexandria, VA: Association for Supervision and Curriculum Development, 1990).

[36]Jack Hassard, *Science Experiences: Cooperative Learning and the Teaching of Science* (Menlo Park: Addison-Wesley, 1990).

TABLE 1.3

DIRECT OR INTERACTIVE
INSTRUCTIONAL BEHAVIORS[a]

Daily Review of Homework	Teachers review the key concepts and skills associated with homework; go over the homework; ask key questions to check for student understanding.
Development	Teachers focus on prerequisite skills and concepts; introduce new ideas, concepts using an interactive approach including examples, concrete materials, process explanations, questioning strategies; check student understanding by using a highly interactive process utilizing questions, and designing a controlled practice activity for individual or group participation; teachers also use a lot of repetition.
Guided Practice	Teachers provided specific time during the lesson for uninterrupted successful practice; teachers use a sustained pace with a lot of momentum; students know that their work will be checked by the end of the period; teacher circulates about the room, checking student work, and asking questions, as needed.
Independent Practice	Teachers assign on a regular basis a homework assignment that is not lengthy and can be successfully completed by the student.
Special Reviews	Effective teachers conduct reviews once a week, preferably at the beginning of week. Focus is on the skills and concepts developed during the previous week; monthly reviews are conducted to review important skills and concepts.

[a]After T. L. Good and J. E. Brophy, *Looking Into Classrooms* (New York: Harper & Row, 1987), pp. 459–495.

neither discovered nor received directly from another person. To help students overcome their naive theories these educators suggest that teaching be organized into a series of stages of learning called the learning cycle. In most learning cycles, the first stage helps students detect and articulate their naive conceptions through exploratory activities; stage two focuses on comparing naive and "scientific" views to develop alternative conceptions; and the third stage provides experiences to encourage students to apply the concepts to new situations.[37]

We will explore other approaches to science teaching. For now, however, inquiry, direct instruction, cooperative learning, and conceptual change teaching should get you started.

[37]See R. Osborne and P. Freyberg, "Learning in Science: The Implications of Children's Science," J. D. Novak and D. B. Gowin, *Learning How to Learn* (Cambridge: Cambridge University Press, 1984).

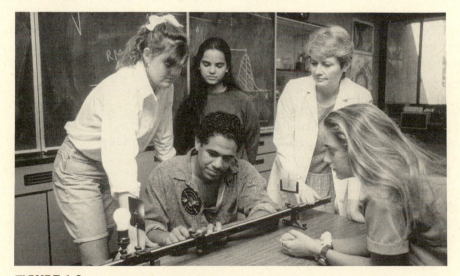

FIGURE 1.9 In Direct/Interactive Teaching, the teacher uses a variety of teaching approaches such as conducting demonstrations, or using audio-visual technologies to help students understand science concepts. © 1990 Tony Freeman.

THE SCIENTIST AND THE STUDENT: TWO CULTURES?

One idea that prevailed during the curriculum reform movement of the 1960s and 1970s was that students were like little scientists: curious, imaginative, interested, and inventive. One idea that has emerged in recent years, however, is that students are quite different than scientists and, indeed, come to science classes with naive theories and explanations for science concepts and phenomena. The assumption made by many science educators that scientists and students are very much alike is questionable and has perhaps contributed to many instructional problems such as motivation, success on standardized tests, and overall performance in science.

Students in middle schools and high schools are not scientists, and we shouldn't be anxious to make them into scientists. They are adolescents, some of whom may choose to be scientists later in life, but most will not. Let's look at some of the differences between scientists and adolescent students and consider some of the implications of these differences.

Some Differences

For starters, most people will not become scientists or engineers. In a typical school with 1000 students in the ninth grade, only 39 will earn baccalaureate degrees in science and engineering, 5 will earn master's degrees, and only 2 will complete the doctoral degree.[38] A more important difference, however, appears when we examine the thinking of scientists and students.

Adolescents are limited in the extent to which they can reason in the abstract, whereas scientists deal with abstractions as commonly as students deal with the concrete. As we will discuss in the next chapter, formal or abstract thinking eludes the majority of middle school and high school students. As some science education researchers point out, scientists work with concepts that have no directly observable circumstances (such as atoms and electric fields) and concepts that have no physical reality (such as potential energy).[39] Students, on the other hand, tend to consider only those concepts and ideas that result from everyday experience. As a result, many students will enter your classroom with misconceptions about scientific ideas, ideas that are firmly held and very difficult to alter.

Another difference between students and scientists has to do with what we might call explanations of concepts and phenomena. According to Osborne and Freyberg, students are not too concerned if some of their "explanations" are self-contradictory, and do not seem to distinguish between scientific (testable, disprovable) and nonscientific explanations. Scientists, on the other hand, are "almost preoccupied with the business of coherence between theories."[40] Osborne and Freyberg also point out that while scientists search for patterns in nature, for predictability, and the reduction of the unexpected, students are often interested in the opposite, thereby becoming interested in looking for the irregular, the unpredictable, and the surprise.

Students' interests, thinking processes, and the way they construct meaning are also limited by their prior knowledge, experiences, cognitive level, use of language, knowledge, and appreciation for the experiences and ideas of others. Scientists' interests, Osborne and Freyberg argue, follow from their participation in the scientific community. Scientists also have available to them a wide range of technical supports enabling them to extend their knowledge base by means of computer networks and data bases, telescopes, electron microscopes, and a common language.

Students and scientists have very different attitudes about science. The more school science students are exposed to, the less their interest in science. For example, in one study, nearly 67 percent of 9 year olds, 40 percent of 13 year olds, and only 25 percent of 17 year olds reported science class to be fun. This pattern persisted when students were asked whether science classes made them feel curious, successful, or comfortable.[41]

Bridging the Gap

How can science education be sensitive to the differences between students and scientists, and in such a way create science programs that nullify the negative trends in attitudes and achievement that have persisted for the past decade?

---

[38] B. G. Aldridge and K. L. Johnston, "Trends and Issues in Science Education," in *Redesigning Science and Technology Education*, ed. R. W. Bybee, J. Carlson, and A. J. McCormack, (Washington, D.C.: National Science Teachers Association, 1984).

[39] Osborne and Freyberg, "Learning in Science," pp. 55–56.

[40] Osborne and Freyberg, "Learning in Science," p. 56.

[41] R. E. Yager and J. E. Penick, "Perceptions of Four Age Groups Toward Science Classes, Teachers, and the Value of Science," *Science Education* 70(4): pp. 355–363.

FIGURE 1.10   High school kids are different than scientists. The odds are that only two out of a 1,000 will earn Ph.D's in science. Science education ought to pay closer attention to the other 98% of the students. (Credits: Top left—Rohn Engh Photography; Top right—Wallowitch)

One place to begin is with pedagogy. Research study after research study has described a picture of the science classroom as a pedagogical monotone. In most classrooms a teach-text-test model prevails. For the majority of students, this model leads to disastrous results. What is needed is greater variety in pedagogy. There are many pedagogical models of teaching that place the student in an active role, as opposed to the widespread practice of students being passive receivers of information. Chapter 7 presents inquiry, conceptual change, direct or interactive, small group, and individualized models of teaching, providing pedagogical varieties for the science teacher.

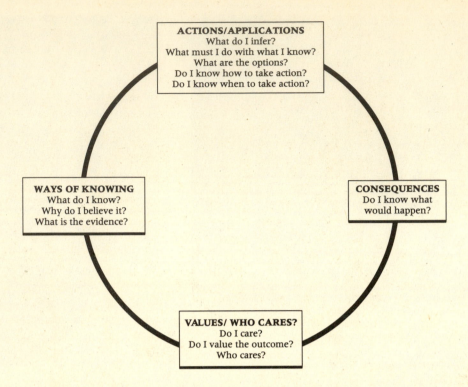

FIGURE 1.11 Mary Budd Rowe's Model of Science Education

Science educators need to reconsider the goals of science teaching, and to take a careful look at its objectives and the concepts that secondary school students are expected to learn. Many science educators suggest that the humanistic and societal aspects of science should be emphasized in the science curriculum. Some suggest that science teaching—and the resulting curriculum—should help students generate ideas about the science-society interface. The interaction between science and society can lead to topics in science teaching that focus on student interests, contemporary scientific, social, and planetary issues, and help students use science concepts and methods in the investigation of these problems.

The emphasis in science teaching is on the teaching of facts and concepts. Very little emphasis is placed on the application of scientific knowledge to societal problems, the consequences of scientific discoveries, or the values undergirding science. Mary Budd Rowe's proposal (Figure 1.11) for a shift in science education incorporates each of these components in a holistic paradigm for a science education program.[42] The sad aspect of this is that even with the attention given to this goal, American students have not done very well on standardized tests, especially when compared with their counterparts in other information-age societies.

As you examine this paradigm, keep in mind that Rowe suggests that for the most part, teachers and texts concentrate on the question "What do I know?" under the Ways of Knowing component. In fact, she points out that a typical high school science text averages between seven and ten new concepts, terms, or symbols per page. Making assumptions about the number of pages in the text, she estimates that students need to learn between 2400 to 3000 terms and symbols per science course. This translates to about 20 concepts per 55 minute period![43]

If these figures are even partially accurate, there is very little time for activities, for thinking about the applications, consequences, or values implicit in science concepts and theories. The implication of this data is that science lesson plans need to incorporate a spectrum of components.

Giving students a broader perspective on science will help bridge the gap separating them from the world of scientists. Most students will not become scientists, but they will become consumers of scientific discoveries and technological inventions, as well as decision makers at the polling booths.

[42]M. B. Rowe, "Science Education: A Framework for Decision-Makers," in L. L. Motz and G. M. Madrazo, Jr., eds., *Third Sourcebook for Science Supervisors* (Washington, D.C.: National Science Teachers Association, 1988), pp. 23–24.

[43]Rowe, "Science Education," p. 25.

THE STUDENTS WE TEACH: WHO ARE THEY?

Our students are adolescents. According to the dictionary, the term adolescence is derived from the Latin word *adolescence*, which means to "grow up." You will teach students in either a middle school, a junior high school or a senior high school who will range in age from 12 to 18. According to psychologists, adolescence is the period of life that is a transition from childhood to maturity. In your pedagogical training you most likely have explored in courses on human growth and development and educational psychology a variety of theories to explain the cognitive, psychosocial, physical, sexual, and emotional development of adolescents. We will explore in the next chapter some aspects of these theories, especially as they impinge on how adolescent students learn, what motivates them, and how to get them interested in learning science. However, it is important to realize that each of the 100 to 175 students you will teach each day is a whole person integrating a constellation of feelings, attitudes, abilities, motivations, physical attributes, and ambitions. It goes without saying that each student is unique. At the secondary education level, however, these students are taking on new roles, are influenced by peer pressure, wonder who they are, and what they will become. As a science teacher, it is easy to put subject matter first because of your love for science and your commitment to teach science and forget that you are teaching students. But science may be regarded as the vehicle that brings you and your students together. You have an opportunity to explore such questions as; How can science teaching contribute to the development of the students we teach? How can science teaching foster the development of healthy persons with positive self-concepts? How can I give the joy I sense about science to my students?

You will come to know your students in the context of a school, some of which will be large urban high schools with student bodies approaching 5000, others small rural schools with only 100 students in each grade. The context of school is important because the institution of the school itself plays a role in the daily life of the student. Let's look at some of the students you will teach and the kinds of environments they will encounter, to introduce the notion that students come first in any discussion of learning, science education, and the profession of teaching.[44]

[44]These scenarios were developed after B. M. Newman and P. R. Newman, *Adolescent Development* (Columbus, Ohio: Merrill, 1986), pp. 330–332.

INQUIRY ACTIVITY   1.4

The Student Comes First

This activity is designed to engage you in an exploration of secondary school students based on several vignettes of students. Your role is to participate with a team of peers in deciding upon some strategies that might interest the students in the vignettes.

Materials

    Student vignettes
    Chart paper
    Marking pens

Procedure

1. Working with a small team of peers, explore each student to discover some of the student's characteristics based on the vignette given here. Your role in the team is to

   a. List as many characteristics of the student as possible.
   b. Identify potential activities, events, and procedures that you think this student will enjoy participating.
   c. Identify potential problem areas for the student. Are there aspects of science class this student may not like? What can be done to mitigate this circumstance?

2. Prepare a profile on each student by summarizing the major characteristics, potential positive activities, and problem areas on a large sheet of newsprint or similar paper.

Minds on Strategies

1. Compare your team's analyses with other teams in your class. How do the analyses compare?
2. Are there any at-risk students in this group, that is, students who potentially would drop out of school? How can teachers make science a positive influence in this person's life?

Vignettes

Pedro.   He attends a large urban high school. He has an academic curriculum and works after school three days a week and on the weekends. He comes to school by car, usually picking up three or four friends on the way. After the half-hour ride to school, Pedro goes to his locker, talks with a few friends, and goes to his first class by 8:00 A.M. His girlfriend, Monica, is in his first class, so he is usually there on time in order to talk briefly with her. Pedro's first class is an ESL class, which he likes because the teacher treats all the students with great respect. The teacher told the students that he would be available after school to help with any language problems.

Pedro does not like math class, yet he feels frustrated because he loved math last year but finds that a lot of time is wasted because of a group of "troublemakers." The teacher is constantly diverted by these students, tending to their misbehaviors. In the regular English class, currently being taught by a student teacher, Pedro is asked to read a poem he wrote aloud. After reading the poem, Pedro is embarrassed and just shrugs his shoulders when asked to explain what he meant by the poem. In biology, the teacher has just begun a unit on amphibians. She announces to the class that in lab this week, lab teams will dissect a frog. Pedro is not too thrilled about this.

Mary.   She is thirteen years old and her family has just moved from New York City to a small town outside a large southern city. She rides the bus to school each morning, getting up at 6:00 A.M. and riding for an hour to reach school by 7:30 A.M. The school, a regional middle school, is in its second year of operation. Mary is a student in one of three eighth grade teams, each of which consists of over 100 students and 4 teachers. When she arrives at school, she goes to the cafeteria to eat breakfast before going to her homeroom. She starts the day with a bowl of cereal, a biscuit and a carton of orange juice.

Her first class, a prechemistry course, is taught by a first-year teacher who has a lot of energy and sometimes surprises the students with a mysterious demonstration. Although she doesn't like the subject of prechemistry she loves coming to this class because her teacher encourages all the students to learn and enjoy science. Her next class is math. All the students in her prechemistry class move en masse to math across the hall. She hates this class. The teacher, who is also on the job for the first time, embarrasses the students by pointing out their mistakes, especially when they are sent to the board to "work" problems. In interrelated arts, the teacher has invited a well-known potter to come to her class to show the students some of his work and how he makes pots. Mary is excited and looks forward to interrelated arts today. Mary's class eats at the first lunch period, which causes her to be hungry every day around 2:00 P.M. Her science teacher has asked for volunteers to form a science club. Mary is not sure whether she will go. She decides to ask two of her friends if they are going. They say they are.

Terrance.   He is the oldest in a family of five children. Both of his parents work, his mother during the day and his father at night. Terrance usually leaves home without breakfast but stops at the "QuickMart" for a sweet roll and a soft drink. Terrance is a very quiet student and tends to keep to himself except for two friends that he sees each day at lunch, and briefly after school while he walks to the bus station to go to work. Terrance reads at the sixth-grade level, and is having a great deal of difficulty with homework assignments in English and in U.S. history. He goes to his homeroom for attendance, and then his first class, general chemistry. His teacher explained that they are using a new book this year, and the emphasis is on chemistry in the community and how chemistry applies to everyday life. In chemistry class, the teacher is explaining the chemistry of digestion, and as he does Terrance's stomach is rumbling. When it rumbles very loudly, a student in the next seat starts to giggle, and pretty soon the back of the class is giggling.

Terrance likes school okay, but he would rather be at work. He is assistant manager of the evening shift of a pizza joint, and he feels very important in this role. He often wishes he was graduated from high school, and gives a great deal of thought to dropping out. His younger brother did.

Joe.   He is a 15 year old in the seventh grade in a junior high school. He is overweight and towers over all the students. He was retained twice in the third grade, and can't wait until next year when he will be able to drop out. Joe goes to bed late each night. He lives with his mother and two older sisters in an apartment in a high rise. Joe is a member of a gang, most of whom live in his apartment building or the ones just adjacent. His gang has not been involved in any violence but regularly meet and smoke dope. Joe knows that his teachers and especially the assistant principal keep an eye on him and his friends. Still Joe has smoked in the boys room and has come to class many times stoned. His first class is life science, and like all the remaining classes, Joe never shows up with his textbook, pencil, or paper.

Joe shows up to school on an average of three or four days a week, and is forever behind in his work. Joe shows some interest when the teacher does a hands-on activity but otherwise disdains reading the text or doing worksheet exercises. The teacher, however, rarely does a hands-on activity, because some of Joe's friends misbehave and cannot be trusted with the teaching materials.

Alicia. She is a senior at a small high school in a midsize city in a Western state. She, like most of the students attending the school, rides the bus. She would like to have her own car, but she can't afford one and her parents refuse to get her one. Alicia is fond of art and language, especially French, and is a member of the drama club. This year she decided to try out for one of the lead roles in *Romeo and Juliet*. In art class, the teacher has agreed to help the drama coach build the set for *Romeo and Juliet*. Alicia offers to make some quick sketches so that they can get an idea of how the different plans would look.

Derek sits down next to Alicia and starts talking about how bad he is going to feel when they leave school in a few months. After class, they go to the student lounge and talk for a while longer. Alicia suddenly feels sad herself and is happy to share her feelings with Derek. The conversation becomes more personal. Derek tells Alicia that he has liked her for a very long time but has been afraid to say anything because Alicia was dating another boy. The bell rings. Alicia and Derek have to go to separate classes. Alicia goes to advanced biology, where the students are giving reports. Her mind wanders to the conversation with Derek. At lunch she does her best to avoid looking at Derek. Derek finds her after school, and talks to her again. He has tears in his eyes and tells her how much he likes her. She tries to comfort him, but nothing helps. She goes home sad, angry, and flattered.

These are only a few of countless scenarios of students in secondary schools. You can add to it from your memory store. As you progress in the process of becoming a science teacher, I hope these scenarios will help you appreciate that students in your class experience a life "outside of science" that will have a significant effect on their learning, just as the theories and models of learning and teaching that will be presented in this book.

THE EFFECTIVE SCIENCE TEACHER: WHO ARE YOU?

As a science teacher, you will have a special role in bridging the gap between the world of science and scientists and the world of students in middle or junior and high schools. Are there characteristics common to teachers who do this effectively? There are two sources of information that will help us with this question. One is the result of the effective-teaching research over the past 25 years, and the other comes from the insight and wisdom of outstanding secondary teachers.

Effective Teachers

In recent years researchers have investigated the relationship between teacher behavior (strategies and methods of instruction) and student performance (conceptual learning, attitudes). Through a technique in which a large number of research studies are synthesized, researchers have found clusters of instruc-

tional strategies and methods that are related to increased cognitive outcomes.[45] At this stage in your study of science teaching, you have probably not mastered these behaviors. Instead, these characteristics will be viewed as advance organizers for our study of effective science teaching.

Individual teachers will vary considerably in their style, and in the specific strategies they use to help students come to enjoy and learn science. However, there appears to be a clustering of broad patterns of teacher behaviors that effective teachers employ. The list that follows has been paraphrased from Borich.[46]

---

[45]See C. A. Hofwolt, "Instructional Strategies in the Science Classroom," in D. Holdzkom and P. B. Lutz, eds. *Research Within Reach: Science Education* (Washington, D.C.: The National Science Teachers Association, 1985), pp. 43–57; and G. Borich, *Effective Teaching Methods* (Columbus, Ohio: Merrill, 1988), especially pp. 1–17.

[46]Borich, *Effective Teaching Methods*, pp. 1–17.

Clarity: Their presentation to the class is clear and understandable. Initial explanations are clear, logical, and easy to follow.

Variety: Teachers who show variety use a variety of behaviors to reinforce students, ask many and varied questions, use a variety of learning materials, equipment, and displays, and use hands-on materials.

Task Orientation: Teachers who spend more time on intellectual content than on procedures or classroom rules tend to have higher rates of achievement.

On-Task Behavior: This refers to the amount of time that students are actually engaged with learning materials and in activities. On-task behavior is closely related to classroom management behaviors of the teacher.

Success Rate: This characteristic is closely related to student self-esteem. Naturally, if students are succeeding at moderate-to-high rates, students are going to feel good about themselves as science learners and have positive attitudes about science. A key behavior here is the teacher's ability to design learning tasks that lead to high success rates but that are not dull or repetitive or viewed as a waste of time.

Using Student Ideas: Acknowledging, modifying, applying, comparing, and summarizing students' comments can contribute to a positive learning environment. Teachers who use student ideas are genuinely interacting with students, thereby enhancing student self-esteem.

Instructional Set: This refers to teacher statements made at the beginning of a lesson, or at transition points in the lesson, that help the students organize what is to come or what has happened before.

Questioning: Teachers can and do ask a variety of questions. Knowing what kinds of questions to ask and when to ask them seems to be important to student learning. Related to questioning is the behavior of "wait time," which refers to the amount of time teachers wait after asking students a question.

Enthusiasm: This humanistic behavior refers to the teacher's vigor, power, involvement, excitement, and interest during a class presentation. Enthusiasm manifests itself in the teacher's use of eye contact, gesturing, movement, supportive and approval behaviors, and a variety of teaching techniques, and in the teacher's love of science.

An Effective Teacher Speaks

There are many effective teachers in the United States. You will read in the Science Teachers Talk sections in the Science Teaching Gazette the comments made by several outstanding science teachers interviewed for this book. An eloquent spokesperson of effective teachers is Eliot Wigginton, one of the best-known high school teachers. Wigginton, who is a secondary teacher in Rabun County, Georgia, is probably best known for his Foxfire books, and community-based, experiential approach to teaching. In his book *Sometimes a Shining Moment: The Foxfire Experience—Twenty Years Teaching in a High School Classroom,* Wigginton grapples with the question, How do we make teaching work? His response was to outline "some overarching truths" about teaching, principles of teaching by which Wigginton differentiates effective from ineffective teachers. Following are brief comments about each of these overarching truths. Wigginton acknowledges that he is constantly searching for ways to answer the question, and says that he tries new approaches, rips apart his lesson plans, and hopes for those moments when things work and his students soar. Examine his list, and compare it to the categories of behaviors that researchers have found to characterize effective teachers. Here in brief are Wigginton's overarching truths about teaching.

They Have a Holistic View of the Subject Matter. This is the characteristic that tends to get students to recall their memorable teachers. "They made the subject come alive" or "She really loved her subject" are some of the things students remember about outstanding teachers. Wigginton claims that effective teachers see the interdependence of their own discipline with all others. They see their subject whole. They are the science teachers who see instantly every major science-related news event. Or as he says, carpet dyes and gymnasium floor waxes and cans of beer become subjects of chemical analysis, and the first spring flowers become targets of botanical scrutiny. These teachers help students relate their subject matter to the whole world, and he goes as far as to say that if there is no way to help students make linkages between this course and the whole world, and relate it to the students' lives, the course should not be offered at all!

They Know How Learning Takes Place. According to Wigginton, the effective teacher understands how learning takes place, knows how to apply the principles of learning, and believes that all students can learn. To Wigginton, this last notion is at the heart of the teaching profession. Teachers who know how learning takes place understand motivation in learning. They have moved away from extrinsic motives (candy, grades, a prize) toward intrinsic motives (natural curiosity, the desire for competence and mastery). They help students make connections between the information they are to learn and their own world. These teachers also know that learning takes place by doing, and that learning begins with *meaningful experiences* and then moves to the text or the teacher, and then on to evaluation, analysis, reflection, and a return to meaningful, hands-on experiences.

They Know Their Students.   Wigginton feels that effective teachers try to bring education and the lives of students together by getting to know them better. He points out that this is a tricky area, because many teachers feel distance should be kept from students—and students may not want to know us. However, Wigginton believes that to make instruction and the curriculum relevant to the students, educators must know their students. He says, for example, that "when I know students reasonably well, I know the extent of the demands I can make upon them; I know something about their talents and abilities and likes and dislikes, and thus I can lead them into educational activities with reasonable hope of success."[47]

They Make Careful Assumptions.   The central idea here is very simple: the best teachers never make negative assumptions about the potential of their students. Wigginton says that too often the disease model of education is at play, wherein students are viewed as defective and it is the job of schools and teachers to fix them. This is in stark contrast to his view that students have a variety of strengths and abilities and that it is the job of the school and teachers to take advantage of them and in the process turn areas of weakness around. As Wigginton says, we make cripples of students on the basis of the assumptions we make about them. As a future science teacher, this is especially crucial, given the negative

attitudes that prevail among students toward science. The evidence from research studies (especially the famous Pygmalion-effect study by Robert Rosenthal) suggests that students who receive attention, have higher goals set for them, and even have more demanded of them, often do advance academically. Students for whom we establish low expectations, and give less attention, often do not advance academically. Teachers' attitudes and the assumptions they make about students can play as important a role in cognitive learning as all the methods, strategies and materials of teaching that we use.

They Understand the Role of Self-esteem.   Effective teachers know that how students feel about themselves foretells how they perceive, react to, and perform in the world. Self-esteem is especially important in science teaching, again because of the negative connotations students have toward the study of science. One of the best remedies—and effective teachers know this—is to plan learning experiences that lead to student success, that build upon the student's dignity and self-worth.

Wigginton explores other characteristics of effective teachers. He suggests that these teachers also recognize their humanness, understand the nature of discipline and control, help students analyze and react to other adults, constantly engage in professional growth activities, and know how to avoid teacher burnout.

As you continue with your study of science teaching, come back to these characteristics resulting from both the science of research and the wisdom of practice.

---

[47]E. Wigginton, *Sometimes a Shining Moment*, p. 216. I recommend this book highly. It is used with all our interns in the middle school and secondary science teacher education program. It is not only inspirational but gives practical suggestions for teaching.

Science Teaching Gazette

Volume One

Case Study 1.
The Student Who Just Can't Relate to This "Physics Stuff"

The Case. A high school physics teacher typically asks students an open-ended evaluation question on each unit exam. On the first exam, the teacher receives this comment from one of the students: "Last year I related to biology so well. I saw things all around me. I just can't relate to this physics stuff. Pushes and pulls; how objects bounce off each other. So it does! So what?"

The Problem. Is this student's "complaint" about physics legitimate? Is relevancy to the students' everyday world something the science teacher should be concerned about? If you were the physics teacher how would you handle this situation? What would you say to the student?   □

Case Study 2.
Kids Are Just Like Scientists

The Case. Northside High School is a technology magnet school (grades 9–12) in a large metropolitan community. The science department consists of 15 teachers, 3 of whom are first-year teachers. Each of the three first-year teachers has been assigned to teach two sections of introductory physical science and three sections of survey biology. The veteran faculty in the science department are very committed to an inquiry approach to science teaching. Mr. Thomas, the science head, at the opening science department meeting, reaffirmed this by saying that instruction should be based on the assumption that "kids are like scientists." He went on to point out that students should be taught to think like scientists, that the

laboratory experiments should reinforce the way scientists do their work, thereby developing in students the same skills that scientists use. One of the first-year teachers, Ms. Jameson, in a private conversation with the other two first-year teachers, disagrees with this philosophy. She claims that some kids simply don't think the way the scientific community thinks and shouldn't be penalized because of it. She says that other approaches should be considered in formulating the underlying philosophy of instruction. One of her ideas is that science instuction should be more application oriented; that science instruction should show students how science relates to their own lives. She wants to discuss these ideas with

Think Pieces

Think pieces are short essays (generally no more than two pages) or posters (no more than one large poster board) that reflect your views (support by references is appropriate) on some topic or subject in science education. Throughout the book a few think pieces will be suggested for consideration, reflection, and action. Here are a few for this chapter:

- Why is inquiry teaching not a common teaching methodology in secondary science classrooms?
- In what ways should the teaching of science reflect the nature of science?
- What are your reasons for wanting to be a science teacher?
- What are the best qualities of a science teacher?
  □

the department head. One of the other beginning teachers suggests that she bring it up at the next department meeting in a week.

At the next meeting the department head reacts negatively to Ms. Jameson's ideas and says that the kids he teaches are quite capable of scientific thinking and that he therefore can't understand why students in her classes wouldn't be capable of this as well.

The Problem. You are the beginning teacher. What would you do in this situation? How do you respond to Mr. Thomas?   □

Research Matters: Teaching Authentic Science by Glen S. Aikenhead[48]

With the pressure to teach authentic science instead of ideal science, what can a science teacher do? Investigations of students' views on the scientific enterprise have explored the following questions:

1. What conditions are necessary for successful learning?

2. How can a teacher evaluate student views?

It is extremely important for a teacher to acquire reliable feedback about his or her own teaching.

Seldom do students pick up authentic images about science from the subtle comments or elements within a science course. Rather, the ideas about the scientific enterprise (its characteristics) and limitations must be the center of attention. Two examples will illustrate this point.

1. If a particular lab is intended to convey that human imagination is involved in scientific model building [e.g., the "black box" lab], students must be asked in the lab to address the role of human imagination; they must discuss it, and they must find it part of their evaluation in the course.

2. If your objective is to teach the distinction between science (the process of understanding natural phenomena) and technology (the process of designing techniques and implements to respond to human needs), projects and problems must be presented to help students distinguish between science and technology.

[48]Glen J. Aikenhead, "Teaching Authentic Science," Research Matters...To the Science Teacher Monograph Series. National Association for Research in Science Teaching. Used with permission.

Students will come to class with their naive ideas [alternative frameworks] about the scientific enterprise—often the conception of ideal science, or scientism. These ideals or mythical notions must be challenged before authentic images can be learned. Simulations, projects, reading assignments, field trips, forums, debates, and especially discussions are all appropriate teaching strategies to help them relearn or reformulate their views.

Most important, the teacher must realize that it usually takes a long time and considerable evidence for students to change their ideas about the nature of science. It may take a full year for students to realize that well-known scientific laws are not truths found in nature but are intellectual constructs. It may take three years before your students develop an accurate view of the methods of authentic science.

On the other hand, ideas that are new or relatively unfamiliar to students are quickly learned. Ideas such as recognizing that the scientific enterprise comprises both public and private science, each employing its own set of values, may be easily assimilated by students. Similarly, students are generally amenable to learning about the social and political contexts of science.

Teachers find that activities that focus on the nature of the scientific enterprise should be introduced early in a course, thus allowing for reinforcement of these ideas during the whole course. Reasonable time taken for such activities does not adversely affect student achievement or traditional science content.

Objectively scored questions do offer objectivity of scoring from the teacher's point of view, but the questions are woefully inadequate in assessing student beliefs. The students' interpretations, however, are clearly evident in their written responses. Student paragraphs, typically two to five sentences in length, are more clearly written when (1) Students are presented with a situation or statement; then (2) asked whether they agree, disagree, or can't tell; and then (3) asked to explain the reasons for their choice. This second point is important because it requires the student to take a position from which to argue. Students will often change their initial choice as they write their explanations. Somewhat surprisingly, similar paragraphs will be written for opposite initial positions.

Teachers trained in science are not comfortable or confident in grading student writing. Here are some guidelines aimed at removing this obstacle. First, familiarize yourself with a range of answers by reading a few responses anticipated to be good and poor. Assign three points to answers that deal with the topic in a sophisitcated way, given the nature of the instructional activity and the maturity level of your class. Two different explanations may each receive three marks, as long as they are logically constructed. Seldom is an answer considered right or wrong but is analyzed as a better or poorer response.

Zero points are assigned to poor or uninformed responses, while one or two points are awarded to more informed responses—those that reflect some degree of realistic understanding. Three points are awarded to answers that are clear, precise, and logical. It is very helpful to

(Continued on p. 34)

Science Is Not Words
by Richard P. Feynman[49]

I would like to say a word or two about words and definitions, because it is necessary to learn the words. It is not science. That doesn't mean just because it is not science that we don't have to teach the words. We are not talking about what to teach; we are talking about what science is. It is not science to know how to change Centigrade to Fahrenheit. It's necessary, but it is not exactly science. In the same sense, if you were discussing what art is, you wouldn't say art is the knowledge of the fact that a 3-B pencil is softer than a 2-H pencil. It's a distinct difference. That doesn't mean an artist gets along very well if he doesn't know that. (Actually you can find out in a minute by trying it, but that's a scientific way that art teachers may not think of explaining.)

In order to talk to each other, we have to have words and that's all right. It's a good idea to try to see the difference, and it's a good idea to know when we are teaching the tools of science, such as words, and when we are teaching science itself.

To make my point still clearer, I shall pick out a certain science book to criticize unfavorably, which is unfair, because I am sure that with little ingenuity, I can find equally unfavorable things to say about others.

There is a first-grade science book which, in the first lesson of the first grade, begins in an unfortunate manner to teach science, because it starts off on the wrong idea of what science is. There is a picture of a dog, a windable toy dog, and a hand comes to the winder, and then the dog is able to move. Under the last

[49]Based on R. P. Feynman. "What Is Science," *The Physics Teacher*, September, 1969, 313–320. Used with permission.

picture, it says "What makes it move?" Later on, there is a picture of a real dog and the question "What makes it move?" Then there is a picture of a motor bike and the question "What makes it move?" and so on.

I thought at first they were getting ready to tell what science was going to be about: physics, biology, chemistry. But that wasn't it. The answer was in the teacher edition of the book; the answer I was trying to learn is that "energy makes it move."

Now energy is a very subtle concept. It is very, very difficult to get right. What I mean by that is it is not easy to understand energy well enough to use it right, so that you can deduce something correctly, using the energy idea. It is beyond the first grade. It would be equally well to say that "God makes it move," or "spirit makes it move" or "movability makes it move." (In fact one could equally well say "energy makes it stop.")

Look at it this way. That's only the definition of energy. It should be reversed. We might say when something can move that it has energy in it, but not "what makes it move is energy." This is a very subtle difference. It's the same with this inertia proposition. Perhaps I can make the difference a little clearer this way: If you ask a child what makes the toy dog move, you should think about what an ordinary human being would answer. The answer is that you wound up the spring; it tries to unwind and pushes the gear around. What a good way to begin a science course. Take apart the toy; see how it works. See the cleverness of the gears; see the ratchets. Learn something about the toy, the way the toy is put together, the ingenuity

of the people devising the ratchets, and other things. That's good. The question is fine. The answer is a little unfortunate, because what they were trying to do is teach a definition of what is energy. But nothing whatever is learned.

Suppose a student would say, "I don't think energy makes it move." Where does the discussion go from there?

I finally figured out a way to test whether you have taught an idea or you have only taught a definition. Test it this way: You say, "Without using the new word which you have just learned, try to rephrase what you have just learned in your own language. Without using the word energy, tell me what you know now about the dog's motion." You cannot. So you learned nothing about science. That may be all right. You may not want to learn something about science right away. You have to learn definitions. But for the very first lesson is that not possibly destructive?

I think, for lesson number one, to learn a mystic formula for answering questions is very bad. The book has some others—"gravity makes it fall," "the soles of your shoes wear out because of friction." Shoe leather wears out because it rubs against the sidewalk and the little notches and bumps on the sidewalk grab pieces

(Continued on p. 35)

(Continued from p. 33)
compose a scoring scheme for each individual question.

Students usually need practice in writing paragraphs about the scientific enterprise. Homework and quizzes are useful places for this to begin. Students who are shy about writing need individual attention and encouragement. English and social studies colleagues may have suggestions for motivating students, as well as comments about scoring schemes and efficient use of marking time.   □

(Continued from p. 34)
and pull them off. To simply say it is because of friction is sad, because it's not science.

I went to MIT. I went to Princeton. I came home, and he (my father) said, "Now you've got a science education. I have always wanted to know something that I have never understood; and so my son, I want you to explain to me." I said yes.

He said, "I understand that they say that light is emitted from an atom when it goes from one state to another, from an excited state to a state of lower energy."

I said, "That's right."

"And light is a kind of particle, a photon, I think they call it."

"Yes."

"So if the photon comes out of the atom when it goes from the exited to the lower state, the photon must have been in the atom in the excited state."

I said, "Well, no."

He said, "Well, how do you look at it so you can think of a particle photon coming out without it having been in there in the excited state?"

I thought a few minutes, and I said, "I'm sorry, I don't know. I can't explain it to you."

He was very disappointed after all these years and years of trying to teach me something that it came out with such poor results.

What science is, I think, may be something like this: There was on this planet an evolution of life in a stage that there were evolved animals, which are intelligent. I don't mean just human beings, but animals which play and which can learn something from experience (like cats). But at this stage each animal would have to learn from its own experience. They gradually develop, until some animal could learn from experience more rapidly and could even learn from another's experience by watching, or one could show the other, or he saw what the other one did. So there came a possibility that all might learn it, but the transmission was inefficient and they would die, and maybe the one who learned it died too, before he could pass it on to others.

The world looks so different after learning science. For example, trees are made of air, primarily. When they are burned, they go back to air, in the flaming heat is released the flaming heat of the sun which was bound in to convert the air into tree, and in the ash is the small remnant of the part which did not come from air, that came from the solid earth, instead.

These are beautiful things, and the content of science is wonderfully full of them. They are very inspiring, and they can be used to inspire others.

Another of the qualities of science is that it teaches the value of rational thought, as well as the importance of freedom of thought; the positive results that come from doubting that the lessons are all true. You must here distinguish—especially in teaching—the science from the forms or procedures that are sometimes used in developing science. It is easy to say, "We write, experiment, and observe, and do this or that." You can copy that form exactly. But great religions are dissipated by following form without remembering the direct content of the teaching of the great leaders. In the same way, it is possible to follow form and call it science, but that is pseudoscience. In this way, we all suffer from the kind of tyranny we have today in the many institutions that have come under the influence of pseudoscientific advisors.

When someone says, "Science teaches such and such," he is using the word incorrectly. Science doesn't teach anything; experience teaches it. If they say to you, "Science has shown such and such," you might ask, "How does science show it? How did the scientists find out? How? What? Where?" It should not be "science has shown," but "this experiment, this effect, has shown." And you have as much right as anyone else, upon hearing about the experiments (but be patient and listen to all the evidence) to judge whether a sensible conclusion has been arrived at.

It is necessary to teach both to accept and to reject the past with a kind of balance that takes considerable skill. Science alone of all the subjects contains within itself the lesson of the danger of belief in the infallibility of the greatest teachers of the preceding generation. So carry on. □

Problems and Extensions

- What are your current conceptions about science teaching? What, in your opinion, does a person need to know to be a good science teacher?
- Interview several science teachers to find out what they like most and least about teaching? What are they doing to improve or change those aspects they like least? Do teachers seem to agree on what they like most about teaching?
- How important is imagination in science? In what ways can imagination be part of the secondary science classroom?
- What has your experience in science been like? Do you see science in the ways Bronowski, Franklin, Kelsey, Feynman, Polyani, or Sagan see it? What is your notion of science?
- Think back to your high school and college experience as a science student. Did you have any teachers that you admired or considered to be outstanding science teachers? What characterized their teaching?

Interviewing a Science Teacher

These are the questions I used to interview teachers in preparation for the Science Teacher Talk sections of future Science Teaching Gazette's. You might want to use some or all of the questions to design and carry out a study assessing local science teachers' views on teaching.

1. If you were to describe to prospective science teachers what you like most about science teaching, what would you say?
2. How do you accommodate students with different learning styles in your classroom?
3. Do you have a philosophy and set of goals that guide your instructional actions with your students? How do you communicate these to your students?
4. Is the inquiry model of teaching important in your approach to science teaching? Why?
5. What strategy of teaching do you find to be most effective with your students?
6. Try to describe your normal teaching method by ranking the following techniques in terms of the frequency with which you use them. (1 means "I use this technique most frequently of all," 2 means "I use this next most often," etc.). Leave blank those you almost never use at all. You may give several items the same number, indicating you use them with equal frequency.

    ____ Lectures that deviate widely from specific text material.
    ____ Lectures that are based closely on the text material.
    ____ Exercises taken from the text and performed by students in class.
    ____ Exercises taken from the text and assigned as homework.
    ____ Exercises I make up myself (or find in nontext sources) and have the students perform in class.
    ____ Exercises I make up myself (or find in nontext sources) and assign for homework.
    ____ Class discussions, normally of the question and answer variety, based closely on text material.
    ____ Open, wide-ranging class discussions that deviate widely from the text and the normal question and answer format.
    ____ Lab periods (for hands-on, scientific experimentation).
    ____ Students working in small groups on research or projects that will be shared with the rest of the class.
    ____ Students working independently on research or projects that will be shared with the rest of the class.
    ____ Students working independently with self-directed study materials other than texts (self-graded workbooks, prepared audio-visual materials, etc.).
    ____ Field trips.
    ____ Class activities and projects that will be shared with an audience outside the classroom (exhibits, publications, plays, debates, etc.).
    ____ In-class visitors from the community.
    ____ Students working independently (or in small teams) on projects specifically designed, with student input, to link real-world experiences in the community with the subject material being studied.
    ____ Other:

7. How do you accommodate students with special needs, such as those with learning or behavior disorders? What have you found to be effective with these students?
8. Do you deal with controversial issues in your classroom? If so, which ones, and how?
9. What tips would you give beginning teachers about planning and preparing lessons?
10. How do you manage your classroom? What is the most important piece of advice you would give a prospective teacher concerning classroom management?
11. How do you evaluate the progress of your students? If you were to evaluate your colleagues, what criteria would you use to judge their teaching?
12. Do you have a favorite science lesson or activity? What is its essence? Why is it your favorite?

RESOURCES

Alic, Margaret. *Hypatia's Heritage: A History of Women in Science from Antiquity through the Nineteenth Century*. Boston: Beacon Press, 1986.

Alic draws on a wealth of biographic and scientific evidence to describe the stories of women whose names have been left out of the history of science, whose work has been suppressed or stolen, and in many cases whose achievements have been denied. The book contains drawings, diagrams, and pictures to illustrate the work of women in science.

Berliner, David C., and Rosenshine, Barak V., eds. *Talks to Teachers*. New York: Random House, 1987.

This is a collection of essays discussing classroom instruction, student motivation and cognitions, teacher expectations, and instructional goals, testing, and planning.

Bronowski, Jacob. *Science and Human Values*. New York: Harper & Row, 1956.

Three essays that describe a central proposition: that the practice of science compels the practitioner to form a fundamental set of universal values.

Bronowski, Jacob. *A Sense of the Future: Essays in Natural Philosophy*. Cambridge, Mass: MIT Press, 1978.

Bronowski explores the philosophy of science through a series of essays on science, imagination, and invention.

Goodfield, June. *An Imagined World: A Story of Scientific Discovery*. New York: Harper & Row, 1981.

What is the nature of scientific discovery? Goodfield explores this question by documenting the processes one scientist went through on the road to discovery.

Gornick, Vivian. *Women in Science*. New York: Simon & Schuster, 1990.

In this book Gornick weaves a story of science out of interviews of a cross section of women scientists in all their diversity, exploring their emotional, intellectual, and physical experiences. The book contains more than 100 brief vignettes, which together help develop and redefine science.

Sagan, Carl. *Broca's Brain: Reflections on the Romance of Science*. New York: Random House, 1979.

This book will give you an overview of Sagan's thinking with tantalizing chapters such as "Can We Know the Universe?" "Reflections on a Grain of Salt," "White Dwarfs and Little Green Men," "Venus and Dr. Velikovsky," and "A Planet Named George."

Wigginton, Eliot. *Sometimes a Shining Moment: Twenty Years Teaching High School*. New York: Doubleday, 1985.

This is Wigginton's classic story of his work as a high school teacher in the north Georgia mountains, and how he developed the Foxfire approach in which students at Rabun County High School researched, wrote, and had published a series of best selling anthologies of the history and experiences of people in the southern Appalachian Mountains. The book provides insights into teaching and a philosophy of education that supports the experiential character of what science education should be today. It is a must-read book.

2

HOW STUDENTS LEARN SCIENCE

Placing a chapter on student learning near the beginning of this textbook reflects a priority shift among secondary science teachers. Increasingly, science educators are paying closer attention to how students learn and are encouraging practicing science teachers to implement the results in the classroom.

In this chapter we will explore some of the current theories that explain how students learn. There is no one theory that we can rely on to explain student learning. Rather, there are several theories that seem to complement each other and taken as a whole provide the science teacher of the 1990s with the most recent ideas and research on how students learn science. Research on learning has moved from an emphasis on behaviorism and Gestalt views, to Piagetian conceptions, and currently to a cognitive science perspective. Cognitive theories are increasingly becoming more inclusive and holistic. Cognitive psychologists are investigating realms of learning that were the dominion of social psychologists, namely student motivation and attitudes toward science. All of this is positive because the science teacher deals with students who are whole.

The students who appear in your classroom have unique ways of learning. Teachers for a long time have recognized this and have devised ways to accommodate students with different learning styles. The chapter concludes with an investigation of this important dimension of learning, namely student learning styles.

PREVIEW QUESTIONS

- What are the trends in achievement and attitudes toward science for secondary students over the past 20 years?
- How important is it to the secondary science teacher to know about learning theory?
- How do behavioral theories explain student learning in science?
- How do cognitive psychologists explain student learning?
- How do social psychologists explain student learning?
- What are behavioral, cognitive and social theories of learning and how do they differ?
- What was the contribution of theorists like Skinner, Bruner, Piaget, Vygotsky, McCarthy and Ausubel to secondary science teaching?
- How do learning styles of students influence learning in the classroom?
- What is metacognition, and how can metacognition help students learn science?

INQUIRY ACTIVITY 2.1

How Do Students Learn Science?

In this chapter you are going to be introduced to a numbers of theories that are used to explain how students learn science. Before you begin your study of this important topic, you will interview practicing science teachers to find out how they think students learn best, and under what circumstances.

Material

Tape recorder

Procedure

1. Working with a team of peers, put together a series of interview questions designed to pique the knowledge base of practicing science teachers about the basic question posed in this chapter: How do students learn science? Some questions that you might consider for the interview follow. You should also refer to the interview questions on page 36 that were used to interview the teachers for the Science Teachers Talk sections of the Science Teaching Gazette.
   a. In your experience, what have you found to be successful ways to help students learn?
   b. Do you believe that a particular theory is more effective than others in explaining how your students learn, for example, behaviorism versus cognitive theories?
   c. How do you motivate students to learn? What are some effective strategies?
   d. What do you think are some of the reasons students have difficulty learning science concepts?
   e. What do you do to help students that are "at risk" or have learning disabilities become interested in and understand science?
2. Once you have a set of questions, contact the teacher you wish to interview. Explain the purpose of the interview and how the information will be used. You should keep in mind human rights issues with regard to interviewing or gathering data on human subjects. Although this is not a research project, you should maintain the confidentiality of the individual, and respect the integrity of this person. You should explain fully why you are doing the interview, and how you will use the results. You should explain that the person's identity will not be used in any discussions with other members of your class, unless permission is granted by the teacher.
3. Visit the school, and conduct the interview. Make sure the teacher understands the purpose of the interview and grants you permission to tape-record the interview, if you plan to do so.

Minds on Strategies

1. What are the main concepts that teachers believe account for student learning? That is, how do practicing science teachers think students learn science?
2. What do science teachers think motivates students to learn science?
3. What are your opinions about how students learn science?

WHAT IS THE STATUS OF STUDENT LEARNING IN SCIENCE?

Since 1969 the National Assessment of Educational Progress (NAEP) has conducted a series (about every four years) of assessments involving nationally representative samples of 9, 13, and in-school 17 year old students.[1] These studies have provided information about trends in science learning among American students. We will examine very briefly some of the results of these studies in order to address the question, What is the status of student learning in science? As you will discover from this chapter, and perhaps even more so from your own experiences, the answer to this question depends on a lot of factors.

[1]I.V.S. Mullis and L. B. Jenkins, *The Science Report Card* (Princeton: Educational Testing Service, 1988).

The NAEP investigated a number of areas of student learning, including achievement in science, attitudes toward science, and experiences in science. We'll touch on each of these. Before we delve into this discussion, you should keep in mind that all test procedures have limitations. A test cannot measure in precise terms what a group of students know. For example, in the results that we will discuss, all questions were designed in a pencil and paper format. There was no performance testing; that is, students were not asked to make observations of real objects, develop inferences based on actual data collection procedures, or design and carry out an experiment. Each test has a built-in bias, and you should be aware of the limitations.

Let's start with a question about learning. Do you think that most 17 year old high school students could answer this question?[2]

Which of the following is the best indication of an approaching storm?
a. A seismogram that is a straight line
b. A decrease in barometric pressure
c. A clearing sky after a cold front passes
d. A sudden drop in the humidity

According to the results of the most recent NAEP study, the odds are that only 41 percent of the 17 year olds would answer this question correctly (correct response: b). Does this surprise you? Let's examine how the NAEP reported its findings, and what the results were.

To report achievement results, the NAEP created five levels of proficiency and used these as a mecha-

nism to examine trends and compare various groups of students. The five areas were defined as follows:

Level 150: Knows everyday science facts
Level 200: Understands simple scientific principles
Level 250: Applies basic scientific information
Level 300: Analyzes scientific procedures and data
Level 350: Integrates specialized scientific information

These levels were computed as the weighted composite of proficiency on five content-area subscales—Nature of Science, Life Science, Chemistry, Physics, and Earth and Space Science. What follows is a description of the nature of each level, including sample items from the 1986 NAEP science test.[3]

Level 150: Knows Everyday Science Facts. Students at this level know some general scientific facts of the type that could be learned from everyday experiences. They can read graphs, match the distinguishing characteristics of animals, and predict the operation of familiar apparatus that work according to mechanical principles. (See Figure 2.1 for sample item.)

Level 200: Understands Simple Scientific Principles. Students at this level are developing some understanding of simple scientific principles, particularly in the life sciences. For example, they exhibit some rudimentary knowledge of the structure and function of plants and animals. (See Figure 2.2 for sample item.)

[2]Mullis & Jenkins, *The Science Report Card*, p. 49.

[3]Mullis & Jenkins, *The Science Report Card*, pp. 4–5. These items were selected from representative samples of the items used on the NAEP test.

Which of the birds pictured below probably lives around ponds and eats snails and small fish?

Heron   Sparrow   Hawk   Swallow

FIGURE 2.1

Why may you become ill after visiting a friend who is sick with the flu?

- ○ The room your friend was in was too warm.
- ○ You ate the same kind of food your friend ate.
- ○ You did not dress properly.
- ● The virus that causes the flu entered your body.

FIGURE 2.2

Level 250: Applies Basic Scientific Information.

Students at this level can interpret data from simple tables and make inferences about the outcomes of experimental procedures. They exhibit knowledge and understanding of the life sciences, including a familiarity with some aspects of animal behavior and of ecological relationships. These students also demonstrate some knowledge of basic information from the physical sciences. (See Figure 2.3.)

Blocks A, B, and C are the same size. Blocks B and C float on water. Block A sinks to the bottom. Which one of the following do you know is TRUE?

- ● Block A weighs more than block B.
- ○ Block B weighs more than block C.
- ○ Block C weighs more than block A.
- ○ Block B weighs more than block A.
- ○ I don't know.

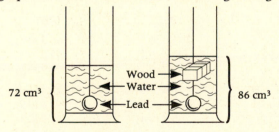

FIGURE 2.3

Level 300: Analyzes Scientific Procedures and Data.

Students at this level can evaluate the appropriateness of the design of an experiment. They have more detailed scientific knowledge and the skill to apply their knowledge in interpreting information from text graphs. These students also exhibit a growing under-

The volume of a block of wood can be found by suspending it in water, as the diagrams above show. What is the volume of the block?

- ● (86 - 72) cm³
- ○ 86 cm³
- ○ (72 - 86) cm³
- ○ (72 + 86) cm³

FIGURE 2.4

standing of principles from the physical sciences. (See Figure 2.4.)

Level 350: Integrates Specialized Scientific Information.

Students at this level can infer relationships and draw conclusions using detailed scientific knowledge from the physical sciences, particularly chemistry. They also can apply basic principles of genetics and interpret the societal implications of research in this field. (See Figure 2.5.)

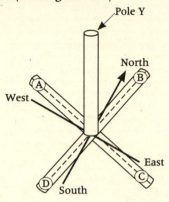

In the central United States at 8:00 a.m. on September 23 it is sunny, and the vertical pole shown in the diagram above casts a shadow. Which shaded area best approximates the position of the shadow?

- ● A
- ○ B
- ○ C
- ○ D

FIGURE 2.5

Trends in Science Achievement

If we compare the results that students showed from 1969 to 1986 (Figure 2.6), it is evident that overall student proficiency and achievement on the test continued to decrease from 1969 to 1982. Although scores increased from 1982 to 1986, the level of performance still remained below the 1969 level.

Let's examine the results for each of the five proficiency levels since 1969. These results give us more insight into student learning because we can talk more specifically. Table 2.1 shows the trends in the percentage of students at or above the five proficiency levels. If we focus on the 13 and 17 year olds, it is clear that secondary students are proficient in knowing everyday science facts (Level 150) and understanding simple scientific principles (Level 200). Beyond these levels, we begin to observe some problems. Although 17 year olds are able to apply basic scientific information, only 53 percent of 13 year olds were proficient in this level (250). Serious learning problems exist at levels 300 and 350. Only one in ten junior high or middle school students and four in ten 17 year olds were able to evaluate the appropriate-

National Trends Average
Science Proficiency, Ages 9, 13, and 17, 1969–70 to 1986 †

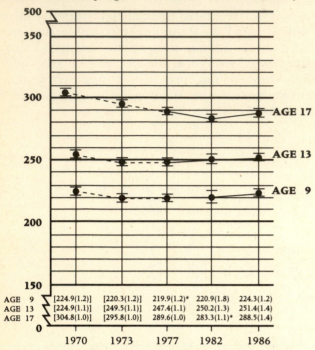

AGE 9	[224.9(1.2)]	[220.3(1.2)]	219.9(1.2)*	220.9(1.8)	224.3(1.2)
AGE 13	[224.9(1.1)]	[249.5(1.1)]	247.4(1.1)	250.2(1.3)	251.4(1.4)
AGE 17	[304.8(1.0)]	[295.8(1.0)]	289.6(1.0)	283.3(1.1)*	288.5(1.4)
	1970	1973	1977	1982	1986

[- - - -] Extrapolations based on previous NAEP analyses.
* Statistically significant difference from 1986 at the .05 level.
   Jackknifed standard errors are presented in parentheses.
†Note: While 9- and 13-year-olds were assessed in the spring
   of 1970, 17-year-olds were assessed in the spring of 1969.

95% CONFIDENCE INTERVAL

FIGURE 2.6   Trends in science academic learning. Mullis and
Jenkins, *The Science Report Card*, p.24

ness of procedures in science or able to interpret results. On level 350, which measured sophisticated knowledge in physical and biological sciences, virtually no 13 year olds and only 7 percent of the 17 year olds were proficient.

The authors of the NAEP study reported results for white, Hispanic, and black students. The largest gains, shown in Figure 2.7, were shown by groups of students considered to be at risk, including black and Hispanic students. However, minority students at ages 13 and 17 still appeared to perform at least four years behind their majority counterparts.

In an analysis of the so-called gender gap, overall proficiency of females was below that of males, and the trend from earlier years continues. According to data (Figure 2.8), the performance gap between 13 year old males and females across the five studies more than doubled, whereas the gap between 17 year old males and females narrowed.

Mullis and Jenkins, the authors of the NAEP report, *The Science Report Card*, summarize the results by making these points:

- Although recent progress has been made, most has occurred at the lower end of the proficiency scale
- Not only is it necessary to increase the average science proficiency of all students in our country, but it is also essential that the percentages of students reaching the higher ranges of proficiency be increased substantially

TABLE   2.1

TRENDS IN THE PERCENTAGE OF STUDENTS AT OR ABOVE THE FIVE PROFICIENCY LEVELS, 1977–1986, FOR STUDENTS AGES 9, 13 AND 17

Proficiency Levels	Age	1977	1982	1986
Level 150	9	93.6 (0.5)[a]	95.0 (0.5)	96.3 (0.3)
Knows everyday science facts	13	98.6 (0.1)	99.6 (0.1)	99.8 (0.1)
	17	99.8 (0.0)	99.7 (0.1)	99.9 (0.1)
Level 200	9	67.9 (1.1)[a]	70.4 (1.6)	71.4 (1.0)
Understands simple scientific principles	13	85.9 (0.7)[a]	89.6 (0.7)	91.8 (0.9)
	17	97.2 (0.2)	95.8 (0.4)	96.7 (0.4)
Level 250	9	26.2 (0.7)	24.8 (1.7)	27.6 (1.0)
Applies basic scientific information	13	49.2 (1.1)[a]	51.5 (1.4)	53.4 (1.4)
	17	81.8 (0.7)	76.8 (1.0)[a]	80.8 (1.2)
Level 300	9	3.5 (0.2)	2.2 (0.6)	3.4 (0.4)
Analyzes scientific procedures and data	13	10.9 (0.4)	9.4 (0.6)	9.4 (0.7)
	17	41.7 (0.8)	37.5 (0.8)[a]	41.4 (1.4)
Level 350	9	0.0 (0.0)	0.1 (0.1)	0.1 (0.1)
Integrates specialized scientific information	13	0.7 (0.1)	0.4 (0.1)	0.2 (0.1)
	17	8.5 (0.4)	7.2 (0.4)	7.5 (0.6)

Mullis and Jenkins, *The Science Report Card*, p. 39.

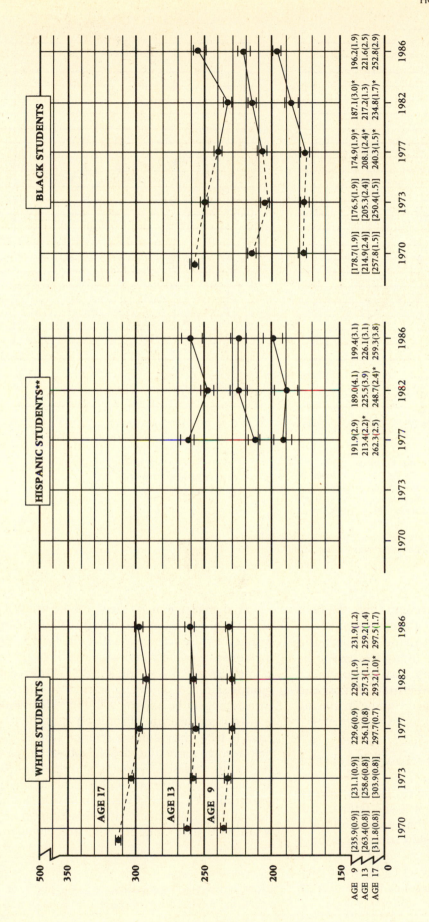

[- - - -] Extrapolations based on previous NAEP analyses.
* Statistically significant difference from 1986 at the .05 level. Jackknifed standard errors are presented in parentheses
** 1970 and 1973 data are unavailable for Hispanic students.
†Note: While 9- and 13-year-olds were assessed in the spring of 1970, 17-year-olds were assessed in the spring of 1969.

FIGURE 2.7 Comparison between White, Hispanic, and Black students ages 9, 13, and 17 in science proficiency, 1969–70 to 1986.† Mullis and Jenkins, *The Science Report Card,* pp. 28-29.

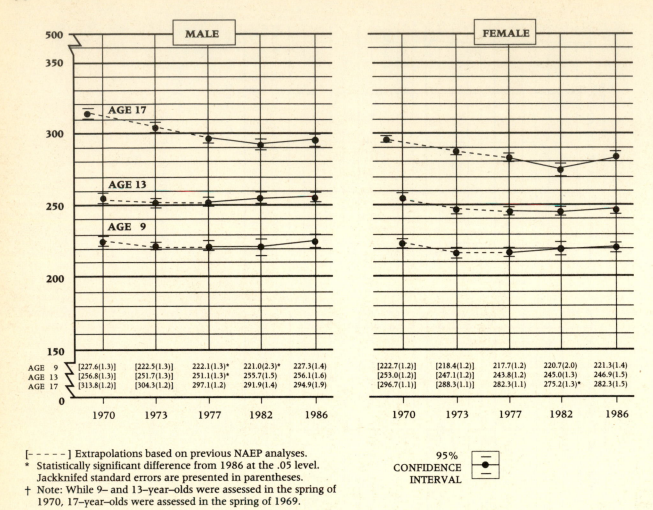

[- - - - -] Extrapolations based on previous NAEP analyses.
* Statistically significant difference from 1986 at the .05 level.
  Jackknifed standard errors are presented in parentheses.
† Note: While 9– and 13–year–olds were assessed in the spring of
  1970, 17–year–olds were assessed in the spring of 1969.

95% CONFIDENCE INTERVAL

FIGURE 2.8 Comparison of trends in average science proficiency by gender of 9, 13, and 17 year olds, 1969–70 to 1986. Mullis and Jenkins, *Science Report Card*, p. 30.

- Students' knowledge of science and their ability to use what they know appear remarkably limited
- Although progress has been made by black and Hispanic students and those living in the south-eastern region of the country, vast performance gaps remain across racial and ethnic groups, and essentially no progress has been made in closing the performance gap between males and females

Trends in Student Attitudes Toward Science

The NAEP studies have also investigated students' perceptions of science. As we will see later in this chapter, some theorists believe that student attitudes play an important role in determining science achievement. In fact, NAEP researchers found that a positive relationship appears to exist between attitudes toward science and proficiency in the subject, especially among eleventh-grade students.[4]

Students have many views on the utility and value of science. For example, when students were asked would they use science as an adult, only 51 percent of the seventh graders and 46 percent of the eleventh grades responded yes. And less than 43 percent of both groups said that knowing science would help them earn a living or be important in their life's work.[5]

Another measure of students' attitude toward science is how they believe science can be applied to help resolve particular global problems such as world starvation, diseases, overpopulation, and birth defects, and such environmental problems as depletion of natural resources, air and water pollution, and destruction of the ozone layer. An interesting finding by the NAEP was that students who perceived numerous

[4]Mullis and Jenkins, *The Science Report Card*, p. 123.
[5]Mullis and Jenkins, *The Science Report Card*, p. 125.

applications of science tended to have higher science proficiency.[6]

In 1977 and 1986, eleventh graders were more likely than seventh graders to believe that science could help alleviate national and global problems. Students were more likely in 1986 than 1977 to agree that science knowledge could help preserve natural resources, reduce air and water pollution, and prevent birth defects. However, students were not as likely to agree that science could solve the world starvation problem.[7]

[6]Mullis and Jenkins, *The Science Report Card*, pp. 128–129.
[7]Mullis and Jenkins, *The Science Report Card*, pp. 129–130.

INQUIRY ACTIVITY   2.2

....

Surveying Students' Knowledge of and Attitudes Toward Science

....

In this activity you will use a survey instrument to find out about student knowledge and attitudes toward science. The instrument will by no means enable you to make conclusive statements about student knowledge and attitudes about science, but it will give you a sampling of their performance and attitudes.

Materials

Science survey instrument (Figure 2.9)

Procedure

1. The survey instrument has two parts: (a) science content and (b) attitudes about science. The science content will consist of the five sample questions from the NAEP test given previously (see Figures 2.1 to 2.5).
2. The attitude part will consist of six attitudinal questions. You might want to have the students explain their response to the attitudinal questions.
3. Seek the permission of the school, the teacher, and the students to whom you will administer the survey instrument. Remember your responsibilities for the human rights of the students you will test. You should explain the nature of the survey to the contact person in the school and then follow the procedures set forth by the school.

Directions: Please answer each of the questions to the best of your ability. Circle your answer for each of the following questions.

Part I.
1. (See Figure 2.1)
2. (See Figure 2.2)
3. (See Figure 2.3)
4. (See Figure 2.4)
5. (See Figure 2.5)

Part II.
6. Knowing science will help me to earn a living.
   Strongly agree
   Agree
   Neutral
   Disagree
   Strongly disagree
7. I will use science in many ways when I am an adult.
   Strongly agree
   Agree
   Neutral
   Disagree
   Strongly disagree

8. Science will be important to me in my life's work.
   Strongly agree
   Agree
   Neutral
   Disagree
   Strongly disagree
9. When you have science in school, do you like it?
   Yes
   No
10. When you have science in school, does it make you feel interested?
    Yes
    No
11. Are there things you learn in science useful to you when you are not in school?
    Yes
    No

FIGURE 2.9   Science Survey Instrument

4. Administer the instrument to a sample of students in one of the following ways:
   a. Ten students in the sixth or seventh grade and in the eleventh grade.
   b. An equal number of boys and girls at any grade level.
   c. A small sample of students enrolled in a biology, a chemistry and a physics course.
5. After administering the instrument, tabulate the results in such a way that you can compare the obvious groups in your study, for example, sixth graders versus eleventh graders, boys versus girls.

Minds on Strategies

1. How did students perform at each of the proficiency levels?
2. What differences in proficiency existed in the groups you compared?
3. How do students perceive science? Are you surprised?
4. How do the results compare with the NAEP results?

STUDENTS AND SCIENCE LEARNING

Paradoxically, one of the significant events in the history of American science education was the launching of Sputnik by the Soviet Union in 1957. From that moment on, the American public, but especially the U.S. Congress, began to ask questions about the state of science education in the United States. Why were the Soviets the first to launch a satellite? Was American science and technology education inferior to Soviet science and technology education? Was the nature of secondary science education in jeopardy? The response of the U.S. Congress was to increase the funding of the National Science Foundation, which led to a plethora of science curriculum projects in the late 1950s, 1960s and 1970s. During this period of science education reform, scientists and science educators began a quest to find ways to improve science learning.

Science learning has at least two dimensions, namely learning about science and how to do science. Champagne and Hornig outline the main elements in these categories and point out that these two views of science learning are quite different. These two views and the subsets of science learning are outlined in Table 2.2.[8] As you can see from Table 2.2, students can learn about the products of scientific inquiry, which includes facts, concepts, principles and theories. The major emphasis in contemporary science programs is on this form of science learning. When newspapers report the proficiency of students in science, they are reporting what students know about this type of science learning. Yet there are many other things that students can learn about. They can learn about the nature of the scientific enterprise, values and attitudes, applications and the risks of science and technology, and science careers.

[8]A. B. Champagne and L. E. Hornig, *Students and Science Learning* (Washington, D.C.: American Association for the Advancement of Science, 1987).

Student learning also includes learning how to do science—learning how to apply the process skills and inquiry skills that are typical of scientific thinking. One of the directions that science education took after Sputnik was to pay attention to the process of science and emphasize inquiry in the development of science teaching materials. The "how to" aspect of science learning is the complement of the "learning about" side of science. Yet even with the emphasis on process and inquiry within the science education community, student learning in science showed disappointing results.

During the last 30 years a number of theories of learning have influenced the way science educators have looked at student learning in science. The early years of science education reform were dominated by the behavioral and developmental theories of learning. The behavioral psychologists, in particular, B. F. Skinner, have influenced science education enormously, as evidenced by the use of behavioral objectives to define student learning and in the classroom management systems that dominate contemporary science teaching. Developmental psychology, led by the work of Jean Piaget, influenced research in science education, especially during the 1960s and 1970s. The impact of Piaget can be seen in the recommendations of science educators to encourage direct experience with learning materials, and to be aware of the students' level of cognitive development.

Recently, however, science education has come under the influence of two other groups of psychological theorists, namely from social psychology and cognitive psychology. In the section that follows, we will examine each of these areas of psychology, how they explain learning, and the implications for science teaching.

TABLE 2.2

WHAT STUDENTS CAN
LEARN IN SCIENCE

They Can Learn About	They Can Learn How To	Processes They Undergo
Knowledge products of scientific inquiry (facts, concepts, principles, theories)	Act upon or apply information (evaluate, manipulate, solve problems)	Internalize values (the utility and risks of science and technology, habits of thought and conceptual skills, and who does science)
The nature of the scientific enterprise (world view, methods, habits of thought, approaches to problems)	Learn (strategies to seek and acquire new information, and to seek and acquire new skills)	Assess self (interest in science, capacity to do science)
Values and attitudes (of the scientific community, society at large, the local community, one's racial or cultural group, one's family)	Produce knowledge (question, test, evaluate)	Make choices (about studying science, science careers, and applying scientific knowledge and skills to daily life)
Applications and risks of science and technology (societal context, personal context)		
Science careers (what scientists do, who scientists are, how scientists get educated)		
Themselves (interest in science, capacity to do science)		

Champagne and Hornig, (1987), *Students and Science Learning*, pp. 3–5. Used with permission.

THEORIES OF
HOW STUDENTS LEARN

You know from your own experience as a student of science that theory is an important aspect of science. Scientists try to unravel and explain the nature of the physical and biological world by developing theories to explain and predict events and phenomena. It is not unusual for two scientists to look at the same set of data and give quite different explanations. Educators and psychologists attempt to explain the nature of student learning just as scientists attempt to explain the nature of the atom, earthquakes, biological change, and so on. And just as there are competing alternative theories in science, there are alternative theories to explain student learning.

What help can theory and associated research be to the beginning science teacher? Aren't theories too impractical for the real world of the everyday secondary science classroom? Perhaps. However, in this book an attempt will be made to show the value of being theoretically oriented and at the same time holding on to a firm practical base. Many science teachers value the work being done by science education researchers. In fact, the National Science Teachers Association (NSTA) has published a multivolume series entitled "What Research Says To The Science Teacher." These volumes describe the results of research on different topics and demonstrate practical applications that can be drawn from this work. Various journals, especially *Science Scope* (for middle school or junior high science teachers) and *The Science Teacher* (for secondary science teachers), include articles on science teaching that demonstrate the practical application of theory and research. And you will find a column in some of the Science Teaching Gazettes entitled "Research Matters...to the Science Teacher." This column is based on brief reviews of research published by the National Association for Research in Science Teaching.

Let's examine the notion of theory and then explore theories of human learning.

The word theory is derived from the Greek word *theoria*, which means vision. A theory of learning is a vision that a psychologist has to explain the complexity of human learning. Theories must be logical and consistent with past observations. They should also predict behavior. Theories that predict behavior and are supported by experimental tests tend to gain credibility. The four theories that we will explore—behavioral learning theory, developmental learning theory, cognitive learning theory and social learning theory—are supported to varying degrees in the literature. As with any theory, learning theories are mod-els of how psychologists think students learn, and they have their limitations, in terms of their ability to explain and predict behavior.

Table 2.3 outlines some of the main elements of each of the theories in terms of

1. How learning is defined
2. The nature of the learner
3. The conditions of learning

Study the chart before you continue, noting the similarities and differences between these three clusters of learning theories. Refer to the chart as you continue with this chapter.

BEHAVIORAL THEORIES OF LEARNING

What do you think the following have in common?

- A teacher says to a student, "I'm proud of you. Your science fair project was outstanding"
- A teacher asks a question. A student answers. The teacher says, "Good answer"

- A biology teacher gives extra credit for students who bring a newspaper clipping on bioethical issues

All of these are applications of behavioral theories of learning. Behavioral theories emphasize overt or observable behaviors in order to influence learning

TABLE 2.3

COMPARISON OF BEHAVIORAL, COGNITIVE, AND SOCIAL LEARNING THEORIES

Learning Theory	Definition of Learning	Nature of Learner	Conditions of Learning
Behaviorism Examples: Operant conditioning (Skinner)	Changes in the overt behavior of the learner as a result of experience	The mind of the student is a black box. Focus is on the interaction between the environment and behavior.	Learning is enhanced through operant conditioning as a result of reinforcement.
Cognitive Theories Examples: Developmental (Piaget) Constructivist Learning Model	Changes in the mental structures that contain information and procedures for operating on that information	Learner constructs knowledge and is actively seeking meaning.	Interacting with the physical world is crucial. What the learner brings to the learning environment and developmental differences in reasoning affect science learning.
Social Learning Theories Example: Cooperative learning	Changes in school achievement as well as in overt behaviors, but also changes in student attitudes and motivation	Learner is a product of social environment and an active participant in it. Emphasis on learner's inner states such as attitude and motivation that effect learner's choice.	Learning occurs as a result of social interactions in formal (school) and informal (family) settings.

Based on information in A. B. Champagne and L. E. Hornig, *Students and Science Learning* (Washington, D.C.: American Association for the Advancement of Science, 1987), pp. 6–13.

and determine if it has occurred. First, we will examine some of these behavioral theories and then identify some principles of behaviorism that can be applied to the classroom.

Conditioning

Conditioning, also referred to as classical conditioning, was one of the first theories of behaviorism. You are probably familiar with the famous experiments by the Russian scientist Ivan Petrovich Pavlov (1849–1936). He found out that a dog's behavior could be conditioned. Here is what he did. A dog, when presented with a piece of meat salivates. Pavlov called the meat an unconditioned stimulus resulting in an unconditioned response (salivation). To condition the response behavior, Pavlov rang a small bell at the same time the meat was presented. After several practice sessions in which the bell and meat were presented simultaneously, the dog eventually learned to salivate when the bell was rung without the meat. In this case the bell is the conditioned stimulus.

According to Hilgard and Bower[9], Pavlov's contribution rested as much on his methodology as on the results of his research. His theorizing and the care with which he explored numerous relationships provided a foundation for further behaviorists.

Connectionism

According to Edward L. Thorndike (1874–1949), the basis of learning is the association between sense impressions and impulses to action.[10] Such an association became known as a connection. Thorndike's theory of stimulus and response became the original S-R psychology of learning. Thorndike theorized that the most characteristic form of learning was trial and error, or learning by selecting and connecting.

Thorndike developed many of his ideas on learning by studying the behavior of animals (cats, dogs, fish, and monkeys) in what he called a problem box. The animal was placed in the problem box and confronted with a situation in which it has to escape from the box or attain food. In the case of the hungry animal trying to escape, it was learning to associate the stimulus (release mechanism) with the response (escape or food). Thorndike developed the "law of effect," which refers to the strengthening or weakening of a connection as a result of its consequences. Thorndike found that rewards strengthened connections but that punishments did not weaken them.

Operant Conditioning

Of all the theories of behavioral learning, operant conditioning probably has had the greatest impact on the science teacher. B. F. Skinner (1904–1990) proposed a class of behavior that is controlled by stimulus events that immediately follow an action. Skinner labeled these operant behaviors because they operated on the environment to receive reinforcement. According to Skinner, once an operant behavior occurs, its future rate of occurrence depends on its consequences. According to Skinner and other modern behaviorists, operant behavior is to be distinguished from responding behavior. Responding behavior involves the reactions of the smooth muscles and glands and includes reflexive reactions such as salivating, secreting digestive juices, shivering, and increased heart or respiratory rates. Operant behavior, on the other hand, involves the striated muscular system (muscles under voluntary control) and results in behaviors such as talking, walking, eating, and problem solving.[11] Responding behavior is controlled by preceding stimuli, as shown in Figure 2.10.

$$US \longrightarrow UR$$

(unconditioned stimulus: lemon juice in mouth)　　(responding behavior: salivation)

FIGURE 2.10　Classical Stimulus-Response Psychology

Operant behavior, on the other hand, is controlled by stimulus events that immediately follow the operant, as shown in Figure 2.11.

$$S \longrightarrow R \longrightarrow S$$

(discriminant stimulus)　(operant response)　(consequent stimulus)

FIGURE 2.11　Operant Conditioning

Skinner designed a special apparatus (others called it the Skinner box) for use with white rats, and later with pigeons. It consisted of a darkened sound-resistant box into which the rat (or pigeon) was placed. The box contained a small brass lever that, if pressed, delivered a pellet of food. Skinner connected the lever with a recording system that produced a

[9]E.R. Hilgard and G.H. Bower, *Theories of Learning* (New York: Appleton-Century-Crofts, 1966).

[10]Discussion based on E.R. Hilgard and G.H. Bower, *Theories of Learning*, pp. 15–47.

[11]W.C. Becker, *Applied Psychology for Teachers: A Behavioral Cognitive Approach* (Chicago: Science Research Associates, 1986), p. 16.

graphical tracing of the rat's behavior. The pigeon box was slightly different; the pigeon pecked for its food at a spot and received grain.[12]

Skinner's work resulted in the development of a number of principles of behavior that have direct bearing on science teaching. Two concepts stand out that have implications for the science teacher, namely, consequences and reinforcement. We will explore these two concepts and then return to them again in Chapter 10, on classroom management.

Skinner found that pleasurable consequences "strengthen" behavior, whereas unpleasant consequences "weaken it." Pleasurable consequences are referred to as reinforcers; unpleasant consequences are called punishers. The teacher who says "Alex, you did such a great job on your laboratory assignment that you can spend the remaining ten minutes working with one of the computer games" is making use of a reinforcer to strengthen classroom work. Let's examine reinforcers a little more carefully.

Behavioral psychologists differentiate between two types of reinforcers, primary and secondary. A primary reinforcer satisfies human needs for food, water, security, warmth, and sex. Secondary reinforcers are those that acquire their value by being related to primary reinforcers or other secondary reinforcers. Secondary reinforcers are the ones that are of greatest value to the science teacher. These reinforcers, which are also called conditioned reinforcers, can be divided into three categories: social, token, and activity.

Social Reinforcers

Social reinforcers are used very effectively by teachers to strengthen desired classroom behavior and learning. Social reinforcers, especially praise, can be a powerful tool for the science teacher. Although Brophy reports that praise is not used very frequently, he did report that most students enjoy receiving some praise and that teachers enjoy giving it.[13] To be effective, praise should be given only when a genuinely praiseworthy accomplishment has occurred. The teacher's praise should be informative, specifying some particulars about the noteworthy behavior or performance to help the student understand his or her successes. And finally, praise should be genuine, sincere, and credible.

Social behaviors can be divided into four clusters: praising words and phrases, facial expressions, near-

ness, and physical contact (Table 2.4) The use of these behaviors is common in many science classrooms.

Token Reinforcers

Token reinforcers are things such as points, gold stars, or chips that can be earned and have a reinforcing effect by pairing them with other reinforcers. Teachers have found the use of tokens very effective in managing student learning and classroom behavior. The use of a point system is especially effective in helping students learn how to manage their behavior, as well as contributing to their success as science learners. Many teachers set up their grading system using a point system; for example, points can be earned for homework, laboratory assignments, projects, quizzes, and tests.

Activity Reinforcers

Activity reinforcers, also referred to as the Premack principle, are a third group of reinforcers that teachers have found effective in the classroom. According to psychologist David Premack, more preferred activities can be used to reinforce less preferred activities.[14] According to the Premack principle, any higher-frequency behavior that is contingent on a lower-frequency behavior is likely to increase the rate of lower-frequency behavior. Thus, the teacher would set up a situation in which students, when they complete the less preferred activity, are permitted to participate in a more preferred activity. In the science classroom, some examples of the Premack principle would be "You may work in the computer game center when you finish cleaning the laboratory," "Those who score over ninety on today's quiz will not have to do homework tonight," or "If all students are in their seats when the bell rings, the class may have three minutes of free time at the end of today's class." These examples will not necessarily work in each situation. The science teacher must determine the preferred activities and then use them to strengthen the less preferred activities.

Theory into Practice

Behavioral theories of learning can be put into practice to the advantage of teachers and students alike. The underlying principle of behaviorism is "reinforce

[12]Many of us dismiss the relevance of operant conditioning, yet many classrooms resemble the Skinner box, in which the students behavior (cognitive, affective, and social) is reinforced with grades, stars, and awards.

[13]J. Brophy, "Teacher Praise: A Functional Analysis," *Review of Educational Research* 51, no. 1 (Spring 1981) pp. 5–32.

[14]D. Premack, "Reinforcement Theory," in D. Levine, ed., *Nebraska Symposium on Motivation* (Lincoln: University of Nebraska Press, 1965).

TABLE   2.4

EXAMPLES OF SOCIAL REINFORCERS
FOR THE SCIENCE CLASSROOM

Praising Words and Phrases	Facial Expressions	Nearness	Physical Contact
Good.	Smiling	Walking among the students	Touching
That's right.	Winking	Sitting in their groups	Patting head
Excellent.	Nodding	Joining the class at break	Shaking hand
That's clever.	Looking interested	Eating with the students	Stroking arm
Fine answer.	Laughing		
Good job.			
Good thinking.			
Great.			
That shows a great deal of work.			
You really pay attention.			
I like that.			
Show the class your model.			
That's interesting Joan. You're doing so well with the microscope.			
That was very kind of you.			

Based on W. C. Becker, *Applied Psychology for Teachers: A Behavioral Cognitive Approach* (Chicago: Science Research Associates, 1986), pp. 24–25.

behaviors you wish to see repeated."[15] According to Robert Slavin the following two principles should guide the use of reinforcement to increase desired behavior changes in the classroom:

1. Determine the behaviors desired from the students and reinforce them when they occur
2. Explain to students the behavior that is desired, and when they show the desired behavior, reinforce the students' behavior and explain why

Science teachers deal with a complex classroom environment, often involving the handling of dangerous materials or doing experiments involving safety issues. Specifying the behaviors that you expect in the laboratory, or whenever students are handling materials, and reinforcing them when they occur will help the students become independent and responsible science learners. Some science teachers post these

[15]R.E. Slavin, *Educational Psychology: Theory into Practice* (Englewood Cliffs, N.J.: Prentice-Hall, 1988), p. 116.

behaviors in the science laboratory after spending time teaching and explaining these behaviors.

Skinner's concept of operant conditioning can be applied to the science classroom in many ways, but three seem clearly the most important, namely in (1) the use of classroom questions and associated techniques, (2) developing a positive classroom climate and (3) the development of programmed teaching materials.

The Use of Classroom Questions

One of the most common teaching behaviors that you will employ is that of asking students questions. Questions can be directed at the whole class, at small groups of students, or at individuals. The technique (developed fully in Chapter 8) involves this sequence:

- The teacher asks a question
- The teacher pauses for at least three seconds (to give students a chance to think of an answer)
- The teacher calls on a student
- The student responds

FIGURE 2.12   The microcomputer can be used as a modern version of Skinner's teaching machine. Tutorial and drill-and-practice programs can be used to reinforce student learning. Courtesy Apple Computer, Inc.

- The teacher responds to the student (choices include praising the student or, using the student idea)

Classroom Climate

Although developed more fully in Chapter 10, Skinner's work can be applied to creating a positive classroom climate by having the teacher respond to student successes rather than failures. For example, rather than pointing out what students are doing wrong, point out what they are doing right. When a student answers a teacher's question with a partially correct response, the teacher should pick up on the correct aspect of the answer to reinforce the student's contribution. Table 2.5 outlines the steps that can contribute to a positive learning environment based on Skinner's concept of operant conditioning.

TABLE   2.5

···

CREATING A POSITIVE CLASSROOM ENVIRONMENT BY MEANS OF OPERANT CONDITIONING

···

Step 1: Analyze the Environment	Step 2: Make a List of Positive Reinforcers	Step 3: Select Sequence of Behaviors to be Implemented	Step 4: Implement Program, Maintain Records of Behavior and Make Changes
Identify positive and undesirable student behaviors receiving reinforcement. What behaviors receive the punishment? What is the frequency of punishment? Have these behaviors been suppressed?	Determine students' preferred activities (students can contribute to this). Consider using punished behaviors as reinforcers. If talking with peers is a disruptive behavior, consider using it (time to talk with peers) as a reinforcer.	Implement a positive reinforcement program. Instead of punishment for tardiness, reward students for being on time.	Make sure classroom rules are clear (see Chapter 10). Make sure students know how to earn reinforcement. Implement reinforcement schedule.

Based on Bell-Gredler, *"Learning and Instruction,"* pp. 105–106.

Programmed Teaching Materials and Computer-Assisted Instruction

Skinner designed teaching machines that controlled the students' progress through a body of material. The teaching machine, usually by means of questions or fill-in-the-blank statements, provided reinforcement for right answers (by confirming them, allowing the student to move ahead). The teaching machine was a vehicle for programming instruction, as well as providing an environment in which students could work at their own rate. Textbook and workbooks were written to teach information about a variety of subjects, especially in science. The textbooks were equipped with a card that could be inserted in a page holder. As students worked through each statement or question, they would slide the card so that the correct answer would appear. Early machines and textbooks were limited in the types of reinforcements they could provide. However, with the development of the microcomputer, not only can a variety of reinforcements be provided (a pleasant sound, a voice), but the software can be programmed to provide a variety of feedback for various responses. Drill and practice, tutorial, and some game software programs are based on the Skinnerian concept of programmed instruction.

Science teachers can make use of Skinner's concept of programmed instruction by providing students the opportunity to work in the microcomputer environment. Drill and practice and tutorial programs are available in most science content areas. Although not the most avant-garde use of the computer, they can help students learn science information efficiently, and with little teacher effort.

Behaviorism has contributed greatly to the teaching of science, but like any theory of learning, it has its limitation and rivals. In the past 20 years there has been an increase in the variety of learning theories to explain student learning in science. Science teachers have available to them the theories proposed by a group of psychologists known as cognitive scientists. These psychologists shift their attention away from observable behaviors and instead focus on skills associated with memory, perception, and conceptual processes, as well as processes related to problem solving, concept discovery, and the use of rules. We now turn our attention to cognitive theories of learning.

COGNITIVE THEORIES OF LEARNING

Cognitive theories of learning had their roots in Gestalt psychology. During the period of time that behavioral theories were being developed, a competing and alternative group of theories was developed by Gestalt psychologists. Unlike behaviorism, Gestalt theory emphasized the importance of mental processes. In Gestalt psychology, the learner reacts to meaningful wholes. According to Gestalt psychologists, learning can take place by discovery or insight. The idea of insight learning was first developed by Wolfgang Köhler; he described experiments with apes in which the apes could use boxes and sticks as tools to solve problems. In the box problem, a banana is attached to the top of a chimpanzee's cage. The banana is out of reach but can be reached by climbing upon and jumping from a box. Only one of Köhler's apes (Sultan) could solve this problem. A much more difficult problem that involved the stacking of boxes was introduced by Köhler. This problem required the ape to stack one box on another and master gravitational problems by building a stable stack. Köhler also gave the apes sticks, which they used to rake food into the cage. Sultan, Köhler's very intelligent ape, was able to master a two-stick problem by inserting one stick into the end of the other to reach the food. In each of these problems the important aspect of learning was not reinforcement but the coordination of thinking to create new organizations (of materials). Köhler referred to this behavior as insight or discovery learning.[16]

Cognitive thinking and research got a boost from the launching of Sputnik. As was mentioned earlier this sparked a massively funded curriculum reform effort in the United States in science and mathematics. The emphasis of the reform was to produce students who could think like scientists through discovery and inquiry learning and active student involvement. This emphasis brought together scientists, teachers, and psychologists. One of the most influential psychologists during this period in science education was Jerome Bruner, director of the Harvard Center for Cognitive Studies.

[16]Discussion based on Hilgard and Bower, *Theories of Learning*, pp. 229–231.

Bruner and Discovery Learning

Because of Jerome Bruner's connection with the National Science Foundation curriculum development projects of the 1960s and 1970s, his thinking had a powerful effect on approaches to science learning. Bruner believed that students learn best by discovery and that the learner is a problem solver who interacts with the environment, testing hypotheses and developing generalizations.[17] Bruner felt that the goal of education should be intellectual development, and that the science curriculum should foster the development of problem-solving skills through inquiry and discovery.

Bruner said that knowing is a process rather than the accumulated wisdom of science as presented in textbooks. To learn science concepts and to solve problems, students should be presented with perplexing (discrepant) situations. Guided by intrinsic motivation, the learner in this situation will want to figure the solution out. This simple notion provides the framework for creating discovery learning activities.

Bruner described his theory as one of instruction rather than learning. His theory has four components, as follows:[18]

Curiosity and Uncertainty

Bruner felt that experiences should be designed that will help the student be willing and able to learn. He called this the predisposition toward learning. Bruner believed that the desire to learn and to undertake problem solving could be activated by devising problem activities in which students would explore alternative solutions. The major condition for the exploration of alternatives was "the presence of some optimal level of uncertainty."[19] This related directly to the student's curiosity to resolve uncertainty and ambiguity. According to this idea, the teacher would design discrepant-event activities that would pique the students' curiosity. For example, the teacher might fill a glass with water and ask the students how many pennies they think can be put in the jar without any water spilling. Since most students think that only a few pennies can be put in the glass, their curiosity is aroused when the teacher is able to put between 25 and 50 pennies in before any water spills. This activity then leads to an exploration of displacement, surface tension, and variables such as the size of the jar and how full the glass is. In this activity the students would be encouraged to explore various alternatives to the solution of the problem by conducting their own experiments with jars of water and pennies.

The Structure of Knowledge

The second component of Bruner's theory refers to the structure of knowledge. Bruner expressed it by saying that the curriculum specialist and teacher "must specify the ways in which a body of knowledge should be structured so that it can be most readily grasped by the learner."[20] This idea became one of the important notions ascribed to Bruner. He explained it this way: "Any idea or problem or body of knowledge can be presented in a form simple enough so that any particular learner can understand it in a recognizable form."[21]

According to Bruner, any domain of knowledge (physics, chemistry, biology, earth science) or problem or concept within that domain (law of gravitation, atomic structure, homeostasis, earthquake waves) can be represented in three ways or modes: by a set of actions (enactive representation), by a set of images or graphics that stand for the concept (iconic representation), and by a set of symbolic or logical statements (symbolic representation). The distinction between these three modes of representation can be made concretely in terms of a balance beam, which could be used to teach students about quadratic equations.[22] A younger student can act on the principles of a balance beam, and can demonstrate this knowledge by moving back and forth on a seesaw. An older student can make or draw a model of the balance beam, hanging rings and showing how it is balanced. Finally, the balance beam can be

[20]Bruner, *Toward a Theory of Instruction*, p. 41.
[21]Bruner, *Toward a Theory of Instruction*, p. 44.
[22]This example was suggested by Bruner in *Toward a Theory of Instruction*, p. 45.

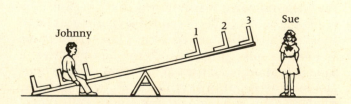

Where should Sue sit to balance the beam?

FIGURE 2.13 The Balance Beam Problem: Here, through an enactive representation (a set of actions), the student can understand the law of moments in a mode that is understandable and recognizable.

[17]J.S. Bruner, *Toward a Theory of Instruction* (Cambridge, Mass.: Harvard University Press, 1967); and M.E. Bell-Gredler, *Learning and Instruction* (New York: Macmillan, 1986).
[18]Bruner, *Toward a Theory of Instruction*. pp. 40–42.
[19]Bruner, *Toward a Theory of Instruction*, p. 43.

described verbally (orally or written), or described mathematically by reference to the law of moments. The actions, images, and symbols would vary from one concept or problem to another, but according to Bruner, all knowledge can be represented in these three forms.

Sequencing

The third principle was that the most effective sequences of instruction should be specified. According to Bruner, instruction should lead the learner through the content in such a way as to increase the student's ability to "grasp, transform, and transfer"[23] what is learned. In general, sequencing should move from enactive (hands-on, concrete) to iconic (visual) to symbolic (descriptions in words or mathematical symbols). However, this sequence will be dependent on the student's symbolic system and learning style. As we will see later, this principle of sequencing is common to theories developed by Piaget, as well as other cognitive psychologists.

Motivation

The last aspect of Bruner's theory is that the nature and pacing of rewards and punishments should be specified. Bruner suggests that movement from extrinsic rewards, such as teacher's praise, toward intrinsic rewards inherent in solving problems or understanding the concepts is desirable. To Bruner, learning depends on knowledge of results when it can be used for correction. Feedback to the learner is critical to the development of knowledge. The teacher can provide a vital link to the learner in providing feedback at first, as well as helping the learner develop techniques for obtaining feedback on his or her own.

Piaget's Developmental Theory

Cognitive theory has perhaps been influenced by no one more than the Swiss psychologist, Jean Piaget. Although Piaget's work began in the early part of the twentieth century, it had little influence in the United States until the 1960s. Piaget's chief collaborator, Barbel Inhelder, attended the 1959 Woods Hole Conference chaired by Jerome Bruner. This conference precipitated interest in Piaget's ideas on cognitive development and led to a great deal of attention

by educators (science educators, especially) and psychologists.[24]

Piaget's theory focuses on the development of thinking patterns from birth to adulthood. To Piaget, learning is an active process and is related to the individual's interaction with the environment. According to Piaget, intellectual development is similar to the development of biological structures, such as those shown in mollusks. Margaret Bell-Gredler describes Piaget's thinking this way.

> He found that certain mollusks, transported from their calm-water habitat to turbulent wind-driven waters, developed shortened shells. This construction by the organism was essential for the mollusks to maintain a foothold on the rocks and thereby survive in rough water. Furthermore, these biological changes, which were constructed by the organism in response to an environmental change, were inherited by some descendents of the mollusks. The organism, in response to altered environmental conditions, constructs the specific biological structures that it needs.[25]

To Piaget, this biological development describes the nature of intellectual development. Intelligence is the human form of adaptation to the environment. Piaget, and other cognitive scientists, theorize that cognitive structures (mental structures) grow and develop through a process of interaction with the environment. How do these mental structures develop?

Development of Mental Structures

According to Anton Lawson and John Renner, both science educators, the most important idea in Piagetian theory is that mental structures (which they nicely describe as mental blueprints) are derived from the dynamic interaction of the organism and the environment by means of a process called self-regulation or equilibration.[26] Lawson and Renner also point out that the mental structure comes not from the organism or the environment alone but from the organism's own actions within the environment. This idea that the individual constructs the mental structure is an underlying principle in all cognitive theories that will be discussed in this chapter.

Let's look at an example of self-regulation in the classroom of a teacher who applies Piaget's theory. The teacher is presenting a unit on air pressure to a group of junior high students. A demonstration is in

---

[23]Bruner, *Toward a Theory of Instruction*, p. 49.

[24]Bell-Gredler, *Learning and Instruction*, p. 221.
[25]Bell-Gredler, *Learning and Instruction*, pp. 193–194.
[26]A.E. Lawson and J.W. Renner, "Piagetian Theory and Biology Teaching," *The American Biology Teacher*, 37 (6): pp. 336–343.

progress. The teacher heats up a metal can containing a small amount of water. After a few minutes steam begins to rise out of the can. The teacher removes the can from the heat source (a hot plate) and caps it. Then the teacher waits. In a few moments the can begins to cave in, bends to the side and falls over. The teacher then asks students to describe and explain what they observed. For many of the students this event is a contradiction or a discrepancy. Discrepancies produce a state of disequilibrium in which you are literally thrown off balance.

According to Lawson and Renner, the students present mental structures are inadequate to explain the crushed can and they must be altered. By means of interaction with the environment (doing experiments and activities on the effects of pressure changes on material objects, for instance), the students can assimilate the new situation and build new structures. Later in this chapter, in the section that deals with conceptual change teaching, you will find out that it is not an easy matter to change students' current mental structures or conceptions.

According to cognitive scientists, there are three additional factors that influence the development of mental structures: experiences with the environment, maturation, and the social environment.

Experience with the environment is essential, since the interaction with the environment is how new structures are made. Piaget distinguishes between two types of experiences, namely concrete and abstract (logico-mathematical). It is important to remind ourselves that knowledge is constructed through experience, but the type of knowledge will be dependent on the type of experiences the individual is engaged in. Concrete experiences are physical experiences in which the student has a direct encounter with physical objects. Piaget suggested that interaction with material objects was essential to the development of thinking. Lawson and Renner suggest that science teachers should use the laboratory to precede the introduction of abstract ideas.[27]

Students need more than experiences with the environment; they also need to interact socially. Here the role of language and verbal interaction in the classroom environment will accelerate or retard cognitive development. The crucial aspect of this factor is that students be given the opportunity to examine and discuss their present beliefs and conceptions. The science teacher should not only provide concrete and abstract experiences with the environment but must provide for social interaction via the use of language.

Small and large group discussions are crucial to the development of cognitive structures.

The third factor facilitating the process of self-regulation is maturation. Piagetian theory is developmental, thereby placing importance on the maturation level of the student.

Cognitive development is a cyclic process involving interaction with new events, materials, properties, and abstractions. Science educators have developed a unique model of learning that is based on Piaget's theory of development.

Cognitive Processes and the Learning Cycle

According to Piaget, the development of new cognitive structures will be the result of three different mental processes: assimilation, accommodation, and equilibration.[28] Cognitive development is the result of the individual's interaction with the environment. The nature of the interaction is an adaptation involving these three mental processes.

Assimilation. Assimilation is the integration of new information with existing internal mental structures. A student who identifies a coarse-grained rock sample as granite is assimilating that rock into his or her schema of rocks. Piaget suggested that assimilation was dependent on the existence of an internal structure so that the new information could be integrated.

Accommodation. Accommodation is the adjustment of internal structures to the particular characteristics of specific situations, events, or properties of new objects. Biological structures, for example, accommodate to the type and quantity of food at the same time that the food is being assimilated. Piaget theorized that in cognitive functioning, internal mental structures adjust to the unique properties of new objects and events.

Equilibration. Equilibration, like assimilation and accommodation, has a biological parallel. The organism, in biological functioning, must maintain a steady state within itself at the same time remaining open to the environment to deal with new events and for survival. In cognitive functioning, equilibration is the process that allows the individual to grow and develop mentally but maintain stability. Piaget suggests, however, that equilibration is not an immobile state but rather a dynamic process that continuously regulates behavior.[29]

---

[27]Lawson and Renner, "Piagetian Theory and Biology Teaching," pp. 336–343.

[28]Bell-Gredler, *Learning and Instruction*, pp. 198–200.
[29]Bell-Gredler, *Learning and Instruction*, p. 200.

This cycle of assimilation, accommodation, and equilibration has been used as a basis for the development of several science teaching cycles called the learning cycle. One of these cycles is shown in Figure 2.14. Three teaching processes—exploration, invention, and discovery (expansion of the idea)—are parallel processes to assimilation, accommodation, and equilibration.[30]

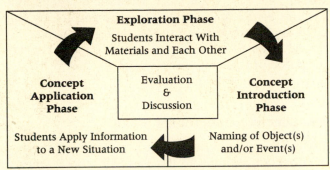

Charles R. Barman, *An Expanded View of the Learning Cycle: New Ideas About An Effective Strategy*, Monograph and Occasional Paper Series #4, Council for Elementary Science International, P.L. Used with permission of C.E.S.I.

FIGURE 2.14   The Learning Cycle

Exploration is an active process involving the students directly with objects and materials. The exploration phase can be open ended, or can be structured by the teacher. The important element is the active engagement of the student for the sake of creating some disequilibrium. During the exploration phase students observe, gather data, and experience new phenomena.

Invention is the phase in the learning cycle that is analogous to accommodation, when new structures are built to integrate new information. Renner calls this phase conceptual invention. The invention process has a high degree of teacher direction. Using the language and experiences of the exploration phase, students invent new concepts with the aid of the teacher. The experiences students had during the exploration phase are used as data for a new structure that is proposed by the teacher. The invention phase is interpretive. Students process new information and modify current conceptions and frameworks to accommodate the new information.

The discovery phase is designed to provide the students with active learning situations in which they can apply, test, and extend the new ideas and concepts. The discovery phase is analogous to equilibration, but like equilibration, it is dynamic. Students, even at this phase, are still in a state of disequilibrium and require further exposure to active learning lessons. The discovery phase allows the student to apply the new ideas to different situations, further reinforcing the development of new mental frameworks.

This is a brief introduction to the learning cycle. You will find more information on the learning cycle idea and its application to science teaching strategies in Chapter 7 and in the section on lesson planning in Chapter 9.

Piaget's Stages of Thinking

Piaget identified four stages or patterns of reasoning that characterized human cognitive development. Piaget viewed these as qualitative differences in the way humans think from birth to adulthood. At each stage the individual is able to perform operations on the environment in order to develop cognitive structures. A summary of the four stages is shown in Table 2.6.

The Sensorimotor Stage

Beginning at birth to about 2 years, the first stage is characterized by perceptual and motor activities. The behavior of children during this stage can be described as nonverbal and is characterized by reflex actions, play, imitating others, and object permanence. Early in this stage of development, if an object that the child has seen is removed from view, the object is forgotten (out of sight, out of mind). However, later in this stage, if a child was playing with an object and it gets hidden from view, the child will look for the object.

The child from the beginning is an agent of his or her own cognitive development. Piaget described the young infant as taking control in procuring and organizing all experiences of the outside world. Young children follow with their eyes, explore things, and turn their heads. They explore with their hands by gripping, letting go, pulling, pushing. We all recognize the exploration of children with their mouths. The child continues its exploration with body movements, and extends this exploration by putting hands, eyes and mouth, into action at once. With these early life explorations, the child, according to Piaget, develops mental schemes or patterns based on these experiences. These experiences, particularly if they are satisfying, will be repeated by the child. Through the processes of assimilation and accommodation, the child builds internal structures; the child adapts to the world.[31]

[30]J.W. Renner and E.A. Marek, *The Learning Cycle and Elementary School Science Teaching* (Portsmouth, N.H.: Heinemann, 1988).

[31]Discussion based on N. Isaacs, *A Brief Introduction to Piaget* (New York: Schocken Books, 1960), pp. 18–22.

TABLE 2.6

PIAGET'S STAGES OF DEVELOPMENT

Stage	Overview
Sensorimotor Period (birth to 1 to 1½ years)	This is the period that is characterized as presymbolic and preverbal. Intellectual development is dependent on the action of the child's senses and on responses to external stimuli. The child is engaged in action schemes such as grasping and reaching for distant objects. Characteristics include reflex actions, play, imitation, and object permanence; nonverbal.
Preoperational Period (2–3 to 7–8 years)	The child's thought is based on perceptual cues and the child is unaware of contradictory statements. For example, the child would say that wood floats because it is small and a piece of steel sinks because it is thin. Characteristics include language development, egocentrism, classification on a single feature, and irreversibility.
Concrete Operational Period (7–8 to 12–14 years)	Logical ways of thinking begin as long as it is linked to concrete objects. Characteristics include reversibility, seriation, classification, and conservation (number, substance, area, weight, volume).
Formal Operational Period (older than 14)	Students are able to deal logically with multifaceted situations. They can reason from hypothetical situations to the concrete. Characteristics include theoretical reasoning, combinatorial reasoning, proportional reasoning, control of variables, and probabilistic and correlational reasoning.

Bell-Gredler, *"Learning and Instruction,"* p. 205.

Near the latter phase of this stage the child's experiences are enriched by means of imaginative play, combined with greatly enhanced exploratory abilities, namely questioning, listening, and talking. These activities lead the child to the next stage of cognitive development, the preoperational stage.

The Preoperational Stage

During the preoperational stage (ages 2 to 7 years), the child's intellectual abilities expand greatly. The child during this stage is able to go beyond direct experience with objects. The preoperational child is able to represent objects in their absence, thereby developing the ability to manipulate in the mind. Thus, the child can engage in activities such as symbolic play, drawing, mental imagery, and language.[32]

Table 2.7 compares how the child's mental abilities change from the beginning of the operational stage to the end of this stage. Note the differences in abilities at the beginning phase, during which the child can classify only on the basis of one characteristic and is not able to conserve, compared with the latter phase of this stage, in which the child can conserve mass, weight, and volume.

The Concrete Operational Stage

The concrete operational stage begins around the age of seven, and extends to the ages of 12 to 14. During this stage of development individuals learn to order, classify, and perform number operations such as adding, subtracting, multiplying, and dividing. They also learn to conserve, develop the ability to determine the cause of events, and apprehend space-time relationships.

Concrete operations means that the child is able to perform various logical operations but only with concrete things. An operation is an action—a manipulation of objects. We might think of an operation as a reasoning pattern. Since most of the students that

[32]T.E. Gross, *Cognitive Development* (Monterey, Calif.: Brooks/Cole, 1985) p. 36.

TABLE 2.7

DIFFERENCES DURING EARLY AND LATE PHASES OF THE PREOPERATIONAL STAGE

Preoperational Stage: Ages 2–4	Preoperational Stage: Ages 4–7 (Called the Intuitive Phase)
Classify on the basis of single property	Able to form classes or categories of objects
Unable to see that objects alike in one property might differ in others	Able to understand logical relationships of increasing complexity
Able to collect things based on a criterion	Able to work with the idea of number
Can arrange objects in a series, but cannot draw inferences about things that are not adjacent to each other	Develops the ability to conserve, for example, mass, weight, volume, continuous quantity, and number

you will teach will be either at the concrete or formal stage of development (or in transition between these stages), we will explore these reasoning patterns in some detail. At the concrete stage there are several reasoning patterns that impinge on science teaching, and ones that will affect students' performance in your classroom. These reasoning patterns include class inclusion, serial ordering, reversibility, and conservation (Table 2.8).

Class inclusion is an important pattern of reasoning in science courses. Class inclusion is a prerequisite for the development of concrete concepts such as animal, plant, rock, mineral, and planet. For example, suppose that you show a student a picture of some plants (carrot, grass, oak tree, cabbage, dandelion). In this task the student is asked to identify which of the pictures is a plant. Most children will readily include the grass in the category of plant, but not tree, carrot, or dandelion (the tree was a plant when it was little, but now it is big, and therefore a tree). As students grow older and their cognitive development gets more sophisticated and their experiences widen, they will develop the ability to include all these objects in the general class of plant.[33]

Students are conservers in the concrete stage. For example, they are able to understand that the quantity of a substance remains the same if nothing is added or taken away. To understand this, show a student a ball of clay the size of a tennis ball. After the student has observed the ball, roll it into a cylinder. Then ask the student if the cylinder (you can call it a snake or a dog) has more, less, or the same amount of clay as the ball. The student that can conserve mass will respond that the cylinder has the same amount of clay. Conservation abilities develop in different areas such as number, mass, weight, and volume.

The reasoning patterns in the concrete operational stage can be explored by administering tasks to individual students. By administering Piagetian tasks, you can develop insights into the reasoning abilities of your students. You might want to do Inquiry Activity 2.3, Piagetian Concrete Reasoning Tasks. To compare

students' concrete and formal reasoning patterns, you might do Inquiry Activity 2.4, The Mealworm and Mr. Short Puzzles.

TABLE **2.8**

....

CONCRETE REASONING PATTERNS

Concrete Reasoning Pattern (Operations)	Explanation and Examples
Class Inclusion	Classifying and generalizing based on observable properties (e.g., distinguishing consistently between acids and bases according to the color of litmus paper; recognizing that all dogs are animals but that not all animals are dogs).
Serial Ordering	Arranging a set of objects according to an observable property and possibly establishing a one-to-one correspondence between two observable sets (e.g., small animals have a fast heart beat while large animals have a slow heart beat).
Reversibility	Mentally inverting a sequence of steps to return from the final condition of a certain procedure to its initial condition (after being shown the way to walk from home to school, finding the way home without assistance).
Conservation	Realizing that a quantity remains the same if nothing is added or taken away, though it may appear different (e.g., when all the water in a beaker is poured into an empty graduated cylinder, the amount originally in the beaker is equal to the amount finally in the cylinder).

Based on R. Karplus et al., *Science Teaching and the Development of Reasoning* (Berkeley, Calif.: Lawrence Hall of Science, 1978), p. 2.3. Copyright © 1977 by The Regents of the University of California. Based in part on research supported by the National Science Foundation. Any opinions, findings, and conclusions or recommendations expressed in this publication are those of the authors and do not necessarily reflect the view of the National Science Foundation.

[33]The example is based on research reported in *Learning in Science: Implications of Children's Science* by R. Osborne and P. Freyberg (Auckland, New Zealand: Heineman, 1985), pp. 6–7.

INQUIRY ACTIVITY **2.3**

....

Piagetian Concrete Reasoning Patterns

....

Piaget developed a number of tasks designed to be administered to individual students. This activity will enable you to administer several tasks to explore the nature of preoperational versus concrete thinking patterns.

Materials

Piagetian tasks, (Figure 2.15) clay, two vials, two candy bars

Procedure

1. Administer these tasks to two or three middle school or junior high age students. You will have to make special arrangements to do this by contacting the school officials. There are certain processes that you should keep in mind when administering these tasks:
   a. Help the student relax by telling the student that you have a few activities or games to play, and that all answers are acceptable.
   b. Accept all answers. Remember that you are trying to understand the student's thinking pattern, and the tasks are a way to discover this. Help the students as much as you can to verbalize their thinking by encouraging them to go on and tell you more.
   c. Give the students time to think after you ask a question.
   d. Always ask the students to justify their answers to the tasks. Again, be patient with the students, by helping them along and even suggesting or paraphrasing what other students may have said and asking the student to agree or disagree with the other students.

2. Gather all the materials needed for each task. When you are administering the tasks, only have in sight the materials pertinent to the current task. You might have all the materials in a small box, removing and returning the materials as needed.

3. Administer the tasks following the procedures outlined in Figure 2.16, which can be found on page 61.

Minds on Strategies

1. Summarize the results using the chart shown in Figure 2.15. *(Continued on p. 62)*

Task	Student Response	Justification	Intellectual Level: (P);Preoperational (Tr);Transitional (C);Concrete
Conservation Mass Weight Volume Continuous Quantity Number			

FIGURE **2.15**

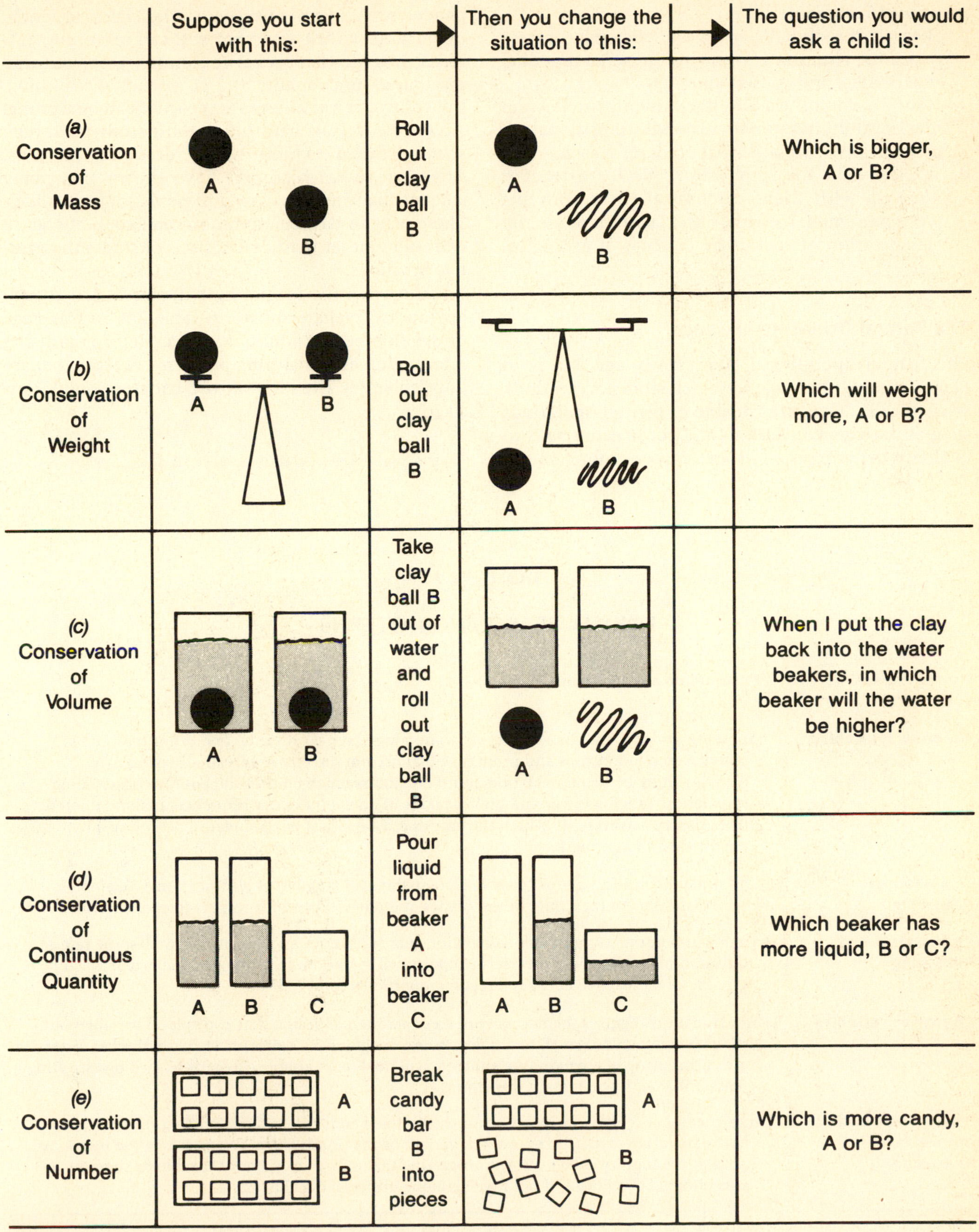
Suppose you start with this:
Then you change the situation to this:
The question you would ask a child is:
(a) Conservation of Mass
A
B
Roll out clay ball B
A
B
Which is bigger, A or B?
(b) Conservation of Weight
A
B
Roll out clay ball B
A
B
Which will weigh more, A or B?
(c) Conservation of Volume
A
B
Take clay ball B out of water and roll out clay ball B
A
B
When I put the clay back into the water beakers, in which beaker will the water be higher?
(d) Conservation of Continuous Quantity
A B C
Pour liquid from beaker A into beaker C
A B C
Which beaker has more liquid, B or C?
(e) Conservation of Number
A
B
Break candy bar B into pieces
A
B
Which is more candy, A or B?

FIGURE 2.16 Concrete Operational Science Tasks. From *Educational Psychology* by N.L. Gage and D. Berliner © 1988 by Houghton-Mifflin. Reprinted by permission of the publisher.

(Continued from p. 60)

2. How did the students do on the following concrete reasoning patterns: class inclusion, serial ordering, reversibility, and conservation?
3. How would you classify the students intellectual level of thinking: preoperational, transitional, or concrete operational? What are your reasons?
4. What would the implications be for a teacher working with these students if these concepts were presented: Newton's first law of motion; the classification of rocks; the concept of geological time?

The Formal Operational Stage

The formal operational stage (over age 14) is in Piaget's theory the stage in which students can think scientifically. They are capable of mental operations such as drawing conclusions and constructing tests to evaluate hypotheses—in short, of performing an ex- panded set of logical operations. The logical or formal operations, which again we will call reasoning patterns, include theoretical reasoning, combinatorial reasoning, functionality and proportional reasoning, the control of variables, and probabilistic reasoning. According to Piagetian theory most students in high school should exhibit these reasoning patterns. However, research studies have shown that many students have not developed these reasoning abilities. These may, in fact, be aspirations and goals of science education rather than descriptions of student's cognitive functions.[34]

The scientific reasoning patterns at the formal operations level are shown in Table 2.9. At this stage of development students are capable of organizing information and analyzing problems in ways that are impossible for a student at the concrete operations stage.

[34]Karplus, *Science Teaching and the Development of Reasoning*, p. 2.5.

TABLE   2.9

FORMAL REASONING PATTERNS

Formal Reasoning Patterns	Explanation and Examples
Theoretic Reasoning	Applying multiple classification, conservation logic, serial ordering, and other reasoning patterns to relationships and properties that are not directly observable. Examples: distinguishing between oxidation and reduction reactions; using the energy conservation principle; arranging lower and higher plants in an evolutionary sequence; making inferences from theory according to which the earth's crust consists of rigid plates; accepting a hypothesis for the sake of argument.
Combinatorial Reasoning	The student considers all conceivable combinations of tangible or abstract items. Examples: systematically enumerating the genotypes and phenotypes with respect to characteristics.
Proportional Reasoning	Stating and interpreting functional relationships in mathematical form. Examples: the rate of diffusion of a molecule is inversely proportional to the square root of its molecular weight; the rate of radioactive decay is directly proportional to its half-life.
Control of Variables	The student recognizes the necessity of an experimental design that controls all variables but the one being investigated. Examples: When designing experiments to find out what factors affect the swing of a pendulum, students will hold one variable constant (e.g., if investigating mass, the length will remain the same).
Probabilistic and Correlational Reasoning	Interpreting observations that show unpredictable variability and recognizing relationships between variables in spite of random variations that mask them. Examples: in the mealworm puzzle (see Inquiry Activity 2.4), recognizing that a small number of specimens showing exceptional behavior need not invalidate the principal conclusion.

Based on R. Karplus, et al., *Science Teaching and the Development of Reasoning* (Berkeley Calif.: Lawrence Hall of Science, 1978), pp. 2.3–2.4. Copyright © 1977 by The Regents of the University of California. Based in part on research by the National Science Foundation. Any opinions, findings, and conclusions or recommendations expressed in this publication are those of the authors and do not necessarily reflect the view of the National Science Foundation. This publication not endorsed by the copyright holder.

INQUIRY ACTIVITY   **2.4**

....

The Mealworm and Mr. Short

....

To help you understand the difference between concrete and formal reasoning patterns, two puzzles (Table 2.10, p. 64) are described, followed by secondary students answers and their explanations of their answers (Table 2.11).

Materials

Paper clips, copies of the puzzles

Procedure

1. First perform the puzzles yourself, writing your answers and explanations as the students did on separate sheets of paper.
2. If you are able to have secondary students complete the puzzles, you can use their responses to respond to the following questions, as well as the students responses included here.

Minds on Strategies

When you have finished completing the puzzles, answer these questions.

1. What differences in reasoning did you find between the students' responses to
   a. The mealworm puzzle
   b. The ratio puzzle
2. What similarities did you observe in the thinking patterns of Norma, David, and Dolores?
3. What similarities did you observe in the thinking patterns of Jean, John, and Harold?
4. Which students are using concrete reasoning patterns? Which are using formal reasoning patterns?
5. What are some characteristics of
   a. Concrete reasoning
   b. Formal reasoning

Concrete Versus Formal Thinking: Implications

While reading through the responses to the two puzzles, you should have noticed that some of the answers were more complete, more consistent, and more systematic than others. Each answer represents the reasoning of the student who wrote it. Yet surprising as it may seem, inconsistent and incomplete answers were just as common as complete and consistent ones. The patterns of thinking that the students show as responses to these puzzle problems are integral aspects of Piaget's theory of cognitive development.

In the mealworm puzzle students using concrete reasoning tend to focus on one variable and exclude the others. The student cannot detect the logic of the experiment that allows one variable to be isolated and separated so that they can be dealt with as distinct causal agents. Formal reasoning patterns are shown when students realize that variables are held constant while only one is allowed to change, as in boxes II or IV, or comparing I with III. The formal thinker considers all possible causal factors to test the hypothesis that light or moisture is responsible for the distribution of mealworms.

In the Mr. Short puzzle, concrete reasoning is recognized by students who simply add the extra amount to the height of Mr. Tall. This type of thinking is a much more direct measure of qualitative difference than is the ratio, which the student makes only by making a correspondence between each individual button and paper clip. Formal thinkers understand that each button corresponds to a certain number of paper clips. Once the ratio is known, the answer is found by calculation.

Concrete thinkers need reference to familiar actions, objects, and observable properties. They use reasoning patterns that include conservation, class inclusion, reversibility, and serial ordering. Their thinking shows inconsistencies between various statements they make, and they often contradict themselves. Most of the students you will teach are concrete thinkers, even though Piaget has suggested that the formal stage of development begins at about age 14. Studies have shown that many adolescents and adults do not develop formal thinking capacities.[35]

Formal or abstract thinkers reason with concepts ($F = ma$), indirect relationships, properties, and theories. They are able to use a wide range of thinking patterns, including combinatorial reasoning, proportional reasoning, control of variables, and probabilistic reasoning. Formal thinkers are able to plan out a course of action, manipulate ideas in their minds, and actively check the validity of their own ideas.

[35]R. Good, *How Children Learn Science* (New York: Macmillan, 1977), p. 141.

TABLE   2.10

THE MEALWORM AND MR. SHORT

The Mealworm Puzzle

An experimenter wanted to test the response of meal-worms to light and moisture. To do this he set up four boxes as shown in Figure 2.17. He used lamps for light sources and constantly watered pieces of paper in the boxes for moisture. In the center of each box he placed 20 mealworms. One day he returned to count the number of mealworms that had crawled to the different ends of the boxes.

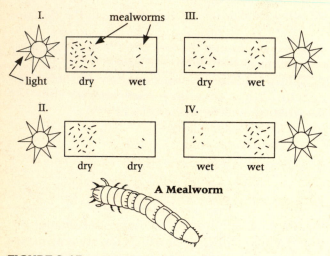

FIGURE 2.17   Mealworm Experimental Boxes

The diagrams show that mealworms respond to (response means move toward or away)

A. Light but not moisture

B. Moisture but not light

C. Both light and moisture

D. Neither light nor moisture

Please explain your choice.

The Mr. Short Puzzle

The drawing in Figure 2.18 is called Mr. Short. We used large round buttons laid side by side to measure Mr. Short's height, starting from the floor between his feet and going to the top of his head. His height was four buttons. Then we took a similar figure called Mr. Tall, and measured it in the same way with the same buttons. Mr. Tall was six buttons high.

Now please do these things:

1. Measure the height of Mr. Short using paper clips in a chain. The height is _____.
2. Predict the height of Mr. Tall if he were measured with the same paper clips _____.
3. Explain how you figured out your prediction. (You may use diagrams, words, or calculations. Please explain your steps carefully.)

FIGURE 2.18   Mr. Short

Based on R. Karplus, et al., *Science Teaching and the Development of Reasoning* (Berkeley, Calif.: Lawrence Hall of Science, 1978), p. 2.3. Copyright ©1977 by The Regents of the University of California. Based in part on research supported by the National Science Foundation. Any opinions, findings, and conclusions or recommendations expressed in this publication are those of the authors and do not necessarily reflect the view of the National Science Foundation. This publication not endorsed by the copyright holder.

CONSTRUCTIVIST LEARNING AND COGNITIVE SCIENCE THEORIES

Science educators, particularly over the past 20 years, using the work of Jean Piaget as a foundation, have developed a number of theories to explain cognitive learning. In general, these researchers are interested in how information is taken in, interpreted, represented, and acted on. These science educators agree with Piaget that knowledge is constructed and theorize that

(Continued on p. 66)

TABLE 2.11

STUDENT RESPONSES TO THE MEALWORM AND MR. SHORT PUZZLES

Science	Mealworm Puzzle Responses	Mr. Short Puzzle Responses
Student 1 (Norma, age 12)	D. "Because even though the light was moved in different places the mealworms didn't do the same things."	Prediction for Mr. Tall: 8 $\frac{1}{2}$ paper clips. Explanation: "Mr. Tall is 8 $\frac{1}{2}$ paper clips tall because when using buttons as a unit of measure he is 2 units taller. When Mr. Short is measured with paper clips as a unit of measurement he is 6 $\frac{1}{2}$ paper clips. Therefore, Mr. Tall is 2 units taller in comparison, which totals 8 $\frac{1}{2}$."
Student 2 (Jean, age 13)	B. "I, II, and III show that mealworms seem to like the light, but in III they seem to be equally spaced. This leads one to believe that mealworms like the dryness, and the reason in pictures III and IV they are by the light is because of the heat that the light produces, which gives a dryness effect."	Prediction for Mr. Tall: 9.2 paper clips. Explanation: "The ratio using buttons of height of Mr. Short and Mr. Tall is 2:3. Figuring out algebraically and solving for x, $\frac{2}{3} = \frac{6\frac{1}{2}}{x}$ gives you 9.2 as the height in paper clips."
Student 3 (David, age 14)	A. "They usually went to the end of the box with the light."	Prediction for Mr. Tall: 9 paper clips tall. Explanation: "I figured it out by figuring that Mr. Short is $\frac{2}{3}$ as tall as Mr. Tall."
Student 4 (John, age 16)	C. "Boxes I and II show they prefer dry and light to wet and dark. Box IV eliminates dryness as a factor, so they do respond to light only. Box III shows that wetness cancels the effect of the light, so it seems they prefer dry. It would be clearer if one of the boxes was wet-dry with no light."	Prediction for Mr. Tall: 9 clips (pencil marks along Mr. Short). Explanation: "I estimated the middle and then one-fourth of Mr. Short. That's about the size of one button. I measured the button with my clips and found 1 $\frac{1}{2}$. So then I counted out six times 1 $\frac{1}{2}$ buttons and got 9."
Student 5 (Dolores, age 17)	C. "In the experiment three mealworms split $\frac{1}{2}$ wet, $\frac{1}{2}$ dry. So it's safe to assume that light was not the only factor involved."	Prediction for Mr. Tall: 8 paper clips tall. Explanation: "If Mr. Short measures 4 buttons or 6 paper clips (2 pieces more than buttons), then Mr. Tall should be 2 paper clips more than buttons."
Student 6 (Harold, age 18)	A. "Because there are 17 worms by the light and there are only 3 by the moisture."	Prediction for Mr. Tall: 9 $\frac{3}{6}$ paper clips tall. Explanation: "Figured it out by seeing that Mr. Tall is half again as tall as Mr. Short, so I took half of Mr. Short's height in clips and added it on to his present height in clips and came up with my prediction."

Based on R. Karplus, et al., *Science Teaching and the Development of Reasoning* (Berkeley, Calif.: Lawrence Hall of Science, 1978), p. 2.3. Copyright ©1977 by The Regents of the University of California. Based in part on research supported by the National Science Foundation. Any opinions, findings, and conclusions or recommendations expressed in this publication are those of the authors and do not necessarily reflect the view of the National Science Foundation. This publication not endorsed by the copyright holder. Note: Students measured a drawing of a "Mr. Short" that was approximately six paper clips tall. Since the drawing you used was slightly larger than a paper clip, your prediction of the height of Mr. Tall in paper clips will differ from the student responses. The reasoning, however, is the same.

students are builders of knowledge structures. These science educators, also referred to as cognitive scientists, have developed a number of alternative models that have direct implication for teaching science to secondary students. The models discussed here represent recent work being done by cognitive scientists. Although these models differ in some respects, they share the following characteristics:[36]

The Importance of Content Knowledge. Cognitive scientists put a lot of emphasis on what they call expert knowledge. They suggest that "experts," say in physics or geology, reason more powerfully about a topic in their respective fields than do novices. In fact one of the things you will find common in the models that follow is an attempt by teachers to find out what the novice learner (your students) know about a topic before teaching them new concepts. Another idea that cognitive scientists have put forth is that learning requires knowledge, yet knowledge cannot be given directly. Students must construct their own knowledge. Yet for this to be done the teacher must provide a learning environment in which students can discuss and question what they are told, investigate new information, and build new knowledge structures. Further, teachers need to provide ways to ascertain what students know, and then find ways to link this knowledge (which quite often is naive) to new knowledge structures.

The Integration of Skills and Content. Because the cognitive approach places the student in the center of learning, the development of thinking skills must be integrated with content knowledge. This is, of course, an idea proposed by many other theorists, especially Piaget and Bruner. It is just as important for students to observe, question, test, and hypothesize, as it is to develop cognitive structures about gravity, electrons, plate tectonics, and carnivores. In fact, without observing, questioning, testing, and hypothesizing, the student has little chance of developing scientific conceptions about these or any other subjects.

The Intrinsic Nature of Motivation. This is a major change in emphasis for cognitive scientists. Typically, motivation has been a subject of study of social psychologists who are interested in attitudes, effort, and attention. Cognitive scientists have realized the importance of developing a learning environment in which students will want to learn. Cognitive scientists, unlike the behaviorists, focus their attention on the intrinsic nature of content and instruction as a

means to motivate students. Cognitive scientists are also learning that students' concept of self can be a contributing factor in motivation. Resnick and Klopfer report that social psychologists have found that motivation is closely related to students' conceptions of intelligence. In one study, if students are helped to understand that intelligence is an incremental ability, rather than a fixed entity, they will stick with a problem and try to use what they have to solve the problem when faced with a challenging or difficult problem. Students who lack such understanding, according to the researchers, might in some cases give up, saying that they lack the intelligence to find the solution.[37] Interesting, thought provoking, challenging, and stimulating approaches to instruction may motivate students more than the positive reinforcers suggested by behavioral psychologists.

The Role of Learning Groups. Cognitive scientists believe that the social setting for learning is crucial. They have found that cooperative problem-solving groups have been effective with students of varying abilities. Resnick and Klopfer suggest that skilled thinkers (the instructor or high-ability students) can demonstrate desirable ways of solving problems, thereby helping students who lack these mental abilities or experiences. The implication for the science teacher is to develop in the classroom open, positive communication patterns, and to place students in small, mixed ability cooperative groups in which social interaction can occur. In either case, cognitive scientists are calling for the formulation of "social learning communities," environments in which questioning, critical thinking, and problem solving are valued. According to cognitive scientists, these learning communities can be critical in helping less able students learn thinking patterns that the more able students possess.

The Constructivist Model

According to the constructivist learning model, a large number of students show up in science classes with lots of ideas about science concepts, many of which are "incorrect" or naive, or are misconceptions. We might think of student misconceptions and naive theories as "alternative frameworks." The aim of science teaching is to "help students overcome naive conceptions or habits of thought and replace them with scientific concepts and principles."[38]

---

[36]Based on L.B. Resnick and L.E. Klopfer, "Toward the Thinking Curriculum: An Overview in Resnick," L.B. and L.E. Klopfer, eds., *Toward the Thinking Curriculum: Current Cognitive Research* (Alexandria, Va.: Association for Supervision and Curriculum Development, 1989).

[37]Resnick and Klopfer, "Toward the Thinking Curriculum," p. 8.
[38]C.W. Anderson and E. L. Smith, "Teaching Science," in V. Richardson-Koehler, ed., *Educators' Handbook: A Research Perspective* (New York: Longman, 1987).

First let's consider the nature of some misconceptions in science, and then describe a theory to improve students' understanding of science.

In a physics or physical science class, the coin toss problem will illustrate the problem of misconceptions. The problem, shown in Figure 2.19 is an application of Newton's laws of motion. A coin is tossed upward in the air, and the student is asked, What are the forces on the coin at point B, when it is moving upward through the air? The typical answer to this problem by students is, "While the coin is on the way up, the force from your hand (Fh) pushes up on the coin. On the way up it must be greater than the force of gravity (Fg), otherwise the coin would be moving down." The expert or physicist's answer is that the only force acting on the coin at point B is the downward force due to gravity (and a small additional downward force due to air resistance). Anderson and Smith report that in one study of college engineering majors, who had a course in high school physics, the percentage of students answering the question correctly after instruction rose from 12 percent to only 28 percent.[39]

In chemistry, an example of a misconception is illustrated when a student, four months into the course, is asked to explain what happens when a nail rusts. His explanation is that "the coldness reacts on it [the nail]....Plastic doesn't rust because coldness doesn't cause the same reaction....Rusting is a breakdown of the iron because it [coldness] brings out the rusting...it [coldness] almost draws it [rust] out, like a magnet...like an attractor it brings it out."[40] Anderson and Smith report that this student consistently said he was satisfied with his explanation, that it made sense to him, and that scientists would explain the

situation in the same way. Misconceptions make sense to the student, and they are firmly held by the learner as a mental structure. This makes things very difficult for the science teacher.

Students hold misconceptions in all areas of science. In a biology class, students were presented with "the amputated finger problem:" If a little girl had an accident and her finger was amputated and she married someone with a similar amputation, what would you predict their children's fingers would look like at birth? Even after instruction in a unit on genetics, students had the following misconceptions: "The finger was cut off too fast for the genes to change"; "The child will probably have a finger missing because the traits of both parents are strong"; and "The lost finger would be inherited from the parents."[41]

Let's look at some alternative frameworks that student's bring to earth science classes. Table 2.12 lists earth science misconceptions in several conceptual areas.

Even after instruction, misconceptions are still prevalent among students in science classes. Teachers who direct their attention to student misconceptions report that a student misconception is "knowledge spontaneously derived from extensive personal experience."[42] Because these misconceptions were derived by students and they make sense to them, the students often hold on to these ideas in spite of alternative (scientific) conceptions. Since it is assumed that knowledge cannot be given directly to the learner, the task of the teacher who focuses on misconception theory is as follows:[43]

1. Help students become dissatisfied with their existing conception
2. Help students achieve a minimal initial understanding of the scientific conception
3. Make the scientific conception plausible to students
4. Show the scientific conception as fruitful or useful in understanding a variety of situations

Lawson and Thompson point out that the intellectual level of the students will affect student understanding in science and may prevent some students from understanding the scientific conception, especially if it is expressed as a theoretical conception.

---

[39]Anderson and Smith, "Teaching Science," p. 87.
[40]Anderson and Smith, "Teaching Science," p. 88.

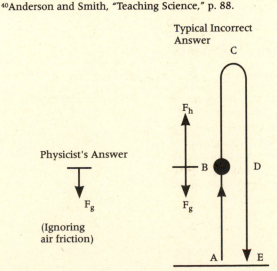

Typical Incorrect Answer

Physicist's Answer

(Ignoring air friction)

FIGURE 2.19 The Coin Toss Problem. Reprinted with permission of the American Physics Teacher and John Clement.

[41]A. E. Lawson and L. D. Thompson, "Formal Reasoning Ability and Misconceptions Concerning Genetics and Natural Selection," *Journal of Research in Science Teaching* 25 (9): pp. 733–746.
[42]Lawson and Thompson, "Formal Reasoning Ability," p. 733.
[43]Anderson and Smith, "Teaching Science," p. 93.

TABLE 2.12

ALTERNATIVE FRAMEWORKS (MISCONCEPTIONS) IN EARTH SCIENCE

Earth in Space	Solid Earth	Biosphere	Atmosphere	Hydrosphere
The Earth is sitting on something.	Mountains are created rapidly.	Dinosaurs and humans existed at the same time.	Rain comes from holes in clouds.	Most rivers flow "down" from north to south.
The Earth is larger than the sun.	Rocks must be heavy.	Humans are responsible for the extinction of dinosaurs.	Rain comes from clouds sweating.	Groundwater "typically" occurs as basins, lakes, and fast flowing streams.
The Earth is round like a pancake.	The earth is molten, except for the crust.	Evolution is goal directed.	The oxygen we breathe does not come from plants.	Salt added to water doesn't change the weight of the water.
Gravity increases with height.	Continents do not move.	Evolutionary changes are driven by need.	Gas makes things lighter.	Glaciers are rapidly created.
Rockets in space require a constant force.	The soil we see today has always existed.	Human beings did not evolve from earlier species of mammals.		
The sun goes around the earth.				
The universe is static, not expanding.				

Excerpted from *Earth Science Education Connection* 1, no. 2 (Winter 1989), p. 3.

They suggest that the concrete student lacks reasoning patterns necessary to evaluate competing hypotheses, for example, the hypothesis of the inheritance of acquired characteristics (the typical misconception) versus the natural selection of genetically acquired characteristics. They may be unable to reject the naive inheritance theory because they do not understand the concept of natural selection and gene transfer. One implication of this study is that teachers should provide students with opportunities to discuss their prior conceptions and carefully compare them with the scientific conceptions. Another and perhaps more important implication is that we need to develop alternative ways of expressing concepts, as has been suggested by Bruner.

Conceptual change research has led to a theory of instruction (The Constructivist Learning Model) in which students' prior conceptions are detected, time is provided for the students to compare their misconceptions with the scientific conceptions, and finally an opportunity is provided for them to use the new conceptions in a variety of learning situations (see Table 2.13). The classroom implication of the constructivist model of teaching will be discussed in Chapter 7.

TABLE 2.13

A MODEL OF CONCEPTUAL CHANGE TEACHING

Before Instruction	Instruction	After Instruction	End Product
Naive conceptions. Identify student misconceptions by means of interviews, having students respond to a few problems based on the central concepts to be taught.	Present information in light of students' misconceptions. Focus on comparing new conceptions with student naive conceptions. Provide opportunities for students to explore new conceptions through laboratory activities, demonstrations, audiovisual aids, and discussions of familiar phenomena. Use questioning strategies and everyday phenomena to help students "test" their new conceptions.	Evaluate the change in students' conceptions. Use questions that were used to assess student misconceptions as a base line for change. Design questions that ask students to justify their ideas.	Scientific conceptions

The Meaningful Learning Model

David Ausubel is a psychologist who advanced a theory that contrasted meaningful learning with rote learning (Table 2.14). In Ausubel's view, to learn meaningfully, students must relate new knowledge (concepts and propositions) to what they already know. He proposed the notion of an advanced organizer as a way to help students link their ideas with new material or concepts. Ausubel's theory of learning claims that new concepts to be learned can be incorporated into more inclusive concepts or ideas. These more inclusive concepts or ideas are advance organizers. Advance organizers can be verbal phrases ("the paragraph you are about to read is about Albert Einstein"), or a graphic. In any case, the advance organizer is designed to provide what cognitive psychologists call the "mental scaffolding to learn new information."[44]

Ausubel believed that learning proceeds in a top-down, or deductive manner. Ausubel's theory consists of three phases: the presentation of an advance organizer, the presentation of a learning task or material, and strengthening the cognitive organization. The main elements of Ausubel's model are shown in Table 2.15.

Concept Mapping for Meaningful Learning

Novak and Gowin have developed a theory of instruction that is based on Ausubel's meaningful learning principles and that incorporates "concept maps" to represent meaningful relationships between concepts and propositions.[45] A cognitive map is a "kind of

[44]Slavin, *Educational Psychology*, p. 186.
[45]J. D. Novak and D. B. Gowin, *Learning How to Learn*, (Cambridge, England: Cambridge University Press, 1984), p. 15.

TABLE 2.14

MEANINGFUL LEARNING CONTRASTED WITH ROTE LEARNING

Type of Learning	Characteristics
Meaningful Learning	Nonarbitrary, nonverbatim, substantive incorporation of new knowledge into cognitive structure
	Deliberate effort to link new knowledge with higher-order concepts in cognitive structure
	Learning related to experiences with events or objects
	Affective commitment to relate new knowledge to prior learning
Rote Learning	Arbitrary, verbatim, nonsubstantive incorporation of new knowledge into cognitive structure
	No effort to integrate new knowledge with existing concepts in cognitive structure
	Learning not related to experience with events or objects
	No affective commitment to relate new knowledge to prior learning

Based on R. E. Slavin, (1988) *Educational Psychology* (Englewood Cliffs: N.J.: Prentice-Hall), pp. 216–217.

TABLE 2.15

AUSUBEL'S MODEL OF LEARNING

Phase I: Advance Organizer	Phase II: Presentation of Learning Task or Material	Phase III: Strengthening Cognitive Organization
Clarify the aim of the lesson.	Make the organization of the new material explicit.	Relate new information to the advance organizer.
Present the organizer.	Make the logical order of the learning material explicit.	Promote active reception learning.
Relate the organizer to students' knowledge.	Present the material and engage students in meaningful learning activities.	

Based on R. E. Slavin, (1988) *Educatonal Psychology* (Englewood Cliffs, N.J.: Prentice-Hall), p. 219.

visual road map showing some of the pathways we may take to connect meanings of concepts."[46] According to Novak and Gowin, concept maps should be hierarchical; the more general, more inclusive concepts should be at the top of the map, and the more specific, less inclusive concepts at the bottom of the map. An example of this hierarchical principle of concept maps is the concept map of oceans drawn by seventh graders shown in Figure 2.20.

The concept map is a tool that science teachers can use to determine the nature of students' existing ideas. The map can be used to make evident the key concepts to be learned and suggest linkages between the new information to be learned and what the students already know. Concept maps can precede instruction and can be used by the teacher to generate a meaningful discussion of student ideas. Following the initial construction and discussion of concept maps, instructional activities can be designed to explore alternative frameworks, resulting in cognitive accommodation.

[46]Novak and Gowin, *Learning How to Learn*, p. 15.

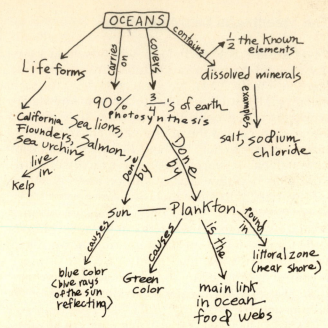

FIGURE 2.20 A concept map prepared from a science textbook by three seventh-grade students working together. Novak and Gowin, *Learning How to Learn*, p. 22. Used with permission.

SOCIAL THEORIES OF LEARNING

Most teaching takes place in groups, and it is therefore imperative that science teachers closely examine the results of research on small group, mixed-ability team learning. At one time or another, students in your classes will be involved with each other doing science laboratory activities, pairing off to answer questions or solve a problem, or working in a small team to prepare a report or make a class presentation. Students interact with each other, and it is important to know how this interaction contributes to student learning. It is also important for the teacher to know how to apply social learning theory to improve student learning and instruction. Enter cooperative learning.

Over the past several years, a major educational innovation has emerged that is affecting classroom learning. Teachers are implementing programs in which students are organized into small groups to accomplish a task, solve a problem, complete an assignment, study for a test, or engage in a hands-on activity.

Cooperative learning is based on the relationships between motivation, interpersonal relationships, and the accomplishment of specific goals. According to social psychology theorists, a state of tension within the individual motivates movement toward the attainment of desired goals. Thus, from this notion, it is the individuals drive to accomplish a desired goal that motivates behavior, whether it be individualistic, competitive, or cooperative.[47]

Cooperative learning theory posits that behavior among individuals in a group is synergistic,[48] that is, the goals of the individuals in a group are linked together in such a way that cooperative goal attainment is correlated positively, or is greater than the separate or individual performance of the group members.[49] This theoretical principle runs through a wide range of cooperative learning models, which will be discussed in detail in Chapter 7 but are alluded to briefly here.

[47]R. T Johnson and D. W. Johnson, "Cooperative Learning and the Achievement and Socialization Crises in Science and Mathematics Classrooms," in A. B. Champagne and L. E. Hornig, *Students and Science Learning* (Washington, D.C.: American Association for the Advancement of Science, 1987), pp. 67–93.

[48]J. Hassard, *Science Experiences: Cooperative Learning and the Teaching of Science* (Menlo Park, Calif.: Addison-Wesley, 1990), p. 18.

[49]Definition is based on D. W. Johnson, et al., *Circles of Learning: Cooperation in the Classroom* (Alexandria, Va.: Association for Supervision and Curriculum Development, 1986).

How does cooperative learning facilitate student learning? There are many points of view on this question. The behaviorist explanation goes like this. Students working in one group compete with other groups that the teacher has established. Students within a group work together to accomplish a task (complete a laboratory report, study together to prepare for a test, complete a science worksheet). Students are placed in a situation in which their success is dependent on the behavior and performance of other students in their group. Success does not necessarily imply a grade, but simply doing well on a competitive task in which one team's performance is rated against other teams' performance. Accordingly, team rewards and individual accountability are essential to achievement. In one of the most widely used models of cooperative learning (Student Teams-Achievement Divisions) student teams study together after being presented information by the teacher. After studying together, students take a test. Test scores are used, along with a system of improvement scores, to chart team recognition.[50]

On the other hand, a cognitive perspective argues that the intrinsically interesting nature of learning tasks combined with the range of abilities and knowledge that students bring to the classroom promotes an environment of learning.[51] Learning tasks that require multiple abilities to accomplish appear to be effective in reducing the domination of group learning by high-ability students. Instead of relying heavily on reading ability, science teachers should design group learning tasks that require reasoning, hypothesizing, predicting, and inductive thinking, and the use of manipulative materials and multimedia sources. According to social psychologists, such tasks "encourage students to modify their perceptions of their own and one another's competence."[52]

A number of social factors affect the success of cooperative learning. As David and Roger Johnson point out, cooperative learning is not having students sit together as they do individual assignments, is not having high-ability students help slower students, and is not assigning a project in which one person does all the work. Cooperative learning is instead based on positive interdependence, face-to-face communication, individual accountability, and interpersonal skills.[53]

STUDENT LEARNING STYLES: IMPLICATIONS FOR TEACHING

Students learn in a variety of ways, and to accommodate these differences teachers have devised a variety of methods and strategies to correlate with these student learning styles. Various strategies have been researched and implemented in the classroom. For example, Rita and Kenneth Dunn have developed a comprehensive approach to learning styles and have found that student learning styles are affected by (1) their immediate environment, (2) their emotionality, (3) their sociological needs, (4) their physical needs, and (5) their psychological processes. Other researchers have explored the dichotomous way the left and right hemispheres of the brain process and interpret information. Some researchers have divided student learning styles into categories, such as Bernice McCarthy. She has devised a system in which four learning styles are identified: innovative learners, analytic learners, common-sense learners, and dynamic learners. In this section we will explore these ideas, and identify some implications for science teaching.

Students, Teachers, and Learning Styles

Learning style pertains to how we learn. To some educators, "one's learning style is a biologically and developmentally imposed set of characteristics that explains why the same lecture, readings, interactions, classroom settings, and teachers affect individuals so differently."[54] Two crucial questions with regard to student learning styles are: In what ways do students differ in their manner of learning? and How do teachers accommodate students with different learning styles? For answers to these questions read the Science Tearchers Talk section in this chapter's Science Teaching Gazette. As you read their comments, note that the teachers allude to a variety of methodologies

[50]R. E. Slavin, *Using Student Team Learning* (Baltimore: The Johns Hopkins Team Learning Project, 1986).

[51]E. G. Cohen, *Designing Groupwork: Strategies for the Heterogeneous Classroom* (New York: Teachers College Press, 1986).

[52]R. Faltus, "Research on Cooperative Small Groups," in *Educational Research and Dissemination Program* (Washington, D.C.: American Federation of Teachers, 1986) pp. 1–25.

[53]Johnson, *Circles of Learning*, pp. 8–10.

[54]R. Dunn and S. A. Griggs, "A Quiet Revolution: Learning Styles and Their Application to Secondary Schools," *Holistic Education Review*, Winter, 1989, pp. 14–19.

in their attempt to increase student motivation and success in the science classroom. Let's explore the theory behind learning styles and then apply learning style ideas to classroom science teaching.

The Psychology of Learning Styles

Some students in your class would rather look at pictures of plants than read about them. You might have a student who prefers discussing questions in a small group to participating in a large group discussion. Another student might prefer to learn chemical nomenclature by matching the chemical symbols (printed on blue index cards) with the names of the elements or compounds (green index cards). We all have preferences in the way we learn. What do we know about learning styles, and how can this be helpful to you as a beginning science teacher?

Discovering Learning Styles

Consider for a moment your own approach to learning. Here are some sample items from an instrument

designed to diagnose student learning styles.[55] Do these describe some of your preferences when it comes to learning?

- I study best when it is quiet.
- I have to be reminded often to do something.
- I really like to draw, color, or trace things.
- I like to study by myself.

Rita and Kenneth Dunn have explored a universe of factors that affect the way students learn. The chart in Figure 2.21 summarizes the variety of elements that fall into one of the following five categories: environmental, emotional, sociological, physical, and psychological. Using the Learning Styles Inventory—a comprehensive approach to assessing students' learning style—researchers have surveyed individuals' styles in each of the 22 areas. The instrument consists of over a hundred preference statements (like the four listed above) that identify students' learning

[55]R. Dunn and K. Dunn, *Teaching Students through Their Individual Learning Styles* (Reston, Va.: Reston, 1978).

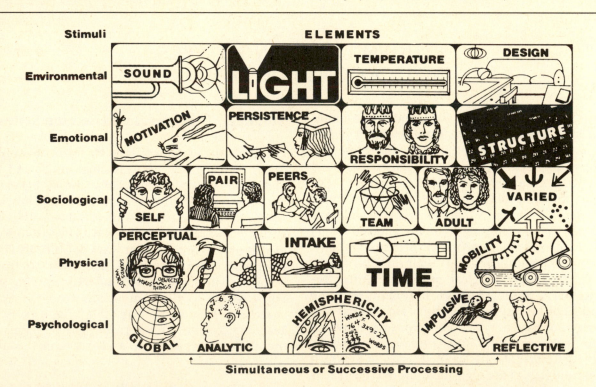

Depiction of the Dunn and Dunn learning styles model as it evolved between 1968 and 1989. Restak and Thies both have reported the biological nature of style. It is believed that the environmental (sound, light, temperature, and design) and physiological (perceptual, intake, time, and mobility) elements are biological in nature, and at least four of those appear to be related to either an analytical (left) or global (right) processing style. The emotional elements (motivation, responsibility/conformity, and structure) and the sociological preferences appear to be developmental—they change with maturation and experience. Persistence often emerges a strong variable among analytics, and the need for short concentration periods interspersed with breaks and the desire to engage in more than one task at a time appears to correlate with strongly global (right) processors.

FIGURE 2.21 Learning Styles Model. From Dunn and Griggs, "A Quiet Revolution," pp. 14-19. Reprinted with permission of Rita Dunn.

preferences.[56] Knowledge of these categories is helpful in understanding the differences in learning preferences of your students. Briefly, here are some comments on the five categories identified by the Dunns and on their implications for science teaching.

Environmental Elements. It shouldn't surprise us that sound, light, temperature, and design affect learning styles. According to Dunn, 10 to 40 percent of students are affected by differences in sound (quiet versus loud), bright or soft lighting, warm or cool temperatures, and formal versus informal seating designs.[57] Science teachers have an opportunity to create a physical learning environment that is appealing to a wide range of students. One of the suggestions that the Dunns make is to "change the classroom box into a multifaceted learning environment." We will explore the classroom learning environment in greater detail in Chapter 9.

Emotional Elements. Motivation, persistence on completing a task, degree of responsibility, and structure (specificity of rules governing work and assignments) constitute emotional elements that affect student learning style.

Physical Elements. There are several physical elements, including perceptual strengths, intake (of food or drink), time (of day), and mobility, that influence learning. Perceptual strengths refers to learning through the different senses. At the secondary school level, greatest emphasis is given to auditory and visual learning. However, secondary teachers who have used electroboards, flip charts, task cards and other manipulatives, have reported increased achievement and interest for the tactile student. Secondary teachers who employed kinesthetic (whole body) activities such as field trips, dramatizing, interviewing, and role playing also reported increases in achievement and interest.[58] Many students also learn better if they are engaged in multisensory learning activities, for example, combining tactile and kinesthetic or visual and auditory.

Sociological Elements. Do students like to learn alone, in pairs, with a small team, or with the whole class? The answer is that students respond to a variety of social groupings and appear to be "unresponsive to a consistent instructional routine."[59] The classroom that provides opportunities not only throughout a science course but within individual lessons for variety in social groupings is paying attention to the sociological needs of the learner.

Psychological Elements. There are a number of psychological factors that psychologists have examined in relation to learning style. Two major ideas emerge in this regard, namely, how learners process information and how learners perceive. Processing information can be viewed as a global process or as an analytical process. Global (processing in wholes) versus analytical (processing in parts) is analogous to right hemispheric thinking versus left hemispheric thinking. Learners appear to perceive either actively or reflectively.

Brain Hemisphericity and Learning Style

In the past quarter of a century considerable attention has been given to what is called brain hemisphericity. According to neurosurgeon Joseph Bogen, brain hemisphericity is the reliance more on one mode of processing than on another by an individual.[60] Roger Sperry, a Nobel laureate in physiology for his work on hemisphericity, explained the nature of hemisphericity this way:

> Each hemisphere...has its own...private sensations, perceptions, thought, and ideas all of which are cut off from the corresponding experience in the opposite hemisphere. Each left and right hemisphere has its own private chain of memories and learning experiences that are accessible to recall by the other hemisphere. In many respects each disconnected hemisphere appears to have a separate "mind of its own."[61]

These early brain researchers found that (1) the two halves of the brain, the right and left hemispheres, process information differently; that (2) in the split-brain patient, there seem to be two different people up there, each with his or her favorite ways of processing information, each with a different mode of thinking; and that (3) both hemispheres are equally important.[62]

[56]Dunn and Griggs, "A Quiet Revolution," pp. 14–19.
[57]R. Dunn, J. S. Beaudry, and A. Klavas, "Survey of Research on Learning Styles," *Educational Leadership*, March, 1989, pp. 50–58.
[58]Dunn, Beaudry, and Klavas, "Survey of Research," pp. 50–58.
[59]Dunn and Griggs, "A Quiet Revolution," p. 18.

[60]J. E. Bogen, "Some Educational Aspects of Hemispheric Specialization," *USCA Educator* 17 (1975): pp. 24–32.
[61]R. W. Sperry, "Lateral Specialization in the Surgically Separated Hemispheres," *The Neuro-Sciences Third Study Program*, F. O. Schmitt and F. G. Warden, eds. (Cambridge, Mass.: MIT Press, 1975), pp. 5–19.
[62]From *The 4MAT® System: Teaching to Learning Styles With Right/Left Mode Techniques* by Bernice McCarthy, copyright 1980, 1987 by Excel, Inc. Used by special permission. Not to be further reproduced without the express written permission of Excel, Inc. Those desiring a copy of the complete work for further reading may acquire it from the publisher, Excel, Inc., 200 West Station Street, Barrington, Il. 60010, 708-382-7272.

Left- Versus Right-Brain Thinking

These neurosurgeons' findings had direct and obvious implications for teaching, but especially for the growing field of student learning styles. Bernice McCarthy, who has applied the results of brain research to the 4MAT model of learning, sees the two hemispheres processing information and experiencing differently. Table 2.16 lists some attributes that she feels make a difference in helping to accommodate students with different learning styles.[63]

One of the arguments that brain researchers make is that school learning emphasizes and favors left-brain learning over right-brain learning. If listening to lectures and relying on the science textbook are left-brain activities, there is evidence to support this argument.[64] For example, teachers who want to increase the number of right-brain activities in their lesson plans, thereby giving right-brain learners more of an opportunity for success, would include such approaches as mind mapping, visualization experiences, imagery, analogies, the use of paradox, role playing, creative writing (yes, in science), demonstrations, experiments, intuitive activities, connecting ideas, and creative problem solving.[65]

Left- Versus Right-Brain: Implications from Research

There is a tendency, as with any theory, to draw simplified interpretations, and so it is with brain functioning and student learning style. One of the major oversimplifications is that rationality is exclusively a left-brain function, and creativity a right-brain function. Evidence supports the idea that both hemispheres play a part in rationality and creativity. There are, however, some results that have powerful implications for you as a teacher. Here are a few.

TABLE 2.16

LEFT AND RIGHT HEMISPHERE INFORMATION PROCESSING

The Left Hemisphere	The Right Hemisphere
Does verbal things	Does visual-spatial things
Likes sequence	Likes random patterns
Sees the trees	Sees the forest
Likes structure	Is fluid and spontaneous
Analyses	Grabs for the whole
Is rational	Is intuitive

Ann Howe and Poul Thompsen report that hemisphericity can play an important role in motivation and science teaching. According to work being done in artificial intelligence, when a person is exposed to some new phenomenon, the first thing that occurs is that in the deep part of the brain he or she gives a preliminary value to it: "Is it interesting or not?" If it isn't, the person doesn't give it any more attention. If it is interesting, after ten seconds or so it enters the right hemisphere, which attempts to make holistic sense of the phenomenon: "What is this all about?" If this succeeds, the information is processed to the left hemisphere, where the brain tries to deal with it analytically. This notion supports the contention that we must pay close attention to the types of tasks that we present to students. Interest is an important aspect of science teaching, and the gatekeeper seems to be the deep recesses of the brain.

Another finding that has implications for teaching has to do with the role of emotion, or feelings. The right hemisphere seems to play a special role in emotion. If students are emotionally involved in an activity, both sides of the brain will participate in the activity, regardless of the subject matter or content.[66]

The two hemispheres are involved in thinking, logic, and reasoning, and in the creation and appreciation of art and music. This contradicts earlier implications that the left brain was the logical side, and the right brain the artistic side.

---

[63]From *The 4MAT® System: Teaching to Learning Styles With Right/Left Mode Techniques* by Bernice McCarthy, copyright 1980, 1987 by Excel, Inc. Used by special permission. Not to be further reproduced without the express written permission of Excel, Inc. Those desiring a copy of the complete work for further reading may acquire it from the publisher, Excel, Inc. 200 West Station Street, Barrington, Il 60010, 708-382-7272.

[64]J. Goodlad, *A Place Called School* (New York: McGraw-Hill, 1984). Goodlad reports that the typical activity in secondary classrooms is lecture. Experiential types of learning activities were rarely, if ever observed by Goodlad's researchers.

[65]From *The 4MAT® System: Teaching to Learning Styles With Right/Left Mode Techniques* by Bernice McCarthy, copyright 1980, 1987 by Excel, Inc. Used by special permission. Not to be further reproduced without the express written permission of Excel, Inc. Those desiring a copy of the complete work for further reading may acquire it from the publisher, Excel, Inc., 200 West Station Street, Barrington, Il. 60010, 708-382-7272.

[66]J. Levy, "Research Synthesis on Right and Left Hemispheres: We Think with Both Sides of the Brain," in R. S. Brandt, ed., *Readings on Research* (Alexandria, Va.: Association for Supervision and Curriculum Development, 1989), pp. 23–28.

INQUIRY ACTIVITY   2.5

Determining Your Hemisphericity

Knowing your own hemisphericity, the tendency to rely on one mode of processing more than another, can be determined by responding to a series of characteristics.

Materials

Right and Left Mode Characteristics (Table 2.17), writing tool.

Procedure

1. In each set, choose the characteristic that is most like you most of the time, and mark it with an *X*. Choose only one answer for each numbered statement.
2. Check your answers by referring to the *4MAT System Workbook: Guided Practice in 4MAT Lesson and Unit Planning.*

Minds on Strategies

1. How would you distinguish between left- and right-mode processing?
2. How can you change a classifying activity (which is left brain) into a right-brain activity?

3. What are some ways to incorporate left- and right-brain processing into science teaching?

Applying Learning Style Concepts to Science Teaching

There are many things you can do to help students learn in the science classroom, and certainly applying what is known about learning styles is a place to begin. There are many sources of information for specific ideas, such as Rita and Ken Dunn's book *Teaching Students through Their Individual Learning Styles: A Practical Approach.*[67] In this section we will consider an approach to learning that incorporates brain research and student learning styles (the 4MAT system) and specific suggestions for teaching students according to their own learning style, and will introduce some tools to help students learn about their own learning (metacognition).

The 4MAT System

4MAT, devised by Bernice McCarthy, is a learning-style system that identifies four types of learners:[68]

[67]McCarthy, *The 4MAT System*, pp. 20–26.
[68]McCarthy, *The 4MAT System*.

TABLE   2.17

RIGHT AND LEFT MODE CHARACTERISTICS

Column A	or	Column B
1. Uses open-ended, random experiments		Uses controlled, systematic experiments
2. Looks at differences		Looks at similarities
3. Controls feelings		Is free with feelings
4. Is a lumper: connectedness is important		Is a splitter: distinctions important
5. Relies primarily on images in thinking and remembering		Relies primarily on language in thinking and remembering
6. Makes objective judgments		Makes subjective judgments
7. Prefers elusive, uncertain information		Prefers established, certain information
8. Problem solves with hunches, looking for patterns and configurations		Problem solves by logically and sequentially looking at the parts of things

1. Imaginative learners
2. Analytic learners
3. Commonsense learners
4. Dynamic learners

Imaginative learners (type 1) perceive information *concretely* and process it *reflectively*. They are sensory oriented as well as reflective. They are imaginative thinkers, and like to work with other people. Their favorite question is, Why? Analytic learners (type 2) perceive information *abstractly* and process it *reflectively*. They are abstract thinkers who think about their creations, which tend to be models and theories. They value sequential thinking. Their favorite questions is, What? Commonsense learners (type 3) perceive information *abstractly* and process it *actively*. They are practical thinkers; they need to know how things work. They are skill oriented, experimenting and tinkering with things. Their favorite question is, How does this work? The dynamic learners (type 4) perceive information *concretely* and process it *actively*. These learners integrate experience and application. They learn by trial and error and are risk takers. Their favorite question is, What if?[69]

How did McCarthy arrive at these four types of learners? McCarthy based her model on the work of David Kolb, who had studied learners and ways in which they perceive and process information.[70] Kolb theorized two continuums as follows:

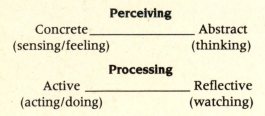

Perceiving

Concrete_____ Abstract
(sensing/feeling)   (thinking)

Processing

Active _____ Reflective
(acting/doing)   (watching)

By combining these dualities, McCarthy developed a system in which four distinct learning styles emerged, each being one of the quadrants in the model (Figure 2.22).

McCarthy's model provides a natural cycle of learning. If you examine the model in Figure 2.23, imagine it as a clock. Learning begins at 12:00 with concrete experience. According to McCarthy, by moving clockwise around the circle, students then experience reflective observation; from this place they move to abstract conceptualization, and finally to active experimentation. In this way, all students are taught in all four ways. Each is comfortable some of the time, while at the same time being stretched to develop other learning abilities.

[69]McCarthy, *The 4MAT System*, pp. 37–43.
[70]McCarthy, *The 4MAT System*, pp. 20–26.

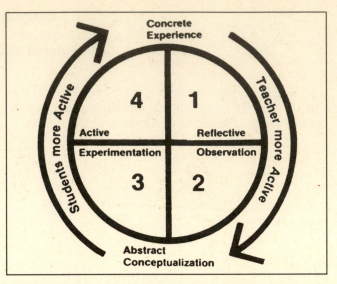

FIGURE 2.22 Types of Learners in the 4MAT System. From *The 4MAT® System: Teaching to Learning Styles with Right/Left Mode Techniques* by Bernice McCarthy, copyright 1980, 1987 by Excel, Inc. Used by permission. Not to be further reproduced without the express written permission of Excel, Inc. Those desiring a copy of the complete work for further reading may acquire it from the Publisher, Excel, Inc., 200 West Station Street, Barrington, IL 60010, 708-382-7272.

Another distinct feature of 4MAT is that each of the four learning styles is integrated with left- and right-brain processing, giving teachers a comprehensive teaching model. Activities for each quadrant are equally divided between left- and right-brain modes. For example, a teacher would begin a learning sequence in quadrant I (concrete/reflective) with a right-mode activity to help the students explore by observing, questioning, visualizing, and imagining. Students would be helped to develop a reason for studying the material. The right-mode activity would be followed by a left-mode activity in which the students reflect on the active-concrete experience they began with. This pattern, alternating between right- and left-brain modalities, is continued in sequence through the remaining three quadrants of the 4MAT model.

In their book *4MAT and Science*, Samples, Hammond, and McCarthy describe science teaching plans, showing how 4MAT can be applied to the science classroom. Plans can be created for any concept or topic in science.

Teaching to Student's Individual Styles

According to some researchers, students who do well in school tend to be the ones who learn either by listening or by reading. The focus on these two senses, especially at the high school level, tends to play havoc with the tactile and kinesthetic learners. Because so much of what happens in classrooms is

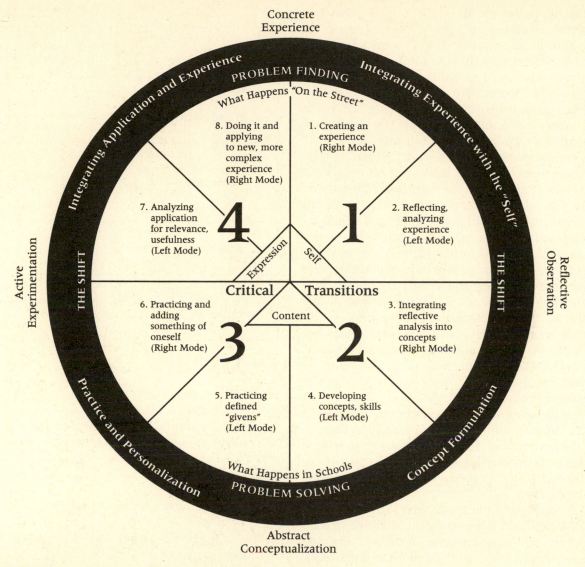

Concrete
Experience

Abstract
Conceptualization

FIGURE 2.23   The 4MAT Learning Cycle. From *The 4MAT® System: Teaching to Learning Styles with Right/Left Mode Techniques* by Bernice McCarthy, copyright 1980, 1987 by Excel, Inc. Used by permission. Not to be further reproduced without the express written permission of Excel, Inc. Those desiring a copy of the complete work for further reading may acquire it from the Publisher, Excel, Inc., 200 West Station Street, Barrington, IL 60010, 708-382-7272.

focused on the auditory and visual modes, students who prefer tactile and kinesthetic modes are actually handicapped. In this section a few suggestions are included to show how these other modalities can be included in science teaching, thereby creating a multisensory approach.

Keep in mind these characteristics: visual learners learn by seeing and imagining; auditory learners learn by listening and verbalizing; kinesthetic learners learn by participating, moving, and talking; tactile learners learn by doing, touching, and manipulating. Also remember that these modes can be combined.

- Bring color into the science classroom: posters, bright bulletin boards, new paint. One technique that is

effective is instead of writing on the chalkboard, write your notes and make drawings with bright marking pens on a flip chart. If you use the overhead projector, write with a variety of colored pens.

- Make tactile learning aids. The Dunns describe a number of tactile aids that can be used repeatedly from class to class and year to year.[71] One example is the task card. This multisensory resource can be used to help students review and check whether they understand material, allows students to work at their own pace or with someone else, and frees

[71]For suggestions on developing tactual and kinesthetic resources see R. Dunn and K. Dunn, *Teaching Students Through Their Individual Learning Styles: A Practical Approach* (Reston, Va.: Reston, 1978) pp. 317–358.

the teacher to work with others. Suppose that you want the students to review the meanings of important ideas and concepts. The teacher would prepare a set of questions with answers (or concepts with associated meanings) and prepare cards as shown in Figure 2.24. The cards are made by cutting oaktag into strips, laminating the strips after writing the information on the cards, and then cutting each card in a unique fashion.

- The computer is a multisensory learning aid. Computers are powerful tools in their own right but can be used to help the tactile and kinesthetic learner. By establishing a computer center in the classroom and providing opportunities for individuals or small groups of students to work in the center with games, tutorials, simulations, or problem solving software, the tactile and kinesthetic learner is given opportunities for personal involvement, manipulation of the keyboard, and movement (to a different place in the classroom).
- Plan an occasional field trip, role-playing sessions and debates, and games in which the students move from one place in the classroom to another and manipulate objects. These activities favor the kinesthetic learner.
- Plan tabletop learning activities in which students handle and manipulate science materials, objects, and specimens. You do not have to wait to go into the lab to favor the tactile learner. It is possible to teach every lesson in such a way that there is some tactile learning.
- Don't forget the auditory learner. Showing films, video, listening to tapes, and hearing music (related to science) are activities that favor the auditory (and visual) learner. Discussion, debates, and question-and-answer sessions favor the auditory learner.

There are many other ways to create a multisensory classroom. These will be presented in Chapters 7 and 8. For now it is important to remember that providing a variety of sensory modes will ensure that each student's learning style is attended to at least some of the time.

Metacognitive Strategies

Have you ever thought about your strengths (and weaknesses too) as a learner? Do you know how you learn? Do you have strategies that you use to learn? These strategies are ways to help students learn about learning and learn about knowledge. These are called metacognitive strategies, and they are playing an increasingly important role in science teaching. We

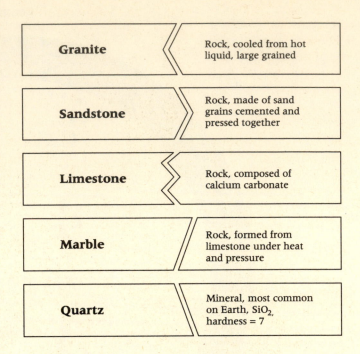

FIGURE 2.24   Sample Geology Task Cards

will briefly discuss them here but will return to them again in Chapter 8.

There are several definitions of metacognition. One view is that it is our ability to know what we know and what we don't know. We might also think of metacognition as the ability:

1. To plan a strategy for producing what information is needed
2. To be conscious of our own steps and strategies during the act of problem solving
3. To reflect on and evaluate the productivity of our own thinking[72]

Teaching metacognitive strategies is a potential new goal for science teachers. Given that student learning styles influence the way students process and perceive information, metacognitive strategies can be useful in helping students understand their unique learning patterns. What are some metacognitive strategies (skills) that students might learn to help them understand their own thinking, thereby increasing their ability to learn science?

Mind Mapping.   Introduced earlier as cognitive mapping, mind mapping is a powerful metacognitive tool. For example, Joseph Novak has reported that high school biology students using concept maps

[72]A. L. Costa, "Mediating the Metacognitive," in R. S. Brandt, ed., *Teaching Thinking* (Alexandria, Va.: Association for Supervision and Curriculum Development, 1989), pp. 120–125.

were more on task in laboratory experiments and reported being very conscious of their own responsibility for learning. Novak has also reported that some teachers are teaching "how to learn" short courses designed to teach students metacognitive strategies. Novak suggests that using cognitive maps as a metacognitive strategy increases meaningful learning over rote learning for students in a variety of science situations.[73]

Illustrating and Drawing. Some learners are visually attuned to looking at things in pictures. There are many opportunities in which students could create an illustration or a drawing to explain their thinking, or to show how they understand a concept.

Brainstorming. "List as many observations of this burning candle as you can." "What are the many ways that one individual can differ from another? What are the many hypotheses to explain the phenomenon?" Brainstorming, a strategy used to help students creatively solve problems, can also be used as a metacognitive strategy. Brainstorming should proceed without censorship. If students are working in a group, all ideas should be accepted. If the students are working alone, they should be told to consider all ideas that "bubble up." Teaching students not to censor their ideas at the beginning of a process is an important metacognitive tool.

Planning Strategies. Students can be shown, prior to an activity (short term or long term) how to go about solving or completing it, special ways that might be helpful for attacking the problem, and any rules and directions to follow (especially if working with equipment, chemicals, or other science materials). Asking students during a learning activity to share their progress or how they are proceeding with the activity or problem enables them to perceive their own thought processes.

Generating Questions and Other Inquiry Strategies. Another metacognitive strategy is to teach students to pose questions regarding some material they will read in their science textbook, homework assignment, research project, or science laboratory investigation. The process of asking questions is the heart of scientific inquiry. Not only does questioning help focus thinking strategies, but the questions themselves show an understanding of the subject matter and can, if students are asked to read information, help them

with comprehension.[74] In Chapter 7, we will explore other science inquiry strategies (processes of science), and it will be shown that these are indeed metacognitive strategies as well.

Evaluating Actions. Some teachers ask students to evaluate what they liked or didn't like, or what were the pluses and minuses of a science learning activity. This process enables students to reflect on and evaluate their actions, and perhaps apply this learning to future actions.

Teaching Capability. Some teachers have a rule in their class: "Outlawed: I can't do it!" Instead these teachers help students focus on what information, materials, or skills are needed to do it. As was mentioned earlier, it is important to teach students that intelligence is not fixed but is a developing ability, based on experience. This position gives students a sense of personal power in that attempting a challenging problem or activity is indeed a way to improve their ability to think.

Communication Skills. Communication skills are not only important to the teacher, but they are an integral metacognitive strategy. In the social learning theory section, it was pointed out that teachers who adopt cooperative learning strategies will need to teach students new social norms and social learning skills. These skills (conciseness, listening, reflecting) are communication skills. An important metacognitive communication skill is reflection. Having students consider other students' ideas, as well as their own, or having students rephrase what they just said are ways of building on and extending ideas.

Journal Keeping. Keeping a diary or log of learning experiences is not new to the science education community. Many scientists have kept logs of their thinking, not only as a record, but more important as a haven for synthesizing and analyzing their thinking. The log is a place where the student can revisit ideas and review thinking processes used in an activity. Combining some of the other strategies, especially mind mapping, illustrating and drawing, and brainstorming, can enhance the quality of logs.

Metacognitive strategies are tools for the science teacher to help students understand their own thinking. Throughout this book you will be introduced to many strategies. Teaching students about their own thinking might be as important as exposing them to the content of science.

---

[73]J. Novak, "The Use of Metacognitive Tools to Facilitate Meaningful Learning," in P. Adley, ed., *Adolescent Development and School Science* (New York: Falmer, 1989), pp. 227–238.

[74]See Costa, "Mediating the Metacognitive," pp. 120–125.

Case Study 1.
A New Approach to Learning

The Case. Lois Wilson, a second year high school biology teacher in a community that has only one high school, took a graduate course in the summer at the local university. In the course, she became extremely interested in a theory of learning called self-directed learning, proposed by Carl Rogers. Self-directed learning provides more freedom for the students in terms of choosing what and how to learn information. Ms. Wilson feels strongly that this "open" method fits with her teaching philosophy better than the more structured approach she was using during her first year of teaching. Prior to the opening of school, Ms. Wilson changes her curriculum plans to reflect the self-directed theory.

She spends the first two weeks of school helping the students become skilled and familiar with self-directed learning. For all of her students, this is a new venture. She planned activities in which students have to make choices regarding objectives, activities, or content. Knowing that students like to work together, she decided to place students in small teams. At the end of the two weeks, she instructs the teams that they should decide and select the activities and content in the first part of the text that would interest them. They should formulate a self-directed plan and carry it out for the remainder of the grading period.

A few weeks later, a rather irate parent calls Mr. Brady, the principal, complaining that her son is wasting his time in Ms. Wilson's class. The parent complains that her son is not learning anything, and she demands a conference with Ms. Wilson.

The Problem. How would you deal with this situation? What would you say to the parent? How do explain your teaching theory to your principal? ☐

Case Study 2.
The Student Who Thought He Failed

The Case. Mr. Wong, a physics teacher in a large comprehensive high school, is known for his innovative approaches to teaching. After attending an in-depth conference and study group on "Implementing Cognitive Theory in the Science Classroom," he decides that he is going to

(Continued on p. 81)

To Understand Is to Invent
by Jean Piaget[75]

The very optimistic outlook resulting from our research on the development of basic qualitative notions, which ought to constitute the foundation of elementary instruction in the sciences, would seem to suggest that a fairly far-reaching reform in this area would help answer societies need for scientists. But this depends on certain conditions that are doubtless those of all intellectual training, although they seem to be particularly important in the various branches of scientific training.

The first of these conditions is, of course, the use of active methods which give broad scope to the spontaneous research of the child or adolescent and require every new truth be learned, be rediscovered, or at least reconstructed by the student, and not simply imparted to him. Two common misunderstandings, however, have diminished the value of the efforts made in this field up to now. The first is the fear (and sometimes the hope) that the teacher would have no role to play in these experiments and that their success would depend on leaving the students entirely free to work or play as they will. It is obvious that the teacher as organizer remains indispensable in order to create the situations and construct the initial devices which present useful problems to the child. Secondly, he is needed to provide counter-exam-

(Continued on p. 82)

[75]Excerpted from J. Piaget, *To Understand Is to Invent* (New York: Viking, 1973), pp. 15–20. Used with permission of Penguin Books USA Inc.

(Continued from p. 80) implement one of the ideas in his teaching approach. At the conference he discovered that determining and helping the students detect their current ideas on the concepts to be taught is an important place to begin instruction. At the conference it was suggested that a simple activity or a demonstration presented to the students would enable the students to demonstrate their ideas verbally and publicly. Mr. Wong planned a demonstration on falling objects. The idea was to let students identify the forces (by making a diagram and labeling it) on the falling object. After doing the

activity and having the students make their diagrams, Mr. Wong carried on a discussion with the class. During the discussion he noticed one of the students was quite upset. The student was embarrassed that he hadn't labeled the diagram "correctly" and felt inferior to the students sitting near him. Mr. Wong noticed that a couple of other students felt the same way.

The Problem. What should Mr. Wong do? What should he say to these students? To the whole class? Should be abandon this new approach?   ☐

Think Pieces

Select one of the following problems (or identify one related to the content of this chapter) and locate some key literature to help you develop a two-page think piece (see Appendix 2.1 for a list of major journals in the field of science education).

* How do Skinner's, Vygotsky's and Piaget's ideas on learning compare?
* What is learning?
* What motivational strategies help students learn science concepts?
* What is conceptual change teaching?
* In what ways do student learning styles influence learning?
* What is the difference between right-brain and left-brain learning? What are the implications of these differences for science teaching?
* What are some metacognitive learning strategies? In what ways can these strategies be helpful to students in a science class?

To find the key literature you should search the Educational Resources Information Center (ERIC) data base, which is on CD-ROM discs in your library. Start the search by identifying the topic of your search. Analyze the topic by listing the keywords in the topic. Then refer to the list of descriptors that can be found in the ERIC thesaurus (see Appendix 2.2 for sample descriptors for science teaching). Going through this process will help you prepare for the search.   ☐

Science Teachers Talk: How Do You Accommodate Students' Varying Learning Styles in Your Classroom?

Ginny Almeder. I accommodate students with different learning styles in my classroom by using different modalities, which include auditory, visual, and tactile components. Each teaching unit is a composite of lecture, written work, large and small group discussion, audiovisual, and laboratory activities. I generally use activities that involve all of the students one way or another. One other thing that I would add is this: There is some flexibility built into participation. For example, following group work students may do an oral presentation or a written presentation using the blackboard. For homework, they may elect to write out their objectives or cross-reference the objectives with the notes. This is a more efficient approach for those students who learn better by listening than by writing. Some students also benefit from re-

versing the teacher-student relationship by working in after-school study groups in which they act as tutors. Some student mentors come to realize very quickly that teaching is a form of learning.

John Ricciardi. I try to plan and construct lesson activities that are constantly in a directional movement or "flow" from one particular learning style to another. Individual learning styles are not fixed, like still pools of water. Maximum brain-mind stimulus is more a style of learning that is symbolized by the water movement in a small country stream…the liquid patterns are observed to be in constant oscillating, pendulating motion. In the classroom, there are, say, 25 different "stream" patterns of thought emanating and synergizing. The

(Continued on p. 82)

(Continued from p. 81)
only real common denominator is that there is a pendulation or "back and forth" learning flow of focal attention. Like the bubbling brook, the brain-mind is constantly jumping here and there, picking and choosing between modalities of information and input such as symbolic, visual, auditory, and kinesthetic. I try to juxtapose my lesson activities to this random mental movement, moving through at least three, and sometimes up to six different instructional modalities within a 50 minute period.

Jerry Pelletier. I approach my class in what I call a multilearning style. Students are given the chance to learn through their various senses. We observe, we listen, we manipulate, we read, we write, we communicate, and we question. In order for a student to understand a scientific concept, they are given these multifaceted lessons. My classroom is set up with heterogeneous groups. When given activities, these groups are asked to work with each other in a cooperative manner. Decisions and questions are produced by the whole group. This allows students to use their own skills and styles in a structured learning environment. Every student participates not only because they are expected to but because the classroom environment is conducive to bringing out the special talents of each student.

Mary Wilde. Accommodating students with different learning styles is easily achieved when teaching science. When introducing a new scientific concept, I begin with a demonstration (visual) to stimulate the thought processes of the child. Through discussion (oral), we, as a group, attempt to explain the concept. After I have aroused the interest, I then proceed to provide more

background information through lecture (oral) and notes (visual). I then try to follow up every new concept with an application activity (tactile), for many of us do not learn until we see and do for ourselves. In order for students to comprehend the most from laboratory activities, I group students heterogeneously by ability and learning style preference. I have much better results and greater achievement if I take into consideration each child's preferred learning style when developing laboratory groups, for the groups function better as a unit.

Bob Miller. I try to vary my presentation of materials depending on the strengths and interests of the students. I offer alternate activities at different levels of difficulty. A large variety of offerings usually results in better work being turned in and possibly it being more creative. ☐

(Continued from p. 80)
ples that compel reflection and reconsideration of over-hasty solutions. What is desired is that the teacher cease being a lecturer, satisfied with transmitting ready-made solutions; his role should rather be that of a mentor stimulating initiative and research.

In physics and the natural sciences the incredible failing of traditional schools till very recently has been to have almost systematically neglected to train pupils in experimentation. It is not the experiments the teacher may demonstrate before them, or those they carry out themselves according to a pre-established procedure, that will teach students the general rules of scientific experimentation—such as the variation of one

factor when the others have been neutralized (*ceteris paribus*), or the dissociation of fortuitous fluctuations and regular variations. In this context more than in any other, the methods of the future will have to give more and more scope to the activity and the groups of students as well as to the spontaneous handling of devices intended to confirm or refute the hypothesis they have formed to explain a given elementary phenomenon. In other words, if there is any area in which active methods will probably become imperative in the full sense of the term, it is that in which experimental procedures are learned, for an experiment not carried out by the individual himself with all freedom of initiative is by definition not an experiment but mere drill with no educational value: the details of the successive steps are not adequately understood.

In short, the basic principle of active methods will have to draw its inspiration from the history of science and may be expressed as follows: to understand is to discover, or reconstruct by rediscovery, and such conditions must be complied with if the future individuals are to be formed who are capable of production and creativity and not simply repetition. ☐

Research Matters: Ideas from a Soviet Psychologist by Anne C. Howe [76]

Over the past two decades most science teachers have been introduced to the discoveries and ideas
(Continued on p. 83)

[76]Anne C. Howe, Ideas from a Soviet Psychologist, Research Matters...To the Science Teacher Monograph Series, National Association for Research in Science Teaching, December, 1990, no. 25. Used with permission.

(Continued from p. 82)
of Jean Piaget. Piaget's ideas now permeate much of our thinking about science teaching and are the basis for some of our most successful curriculum projects and teaching strategies.

Another psychologist whose ideas should become better known to science teachers is Lev Vygotsky. Vygotsky was a Soviet psychologist who was born in the same year as Piaget but died in his thirties. His work was ignored in the United States for many years after he died but now we are beginning to see that he had some very interesting and stimulating ideas that may be relevant to science teaching.

Vygotsky, like Piaget, believed that the learner *constructs* knowledge; that is, that what we know is not a copy of what we find in the environmnent but is, instead, the result of *thought and action* [emphasis mine]. Although both of these thinkers focused on the growth of children's knowledge and understanding of the world around them, Piaget placed more emphasis on the role of teaching and social interaction in the development of science concepts and other knowledge. He also sought to show that language plays a central role in mental development.

Mental Development. Vygotsky believed that development depends on both natural or biological forces and social and cultural forces. Biological forces produce the elementary functions of memory, attention, perception and stimulus-response learning but social forces are necessary for the development of the higher mental functions of concept development, logical reasoning and judgment. A main difference between lower, or elementary, and higher mental functions is a shift from outside control to inner control. Through social interaction the child gradu-ally assumes more responsibility and becomes more self-directed and autonomous.

In practical terms this means that what children can do and learn in school depends, at every age, on such things as their attention span, their ability to remember and other biologically determined factors, but this is only part of the story. What they can do and learn also depends on the interaction that takes place between and among children and between children and adults, especially their parents and teachers. It is through social interaction, including the interaction with teachers, that children become conscious of their basic mental functions and able to use them in their growth toward self-control, self-direction and independent thinking and action.

In a series of experiments on the development of scientific concepts, Vygotsky and his students studied the differences between concepts acquired outside the school, which he called "spontaneous" concepts and those learned from instruction in school, which he called "scientific." Working with second and fourth grade children, he found that the children could explain and articulate the concepts learned in school better than they could those learned from everyday experience. In contrast to spontaneous concepts, those taught in school are defined, discussed, thought about and tied in with other concepts; because of this it becomes possible to generalize and to build a system of concepts.

Both Vygotsky and Piaget were interested in children's mental development and both conducted their research by asking children questions about rather ordinary things but, as Vygotsky pointed out, there was an important difference in their views. While Piaget assumed that development and instruction were separate, Vygotsky focused on the interaction between them. He believed that instruction leads to mental development and that further development makes higher level instruction possible.

Interaction as a Force in Development. Vygotsky's ideas were developed through observation and study of children in the context of their daily lives, including school and family. His writings emphasize the importance of the social interaction that takes place between children and adults or older peers as the main force in cognitive, or mental, development.

How can you use these ideas in teaching science? By applying the same general principles, the teacher sets tasks that are just beyond what the pupils can do on their own but are attainable with help. We might think of this as the teacher building a scaffold for the learner to climb to a higher level. At first the learner depends on the teacher to show him/her what to do but gradually the learner masters the task and gains control over a new function or concept.

Teachers can provide a classroom environment that makes it possible, even probable for pupils to master new skills, gain new knowledge or improve understanding of things already known.

Here are some practical suggestions for teachers to implement these ideas:

1. The teacher models the behavior or skill that is to be taught or encouraged until the pupils can internalize the behavior. For example, if you use the metric system consistently and without even making a comment about it, the children will accept it as the way things are measured in science.

(Continued on p. 84)

(Continued from p. 83)

2. Peer tutoring is incorporated into the teaching strategies used in the classroom. Pupils have to be taught how to be good tutors and to accept tutoring but this is a very powerful teaching method when well used.

3. Cooperative learning is used as a regular instructional strategy. Cooperative learning methods, which are gaining popularity all over the United States today [and other countries, including the Soviet Union], take advantage of pupil interaction as a means of promoting both academic and social learning.

Vygotsky's ideas call for a classroom where active exchanges between children themselves and between children and teacher are an ongoing part of daily life. In this setting the teacher sets tasks that are just beyond the learners' current levels of competence and provides the help that learners need to reach higher levels. The help may take various forms but an important aspect will be opportunities for children to work together, to give and receive verbal instructions, to respond to peer questions and challenges and to engage in collaborative problem solving. ☐

Problems and Extensions

- Suppose that you could interview one of the following theorists: Piaget, Skinner, Dunn, McCarthy, Pavlov, Bruner, Vygotsky, or Ausubel. How would the theorist you choose respond to the following: (1) Students learn best when they…(2) What is learning? (3) What is the best way to teach science? (4) How would you motivate reluctant learners?

- Visit one or two secondary science classrooms to find out if there is any evidence of the following learning theories being put into practice: behavioral theories, cognitive theories, and social learning theories. Cite examples of classroom practice that are related to the theories you observe.

- Student attitudes in science have been slipping in recent years. What are some ways to turn this trend around? Collaborate with a small group and brainstorm as many ideas as you can think of. Rank-order the ideas, and then present them to the class. Ask members of the class if they have witnessed any of the recommendations your group made.

- Select a chapter from a secondary science textbook. Examine the chapter in light of the theories of learning that you have studied in this chapter. Is there any evidence of behaviorism in the chapter? Cognitive learning theory? Social learning theory?

- Design a model of learning using what you have learned in this chapter and from your previous courses and experiences. Your model should be a visual display and should reflect your beliefs and knowledge about student learning.

- Become one of the following characters, and role-play a discussion about improving learning in the science classroom. The fundamental question is, What are the best ways to improve student achievement and attitudes in science? The characters include Piaget, Skinner, Pavlov, Dunn, McCarthy, Ausubel, and Vygotsky.

- How do you suppose the following scientists would explain learning, and what recommendations do you think they would make about contemporary science teaching: Albert Einstein, Rosalyn Franklin, Marie Curie, Linus Pauling, Stephen Jay Gould, Issac Azimov, and Benjamin Banaker.

APPENDIX 2.1
SCIENCE TEACHING JOURNALS

The Biology Teacher
Journal of Chemical Education
Journal of College Science Teaching
Journal of Earth Science Teaching
Journal of Research in Science Teaching
The Physics Teacher

School Science and Mathematics
Science and Children (elementary)
Science Education
Science Teacher (high)
Science Scope (middle)

APPENDIX 2.2
ERIC THESAURUS:
SOME DESCRIPTORS FOR
SCIENCE EDUCATION

(Note: These are the main categories in the science teaching section of the ERIC thesaurus. There are hundreds of other descriptors in the thesaurus that you can use to narrow your search. Use these descriptors to develop your own problems for writing a think piece.)

SCIENCE ACTIVITIES
SCIENCE AND SOCIETY
SCIENCE CAREERS
SCIENCE CLUBS
SCIENCE COURSE IMPROVEMENT PROJECTS
SCIENCE CURRICULUM
SCIENCE DEPARTMENTS
SCIENCE EDUCATION
SCIENCE EDUCATION HISTORY
SCIENCE EQUIPMENT
SCIENCE EXPERIMENTS
SCIENCE FACILITIES
SCIENCE FAIRS

SCIENCE HISTORY
SCIENCE INSTRUCTION
SCIENCE INTERESTS
SCIENCE LABORATORIES
SCIENCE MATERIALS
SCIENCE PROGRAMS
SCIENCE PROJECTS
SCIENCE SUPERVISION
SCIENCE TEACHERS
SCIENCE TEACHING CENTERS
SCIENCE TECHNOLOGY AND SOCIETY
SCIENCE TESTS
SCIENCE UNITS
SCIENTIFIC AND TECHNICAL INFORMATION
SCIENTIFIC ATTITUDES
SCIENTIFIC CONCEPTS
SCIENTIFIC ENTERPRISE
SCIENTIFIC LITERACY
SCIENTIFIC OBSERVATION

RESOURCES

Bell-Gredler, Margaret E. *Learning and Instruction: Theory into Practice.* New York: Macmillan, 1986.

This is without question one of the best books to provide you with an overview of all the theories of learning and instruction. Bell-Gredler writes not only about the theories of learning but provides concrete examples of how to implement the theories into instructional strategies. If you want to know about the theorists (Piaget, Bruner, Gagne, Bandura, Weiner), this is the book.

Bybee, Rodger W., and Robert B. Sund. *Piaget for Educators,* 2d ed. Prospect Heights, Ill.: Waveland Press, 1990.

This is the essential book for science teachers wishing to know how to apply Piaget's ideas to the classroom. The book gives details of Piaget's life and explores in detail his theories and relates them to practical applications.

Champagne, Audrey B., and Leslie E. Hornig. *Students and Science Learning.* Washington, D.C.: American Association for the Advancement of Science, 1987.

This book contains papers from the 1987 National Forum for School Science. The book presents recent ideas about the practical application of learning theories, cognition, and science teaching, and intervention programs in math and science for minority and female middle school students.

Duckworth, Eleanor. *The Having of Wonderful Ideas and Other Essays on Teaching and Learning.* New York: Teachers College Press, 1987.

This book presents examples of learner-centered teaching activities engaging learners in mind, sense of self, sense of humor, range of interests, interaction with other people, and problem solving.

Good, Ronald G. *How Children Learn Science: Conceptual Development and Implications for Teaching.* New York: MacMillan, 1977.

This is one of the most thorough treatments of how students learn from the standpoint of developmental psychology and the work of Jean Piaget. Good translates Piaget's work into understandable examples for the science teacher.

Lawson, Anton E., Michael R. Abraham, and John W. Renner. *A Theory of Instruction: Using the Learning Cycle to Teach Science Concepts and Thinking Skills.* NARST Monograph No. 1, National Association for Research in Science Teaching, 1989.

This monograph traces the origins of the learning cycle in science instruction, describes in detail what the learning cycle is, and provides good examples to illustrate the application of the learning cycle in science instruction.

Mullis, Ina V. S., and Lynn B. Jenkins. *The Science Report Card: Elements of Risk and Recovery.* Princeton, N.J.: Educational Testing Service, 1988.

This report summarizes the trends in achievement, attitudes, and opportunities for science learning based on the 1986 national assessment.

Slavin, Robert E. *Educational Psychology: Theory into Practice.* Englewood Cliffs, N.J.: Prentice Hall, 1988.

If you took a course in educational psychology, you may have used this book. It provides a complete introduction to the field of educational psychology in a very readable and understandable way. If you need to track down a concept or term in educational psychology, refer to this book.

CHAPTER

3

THE GOALS AND HISTORY OF SCIENCE EDUCATION

We are just a few years from the year 2000, the beginning of a new millennium, and perhaps the beginning of a new era in human history. Every student you teach will be adults in the next century. Is the science education these students are receiving adequate to help them meet the challenge and expectations of the twenty-first century? Educators have used the year 2000 as a reference point to evaluate current practices, warn people of ecological, population, and economic disasters, and make suggestions for what education should be. As we approach the year 2000, it is even more evident that there is the need to evaluate the goals of the science curriculum, to reflect on the past emphases in science education, and some would say, make bold recommendations for change in the teaching of science.

For example, in the 1980s there were a number of studies that compared the academic performance of U. S. students in science with their counterparts in Japan, Germany, Sweden, the Soviet Union, and some other countries. In every comparative study the United States ranked as one of the lowest in science performance. In studies comparing science students' performance over the period 1968–1986, the general trend was a lowering of student cognitive achievement and diminished attitudes toward science. Economic disadvantage is likely to result from this.

In the 1980s the scientific community and governments around the world began to recognize the importance of developing a global perspective for the solution of ecological-environmental problems, the reduction of nuclear arms, the ethics involved in the use of high technology, and the drug epidemic that has especially ravaged North and South American societies. These and other global problems signaled the advent of a new and imperative way to think, namely globally or holistically.

Another significant event in the late 1980s was the democratization of Eastern Europe and the Soviet Union. No one could have predicted these events, but the implications of these dramatic changes provide a new alliance of democratic thinking.

The world of the 1990s and beyond will be a world dominated by global interdependence in a variety of areas including the economy, politics, science and technology, individual and societal security, and the environment. According to many observers, acts of environmental destruction may pose a greater threat to national security than the military aggressiveness of other nations. This is especially plausible in light of the changes in Eastern Europe.

Economic security will depend on each nations' ability to compete in the global marketplace. Technological and scientific changes and advances will continue at breathtaking leaps, posing greater problems for individuals, societies, and governments, while offering further hope of alleviating some of the world's pressing problems.[1] Yet it is more than competition. The decade of the 1990s and beyond must be based on agreement and cooperation if we are to solve the real problems facing the planet today, such as alleviation of world hunger, reducing environmental threats, and creating a sustainable world society.

The world of science education—in this country and around the world—will have to grapple with

---

[1]See Lee F. Anderson, "A Rationale for Global Education," in Kenneth A. Tye, ed., *Global Education: From Thought to Action* (Alexandria, Va.: Association for Supervision and Curriculum Development, 1990).

how it can contribute to the perceived needs of individuals, societies, and the planet as a whole.

This chapter will focus on the goals of the science curriculum. We will explore changes in the goals of science education during this century but will pay particular attention to the contemporary nature of the science curriculum, and to reports and recommendations issued by a number of groups, commissions, and professional societies in the 1980s, and their implications for science as we approach the year 2000.

PREVIEW QUESTIONS

- What are goals and how are they formulated?
- What factors or forces affected the goals of science education for the 1990s?

- How and why have the goals of the science curriculum changed in this century?
- What are the origins of modern science education?
- What was the progressive education movement and how did it influence the science curriculum?
- What were the characteristics of the science curriculum during the golden age of science education?
- What changes brought an end to the golden age of science education?
- What are the contemporary trends in science education?
- What will science education be like for the 1990s and into the next century?

....
THE PHILOSOPHY AND GOALS
OF SCIENCE EDUCATION
....

What are the goals of science teaching? Is it possible to agree on goals for science teaching? Who determines the goals of science teaching? Have the goals of science teaching changed during the twentieth century? These are a few of the questions that we'll explore in this section to help us gain insight into an important aspect of science education, namely, the goals of science teaching.

Goals are ends toward which we direct our attention. They tend to be reflective statements expressing an individual's or a group's philosophical perspective on the broad aims or objectives of education. For example, consider this goal:

> *Science education should prepare individuals to utilize science for improving their own lives and for coping with an increasingly technological world.*[2]

This broad aim or goal, which reflects a recent emphasis in science education and is one that many recommend for the 1990s and beyond, highlights how science education should contribute to the personal needs of individuals as well how to use science to make everyday decisions. Goals should have relevance to students lives.

When we speak about the goals of science education (or for that matter, the goals of education in gen-

eral) it is important to keep in mind that the United States is the only highly technical, industrialized country that does not have a national science curriculum. Education is set in a decentralized system, with individual states and school districts determining the goals and the curriculum for their students. As you study the material in this and the next two chapters on the science curriculum, ask yourself questions such as, What are the implications of a decentralized educational system on goal setting? What are the pros? What are the cons?

Goals represent the desired or hoped for directions for science teaching. What are the desired directions of science education for the 1990s and beyond? To answer this question, we will examine the recommendations of a number of commissions, professional organizations, and convened groups. The science education community has been and will be influenced by these reports and recommendations. The impact will influence, among other things, the nature of textbooks, curriculum development efforts at the national and local levels, and teaching practices over the next decade and beyond.

The goals of science education in the future will be very different from the goals of the 1960s, which shaped science education curriculum, instructional practices, and textbooks into the 1990s, in that they will encompass science competence for all students, regardless of sex, race, and economic status. There will be disagreement among science teachers about what goals are of the most worth. For example,

---

[2]This statement is one of four goal clusters identified in Project Synthesis. See Norris C. Harms and Robert E. Yager, *What Research Says to the Science Teacher, vol. 3* (Washington, D.C.: National Science Teachers Association, 1981).

FIGURE 3.1 In the elementary science curriculum students should be involved in hands-on, inquiry oriented activities. The elementary science program should develop scientific, technological, and health knowledge, as well as develop positive attitudes about learning science. © Paul Conklin.

some science educators raise the question, 'Why should all students be taught to think like scientists?' Some science educators claim that instead of valuing only the mode of thinking practiced by the scientific community, science education should value the abilities and cultural relevance of diverse modes of thinking that can contribute to the solution of problems.[3]

One of the trends in the formulation of science education goals is the emphasis on citizenship education, that is, science should be of value to people as individuals, workers, and citizens. Alan Voelker describes this direction this way:

> Science education literature indicates that the main reasons for school science are (1) to provide background for citizenship, (2) to provide background for those entering occupations or careers oriented toward science and technology, and (3) to contribute to the preparation of scholars. Science teaching in elementary and secondary schools should emphasize citizenship education. Schools, the province of all citizens, can best serve future scholars and science-related career aspirants in an educational atmosphere where scientific literacy for all citizens is given top priority.[4]

Voelker points out that if the science curriculum is to become truly responsive and responsible to citizens in a scientifically and technologically oriented world, the concerns of all citizens need to be elevated to top priority. The science curriculum that was designed for aspiring scientists and engineers, with its heavy emphasis on the scientific process, may have to give way to new forms of science curriculum that advocate a variety of ways and methods of getting involved in science education. This orientation toward citizenship has its roots in reports issued in the latter part of the nineteenth century but was most notably developed in the 1930s when science educators insisted that science education should be an integral part of general education in a democratic society.

Champagne and Hornig suggest that many goals have been proposed for the science curriculum of the future, including the following:

- The development of a productive work force that will maintain economic prosperity and security (a nationalistic, economic goal)
- The development of a literate citizenry that is knowledgeable about scientific and technological issues and able to make informed decisions in their public and private lives (scientific literacy goal)
- The widespread adoption of the intellectual style of scientists, which is equated with better thinking ability (an academic or discipline-centered goal)
- The development of the ability to apply social, ethical, and political perspectives to interpretations of scientific information (an application goal)[5]

These and other goals will be proposed. However, the tide has turned in favor of developing science curriculum programs whose goals are solidly based on the interaction of science and citizens. Gone are the days (at least for now) when science was viewed as a pristine discipline, with its underlying assumptions and tacit way of knowing, to be learned by every school child as if he or she were a little scientist.

What do you think should be the goals of the science curriculum? Read ahead and participate in the following inquiry activity.

[3]See R. A. Cohen, "A Match or Not a Match: A Study of Intermediate Science Teaching Materials," in A. B. Champagne and L. E. Hornig, eds., *The Science Curriculum* (Washington, D.C.: American Association for the Advancement of Science, 1987), pp. 35–60.

[4]A. M. Voelker, "The Development of an Attentive Public for Science: Implications for Science Teaching," in R. E. Yager, ed., *What Research Says to the Science Teacher, vol. 4* (Washington, D.C.: National Science Teachers Association, 1982), pp. 65–79.

[5]Audrey B. Champagne and Leslie E. Hornig, *The Science Curriculum* (Washington, D.C.: American Association for the Advancement of Science, 1987), pp. 1–12.

INQUIRY ACTIVITY   **3.1**

The Goals of the Science Curriculum

What do you think teachers emphasize in their courses? What do you think should be emphasized? This inquiry activity will give you an opportunity to find out and compare your opinions with practicing science teachers' opinions.

Materials

Index cards, paper, writing instrument

Procedure

1. Write the following goals on individual index cards and use the cards while you go through some decision-making procedures.
   a. Become interested in science
   b. Learn basic science concepts
   c. Prepare for further study in science
   d. Develop inquiry skills
   e. Develop a systematic approach to solving problems
   f. Learn to effectively communicate ideas in science
   g. Become aware of the importance of science in daily life
   h. Learn about applications of science in technology
   i. Learn about the career relevance of science
   j. Learn about the history of science
   k. Develop awareness of safety issues in the laboratory
   l. Develop skill in laboratory techniques
2. First consider what you think. If you were teaching a science course in grades 7 to 9, which goal would you emphasize the most? Rank-order the remainder of the goals from most emphasized to least emphasized. Write your sequence on a sheet of paper.
3. Shift your attention to a high school course. Rank-order the goals from most emphasized to least emphasized at the high school level.
4. Finally, group the goals into clusters or categories. What criteria did you use? Is it easier to consider the clusters of goals rather than the entire list of goals?

Minds on Strategies

1. The results of the survey show that secondary science teachers consider the most important goals to be learning basic science concepts, developing a systematic approach to solving problems, and becoming aware of the importance of science in daily life.[6] How do these results compare with your results?
2. Are there differences between middle or junior high science goals and high school science goals? Should there be?

[6]The results for each goal, given as a percentage for grades 7 to 9 and 10 to 12, respectively, are as follows: goal 1:52, 47; goal 2:85, 85; goal 3:52, 56; goal 4-63, 57; goal 5:62, 67; goal 6:42, 49; goal 7:67, 60; goal 8:40, 40; goal 9:28, 32; goal 10:12, 11; goal 11:50, 53, goal 12:46, 57. Based on Iris R. Weiss, "The 1985–1986 National Survey of Science and Mathematics Education," in *The Science Curriculum*, Audrey B. Champagne and Leslie E. Hornig (Washington, D.C.: American Association for the Advancement of Science, 1987), pp. 225–233.

SCIENCE EDUCATION REPORTS INFLUENCING THE FUTURE

During the 1980s science educators reflected on the philosophy and goals that should guide science programs and the science curriculum into the twenty-first century. It was a period marked by numerous reports and proposals (over 300) by convened committees funded by various public and private granting agencies, as well as by a number of scientific organizations. This period of reflection was brought on by what some citizens in the United States perceived as a "crisis" in (science) education. One report claimed that the nation was at risk because of the "rising tide of mediocrity that threatens our very future as a nation and a people."[7] The report went on to say that "if an unfriendly foreign power had attempted to impose on America the mediocre educational performance that exists today, we might well have viewed it as an act of war." Strong statements, indeed. Yet a host of reports appeared that made similar charges. Later in the chapter we'll examine these reports in more detail.

Most of the organizations within the community of science education issued reports making recommendations for science teaching in general, and depending on the organization, recommendations for one aspect of science education, for example, biology, chemistry, or earth science. Organizations typically do this from time to time, especially during periods of crisis. For example, one report issued by the National Science Teachers Association, called *Project on Scope, Sequence and Coordination (SS&C)*, has proposed a new organization of the science curriculum. The Committee on High-School Biology Education of the National Research Council issued a report in 1990 entitled *Fulfilling the Promise: Biology Education in the Nations Schools.*[8] The American Geological Institute developed a report on the teaching of earth science entitled *Earth Science Education for the Twenty-first Century.*[9] Another report, *Professional Standards for Teaching Mathematics*, issued by the National Council of Teachers of Mathematics, stresses the connection between mathematics, science, and computers.

For now, let's focus our attention on four reports that have had, and will continue to have, a powerful impact on the direction of science education. We will briefly examine the following reports, their recommendations and implications for goals of the science curriculum:

> *New Designs for Elementary School Science and Health* (1989), by the Biological Sciences Curriculum Study (BSCS)
> *Project Synthesis* (1981), by the National Science Teachers Association (NSTA)
> *Project 2061: Science for All Americans* (1989), by the American Association for the Advancement of Science (AAAS)
> The Scope, Sequence, and Coordination Project (1990, NSTA)

Further, we will also draw some conclusions based on the implications of recent research on cognitive science and problem solving on the goals of the science curriculum.

New Designs for Elementary School Science and Health

We'll start with elementary science. Although you are going to be a secondary science teacher, each of the students you teach will have experienced an elementary school science curriculum, and therefore it is important for you to be aware of developments and changes at the elementary level as well. The science curriculum should be perceived as a K–12 integrated and articulated process, and many school districts and state departments of education are moving in this direction.

New Designs for Elementary School Science and Health is a collaborative report developed by International Business Machines (IBM) and the Biological Sciences Curriculum Study (BSCS). The *New Designs* report makes specific recommendations for K–6 science, technology, and health.

The report contains a framework—a foundation and structure—for the curriculum and instruction at the elementary school level. It represents the most advanced set of recommendations regarding science, technology, and health at the elementary school level, and therefore is an important document that will influence elementary science in the years ahead. The framework adopts an inquiry approach to science, bases its structure on major concepts and skills for science, technology, and health, and promulgates an instructional model based on a constructivist learning theory (see Chapter 2).

The BSCS's program, *Science for Life and Living: Integrating Science, Technology, and Health*, is a K–6 program whose fundamental goal is as follows:

> Children should learn about science, technology and health as they need to understand and use them in their daily life and as future citizens. Education in the elementary years should sustain children's natural curiosity, allow children to explore their environments, improve the children's explanations of their world, help the children to develop an understanding and use of technology, and contribute to the informed choices children must make in their personal and social lives.[10]

[7]National Commission on Excellence in Education, *A Nation at Risk* (Washington, D.C.: U.S. Department of Education, 1983).

[8]National Research Council, *Fulfilling the Promise: Biology Education in the Nation's Schools* (Washington, D.C.: National Academy Press, 1990).

[9]National Center for Earth Science Education, *Earth Science Education for the Twenty-first Century* (Alexandria, Va.: American Geological Institute, 1990).

[10]Biological Sciences Curriculum Study (BSCS), *New Designs for Elementary School Science and Health* (Dubuque, Iowa: Kendall/Hunt, 1989), p. xxii.

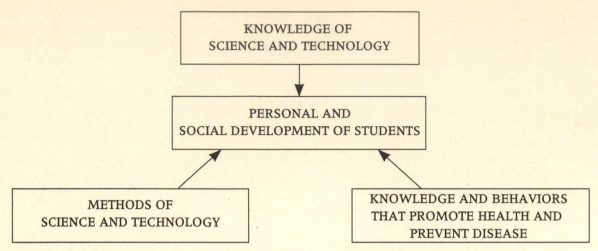

FIGURE 3.2   Relationships among science curriculum goals (BSCS, *New Designs for Elementary School Science and Health*, used with permission).

The report outlines the strategies curriculum planners might use for developing the curriculum for the 1990s and beyond, basing their plan on the BSCS elementary science program.

Through a series of commissioned papers and panel reports, as well as surveys and reviews of the literature, *New Designs* made recommendations for elementary science in three areas:

1. A curriculum framework for elementary science and health
2. An instructional model for contemporary elementary school science
3. The integration of technology and elementary school science and health

The Curriculum Framework

The developers of this report take the position that science should result in the scientific and technological literacy of citizens. Specifically, they assert that "science education programs should help students to (1) fulfill basic human needs and facilitate personal development, (2) maintain and improve the physical and human environment, (3) conserve natural resources, and (4) develop greater social harmony at the local, regional, national and global levels."[11]

Curricula for elementary science should be based on a set of goals that answer the question, What should education in the proposed science curriculum achieve? Here are the BSCS proposals:

- Science education should develop fundamental scientific, technological, and health knowledge
- Science education should develop a fundamental understanding of, and ability to use, the methods of scientific inquiry

[11]BSCS, *New Designs*, p. 27.

- Science education should prepare citizens to make responsible decisions concerning social issues that relate to science, technology, and health
- Science education should contribute to a fulfillment of the students' personal needs
- Science education should inform students about careers in science, engineering, and health
- Science education should develop behaviors that promote health and prevent disease[12]

The relationship between the goals for the *New Designs* curriculum framework is shown in Figure 3.2.

The developers also based their curriculum framework on several characteristics of students:[13]

1. Students have motivation
2. Students have developmental stages and tasks that influence learning
3. Students have different styles of learning
4. Students have explanations, attitudes, skills, and sensibilities about the world

The Instructional Model

Basing their work on an earlier proposal of a learning cycle by Karplus (see Chapter 2), the *New Designs* report proposed an instructional model on a constructivist view of learning. As developed in the last chapter, the constructivist view posits that students redefine, reorganize, elaborate, and change their initial concepts through interaction between themselves and their environment.

In the *New Designs* report, a five stage model of learning is outlined that will be used as the learning cycle in elementary science. The essential elements of each stage of the learning cycle are shown in Table 3.1

[12]BSCS, pp. 25–30.
[13]BSCS, pp. 30–32.

TABLE 3.1

INSTRUCTIONAL MODEL OF THE BSCS ELEMENTARY SCIENCE PROGRAM

Stage of Learning	Function and Nature of Student Activity
Stage 1: Engage	These activities mentally engage the student with an event or a question. Engagement activities help the students make connections with what they already know and can do.
Stage 2: Explore	The students work with each other, explore ideas together, usually through hands-on activities. Under the guidance of the teacher, they clarify their understanding of major concepts and skills.
Stage 3: Explain	The students explain their understanding of the concepts and processes they are learning. Teachers clarify their understanding and introduce and define new concepts and skills.
Stage 4: Elaborate	During these activities the students apply what they have learned to new situations, and they build on their understanding of concepts. They use these new experiences to extend their knowledge and skills.
Stage 5: Evaluate	The students assess their own knowledge, skills, and abilities. These activities also focus on outcomes that a teacher can use to evaluate a student's progress.

Based on *New Designs for Elementary School Science and Health*, Biological Sciences Curriculum Study. (Dubuque, Iowa: Kendall/Hunt, 1989), pp. 36–39.

The Integration of Technology and Elementary Science

New Designs takes the position that educational computing will play a significant role in science education programs, and that teachers will find the use of this technology as easy as using home appliances, provided that in-service training and appropriate software are made available. For technology to be integrated into the teaching of elementary science, the microcomputer must become an integral part of the instructional process.

One of the key notions suggested by the BSCS project is the creation of a technology-oriented learning environment. It goes without saying that integrating the microcomputer into science teaching will require the reformulation not only of the curriculum but of the learning environment. Figure 3.3 on the next page, is a schematic of the technology-oriented classroom. Note that in addition to the MBL (micro-computer-based laboratory) station, the classroom would contain computer stations for courseware and word processing, an interactive video station, and stations for listening, writing, and manipulative materials. Also note that students would be organized into small cooperative groups.

The *New Designs* report suggests that science, health, and technology can be integrated to create an elementary science program that fuses some of the best of earlier programs with new advances in cognitive psychology, cooperative learning, and microcomputer technology. The elementary science goals of the future will reflect an integrated, interdisciplinary model, one in which students are placed in environments in which they not only will learn about the world around them but will learn about themselves as well.

We will now turn our attention to three other reports that made recommendations not only about elementary science but about middle and high school science as well. Let's look at the recommendations made by Project Synthesis and then examine the ideas proposed by Project 2061, and NSTA's Scope, Sequence and Coordination Project.

Project Synthesis

Project Synthesis was an effort to examine the state of science education by studying a number of research studies funded by the National Science Foundation.[14] The studies included *The Status of Pre-College Science, Mathematics and Social Science Education: 1955–1975, Case Studies in Science Education*, and *1977 National Survey of Science, Mathematics and Social Education*. Project Synthesis also examined the third set of results of the National Assessment of Educational Progress. In addition to these reports the staff of Project Synthesis also examined journal articles in science education and analyzed the most widely used textbooks because of their influence on science education. The synthesis group was comprised of 23 science educators, who were divided into five subgroups: biological sciences; physical sciences (including earth science); inquiry; science, technology, and society; and elementary science.

Project Synthesis attempted to develop a set of goals that were broad and could be generally accepted, had meaning to a number of audiences, were unbiased, were limited in number, had unifying fea-

[14]Harms and Yager, *What Research Says to the Science Teacher*.

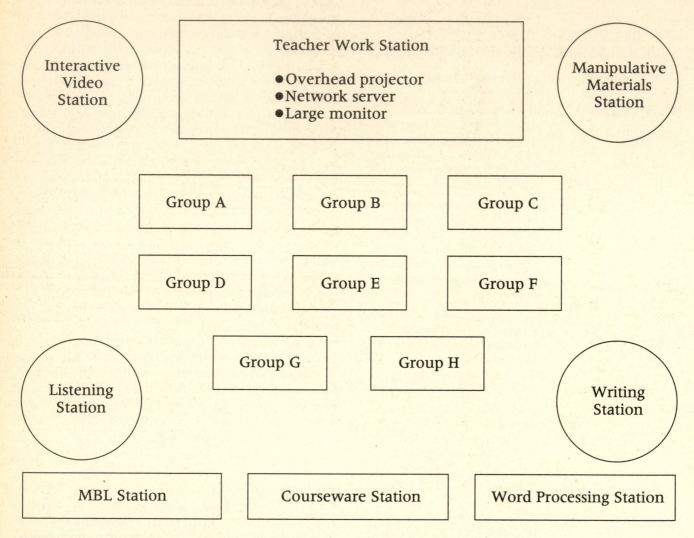

FIGURE 3.3   The Technology-Oriented Classroom (BSCS, *New Designs for Elementary School Science and Health*, used with permission.

tures, could lead to operational definitions in terms of student objectives, and finally could be useful for policy statements. The project adopted the term "goal cluster" to define the major goals, and in collaboration with the leaders of the five groups, the project staff identified four goal clusters in which student outcomes could be grouped as follows: the individual, societal issues, academic preparation, and career choice. The four goal clusters of Project Synthesis are presented in Table 3.2.

Using the goal clusters as an organizational structure, desired states were generated for biology, physical science, inquiry, science, technology, and society, and elementary science; a synthesis was made of the actual states of science education, and finally recommendations were made based on the discrepancies between the desired and actual states of these areas.

Biology

The findings and recommendations of the biology group reflect the general considerations of the other groups. The biology group recommended that the desired biology program should focus on humanity as a part of nature by suggesting the following goals:

- To understand human beings as a distinctive organism; to appreciate the universal human need to be in touch with our own nature, and all of nature
- To learn to live in harmony with nature and to minimize the dissonance between social and natural environments[15]

The biology group felt that the goals of biology should result in courses, topics, or modules that em-

[15]See P. Hurd, "Biology Education," in Harms and Yager, *What Research Says*, p.13.

TABLE 3.2

PROJECT SYNTHESIS
GOAL CLUSTERS

Goal Cluster	Focus	Statement
I	Personal needs	Science education should prepare individuals to utilize science for improving their own lives and for coping with an increasingly technological world.
II	Societal issues	Science education should produce informed citizens prepared to deal responsibly with science-related societal issues.
III	Academic preparation	Science education should allow students who are likely to pursue science academically as well as professionally to acquire the academic knowledge appropriate for their needs.
IV	Career education and awareness	Science education should give all students an awareness of the nature and scope of a wide variety of science and technology-related careers open to students of varying aptitudes and interests.

Based on Norris C. Harms and Robert E. Yager, *What Research Says to the Science Teacher*, vol. 3. (Washington, D.C.: National Science Teachers Association, 1981).

phasized environmental or ecological studies, human anatomy and physiology, health (especially related to alcohol, drugs, tobacco and disease), and future studies in terms of the quality of life. The biology group pointed out that even though science educators recommended (in the literature) that science teaching should emphasize a broader perspective, emphasizing the relationship of science and technology, societal and cultural dimensions, personal and humanistic values, decision-making and problem-solving skills, the current state of biology teaching emphasized vocabulary and the narrow use of student objectives (normally content oriented). Very little attention was given to general goals such as inquiry, problem-solving, or the nature of human beings.[16]

Physical Science

Each of the subgroups found discrepancies between the desired state and the actual state of science teaching. The physical science group, for instance, found that course content was narrow, with little emphasis on the relationship of the content of the science to personal or social issues. They

found that most of the emphasis was placed on goal cluster III (academic preparation), and recommended that physical science courses be modified to provide experiences related to personal needs, societal issues and career awareness. They recommended a greater emphasis on interdisciplinary programs, designing science programs that would enable students to pursue topics of interest to them, and the inclusion of socially relevant problems in physical science.

Inquiry

The inquiry group reported that the desired state of inquiry was rare. This group defined the desired state of inquiry to include emphasis on three themes: science process skills, nature of scientific inquiry and general inquiry processes. An interesting finding of this group was that although science teachers favor the use of inquiry practices, there was very little inquiry in actual classrooms. This group pointed out that many teachers found inquiry methods difficult to manage, and that equipment and materials were not readily available. The inquiry group, after considering many alternatives, recommended a reformulation of the traditional views

16Hurd, pp. 12–32.

FIGURE 3.4 Project Synthesis recommended that student learning in science should focus not only on academic preparation, but personal needs, societal issues, and career awareness, as well. © Myrleem Ferguson.

about teaching scientific inquiry. They recommended that all student outcomes with respect to inquiry should be responsive to individual differences, personal goals and community wishes.

Science/Technology/Society

The science/technology/society (STS) group examined science education goals in relationship to the interface of science, technology, and society. They identified several areas of concern that need attention in the STS domain: energy, population, human engineering, environmental quality, utilization of natural resources, national defense and space, sociology of science, and effects of technological development.

The work of Project Synthesis was fundamentally carried out by science educators. We will return to other recommendations of Project Synthesis when we examine the middle/junior high and high school science curricula in the next chapter. We now turn our attention to Project 2061, which was published in 1989, and represents the recommendations of the prestigious American Association for the Advancement of Science (AAAS). Many of its recommendations were proposed by scientists, although the report itself was written by science educators.

Project 2061: Science for All Americans

Project 2061: Science for All Americans (2061 is the year of the next arrival of Halley's Comet and the period of the next human life span) is a massive project to examine the goals of science education, develop experimental programs in selected school districts around the nation, and eventually use the results of the project to transform the face of science education. Quite a challenge.

Science for All Americans is a report written by the Project 2061 staff of the American Association for the Advancement of Science (AAAS). According to the AAAS project staff, it is suggested that the "terms and circumstances of human existence can be expected to change radically during the next human life span. Science, mathematics and technology will be at the center of that change—causing it, shaping it, responding to it. Therefore, they will be essential to the education of today's children for tomorrow's world."[17]

Project 2061 believes that science for all americans is about scientific literacy. To Project 2061, a scientifically literate person is one who

- Is aware that science, mathematics and technology are interdependent human enterprises with strengths and limitations
- Understands key concepts and principles of science
- Is familiar with the natural world and recognizes both its diversity and unity
- Uses scientific knowledge and scientific ways of thinking for individual and social purposes[18]

Although the report points out that most Americans do not understand science, it takes a point of view that goes beyond individual self-fulfillment and

[17]*Science for All Americans: A Project 2061 Report on Literacy Goals in Science, Mathematics, and Technology,* American Association for the Advancement of Science, (Washington, D.C., American Association for the Advancement of Science, 1989).
[18]*Project 2061,* p. 20.

FIGURE 3.5 Project 2061: Science for All Americans Logo of Project 2061. Used with permission.

the national interests of the United States. As it points out, more is at stake. The serious problems that humanity faces are global: unchecked population growth, acid rain, the shrinking of tropical forests, the pollution of the environment, disease, social strife, extreme inequities in the distribution of the earth's wealth, the threat of nuclear holocaust, and many others. In this context scientific literacy takes on a global perspective. The scientifically literate citizen in a global world would realize the potential of science in the following ways:[19]

- Science can provide humanity with the knowledge of the biophysical environment and of social behavior to effectively solve global and local problems
- Science can foster the kind of intelligent respect for nature that should inform decisions on the uses of technology
- Scientific habits of mind can be of value to people in every walk of life to solve problems that involve evidence, quantitative considerations, logical arguments and uncertainty
- The knowledge of principles related to the nature of systems, for example, can give people a sound basis for assessing the use of new technologies and their implications for the environment and culture
- Knowledge of technology, especially continuous development and creative uses, can help humanity deal with survival and work toward a world in which humanity is at peace with itself and its environment

Project 2061 is divided into three phases, each with its own actions and goals: (I) content identification, (II) educational formulation, and (III) educational transformation.

The purpose of phase I is to build a rationale for science education and develop an outline of what science, mathematics, and technology content ought to be included in an education for elementary and secondary curriculum. The *Science for All Americans* report included recommendations on the nature of science, mathematics, and technology, the physical setting, the living environment, the human organism, human society, the designed world, the mathematical world, historical perspectives, common themes, and habits of mind.

Five panels were organized, and each developed reports outlining the subject matter that should constitute the essence of literacy in science, mathematics, and technology. The panel reports that are of most interest to science teachers include *Biological*

and Health Sciences, Physical and Information Sciences and Engineering, and *Technology.*

Physical Science

The panel on physical and information sciences and engineering recommended that the task of teaching the physical and information sciences and engineering in elementary and secondary schools can be facilitated by focusing on key unifying concepts. They identified four key unifying concepts—materials, energy, information, and systems—and several other unifying concepts including equilibrium, time rate, conservation, efficiency, uncertainty, risk, cost-effectiveness, and benefit analysis. They then identified key specific concepts in physics, chemistry and earth, planetary, and astronomical sciences, information science and computer science, and engineering. Table 3.3 shows the key unifying concepts for physics, chemistry, earth science, information science, and engineering.

Biology

The biological panel suggested that citizens should come to know biology in order to understand themselves as a product of evolution, and as single individuals in an ecological scheme. The panel also felt that biology "teaches rules to live by." The topics that the panel included in its report show the emphasis that biology in the elementary and secondary curricula should take: human biology, the evolution of diverse life-forms, environmental biology, and human ecology. The panel posed a number of questions that they hoped 18 year olds would be able to answer. Some of these questions include, How does the human organism work and what does it take to keep it healthy? How are humans like and unlike other living organisms? What determines the productivity of an ecosystem? How does accumulating pollution affect humankind's future?

Technology

The technology panel defined technology as the application of knowledge, tools, and skills to solve practical problems and extend human capabilities. The panel highlighted the notion that technology is part of social progress, as well as a technical process. Technology education should emphasize the interface between technology and society. They pointed out that citizens must be able to develop, articulate, and illustrate how technology affects society and how society affects technology.[20]

---

[19]*Project 2061*, pp. 12–13.

[20]*Biological and Health Sciences, A Project 2061 Panel Report* (Washington, D.C.: American Association for the Advancement of Science, 1989).

TABLE   3.3

KEY UNIFYING CONCEPTS IN SCIENCE, TECHNOLOGY, AND ENGINEERING

Physics and Chemistry	Earth, Planetary, and Astronomical Sciences	Information Science and Computer Science	Engineering
All physical and chemical phenomena are governed by a few basic interactions.	Our universe has an enormous number of galaxies.	Information is the meaning attributed to data.	A key ingredient of the engineering process is the ability to plan and manage a project.
The quantum principle: on a microscopic-length scale, many physical quantites—such as electric charge, mass, and energy—are found in tiny fixed units called quanta; atoms gain and lose energy in fixed quantum units.	The Sun is one of the many stars within the Milky Way Galaxy; Earth is one of the planets of the Sun.	Different kinds of information can be derived from the same data.	Modeling conceptualizes the problem to be solved and the solution itself, formulating them as much as possible in quantitative terms.
The behavior of simple static and moving systems can be explained using the laws first laid out by Isaac Newton.	Earth is a nearly spherical rotating body; its dimensions, motion, and position relative to the Sun govern our lives.	Information can be expressed in many forms and can be represented in analog (continuous) or digital (discrete) formats.	Every design or system has constraints that must be identified and taken into account.
Intuitive ideas of space, time, energy and mass fail at great distances and when speeds approach the speed of light.	Forces deep within the earth, acting over geologic time, have caused continents to move, rupture, and collide.	Information generally degrades during transmission or storage.	Optimization endeavors to determine the best possible solution to a problem under its relevant constraints.
Electromagnetic radiation, of which light is an example, occurs in a very large range of wavelengths; such radiation is emitted and absorbed as particles or bundles of energy.	Life on Earth arose through natural processes several billion years ago.	All systems, both natural and human-made, are internally coordinated by processes that convey information.	The design is the core of the engineering process.
Electrical phenomena can be understood in terms of the behavior of the electric charge.	The oceans and atmosphere are very large, buffered chemical, biological and geological systems.	Information is more useful when it is represented by orderly collections of symbols called data structures.	Each design has its side effects.
For many purposes, matter can be viewed as being composed of atoms and molecules that have well-defined sizes, shapes, structures, compositions, and energy contents.	Climate is the long-term expression of the movements of masses of fluids in the ocean and atmosphere.	Procedures can be formalized as algorithms.	The artifacts created by design cannot be allowed to perform unassisted; they require operation supervision, maintenance and repair.
Atomic nuclei undergo changes.	Waves, wind, water and ice are agents of erosion and deposition that sculpt the earth's surface to produce distinctive landforms.	Computing machines are constructed from simple components.	We must constantly be alert to the possibility of engineering failure; most failures occur at the interface between systems; failures can occur in all systems, even well-maintained and supervised ones.
	Humans need and use many substances found naturally on the earth's surgace and in its interior.	All general-purpose computing machines are fundamentally equivalent.	
	All places on Earth have characteristics that give them meaning and character and that distinguish them from each other.	To ensure that an information system will be successful in the real world, the design must include both logical rigor and an understanding of social forces, cultural beliefs and economic realities.	

From George Bugliarello, *Physical and Information Sciences and Engineering: Report of The Project 2061 Phase I Physical and Information Sciences and Engineering Panel.* (Washington, D.C.: American Association for the Advancement of Science, 1989). Used with permission of AAAS.

The technology group identified several fields of technology that should provide the framework for technology education: materials, energy, manufacturing, agriculture and food, biotechnology and medical technology, environment, communications, electronics, computer technology, transportation, and space. The panel report outlined the nature of each of these technology fields and identified suggested experiences for students. For example, activities suggested by the panel in the communications area included having students make simple devices that can be used in communications, from historical gadgets (such as a carbon microphone or simple telegraph) to modern electronic circuits. Students should also be encouraged to undertake imaginative projects, such as inventing ways of communicating with people in remote lands or searching for information from outer space that might reveal life there.[21]

The second phase of the project is an extension of phase I of Project 2061. In the second phase, the focus of attention will be to develop, in five school districts across the nation, alternative K–12 curriculum models for education in science, mathematics, and technology.[22] Development teams will design curriculum plans for science, mathematics, and technology and then field-test and evaluate the results. The second phase will also include creating a set of blueprints for reforming the other components of education that complement curriculum reform, increase the pool of educators and scientists able to serve as experts in the school curriculum reform, and foster public awareness of the need for reform in science, mathematics, and technology education.

Phase III of Project 2061, referred to as the educational transformation phase, will be a "highly cooperative, nationwide effort which will mobilize resources, monitor progress, and, in general provide direction and continuity of the effort."[23] The goal of the third phase is to use the products of phases I and II to broaden the scope, and raise the quality of education in science, mathematics, and technology.

The overall and general direction of Project 2061 is multifaceted but includes the following:[24]

- To ensure scientific literacy, curricula must be changed to reduce the sheer amount of materials
- Rigid subject-matter boundaries should be weakened or eliminated

- More attention should be paid to connections between science, mathematics and technology
- The scientific endeavor should be presented as a social enterprise that strongly influences and is influenced by human thought and action
- Scientific ways of thinking should be fostered
- Teaching related to scientific literacy needs to be consistent with the spirit and character of scientific inquiry and with scientific ideas
- Teaching should begin with questions about phenomena rather than with answers to be learned
- Students should be actively engaged in the use of hypotheses, collection and use of evidence, and the design of investigations and processes

The Scope, Sequence and Coordination Project

The Project on Scope, Sequence, and Coordination (SS&C) was initiated by the National Science Teachers Association as a major reform effort to restructure secondary science teaching.[25] One major feature of this reform proposal is the elimination of tracking of students (survey, general, and advanced courses, for example), and its replacement with a science program in which all students study science in a well-coordinated science curriculum teaching each year physics, chemistry, biology, and earth and space science. An outcome of this concept is a school science curriculum that provides for "spacing" the study of the sciences over grades 6 to 12.

SS&C has also integrated the outcomes of Project 2061 as the goals for the science curriculum. Although these are separate projects, each having its own staff, infrastructure, and dissemination centers, both projects "contend that less content taught more effectively over successive years will result in greater scientific literacy of the general public."[26]

According to the NSTA, the fundamental goal of SS&C is to make science understandable by essentially all students. To do this, the project emphasizes that students should become actively engaged with experiencing phenomena rather than the customary approach of naming phenomena. The outcome of this concept is that the reformed science curricula will contain a greatly reduced number of topics and "their accompanying baggage of facts and terminology."

SS&C Curriculum Models
One might assume that the SS&C proposal will

[21]*Technology, A Project 2061 Panel Report* (Washington, D.C.: American Association for the Advancement of Science, 1989), p. 23.
[22]*Project 2061*, p. 161.
[23]F. J. Rutherford, A. Ahlgren, and J. Merz, "Project 2061: Education for a Changing Future," in Champangne and Hornig, *The Science Curriculum*, p. 62.
[24]*Project 2061*, p. 5

[25]See *Scope, Sequence, and Coordination of Secondary Science: A Rationale* (Washington, D.C.: National Science Teachers Association, 1990).
[26]*Scope, Sequence, and Coordination*, p. 1.

lead to a single curriculum model. The evidence so far is that a number of models have been developed rather than a single approach. Through a series of science curriculum grants by SS&C to sites in Texas (Houston), California, North Carolina, Iowa, and Puerto Rico, curriculum reform efforts were underway in hundreds of schools. A number of different models of curriculum reform are possible based on the central tenets of SS&C. For example, the model described in the rationale statement of SS&C suggested a science curriculum that moved from the concrete to the abstract from grades 7 to 12 (Table 3.4)

The general concept to keep in mind with this revised model is that students would begin their secondary science experience by having concrete experiences with phenomena prior to naming or symboling them. Students would engage in the exploration of phenomena to construct concepts based on personal experience. As the students mature and develop a repertoire of science concepts at the concrete and conceptual level, they would move to increasingly more complex and abstract concepts and experiences.

A number of approaches are possible. For example, in the Houston model a "block" approach in which a sequenced collection of laboratories focusing on a series of coordinated concepts from biology, chemistry, earth and space science, and physics has been adopted. An example of a block, "Floating and Sinking," will help illustrate the Houston approach to curriculum:

What is density and how can it be experienced directly? How is the density of a solid or a liquid measured? Why do some things float while others sink? Before the block was over, students discussed density (physics), solutions (chemistry), oceans (earth science), and marine organisms (biology). These and other topics led students to experience the effect of density on familiar objects for themselves and they more easily learned the concepts involved.[27]

In the Iowa project of SS&C, the emphasis is on STS. In California, referred to as the One Hundred Schools Project (actually there were over 250 schools at the last count), a bottoms-up approach, with strong involvement of local teachers and local networks, has resulted in a wide range of curriculum change. According to the California developers, one aim is to create a challenging, nontracked science program, especially for schools with high minority populations. In Puerto Rico, the emphasis is on integrated blocks, that is, blocks not only in English and Spanish but integrating science and mathematics.

SS&C Principles

On what principles are reform efforts based? What are some of the guiding principles for reforming science curriculum according to SS&C? Examine the following list and think about the implications of

[27]"Houston Develops All New Classroom Blocks," *Currents*, March/April, 1991, p. 3.

TABLE 3.4

MODEL OF A REVISED SCIENCE CURRICULUM FOR GRADES 7 THROUGH 12 IN THE UNITED STATES

	Grade Level						
	7	8	9	10	11	12	
							Total Time Spent
	Hours per Week by Subject						
Biology	1	2	2	3	1	1	360
Chemistry	1	1	2	2	3	2	396
Physics	2	2	1	1	2	3	396
Earth / Space Science	3	2	2	1	1	1	360
Total Hours per Week	7	7	7	7	7	7	
Emphasis	Descriptive; phenomenological		Empirical; semi-quantitative		Theoretical; abstract		

Scope, Sequence, and Coordination of Secondary Science: A Rationale (National Science Teachers Association, 1990). Used with permission.

these statements on the development of science lesson plans or units of instruction.[28]

1. The four basic subject areas, biology, chemistry, earth and space science, and physics are addressed each year, and the connections between them are emphasized

2. The coordinating themes identified by Project 2061 are used as unifying threads between the disciplines: systems, models, constancy, patterns of change, evolution, and scale

3. Science is shown to be open to inquiry and skepticism, and free of dogmatism or unsupported assertion by those in authority. The science curriculum promotes student understanding of how we come to know, why we believe, and how we test and revise our thinking

4. Science should be presented in connection with its applications to technology and its personal and societal implications

5. Students should have the opportunity to construct the important ideas of science, which are then developed in depth through inquiry and investigation

6. Vocabulary is used to facilitate understanding rather than as an end in itself. Terms are presented after students experience the phenomena

7. Texts are not the source of the curriculum but serve as references. Everyday materials, laboratory equipment, video software, and other printed materials such as reference books and outside reading provide a substantial part of the student learning experience

8. Lessons provide opportunities for skill building in data collection, graphing, record keeping, and the use of language in verbal and written assignments

9. Of particular importance is that instruction enhances skepticism, critical thinking, and logical reasoning skills. Thinking and reasoning skills such as controlling variables and drawing inferences need to be fostered

[28]The list is based on *Scope, Sequence, and Coordination*, pp. 8–9.

THE INFLUENCE OF RESEARCH ON THE GOALS OF SCIENCE TEACHING

You will discover that the science education community is increasingly interested in correlating the results of research with the practice of science teaching. For example, the National Science Teachers Association has published a series of monographs entitled *What Research Says to the Science Teacher*. One of the reasons for this series is the difficulty that has existed in trying to connect research to practice. It is fitting that the last volume of the 1980s published in this series was entitled *Problem Solving*. The monograph contained a series of papers designed to help the science teacher focus on the important goal of helping students solve problems in the disciplines of biology, chemistry, earth science, and physics.[29]

Why should problem solving be considered an important goal of science teaching as we approach the twenty-first century? As the researchers point out, problem solving enables teachers to give students the opportunity to integrate thinking skills to solve a variety of problems. For example, the earth science teacher can rely on the concreteness of earth science phenomena, and the fact that students can connect directly to issues and problems dealing with the land, sky, water, and air. Interesting problems, having global consequences yet involving local actions, can help students explore problems that affect them directly. This approach to problem solving ties in directly to the goal that science should be of value to people as individuals, workers, and citizens.

The National Association for Research in Science Teaching has published a series of documents entitled *Resrach Matters...To the Science Teacher* on a range of topics. These two-to-three-page papers are like consumer reports in which researchers translate the results of studies into practical-consumer-like reports. Some of these, such as "Pupil Behavior and Motivation in Eighth Grade Science" and "Encouraging Girls in Science Courses and Careers," are found in a column of the Science Teaching Gazette.

Based on their research, researchers in the cognitive science tradition have also suggested some directions for the future science curriculum. Here are

[29]See *What Research Says to the Science Teacher: Problem Solving, vol. 5* (Washington, D.C.: National Science Teachers Association, 1989).

several directions that cognitive researchers feel are direct implications of their research:[30]

1. *The goals of the science curriculum should be redefined and broadened to reflect technological advances and societal needs.* Their position is that science education should focus its attention on creating environments that foster science for all citizens rather than on science for future careers in science

2. *The goals of the middle-grades curriculum (grades 4 to 8) should focus on students' concerns about the impact of science on society.* For example, science programs at this level that involve students in environmental education activities, in which they explore on the local level connections between science and society, would capitalize on students interest in these topics. Having students investigate pollution, recycling, causes of disease, waste disposal, and the impact of power plants on the environment are examples of promising topics. The middle years are significant and influential to students. It is the time when students lose interest in science, and it is imperative that science programs be designed to capitalize on their interests and motivations

3. *Less is better.* Researchers think science courses should cover less topics and go into the topics that are included in more depth. By going into selected topics in depth, teachers would be able to plan in-depth problem-solving activities. Currently teachers must skip quickly from one topic to another, giving very little attention to real understanding of science concepts and problem solving

4. *The content of the science disciplines should be integrated.* Project 2061, SS&C and the *New Designs* report cited earlier recommended an interdisciplinary approach to the science curriculum. Students should be helped to make connections between the fields of science by being engaged in activities that require them to link one discipline to another. Except for elementary science, the science curriculum in the United States is organized in such a way that students study only one science discipline each year, reinforcing the separateness of science fields. As we will see in chapter 5, students in a number of other countries are introduced to physics, biology, and chemistry early in their secondary education, and continue to study *each* of these subjects for as many as five years

There are many other implications for science teaching based on the work of science education researchers. We have identified a number of implications that have direct bearing on the goals and nature of the curriculum. In Chapter 2, a number of implications were identified related to cognitive development, student learning styles, and metacognitive strategies. As you continue your study of science teaching, it is important to make connections between research and science teaching practice.

---

[30]The implications that follow are based on a paper describing a planning conference on research and science teaching held at and sponsored by the Lawrence Hall of Science, Berkeley, California, and the National Science Foundation. See M. C. Linn, "Establishing a Research Base for Science Education: Challenges, Trends and Recommendations," *Journal of Research in Science Teaching* 24, no. 3 (1987): 191–216.

TWENTIETH–CENTURY SCIENCE EDUCATION: A HISTORICAL PERSPECTIVE

Have you ever wondered what science teaching was like 100 years ago? What did teachers emphasize in their lessons? What did educators believe science could contribute to the education of students? How is teaching today related to the past? What can we learn about contemporary teaching by reaching into the past?

The historical perspective is important from the standpoint of trying to understand contemporary trends and changes in science teaching. A great many events, developments, and reports have contributed to shaping the goals and nature of the K–12 science curriculum during the twentieth century. In this section we will go back in time and explore briefly some of these forces that have affected science education. Dividing the history of science education into time units is an arbitrary process; therefore, for convenience I have identified the following phases:

Phase I: The beginnings of modern science education, pre-1900–1930

Phase II: Progressive education and science education, 1930–1950

Phase III: The golden age of science education, 1950–1977

Phase IV: Textbook controversies and the back to basics movement, 1977–1983

Phase V: A nation at risk, the 1980s

Phase VI: Science for all, toward the year 2000 and beyond

INQUIRY ACTIVITY 3.2

Icons of Science Education

The photograph of Clarence Darrow and William Jennings Bryan at the 1925 trial of John T. Scopes (the science teacher accused of breaking the state law against teaching evolution) is surely one of the icons of the history of science education. What are some other icons of the history of science education? In this inquiry activity you will work with a collaborative group to explore one of the phases of science education (as identified in this chapter) in order to identify the icons of that phase.

ence education history to investigate. The team's task is to assemble icons—images, representations, artifacts, pictures—of the period under investigation. The icons can be presented on a large poster or as a series of individual exhibits.

Minds on Strategies

1. What are some icons of science education? Compare the icons from the different periods of science education history. How has science education changed?
2. What are the important issues in science education today?
3. What are some of your predictions of science education for the year 2061?

TABLE 3.5

FIGURE 3.6 The famous Scopes Trial as an icon of science education. The matter of teaching evolution in the public schools has remained an issue for science educators. © AP/Wide World Photos.

Materials

Text material
Reference materials on science education history (see references in this chapter)
Poster board, art supplies

Procedure

1. The class will inquire into the various phases of the history of science education. The questions in Table 3.5 are designed to be used to explore any phase in the history of science education.
2. Each team should choose a different period in sci-

INQUIRY INTO THE HISTORY OF SCIENCE EDUCATION

Phases of Science Education History	Inquiry Questions
I. The beginnings of modern science education, pre-1900–1930	What social forces influenced science education?
II. Progressive education and science education, 1930–1950	What reports or commissions influenced the goals of science teaching?
III. The golden age of science education, 1950–1970	What were the desired goals of science teaching?
IV. Textbook controversies and the back-to-basics movement, 1975–1983	What was the focus of the science curriculum?
V. A nation at risk, the 1980s	
VI. Science education for all, toward the year 2000 and beyond	

Phase I: The Beginnings of Modern Science Education, Pre-1900–1930

The roots of what we might call modern science education can be traced to the latter part of the nineteenth century; many of the recommendations and philosophies proposed in these early years have influenced science education, even to today. Several important ideas that form the foundation for modern science education had their origins during this period. These include the organization of the science curriculum, the methodology of inquiry, nature study, elementary science, and the social goals of science teaching.

In 1895 the National Education Association appointed a committee of 12 university scientists and an equal number of high school teachers. Four years later the committee recommended a K–8 and 9–12 science program as follows:[31]

Elementary School (K–8)
  Nature study (two lessons per week)
High School (9–12; four lessons per week)
  Grade 9: physical geography
  Grade 10: biology; botany and zoology; or
        botany or zoology
  Grade 11: physics
  Grade 12: chemistry

The main purposes of teaching science during this early period were:

• Formal discipline
• The teaching of facts and principles
• College preparation

There was some but limited emphasis on the skills and methods of science, and very little emphasis on scientific method, scientific attitudes, appreciations, or the social implications of science

Elementary School Science

At the elementary school level, two contrasting approaches to science education dominated the curriculum, namely nature study and elementary science. Nature study was a child-centered approach to teaching that focused on helping students develop a love of nature. The content focused primarily on the study of plants, animals, and ecology, with teachers emphasiz-

ing the study of the local environment. Many of the progressive schools in America adopted the nature study approach. Study guides, teachers' resource books, and lesson plans were written and distributed describing the approach of the nature study advocates. Because of its progressive nature, the nature study movement was associated with an educational concept known as teaching the whole child. Nature study was an interdisciplinary approach, and science was seen as an integral part of art, language, and literature. Lawrence Cremin, in his book *The Transformation of the School*, says this about the nature study movement:

> Science was begun in the form of nature study, and under the brilliant leadership of Wilbur Jackman, the children conducted trips through neighboring fields and along the lake shore. They made observations, drawings, and descriptions, thus correlating their work in science with their studies in language and art.[32]

Nature study was the dominant approach to science in the elementary school, and its program reached its peak in the period 1900–1910. Elements of the nature study movement still persist today. But as a movement it faded away in the 1930s. Individual "nature study units" were integrated into the dominant approach to science in the elementary school, and as a movement it reappeared in the late sixties and seventies. In the language of the 1990s, nature study is analogous to the environmental education movement, which will be discussed in Chapter 6. Elementary science, however, moved into the classroom, and the content was broadened to include physical science.

The alternative approach to science in the elementary school was the elementary science movement, which was defined and outlined by Gerold S. Craig.[33] In a study he did on the curriculum for elementary science he recommended a science program that was broad in scope, including the life and physical sciences. He devised a continuous program, K through 8, emphasizing an articulated unified program aimed at developing an understanding of significant ideas in science. The content of elementary science was developed by panels of parents and educators who identified important ideas and generalizations that should be emphasized grade by grade. This method, and indeed the results of study, still dominate the content of elementary science textbooks and state science curriculum guides, even today. His recommen-

[31]P. D. Hurd, *Biological Education in American Schools, 1890–1960* (Washington, D.C.: American Institute of Biological Sciences, 1961), pp. 16–17.

[32]L. A. Cremin, *The Transformation of the School* (New York: Vintage, 1964), p. 133.

[33]G. S. Craig, *Certain Techniques Used in Developing a Course of Study in Science for the Horace Mann Elementary School* (New York: Columbia University Press, Teachers College, 1927).

dations became the basis of elementary science text programs around the country. Science readers were developed for each grade, and consequently elementary science became more abstract in comparison with the nature study approach, which valued first-hand experiences in nature by the students. For all practical purposes this approach remained the same until the development of the elementary science reform projects of the 1960s and 1970s.

Secondary School Science

The secondary science curriculum during much of the period up to 1920 consisted of one-semester courses in many different subjects, such as astronomy, geology, physical geography, botany, zoology, and physiology, in the first two years of high school. Later during this period the organization of the science curriculum (9–12) evolved to general science, biology, chemistry, and physics as one-year courses. This pattern, for all practical purposes, has persisted to the present.

High school teachers, near the latter part of this phase, shifted their goals to include process, attitudes, and application. Some major goals proposed during this period included scientific thinking and understanding of the scientific method, development of attitudes toward science, and increased emphasis on the practical application of scientific knowledge.

A phenomenon that had an impact on the science curriculum was the junior high movement. To provide a transition from the elementary to high school environment, a 6-3-3 organization of schools was proposed. Elementary science was limited to grades K through 6. General science had replaced physical science as the "terminal" course for students not completing high school. With the advent of the junior high school, general science was pushed down to the seventh and eighth grades, and became the dominant science offering at this level for many years. Although the junior high school was designed to provide a transition from elementary school to high school, its curriculum took on the appearance and character of the high school. Teachers were organized into subject matter departments, and the course offerings reflected the high school curriculum.

Roots of Inquiry

Now let's shift our attention away from the content of the early science, and take a look at the nature of inquiry in science teaching. What were its roots? When did it emerge in science teaching?

According to Carlton Stedman, inquiry and experimentation, as currently understood and practiced, were resisted by most practicing scientists for philosophical reasons until the mid 1800s.[34] Their place in science teaching has taken even longer. In 1842 the American Association for the Advancement of Science was organized as a result of a conference in Boston of the American Association of Geologists and Naturalists. The geology group essentially was transformed into the AAAS. The AAAS joined forces with the forerunner of the National Education Association, and both proposed the idea of a national university and a school for advanced training in science. However, because they were concerned with basic research and more uniform curricula, the time was still not receptive for inquiry.[35]

Much of the early innovation in science education had its beginnings at the university level, especially in the work of several college teachers. Stedman highlights the work of three early scientists who influenced science teaching.

Benjamin Silliman (1779–1864) developed Yale's first chemistry lab and is credited with teaching the first course in geology in the United States. He believed in using visual aids in teaching and illustrated his lectures with five-foot-square pictures.[36]

Louis Agassiz' claim to fame as a scientist was his early work on glacial theory. But Agassiz was also an outstanding educator who blended Pastalozzi's object lessons (having the student experience objects using the senses), discovery learning, and active involvement. As a teacher, Agassiz had students work in his lab to observe specimens first hand, thereby gaining knowledge experientially. Stedman reports that Agassiz gave students over 1000 specimens and had them separate the specimens into species. Upon examining their work he praised their accomplishments and discoveries.[37]

Asa Gray, a colleague of Agassiz' at Harvard, was a biologist who, according to Stedman, was the "first citizen of Darwin." Gray, through public speaking and writing tried to reduce the conflict between evolution and religious views, a conflict that is still with us. As a teacher, Gray tried to help students perceive his subject (botany) as a whole related system.[38]

Silliman and Agassiz attempted to make learning meaningful (at a time when the classics flourished and mental discipline was stressed) by encouraging laboratory experiences and practical experience. It is here that the roots of inquiry teaching lie. But it took

[34]Carlton H. Stedman, "Fortuitous Strategies on Inquiry in the 'Good Ole Days,'" *Science Education* 71 (5): 657–665.

[35]Stedman, "Fortuitous Strategies," 657–665.

[36]Stedman, "Fortuitous Strategies," 657–665.

[37]Stedman, "Fortuitous Strategies," 657–665.

[38]Stedman, "Fortuitous Strategies," 657–665.

FIGURE 3.7 An early science teaching laboratory. North Wind Archives.

many years for the roots of inquiry to grow into a dominant philosophy in science teaching. This occurred in the late 1950s and 1960s with the development of the science reform projects.

Phase II: Progressive Education and Science, 1930–1950

Several reports influenced science teaching during this period. In 1932 *A Program for Science Teaching* was published by the National Society for the Study of Education, which emphasized a continuous 12 year curriculum, the organization of science courses around general principles, generalizations or "big ideas" in science, and the inclusion of the method of science as an integral part of science. Although the term "inquiry" was not used, the implication was that science teaching should constitute more than the teaching of facts and should involve students in the methods of science—observing, experimenting and hypothesizing.

The authors of *A Program for Science Teaching* proposed that science education should contribute to the major aim of education, namely "life enrichment through participation in a democratic social order." In fact the report suggested that the utility of subject matter be measured by the degree to which it reaches the interests and relates to the well-being of students.[39]

The authors also recommended a K–12 science curriculum based on the major generalizations of science. They recommended adding more physical science to the elementary program and suggested general science for grades 7 to 9. They suggested that content should include problems that are related to the student's world and that would enable the students to use the methods of science.[40]

A second report of this period was *Science in General Education*, published by the Progressive Education Association (PEA) in 1938. This report focused attention on the needs of students and recommended a program that looked at the psychology of the learner. The report also recommended a greater correlation between science and everyday living. The main aims of science, according to this report, included[41]

1. Acquiring understanding in science as distinguished from information
2. Developing the ability to think
3. Developing particular skills or abilities related to problem solving
4. Developing certain attitudes and dispositions useful in problem solving

The progressive education movement was a movement that provided an alternative approach to the tra-

---

[39]Hurd, *Biological Education*, pp. 57–58.

[40]Hurd, *Biological Education*, pp. 57–58.
[41]Commission on Secondary School Curriculum, *Science in General Education* (New York: Appleton-Century, 1938).

ditional school. Dewey suggested that the progressive education movement appealed to many educators because it was more closely aligned with America's democratic ideals. Dewey put it this way:

> One may safely assume, I suppose, that one thing which has recommended the progressive movement is that it seems more in accord with the democratic ideal to which our people is committed than do the procedures of the traditional school, since the latter have so much of the autocratic about them. Another thing which has contributed to its favorable reception is that its methods are humane in comparison with the harshness so often attending the policies of the traditional school.[42]

Two aspects of the progressive education movement that affected all of education but perhaps appealed most to science teachers, was the movement's notion of the child-centered curriculum, and the project method. Both of these ideas have survived, even to this day, and over the years were given different degrees of emphasis. For example, in the late 1960s and 1970s, the child-centered curriculum was represented in the humanistic education movement (sometimes known as affective education). The humanistic ideas of the present day were similar to the progressive ideals of the 1930s.

The child- or student-centered approach is a major paradigm implying beliefs about the nature of learning, the goals of education, and the organization of the curriculum. Emphasis on student-centeredness has waxed and waned in American education to the present day. The PEA represents one of the first and most important advocates of student-centeredness.

The PEA sparked the development of a number of experimental schools that embodied the philosophy of the progressive educators. Science in the progressive schools was an opportunity to involve students directly with nature or with hands-on experiences with science phenomena, and to relate science to not only the emotional and physical well-being of the child but to the curriculum as a whole. There is a rich literature on this movement that describes innovative child-centered programs such as the organic school, Dewey's Schools of To-Morrow, the Gary (Indiana) plan, the Lincoln School, and the Parker School.[43]

The project method, which had not been a foreign idea to science teachers, was given great impetus by the progressive education movement. The project method was described by William Kilpatrick, a professor at Teachers College of Columbia University.

Glatthorn describes Kilpatrick's approach as follows:

> Any meaningful experience—-intellectual, physical, or aesthetic—could serve as the organizing center of the project, as long as it was characterized by purpose. And for every project, there were four steps: purposing, planning, executing, and judging. While it might be necessary from time to time for the teacher to make suggestions about each of these four stages, it was preferable, from Kilpatrick's viewpoint, for the child to initiate and determine each stage. And these projects, strung together, became the curriculum.[44]

Near the end of this period there was an increasing emphasis on the importance of science in general education. The report *Science Education in American Schools* (1947) by the National Society for the Study of Education summarized the Educational Policies Commission Report and the Harvard Committee on General Education with the following statements:

- Science instruction should begin early in the experience of the child
- All education in science at the elementary and secondary levels should be general. Even for students going to college, general courses in biological science and in physical science [according to the Harvard report] "should make a greater contribution to the student's general education and his preparation for future study than a separate one-year course in physics and chemistry." The document of the Educational Policies Commission goes even further in its recommendations for reorganization of high school science courses
- The development of competence in use of the scientific method of problem solving and the inculcation of scientific attitudes transcend in importance other objectives in science instruction[45]

Phase III: The Golden Age of Science Education, 1950–1977

During the period 1950–1977 science education witnessed a massive curriculum movement (the course content curriculum project movement) that no other period has witnessed. It was a period of federal intervention in education and of the expenditure of huge sums of money on curriculum development and teacher training. It was, as many coined it, the golden

[42]J. Dewey, *Experience and Education* (New York: Collier, 1938), pp. 33–34.
[43]See Cremin, *The Transformation of the School.*
[44]A. A. Glatthorn, *Curriculum Leadership* (Glenview, Ill.: Scott, Foresman, 1987), p. 39.
[45]National Society for the Study of Education. Forty-sixth Yearbook of the National Society, *Science Education in American Schools* (Chicago: University of Chicago Press, 1947).

age of science eduation. Why did this happen? What were the forces influencing this movement?

Paul DeHart Hurd made an interesting point when he said that "at some point an awareness emerges: the present curriculum is no longer serving the needs of either student or society."[46] And something needs to be done. The desire was to *reform not revise* the nation's science and mathematics curriculum.

After World War II the advances in science and technology increased at an enormous rate. These advances were described in a series of stages—the first, the atomic age, then the age of automation, then the space age, and the computer age. These changes occurred at a very rapid pace and there was a growing concern that America was not producing enough scientists and engineers to meet this need.[47]

The early 1950s has been characterized as an era in which American anxiety often bordered on hysteria. America was in the midst of the cold war with the Soviet bloc nations. The McCarthy hearings were in full swing and there was a massive retaliation against communist aggression abroad. England points out that many Americans of "normally placid temperament became convinced that Moscow was directing a conspiracy, reaching around the globe, to bury Western democracy, and their fears were intensified by the speed of the Russian development of nuclear weapons. Belief in a substantial margin of American superiority began to crumble."[48]

The National Science Foundation (NSF), which had been created in 1950, took the leadership in addressing the problem of personnel shortages in science and engineering, and the concern that high school science courses were inadequate in light of the rapid changes in science and technology, that many science teachers needed more training in science and improved methods of teaching, and that the textbooks being used were outmoded, and needed to be changed.

Even with the growing concern for science education in the schools, the NSF showed reluctance in the beginning to support high school science. Support would be directed at colleges and universities.[49] Partly due to limited funds, NSF didn't think it could make a dent in the nation's schools. However, in 1953 the first two "NSF Summer Institutes" were held at the University of Colorado for college mathematics teachers, and at the University of Minnesota for college

physics and a companion group of high school physics teachers. Deeming the institute a success, NSF moved into the institute business and sponsored 11 in 1954, and 27 in 1955 (two of these were year-long institutes for high school mathematics and science teachers).[50]

In 1956 NSF funded a group of physicists at the Massachusetts Institute of Technology under the direction of Professor Jerrold R. Zacharias. Known as the Physical Science Study Committee (PSSC), they outlined in 1956–57 the ideas for a new high school physics course and in the summer of 1957 assembled 60 physicists, teachers, apparatus designers, writers, artists, and other specialists to produce a pilot version of the PSSC physics course (text, teacher's guide, lab manual, equipment, and later films and books written by scientists).[51]

Oddly enough, as late as the fall of 1957, the NSF was thinking of reducing its institutes for science teachers. The NSF-sponsored teacher institutes had come under attack. Some people were concerned that the Academic Year Institutes were aggravating the shortage of teachers by taking science and mathematics teachers (often the best) out of the classroom. It was also known that many high school teachers showed little interest in applying for refresher courses in science (even with tuition paid and a stipend).[52]

Then on October 17, 1957, everything changed.

The Sputnik Watershed

The Soviet launch of Sputnik into an orbit about the Earth was a shot that had reverberating effects on American science and mathematics education like no other event in this century. According to England, "the launching of Sputnik changed the outlook dramatically. It brought another big boost in the institute budget and a chance to try out a variety of other projects, some of them reaching down into the elementary school."[53]

Sputnik created an intellectual climate that fostered the adoption of new courses of study by the nation's schools, spurred Congress to ensure funding for course improvement projects and summer institutes to train teachers in the new courses of study in science and mathematics.

The golden age of science education came about as a response to the apparent superiority of Soviet science and technology, yet its roots lie prior to Sputnik with the perceived crisis in the quality of secondary science teachers, courses, textbooks, and teaching materials.

[46]P. D. Hurd, *New Directions in Teaching Secondary School Science* (Chicago: Rand McNally, 1970), pp. 1–2.

[47]Hurd, *New Directions*, p. 2.

[48]J. M. England, *A Patron for Pure Science: The National Science Foundation's Formative Years, 1945–57* (Washington, D.C.: National Science Foundation, 1982), p. 248.

[49]England, *A Patron for Pure Science*, p. 238.

[50]England, *A Patron for Pure Science*, p. 238.

[51]Uri Haber-Schaim et al. *Physical Science Study Committee Physics* (Boston, Mass.: D.C. Heath, 1971) p. 627.

[52]England, *A Patron for Pure Science*, p. 237.

[53]England, *A Patron for Pure Science*, p. 247–248.

Reform Projects

More than $117 million was spent on over 53 separate course improvement projects during the years 1954 to 1975.[54] The curriculum projects developed during the golden age of science education were characterized as follows:

1. The reformers were typically scientists, specialists in physics, chemistry, biology, or earth science. Their views of what elementary and secondary science should be were colored by their research interests and the nature of science.[55] The reformers criticized contemporary science programs by stating that although "biology, chemistry and physics is taught, there is very little evidence of the science of these subjects presented."[56]

2. The criteria for the selection of content came from the discipline (as opposed to earlier criteria such as meeting the needs of the child). Thus the "content of science, with its concepts, theories, laws and modes of inquiry is given."[57] The new science curricula were also based on the assumption that science was a way of knowing. The learning activities in the new curricula were designed to help students experience science as a way of knowing and therefore engaged them in "experimenting, observing, comparing, inferring, inventing, evaluating, and many other *ways* of knowing."[58] (Italics mine.)

3. According to the developers of the projects, students would be motivated instrinsically. The interesting nature of the science activities would themselves motivate students. This idea was fully developed by Jerome S. Bruner in his book *The Process of Education* (Cambridge, Mass.: Harvard University Press, 1960). The book was a report of a conference among scientists and educators at Woods Hole in 1959.

4. The goals of science teaching as they developed with these reform projects were very different from science teaching goals in the past and are in stark contrast to what science educators are advocating for the 1990s and beyond. Hurd describes the goals of the curriculum reform projects in this way:

> What is expressed is more a point of view or rationale for the teaching of science. This point of view lacks the societal orientation usually found in statements of educational goals. The "new" goals of science teaching are drawn entirely from the disciplines of science. Social problems, individual needs, life problems and other means used in the past are not differentiated for local schools or for individuals. If the discipline of science is to be the source of educational goals, then differences are not possible without violating the structure of science. To do otherwise would be teaching un-science, a major criticism of the traditional curriculum. Thus we find the reformers define educational goals within the framework of science as they are reflected in the separate disciplines. Their logic is simple and straightforward: this is what we know with some degree of reliability; this is how we find out about what we know; and this is how it all fits into the big picture—*conceptual schemes* [italics mine]. Now then, whatever personal or social problems involving science are to be solved must begin with authentic concepts and with fruitful processes of inquiry.[59]

[54]J. Sealy, "Curriculum Development Projects of the Sixties." R & D Interpretation Service Bulletin, Science. (Charleston, W. Va.: Appalachia Educational Laboratory, 1985).
[55]Hurd, *New Directions*, p. 15.
[56]Hurd, *New Directions*, p. 27.
[57]Hurd, *New Directions*, p. 31.
[58]Hurd, *New Directions*, p. 31.

[59]Hurd, *New Directions*, p. 34.

INQUIRY ACTIVITY   3.3

Course Improvement Projects and Traditional Science Courses: How Were They Different?

The course improvement projects developed during the golden age of science education had an enormous impact on science teaching during that period and have influenced the science education of today. In this inquiry activity you will compare and contrast a course improvement project science textbook and a commercial science textbook of the same period.

Materials

Course improvement project textbook
Commercial science textbook (same subject and time period of the course improvement textbook)

Procedure

1. Working with a team, select one text or science program from the list of course improvement projects (see the course in improvement projects in chapters 4–5.). Find a commercial text or science program in the same content area. The class should make sure that teams select programs from the elementary, middle or junior, and high school levels.
2. Make a chart using your own criteria to compare and contrast the texts. You might consider the following questions for your criteria:
   a. Who are the authors of the textbook? (Scientists, science educators, science teachers?)
   b. What are the goals of the course?
   c. How is the content organized in the textbooks?
   d. What emphasis is laboratory work given?
   e. To what extent is the application of science to society emphasized?

Minds on Strategies

1. What are the fundamental differences between the "traditional" science programs and the course improvement project science programs?
2. Jerome Bruner was one of the psychologists associated with the early course improvement projects. How were his theories manifested in these programs? (Refer to Chapter 2 for Bruner's ideas).
3. What do you think were some of the problems associated with implementing the course improvement project programs?

Phase IV: Textbook Controversies and the Back-to-Basics Movement, 1975–1983

Nothing lasts forever! The course content improvement projects, which were propelled by the nation's desire to improve science and mathematics, came under attack in the mid 1970s. There were two ideas that surfaced during this time that had an impact on science education. One was a movement that became know as the back-to-basics movement, and the other had to do with the questioning of science textbooks by individual citizens, textbook watchers, and religious groups (especially fundamentalist Christians).

The back-to-basics movement was not only a reaction to the current wave of course content improvement projects but also reflected antagonism toward the progressive movement in education, which surfaced in the late 1960s and 1970s as the humanistic education movement. The conservative movement "labeled progressive schools 'anti-intellectual playhouses' and 'crime breeders,' run by a 'liberal establishment.'"[60] One of the projects funded by NSF came under furious scrutiny by certain conservative congressmen, most notably Representative John Conlan of Arizona. According to Nelkin, citizen groups in Arizona complained to Conlan (then a state senator) about Man: A Course of Study (MACOS). Conlan's staff investigated the complaints, and eventually Conlan took steps to stop appropriations for MACOS "on the grounds of its 'abhorrent, repugnant, vulgar, morally sick content.'"[61] Others criticized MACOS. Nelkin claims that the Council for Basic Education objected to MACOS for its emphasis on cultural relativism, and its lack of emphasis on skills and facts. Even liberal congressmen got on the anti-MACOS bandwagan because of their desire to limit the executive bureaucracies, such as NSF, and for "their resentment of scientists, who often tended to disdain congressional politics; and above all, the concern with secrecy and confidentiality that followed the Watergate affair."[62]

The MACOS controversy brought the issue of censorship into the public arena. However, to avoid the claim of censorship, which probably would not have been acceptable to many in Congress, Conlan focused on the federal government's role in implementing MACOS, as well as all other NSF-funded curriculum projects. One issue that surfaced was "the marketing issue—the concern that the NSF used taxpayers' money to interfere with private enterprise."[63] Along with this was the position expressed by such conservative writers such as James Kilpatrick, who attacked NSF science programs as "an ominous echo of the Soviet Union's promulgation of official scientific theory."[64] The tenor of the times was quite clear: "resentment of the 'elitism' of science reinforced concern that NSF was naively promulgating the liberal values of the scientific community to a reluctant public."[65] The result: on April 9, 1975, Congress terminated funds for MACOS; further support of science curriculum projects was suspended, and the entire NSF educational program came under review.

The federal funds that would have been used to support the NSF curriculum projects were used for

[60]D. Nelkin, *Science Textbook Controversies and the Politics of Equal Time* (Cambridge, Mass.: MIT Press, 1977), p. 41

[61]Nelkin, *Science Textbook Controversies*, p. 112.

[62]Nelkin, *Science Textbook Controversies*, p. 112.

[63]Nelkin, *Science Textbook Controversies*, p. 114.

[64]Nelkin, *Science Textbook Controversies*, p. 114.

[65]Nelkin, *Science Textbook Controversies*, p. 114.

research and college science programs. The NSF did, however, fund three large studies to answer the charges that science and mathematics education had not improved as a result of the course content curriculum projects.[66] The three studies, which were mentioned earlier in this chapter, were (1) a review of the science education research between 1954 and 1974;[67] (2) a demographic study that assessed such factors important to science educators as enrollments, offerings, teachers, and instructional materials;[68] and (3) case studies of science teaching.[69]

Because of the volume of data provided by the three status studies, the NSF funded a fourth study, called Project Synthesis (discussed at length earlier in the chapter). Project Synthesis recommended expanding the goals of science education to include not only content (science as preparation for further study) but also science for meeting personal needs, resolving societal problems, and for career awareness.

To recoup its position as an active force in science education, the NSF prepared a report in 1980 entitled *Science and Engineering Education for the 1980s and Beyond* for the Carter administration. Unfortunately (for NSF) Carter was defeated, and the new president, Ronald Reagan, rejected the reports recommendations and tried to eliminate the science education section of NSF, and during the early 1980s the influence of the NSF in science education was limited to college faculty improvement and graduate student fellowships in the basic sciences.[70]

Another wave of controversy occurred during this period, and that was the teaching of evolution in the public schools, and other science concepts and ideas that touched on beliefs, religion, values, and morals (e.g., teaching human sexuality, human reproduction, and birth control in biology classes). But it was Darwin's evolutionary theory that resulted in court cases and laws being passed to regulate the teaching of evolution (such as giving equal time to "creation science" if "evolution science" was presented in a science class).

A phenomenon that reached its pinnacle during this time was the general scrutiny of textbooks, especially in biology, earth science, social studies, and literature. In the 1960s, when BSCS, whose textbooks emphasized Darwin's theory of evolution (in contrast with many high school biology texts at the time), submitted its books for state adoption in the lucrative market of Texas, serious trouble surfaced. The Reverend Ruel Lemmons led a protest (that reached the Texas governor's office) to get the BSCS textbooks banned, claiming the books were pure evolution, completely materialistic, and atheistic.[71] The books were not banned, but there were changes made to lighten their evolutionary emphasis. Nelkin reports that BSCS had to specify that evolutionary theory was a theory, not a fact, and that it had been modified, not strengthened, by recent research.[72]

This was just the tip of the iceberg. A group of fundamentalists began to develop a world view of creation based on the story in Genesis in the Bible. The creationists rejected the notion of a 5 billion year old Earth, instead claiming that biological life began approximately five to six thousand years ago. One of the forceful voices in the creation science movement was Henry Morris. In an article published in the *American Biology Teacher*, he set out the differences, from his point of view, between the creation model and the evolutionary model.[73] Essentially the creationists "theorized" that all living things were brought about by the acts of a creator. The evolutionary model proposed that all living things were brought about by naturalistic processes due to properties inherent in inanimate matter. The creationists, in their literature, set the creation model alongside the evolutionary model and insisted that good science education would provide alternative views on the same topic and allow students to evaluate them to form their own position.

In 1969 the California Board of Education modified the *Science Framework on Science for California Schools* so that the theory of creation would be included in textbooks. The *Science Framework on Science for California Schools* sets forth the guidelines for the adoption of science textbooks (currently over $40 million are spent on science books in California during the science adoption year). Vernon Grose (an aerospace engineer) wrote and presented a document arguing that evolutionary theory was biased and should be taught only if alternative views were presented. He convinced the board of education to mod-

[66]R. E. Yager, "Fifty Years of Science Education, 1950–2000," in *Third Sourcebook for Science Supervisors*, L. L. Motz and G. M. Madrazo, Jr., eds. (Washington, D.C.: National Science Teachers Association, 1988), pp. 15–21.

[67]S. L. Helgeson, P. E. Blosser, and R. W. Howe, *The Status of Pre-College Science, Mathematics, and Social Science Education: 1955–75* (Columbus, Ohio: Center for Science and Mathematics Education, The Ohio State University, 1977).

[68]I. R. Weiss, *Report of the 1977 National Survey of Science, Mathematics, and Social Studies Education* (Washington, D.C.: Center for Educational Research and Evaluation, U.S. Government Printing Office, 1978).

[69]R. E. Stake, and J. Easley, *Case Studies in Science Education* (Urbana, Ill: Center for Instructional Research and Curriculum Evaluation, University of Illinois, 1978).

[70]Yager, "Fifty Years of Science Education," p. 18.

[71]Nelkin, *Science Textbook Controversies*, p. 29.

[72]Nelkin, *Science Textbook Controversies*, p. 29.

[73]H. Morris, "Creation and Evolution," *American Biology Teacher*, March, 1973.

ify its position on the teaching of evolution; the board inserted the following statement into the framework:

> All scientific evidence to date concerning the origin of life implies at least a dualism or the necessity to use several theories to fully explain relationships.... While the Bible and other philosophical treatises also mention creation, science has independently postulated the various theories of creation. Therefore, creation in scientific terms is not a religious or philosophical belief. Also note that creation and evolutionary theories are not necessarily mutual exclusives. Some of the scientific data (e.g. the regular absence of transitional forms) may be best explained by a creation theory, while other data (e.g., transmutation of species) substantiate a process of evolution.[74]

The "evolutionists were incredulous that creationists could have any influence."[75] A number of individuals and groups such as the National Association of Biology Teachers (NABT), the National Academy of Science, and the Academic Senate of the University of California protested and lobbied against the state board's ruling. The solution to the creation-evolution issue resulted only after the state board had received numerous complaints about the earlier decision. In 1972 the California Board of Education decided to approve a statement prepared by its curriculum committee by proposing neutrality in science textbooks. Dogmatic statements in science books would be removed and replaced with conditional statements. Textbooks dealing with evolution would have printed in them a statement indicating that science cannot answer all questions about origins, and that evolution is a theory, not a fact. Some textbooks, even in the 1990s, contain statements to this effect, usually printed on the inside cover. The effect of this policy change prevented textbook publishers from having to include in science books a section on "creation science." The board's decision, which was called the Antidogmatism Statement, caused publishers to rethink the way they were presenting science information in textbooks.

By the late 1970s the emphasis in education had shifted to the "basics," and the literature was replete with back-to-basics slogans: for example, schools need to attend to basic skills. The place of science in the curriculum lost the priority it had had in the 1960s. The interest now was on teaching the basic skills of reading, mathematics, and communication.

Phase V: A Nation at Risk, the 1980s

Some point to the year 1983 as being similar in some respects to the year of Sputnik, 1957. Two reports were issued by very prestigious commissions, which marked another turning point in the nation's perception of education (actually, over 300 reports were issued during the mid 1980s calling for reforms in education). In 1957 the perceived concern was with the nation's inability to prepare scientists and engineers to meet the challenge of the space age. In 1983 the perceived concern was with America's ability to compete in the world economically. In the report *A Nation at Risk* by the National Commission on Excellence, established by the secretary of education to report to the American people on the quality of America's education, the authors (only 1 of 14 being a teacher) made this statement:

> If an unfriendly foreign power had attempted to impose on America the mediocre educational performance that exists today, we might well have viewed it as an act of war. As it stands, we have allowed this to happen to ourselves. We have even squandered the gains in student achievement made in the wake of the Sputnik challenge. Moreover, we have dismantled essential support systems which helped make those gains possible. We have, in effect, been committing an act of unthinking, unilateral educational disarmament.[76]

The report goes on to say that the United States is being overtaken as the leader in science, technology, commerce, industry, and innovation by Japan, as well as some other nations, and that this is most likely due to the "rising tide of mediocrity in our schools."[77]

The commission supported their statement of risk by citing results on various achievement test scores. For example, they reported that international comparisons of student achievement completed a decade ago reveal that on 19 academic tests American students were never first or second and, in comparison with other industrialized nations, were last several times. They also cited falling scores on standardized tests such as the College Boards and the SATs.[78] It should be pointed out that some took issue with the commission for citing the poor performance of American students on international comparisons. The

---

[74]Nelkin, *Science Textbook Controversies*, p. 83.
[75]Nelkin, *Science Textbook Controversies*, p. 85.

[76]National Commission on Excellence in Education, *A Nation at Risk*, p. 5.
[77]National Commission on Excellence in Education, *A Nation at Risk*, p. 5.
[78]National Commission on Excellence in Education, *A Nation at Risk*, pp. 8–9.

FIGURE 3.8   Science education in the 1990s will attend to global environmental issues such as air pollution, global warming, ozone depletion, and deforestation.

director of the report of the International Project for the Evaluation of Educational Achievement warned against using the data to make generalizations about individual countries' educational system. He also stated that "the best American students are comparable in achievement to those of other advanced nations." He went on to say that the comprehensive systems, such as in the United States, result in bringing more people into the talent pool.[79]

The report nevertheless, had an impact on education in general, and on science education specifically. Using the language of the back-to-basics movement, the commission recommended that all students seeking a diploma from high school be required to take a curriculum, which they called the five new basics, consisting of (1) four years of English, (2) three years of mathematics, (3) three years of science, (4) three years of social studies, and (5) one-half year of computer science (college-bound students should take two years of a foreign language). With regard to high school science they recommended a curriculum that would provide all graduates with the following:[80]

1. The concepts, laws, and processes of the physical and biological sciences
2. The methods of scientific inquiry and reasoning
3. The application of scientific knowledge to everyday life
4. The social and environmental implications of scientific and technological development

They also said that science courses must be revised and updated for both the college-bound and those not going to college. With regard to computer science, students should (a) understand the computer as an information, computation, and communication device, (b) use the computer in the study of the other basics and for personal and work-related purposes, and (c) understand the world of computers, electronics, and related technologies.[81] Among the other recommendations, the commission suggested raising college entrance requirements, administering standardized achievement tests at major points in students' careers and upgrading textbooks by having states evaluate texts on the basis of their ability to present rigorous and challenging material clearly. The report also recommended that American students stay in school longer each day (from six hours to a seven hour day) and extending the school year to at least 200 days per year (then 180), but preferably to 240 days.

The report did give state and local district science educators the "rationale" to require more science, and indeed many states passed laws requiring more science for graduation. There was also an increased effort to require more time on task, and some states required science teachers to log in the number of hours students were engaged in science laboratory activities. (Georgia, for example, requires that 25 percent of instructional time be spent doing laboratory activities.)

Another, and perhaps more significant report for science education was *Educating Americans for the Twenty-first Century*, issued by the National Science Board. The report is very clear about the main goal for improving science, mathematics, and technology education. The authors of the report state the following as the one basic objective:

[79]For a full analysis and critique of various commission reports of 1983, see D. Tanner, "The American High School at the Crossroads," *Educational Leadership*, March, 1984, pp. 4–13.

[80]National Commission on Excellence in Education, *A Nation at Risk*, p. 25.

[81]National Commission on Excellence in Education, *A Nation at Risk*, p. 26.

The improvement and support of elementary and secondary school systems throughout America so that, by the year 1995, they will provide all the nation's youth with a level of education in mathematics, science, and technology, as measured by achievement scores and participation levels, that is not only the highest quality attained anywhere in the world but also reflects the particular and peculiar needs of our nation.[82]

Some critics have charged that the report is overly nationalistic, and was reminiscent of the era of the cold war and the space race.[83] The report also uses the language of the back-to-basics movement, stating that education must return to the basics, but the basics of the twenty-first century are not only reading, writing, and arithmetic but include communication, higher-order problem-solving skills, and scientific and technological literacy.[84] The report also has been criticized for its failure to address the social implications of science, and the importance of science education in general education. Further critics have charged that the board ignored the shortcomings of the course content curriculum projects of the 1960s and 1970s, and instead recommended that the National Science Foundation be the leader in curriculum development efforts by promoting the development of new curricula.[85]

Even with these criticisms, the report has had an impact and has been used by a number of groups as a rationale for charting "new" curricula, and for seeking funding. *Educating Americans for the Twenty-first Century* suggested that the K–12 science curriculum be revamped, and that science and technology education at the elementary and secondary levels should result in the following outcomes:

- The ability to formulate questions about nature and seek answers from the observation and interpretation of natural phenomena
- The development of students' capacities for problem solving and critical thinking in all areas of learning
- The development of particular talents for innovative and creative thinking
- An awareness of the nature and scope of a wide variety of science- and technology-related careers open to students of varying aptitudes and interests
- The basic academic knowledge necessary for advanced study by students who are likely to pursue science professionally

- The scientific and technical knowledge needed to fulfill civic responsibilities and improve the students' own health and life and their ability to cope with an increasingly technological world
- The means for judging the worth of articles presenting scientific conclusions

The report also recommended a K–12 curriculum plan as follows:

K to 6.   The emphasis should be on phenomena in the natural environment, with a balance between biological and physical phenomena. The curriculum should be integrated with other subjects and should be implemented with hands-on activities.

Grades 7 to 8.   The focus in the middle school should be on the biological, chemical, and physical aspects related to the personal needs of adolescents. The curriculum should also focus on the development of qualitative analytical skills. Experimentation, science texts, and community resources should be used in instruction.

Grades 9 to 12.   Biology should be presented in a social and ecological context. Topics should include health, nutrition, environmental management, and human adaptation. The biology curriculum should be inquiry oriented, and problems should be selected in a biosocial context involving value or ethical considerations. Chemistry should emphasize the social and human relevance of chemistry, and should include topics from descriptive and theoretical chemistry. Physics courses should be designed for a wide variety of students and should be built upon students' earlier experiences in physics. The emphasis in grades 9 to 11 should be on the application of science and technology. Schools should offer discipline-oriented career preparation courses in grades 11 to 12, in which several disciplines would be taken in each year.

Because of these and other reports,[86] action to modify and change school programs was set into motion at the national, state, and local levels. At the national level, the National Science Foundation received appropriations from Congress to once again develop curriculum materials for the schools. During the eighties and continuing into the nineties, the NSF

[82]National Science Board, *Educating Americans for the Twenty-first Century* (Washington, D.C.: National Science Foundation, 1983), p. 5.
[83]Tanner, "The American High School," p. 8.
[84]National Science Board, *Educating Americans*, p. v.
[85]Tanner, "The American High School," p. 8.

[86]Additional reports issued from 1982 to 1984 include: Task Force on Education for Economic Growth, *Action for Excellence: A Comprehensive Plan to Improve Our Nation's Schools* (Denver, Colo.: Education Commission of the States, 1983); J. I. Goodlad, *A Place Called School* (New York: McGraw-Hill, 1984); E. L. Boyer, *High School: A Report on Secondary Education in America* (New York: Harper & Row, 1982); M. Adler, *The Paideia Proposal* (New York: Mac-millan, 1982); College Examination Board, *Academic Preparation for College: What Students Need to Know and Be Able to Do* (New York: The College Board, 1983).

funded a number of curriculum projects, especially at the elementary and middle school levels. Many of these funded projects were joint efforts involving the public and private sectors. In nearly all of these projects, a publisher or a commercial enterprise teamed up with a university group or science education center (such as the BSCS) to develop science curriculum materials. Some of these projects will be discussed later in this chapter.

Science education in the eighties was also a time of reflection and recommendations for science education toward the year 2000. Earlier in the chapter, we looked at the work and recommendations of Project Synthesis. Project Synthesis represents the most comprehensive approach to the analysis of science education and science curriculum goals.

In 1982 the National Science Teachers Association began its Search for Excellence in Science Education program. This program, which was begun by then NSTA president Robert Yager, was designed to identify exemplary programs in science education. The criteria used to identify exemplary programs were the desired states identified by the Project Synthesis subgroups (elementary science, biology, physical science, science, technology, and society, and inquiry). Several volumes in the Search for Excellence Series have been published by the NSTA as follows:

Volume 1 (1982)
Science as Inquiry, Elementary Science, Biology, and Physical Science
Volume 2 (1983)
Physics, Middle School/Junior, and Non-School Settings
Volume 3 (1984)
Energy Education, Chemistry, and Earth Science

Additional monographs in the series included *Teachers in Exemplary Programs: How Do They Compare? Centers of Excellence: Portrayals of Six Districts,* and *Exemplary Programs in Physics, Chemistry, Biology, and Earth Science.* We will discuss some of these programs in the next two chapters.

Phase VI: Science for All, Toward the Year 2000 and Beyond

This chapter began with an examination of goals and objectives for the science curriculum with an eye to the future. We discussed the results of Project Synthesis, Project 2061, *New Designs for Elementary School Science and Health,* Scope, Sequence, and Coordination Project, and cognitive research, and identified some of the implications for the science curriculum goals.

As we move toward the year 2000 and beyond, the goals of science education will be influenced not only by the foregoing, but by several other factors. Here, very briefly, are some of those factors that will affect science education in this decade and the next. Although recent research has shown that the science curriculum has not changed very much in the last 15 years, these issues will be debated, described, and put into practice in individual classrooms, school districts, and states. We must keep in mind that this country does not have a national curriculum and that much of the action in curriculum development occurs in individual school districts and classrooms.

Multicultural Education Programs

School districts are taking a proactive position in infusing multicultural programs into their schools. Portland, Oregon, and the states of Minnesota and Iowa have developed programs whereby multicultural education is part of the entire curriculum, not simply a unit taught here or there.

In 1989 a Multicultural Science Education Division was established by the Board of Directors of the National Science Teachers Association. The first director of this new NSTA division was elected in 1990. The Association for Multicultural Science Education (AMSE) was also formed in 1989 to deal with issues related to multicultural education. The first organizational meeting of the AMSE was held in Atlanta at the 1990 NSTA annual meeting. The new association's purposes are as follows:

1. To explore issues and initiate programs relevant to teaching science to students of diverse backgrounds and motivate them to consider science as a career
2. To awaken the interest of nonwhite and female students in science
3. To promote programs that make science teachers successful in a culturally diverse classroom
4. To explore changes that need to be made in the science curriculum to meet the needs of all students
5. To identify, recruit, and fully involve current teachers of all minorities in becoming important forces in science education

Several of the NSF funded projects of the late 1980s and early 1990s were aimed at developing science education programs in multicultural environments. Several of these will be discussed in the next chapter.

Global Environmental Issues and Problems

The major conferences held during the 1990s will have to do with the environment and the economic impact environmental problems will have on all nations. Science education programs will attend to a number of global environmental issues such as air pollution, global warming, ozone depletion, and deforestation. Science education will have to address the global environmental problems head on in the nineties. Earthday 1990 (April 27) marked the twentieth anniversary of the environmental education movement and represented a call for action among governments, action groups, and citizens to deal with the fate of the earth's environment. The environmental theme will permeate science curriculum development at the local, state, national, and international level—in short, globally.

The Science-Technology-Society Interface

The science-technology-society interface, which will be developed more fully in Chapter 6, is the bridge that science teachers will use to connect the ideas of science with their utility in society. It will be the vehicle to humanize science teaching and make the content of science relevant to students in middle and high schools. The science-technology-society (STS) theme makes sense as an organizing principle for science teaching, especially in light of the recent history of science curriculum development. If you recall, the curriculum reform of the 1950s and 1960s was discipline centered, and this approach was found to be ineffective with most students. The STS theme forces us to ask questions such as, How does science relate to the world of the student? How can science contribute to the fulfillment of a healthy life and environment? What is the relationship between humans and the environment? How can science be in the service of people?

The STS theme has the intrinsic appeal of forcing us to stay current, and to relate science to the students' world today and dreams of tomorrow. It also has the structure of incorporating some of the other major forces that are influencing education today, namely multicultural programs, interdisciplinary thinking, conceptual themes, and global environmental issues.

Key Concepts and Conceptual Themes

If you recall, the notion of "big ideas" was put forward in the 1930s and was an important construct during the curriculum reform era of the late 1950s and 1960s. The major themes of science have also been called overarching truths and unifying constructs. Project 2061 (published in 1989) recommends that science content be organized around conceptual themes. Because of the prestigiousness of the AAAS, many school districts and states will adopt this notion.

Conceptual themes, however, have support in the recent research in cognitive science. One of the recommendations that has been made is to reduce the amount of content in science courses. Organizing courses around a reduced list of conceptual themes is right on target. For example, the 1989 Science Framework for California Schools has identified six themes for its K–12 framework as follows:[87]

Energy
Evolution
Patterns of change
Stability
Systems and interactions
Scale and structure

Accordingly, the main criterion of a good theme is its ability to integrate facts and concepts into overarching constructs. Themes have the characteristic of linking content between the disciplines, therefore being transdisciplinary. But they also have the quality of helping students link science with other disciplines in the curriculum, such as mathematics, history, economics, political science, language arts, and literature.

Interdisciplinary Thinking and Planning

Key concepts and conceptual themes are the structures needed to link thinking across the disciplines, for example, biology and physics or geology and economics. Interdisciplinary thinking is a strategy that will enable students to see the relevance of science content, especially when science is linked with the issues of the day. For example, the field of sports physiology is not only a way of linking biology, health and physics but also a way of connecting science with economics, management, and sports.

Middle schools are organized to put interdisciplinary thinking and planning into action more so than junior high schools or senior high schools because of the organization of teachers into multidiscipline teams and because in many middle schools, these teams have time allotted in the school day for planning. In many middle schools, interdisciplinary units of study are developed by the mathematics, science, social studies, and language arts team. In some high schools, the science teacher plans with the social studies teacher to investigate the historical aspects of a particular environment or community. In other settings, the mathematics and science teacher team up to help students integrate topics in mathematics with topics in biology, earth science, chemistry, and physics.

[87]California Department of Education, *Science Framework for California Schools*, 1989, p. 17.

Case Study 1.
Rehashing the Sixties

The Case. The report *Science for All Americans,* issued in 1989 by the American Association for the Advancement of Science, outlines recommendations for the improvement of science in the nation's schools. Newsome Wave, a former high school science teacher but now a school superintendent, is a sharp critic of the report. He asserts that the AAAS has created a warmed-over version of recommendations that are similar to the reform proposals of the 1950s and 1960s. In an article published in a major Los Angeles newspaper, he said that "this report (*Science for All Americans*) reinforces the elitism of the scientific establishment and fails to deal with the needs of all the students who pass through our schools. How can a report that has scientists, who for the most part are remote from and outside the school environment, identify what should be taught in the schools, be taken seriously? One of the lessons we learned from the curriculum reform projects of the sixties was they addressed the needs of a very small part of the school population. The AAAS seems headed the same direction that cursed the sixties science reform project. Haven't we learned that national curriculum projects simply can't meet the needs of the diverse population of school students, let alone the diversity of the science teaching force."

Reginald Regis, the coordinator of science of a highly populated western state and an advocate of the AAAS's effort at science curriculum reform, wrote a blistering rebuttal to Wave in the journal *New Science*. Regis pointed out that Wave failed to mention that the AAAS has developed a broad program that includes the involvement of cooperating school

(Continued on p. 118)

Case Study 2.
Divine Intervention

The Case. Mr. John Moore is a biology teacher in a small community about 50 miles from a large metropolitan area in the midwestern region of the country. He has been teaching biology to ninth and tenth graders for seven years in this school district. The state he teaches in adopts new science textbooks every five years, and this year it is time to review and make final selections for secondary science books. Mr. Moore was asked by the district science supervisor to chair the six-member biology textbook adoption committee. The district procedures include

(Continued on p. 118)

Think Pieces

- What should the goals of the science curriculum be for the 1990s?
- How do the recommendations of Project Synthesis, Project 2061 and SS&C Project compare? What are the similarities and differences?
- Write an article for the newspaper using one of the following headlines:

  Science Educators Create New Goals for Teaching: New Hope for the Future

  Science Teacher Designs New Course: Career Awareness Biology

  Project 2061 Approved by the School Board

  Multicultural Science Curriculum Implemented   ☐

Back to the Future with Science Education
by Larry Loeppke[88]

Twenty-five years ago the context for changing educational materials and practices was clear. Our country's goal was to secure peace and national prosperity through global political and economic dominance. As a technological society in full bloom, we were developing new nuclear submarines, new B-52 bombers, lunar landing modules, and an interstate highway system. Classroom pedagogy reflected the scientific foundation of our technology. The application of scientific inquiry began to be used as a teaching technique in the public schools, thanks to new curricula developed with National Science Foundation funds by such groups as the Biological Sciences Cur-

(Continued on p. 118)

(Continued from p.117)
districts. He also pointed out that although panels comprised of scientists and university officials did write the unifying concepts, educators reviewed the panels' recommendations.

Wave and Regis have agreed to appear together at the annual conference of the National Science Teachers Association to debate the issues surrounding each educator's point of view.

The Problem: First consider your position in this case. Is the AAAS report a rehash of the wishes of the sixties? What do others in your class think? Take the position of either Mr. Wave or Mr. Regis. What are the facts that support your position? What are your arguments? Prepare yourself for a debate with your opponent. ☐

(Continued from p.117)
placing the textbooks in all the district's school libraries and three of the library branches so that parents and interested citizens can review the books that are being considered for adoption.

During the review process, Mr. Moore receives a letter from an irate parent who objects to some of the content in the biology books, especially the treatment of evolution. The parent, Mr. Alan Hockett, is an engineer with a Ph.D. in chemical engineering from a prestigious university in California and claims that creation theory is as likely a scientific hypothesis as evolution. He points out in the letter that neither theory can be supported by observable events nor can be tested scientifically to predict future events, nor are they capable of falsification. He claims that not to give students opposing "scientific views" is indoctrination. He says that "equal time" should be given

to creation science if evolution is taught in the biology curriculum. He demands to meet with the committee, and ends his letter by saying that he will go to the school board if he is not satisfied with the committee's responses to his claims.

The Problem. Should the committee meet with Mr. Hockett? What should the committee do to prepare for the meeting? How will this affect the adoption process? What position should the biology committee take? ☐

(Continued from p.117)
riculum Study (BSCS), the Earth Science Curriculum Project, the Physical Science Study Committee, and the American Chemical Society (ACS).

For the young people of the early sixties, the relevance of science was fairly clear. New scientific miracles, such as Corning Ware, were in use in the family kitchen and discussed at the dinner table. Entertainment shows such as "Star Trek" convinced us that ionic bonding, plasma physics, and other such mysteries were going to be a part of our daily lives. Students were fascinated by the impact science would have on their careers; someday, as members of the work force, they too would be bringing us such new and exciting technological miracles as remote control television and microwave ovens.

Over the years, however, we became accustomed to miracles. Science became simply another part of our economic enterprise. We described rock bands as "awesome," not laser beams at the grocery checkout. We came to attention when a space shuttle exploded, not when it flew. We became wired to the rest of the world through new global communications technology, then suffered the fate of novices on the global economic stage. As we lost interest

in the mystery and novelty of scientific inquiry and technological innovation, we lost our sense of their immediate relevance and importance. And then we lost our memory of scientific facts. Educators who deal in study skills and learning theory would put it this way: without a "significant context" there is no long-term memory of content.

When scientific literacy began to drop in the seventies, the "education industry"—public and private entities who develop, publish, and teach our science curricula—attempted to tackle the problem by increasing the number of facts students were expected to learn. Most of the textbooks that have been used in classrooms ever since exemplify this "content" solution. They are fact filled because market research shows that educators want them to be. And educators want textbooks full of facts because student assessments measure how many facts students know, not how much knowledge they can apply. Textbooks now not only include a growing number of facts but are also written at oversimplified reading levels, come with ancillary materials that have the effect of reducing the teacher's individualized input, and generally have the effect of replacing scientific inquiry with memorization. Meanwhile, students have been left to fend for themselves when it comes to figuring out why what they are memorizing is personally relevant and applicable to their lives.

Fortunately, as we approach the new century, educators are involved in a great "rethink" about the process of education. And it seems that science educators have now developed a fairly clear—though still very general—consensus on two broad areas of change: (1) Science teaching should be the guided but direct application of sci-

(Continued on p.119)

(Continued from p. 118) entific inquiry by the student in a hands-on fashion; (2) Science teaching should be carried out in a context that is personally and socially relevant to the student in order to fan the spark of curiosity generated by scientific inquiry.

With this foundational agreement, it is simply unnecessary for us to enter the twenty-first century with the kind of mediocre textbooks, ineffective pedagogy, and ill-prepared graduates we have tolerated over the past 15 years. Our schools, our universities, and our educational businesses know where we need to go. Now we must manage our resources wisely and cooperatively to get there.

The trends that will shape science education for the twenty-first century are increasingly evident. To some degree, the question of where we are going has already been answered. We are moving toward the teaching of science as a thinking process—a way of "knowing and reknowing" our world as it changes before us. The question of how this teaching can be done is also in the process of being answered. New curricula, new textbooks, and teacher training are all in the works. How widespread will these changes be? How lasting? These are the questions we must ask ourselves now. And the answers depend largely on the willingness of those of us engaged in science education to rediscover that science is the channeling of wonder into a learned process of real-world investigation. □

[88]Based on L. Loeppke, "Back to the Future with Science Education," *Science Books & Films* 24, no. 4 (March/April 1989) 194, 262. Used with permission.

Case Study 3. New Science Goals— Just Another Fad?

The Case. Miss Jennifer Harris is the chair of the science department of Block High School in an urban school district in the northeast part of the country. It is her first year as chair, but her tenth year of teaching at Block. The science department has agreed to implement in all courses a new emphasis by incorporating the goals proposed by Project Synthesis. At a meeting of the science department, Miss Harris explains that in each science course, content will be emphasized in terms of

• Personal needs
• Societal issues
• Academic knowledge
• Career education

The teachers agree that all course syllabi should reflect these "goal clusters," and they agree to rewrite them to show this change.

About a month into the term, a student in Miss Harris's first-period class hands her a note from the student's father. The father is furious that these new science goals are being implemented. He is sick and tired of these new education fads, which seem to come and go and never produce any positive results. He wants to know why time is being spent in science classes on personal needs and career education. Can't that be done by the school counselor?

The Problem. What should Miss Harris do? How should she respond to her student's parent? Are these goals simply a new education fad, or are they grounded in defensible educational practice? □

Science Educators Should Teach about the Interaction of Science and Society by Robert Yager[89]

Recently there have been several reports concerning major shifts in the goals of science education. Helgeson, Bloser, and Howe reported that the goals of science education at the secondary level are in a period of major transition. An NSTA working paper, entitled *Science Education: Accomplishments and Needs*, identified several goals for science education for the eighties. An analysis of the paper revealed general agreement among the leadership of five groups of science educators that an increased focus upon science and society or science in society was needed. Harms and the Project Synthesis research team reported that the science and society focus is basic if K–12 science teaching is to reach a more desired state. Several college courses have been developed with such an emphasis on the interrelationships between science and society.

These developments of the past five years indicate the centrality of the science-society interface for science education as a discipline. All other goals—those that have characterized the field for the past 50 years—probably can be subsumed by the science-society goal. This new goal may provide a major justification for the study

(Continued on p. 120)

(Continued from p. 119) of science in K–12 settings and general education science requirements in college.

Defining the discipline of science education as the study of the science-society interface removes the restriction that science education is a school or collegiate program. At the same time, it does not exclude such settings as places that the interface may be effectively considered. Such a designation provides parameters for research efforts, curriculum planning, and educational programs.

Science education is defined, then, as the discipline concerned with the study of the interaction of science and society—that is, as the study of the impact of science upon society as well as the impact of society upon science. Their interdependence becomes a reality and the interlocking concept of the discipline. Research in science education centers upon this interface.

An analogy may help explain such a definition. It may help elaborate upon the advantages of such a view for the discipline. The discipline of science education, when defined as the interface between science and society, may be likened to the cell membrane which surrounds a living cell, separating the living material from its surroundings. The membrane is a dynamic one through which all material must enter and exit the cell itself. Studying the process and the factors controlling such movement, the direct involvement of the membrane in the actions can be used as a parallel in terms of science education and its role in assisting society to understand and to use science while also assisting professional scientists to understand and to affect society.

When one uses the cell membrane as an analogy, the importance of the discipline to society as a whole, to the entire scientific community, and to the future of humanity is apparent. Many biologists consider the membrane the most vital aspect (structure and function) of life itself. Similarly, such a definition of science education gives it a primary role in today's world. Defining science as "the discipline concerned with the study of the interaction of science and society" provides clear justification for science education by making it clear that science education is a vital link to the future of mankind. ☐

[89]Based on R. Yager, "Defining the Discipline of Science Education," *Science Education* 68, no. 1 (1984): 35–37. Used with permission.

Another View of Science Education by Ron Good, J. Dudley Herron, Anton E. Lawson, and John W. Renner[90]

The current "crisis" in science education has caused a flurry of activity at many levels. This brief article looks critically at the attempts by certain persons to redefine science education as the interface between science and society. We hope a continuing open dialogue on this issue will encourage the widest possible participation by those knowledgeable about and interested in public science education.

If the area of study called science education is a discipline, those practicing it should agree what it is. To develop an understanding of what science education is, the name needs to be analyzed and that analysis, we believe, consists of first describing science. Several of those descriptions follow:

> The object of all science is to coordinate our experiences and bring them into a logical system (Albert Einstein).
> The task of science is both to extend the range of our experience and reduce it to order (Niels Bohr).
> Science is built up with facts as a house is with stone, but a collection of facts is no more science than a heap of stones is a house (Jules Henri Poincaré).

We believe that the foregoing descriptions of science were synthesized into a single, excellent definition by science historian Duane Roller when he said, "Science is the quest for knowledge, not the knowledge itself."

We accept Roller's definition of science; science education, then, involves the process of educating students in how to quest—or search—for knowledge. That description, however, is not unique to the natural sciences until we specify that those searches for knowledge are made in the biological and physical worlds. Science education then becomes the discipline that bears the responsibility of leading students to learn how to search for knowledge in the biological and physical worlds. There is little doubt in our minds that once how to conduct a quest is learned, it will be used by students in other content areas *if* they are allowed. Of course, the process of science generates declarative knowledge; therefore, that declarative knowledge also falls under the rubric of science education. We, therefore, choose to define science education as the discipline devoted to discovering, developing, and evaluating improved methods and materials to teach science, that is, the quest for knowledge, as well as the knowledge generated by that quest. ☐

[90]Based on R. Good, J. D. Herron, A. E. Lawson, and J. W. Renner, "The Domain of Science Education," *Science Education* 69, no. 2 (1985): 139–141. Used with permission.

Problems and Extensions

- Design a K–12 scope and sequence of science goals. The scope and sequence should delineate elementary, middle, and high school, and show the connection and articulation of goals across the three grade groupings.
- Find or create examples of science activities that illustrate each of the following goal clusters: personal needs, societal issues, academic preparation (process and content of science), and career education.
- Find two textbooks in your field (e.g., biology, chemistry, earth science, or physics), and examine them in light of the Project Synthesis goal clusters. To what extent is each goal cluster present? What is the relative emphasis of each goal cluster? Were there other goals apparent in the textbooks? What is your overall assessment of the two books with respect to goal emphasis? Which book would you want to use?
- Examine one textbook in one subject area from at least two different eras of science teaching in terms of the Project Synthesis goal clusters. To what extent is each goal cluster present? What is the relative emphasis of each goal cluster? Were there other goals apparent in the textbooks? What is your overall assessment of the two books with respect to goal emphasis? Which book would you want to use?
- Design a time line of events in the history of science teaching in the last two centuries. Plot the events on adding machine tape or some other material.

RESOURCES

Archambault, Reginald D. *John Dewey on Education*. New York: The Modern Library, 1964.

The book represents a collection of Dewey's major writings on education, together with certain basic statements of his philosophical position that are relevant to understanding his educational views.

Cremin, Lawrence A. *American Education: The Metropolitan Experience, 1876–1980*. New York: Harper & Row, 1988. This is the final volume of Cremin's history of American education. The book covers the leading educational theorists of the era; it traces the development of educational programs in schools and college, and it portrays the role of education in the lives of individual Americans.

DeBoer, George E. *A History of Ideas in Science Education: Implications for Practice*. New York: Teachers College Press, 1991.

If you need to write a report on any aspect of the history of science education in the United States from the mid-nineteenth century to today, this is the book to get. DeBoer's book traces the history of ideas in science education by examining the works of scientists, researchers, professors, and classroom teachers to put together an account of science education that exists nowhere else.

Duschl, Richard A. *Restructuring Science Education*. New York: Teachers College Press, 1990.

This book reviews the last 30 years of science education and in the process clarifies the philosophical orientation that is most prevalent in today's school science programs. It also suggests that science teachers should consider the structure and development of theories as an organizing framework for curriculum and instruction.

Fulfilling the Promise: Biology Education in the Nation's Schools. Washington, D.C.: National Academy Press, 1990.

This book presents a vision of what biology education in our schools could be as well as specific practical recommendations. The book traces the development of biology education in the nation's schools to its present state; problems that have impeded the improvement of biology education; how to prepare biology teachers; and what is needed to improve biology education.

Hurd, Paul DeHart. *New Directions in Teaching Secondary School Science*. Chicago, Ill.: Rand McNally, 1970.

This book examines trends in secondary school science during the golden era of science education, and the perspective they provide for innovative change.

James, Robert K., and V. Ray Kurtz. *Science and Mathematics Education for the Year 2000 and Beyond*. Bowling Green, Ohio: School Science and Mathematics Association, 1985.

A collection of articles responding to a central theme: "In a world which is increasingly scientific and technological, what science and mathematics knowledge and skills will be of greatest value?" The book also contains the reactions of a wide range of in-service practitioners to the suggested curricula.

Harms, Norris C., and Robert E. Yager. *What Research Says to the Science Teacher*, Vol. 3. Washington, D.C.: National Science Teachers Association, 1981.

This is the report of Project Synthesis with chapters devoted to the overall purpose, organization, and procedures used, to biology education, physical science, inquiry, elementary school science, and science-technology-society.

Murnane, Richard J., and Senta A. Raizen, eds. *Improving Indicators of the Quality of Science and Mathematics Education in Grades K–12*. Washington, D.C.: National Academy Press, 1988.

A report of the Committee on Indicators of Precollege Science and Mathematics Education (established by the National Research Council). The report discusses recommendations for improved ways of monitoring the condition of education in the fields of science and mathematics.

Nelkin, Dorothy. *Science Textbook Controversies and the Politics of Equal Time*. Cambridge, Mass.: MIT Press, 1977.

This book discusses and analyzes the creationism movement and its expression of basic and widespread criticism of science and its pervasive influence on social values.

CHAPTER
4
THE SCIENCE CURRICULUM: MIDDLE SCHOOL PATTERNS AND PROGRAMS

In the next two chapters we will explore the science curriculum in the secondary schools by focusing first on the curriculum of middle and junior high schools, and then on the curriculum of the senior high school, as well as science curriculum perspectives in other countries. You have discovered from the last chapter that the science curriculum has evolved over the past century as a result of social, political, economic, and educational forces. Focusing on the science curriculum early in your study of science teaching will give you the big picture—a holistic view—of science in the school curriculum. From this vantage point, the methods, strategies, development of lesson plans, and management techniques can be viewed in the context of this larger framework.

We will begin our study of the curriculum by examining the patterns—past, present, and future—of the curriculum in the middle or junior high years. The middle or junior high curriculum, which follows the elementary science curriculum, is the crucial part of the K–12 science program. All students study some science during these years; this is the time during which students are encouraged or discouraged from developing a positive relationship with science. What are the middle or junior high curriculum patterns? What do you think should be emphasized in science for preadolescent students? What will help students become scientifically literate citizens in the twenty-first century?

PREVIEW QUESTIONS

- What is meant by the term *curriculum*?
- What were the characteristics of the junior high science curriculum reform projects of the 1960s?
- What is the desired or ideal middle school science program?
- How do middle schools differ from junior high schools?
- What are the characteristics of exemplary middle school science programs?

THE CURRICULUM

First, what is the curriculum? The curriculum is more than a textbook, a series of textbooks, or a curriculum guide. The curriculum includes the goals, objectives, conceptual themes, day-to-day learning activities, ancillary materials (films, videos, computer programs), and evaluation. You might think of the curriculum from three vantage points; the experience of the student, of the teacher, and of an observer. It is quite possible for the curriculum to be perceived one way by the teacher, experienced quite differently by the student, and reported in a third manner by an observer. Curriculum takes place in an environment, and no matter what the environment is, it will influence the curriculum in a variety of ways. For the sake of communication we will use the following definition of curriculum:

The curriculum is the plans made for guiding learning in schools, usually described in documents (textbooks, curriculum guides, course syllabi, lesson plans) of several levels of generality, and the execution of those plans in the classroom, as experienced by learners; those experiences take place in a learning environment (classroom, laboratory, outdoors) which also influences what is learned.[1]

As you read about and discuss the science curriculum, realize that the science curriculum can only be fully appreciated in the context of real schools, teachers, and students.

In the study of the science curriculum, the impact of the teacher cannot be overlooked. Even with descriptions of course outlines for year-long courses, or school system curriculum guides for K–12 programs, curriculum is in actuality a dynamic concept involving interaction patterns among teachers, students, and learning experiences. Thus, these sections include descriptions of a few exemplary science curriculums described by practicing science teachers. As you explore these sections, and as you visit schools and observe the science curriculum in action, ask yourself, How does the teacher influence the curriculum? How will you influence the curriculum? How do the students influence the curriculum? How does your philosophy of education influence the curriculum?

[1]Based on A. A. Glatthorn, *Curriculum Leadership* (Glenview, Ill.: Scott, Foresman, 1987).

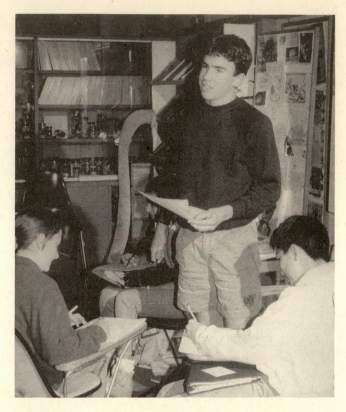

FIGURE 4.1 The science curriculum: a dynamic process involving the interaction of teachers and students in learning activities. © 1990 Tony Freeman.

What kind of science curriculum will you create in your classroom?

INQUIRY ACTIVITY   4.1

Exploring Science Curriculum Materials

It is important to become familiar with science curriculum materials, the textbooks as well as the ancillary materials. In this inquiry activity you will investigate two textbooks representing the same curriculum field, for example, earth science, biology, chemistry, or physics. You can examine either two contemporary books, or choose a contemporary book and compare it with one that was developed in the same discipline during the golden age of science education. Your investigation will be designed to identify and compare differences and similarities in the curriculum materials you explore.

Materials

Two science textbooks
Optional: ancillary materials (tests, lab manuals, teacher's guide, audio-visual materials, computer software)

Procedure

Analyze each text using the chart shown in Figure 4.2. A rating scale on a five point spread will allow you to gather some data for comparative purposes. If other individuals or groups chose the same books, you will be able to compare your interpretations.

Minds on Strategies

1. To what extent are the Project Synthesis goals met in the curriculum materials you investigated?
2. How are contemporary curriculum materials different than those produced during the golden age of science education? (You may have to collaborate with others who studied books in this era.)
3. Is the science-technology-society theme evident as an integral part of the curriculum materials? How is this accomplished?

Evaluation Criteria	Textbook 1 _____					Textbook 2 _____				
	Poor				Outstanding	Poor				Outstanding
	1	2	3	4	5	1	2	3	4	5
Goals: (Project Synthesis)										
1. Evidence is present that science should prepare individuals to utilize science to improve lives in a technological world.										
2. Evidence is present that science should produce informed citizens prepared to deal with science-related societal issues.										
3. Evidence is present that science should provide adequate knowledge for those pursuing science careers.										
4. Evidence is present that science should give all students an awareness of the nature and scope of a wide variety of science and technology careers.										
Content of Text										
1. Accuracy 2. Depth 3. Breadth 4. Processes of Science 5. Nature of Science 6. Understandability 7. Other										
Laboratory Investigations										
1. Process oriented 2. Integral part of curriculum 3. Other										
Features										
1. Design of the book 2. Use of photography and graphics 3. Use of learning tools for students 4. Readability 5. Interest 6. Other										
Ancillaries										
1. Laboratory manual 2. Study aids 3. Computer software 4. Films, videos, filmstrips 5. Other										
Summary										
1. Overall Point Total 2. Strengths 3. Weaknesses										

FIGURE 4.2 Curriculum Evaluation Chart

MIDDLE SCHOOL AND
JUNIOR HIGH SCIENCE CURRICULA

Up until the late 1960s, the dominant attendance pattern for early adolescent students (ages 11 to 14) was the junior high school, typically grouped in grades 7, 8, and 9. The first middle school opened in Bay City, Michigan, in 1950.[2] The middle schools didn't emerge on the American school scene until the 1960s, and then grew rapidly in the 1970s and 1980s. The middle schools dominate the educational organization of early adolescents today, and in most cases comprise grades 6, 7, and 8 (some arrangements of middle schools also include 7 to 8, or 7 to 9).

As was discussed earlier, the junior high school emerged in the period 1910–1920. For many years the science curriculum of the junior high school consisted of three years of general science. General science programs generally included units at each grade level in physics, chemistry, biology, geology, meteorology, and astronomy.

In the 1960s several curriculum reform projects were developed that influenced the junior high science curriculum. Because the junior high was the typical pattern in American school systems for the early adolescent, these reform projects were geared for junior high schools, not middle schools. Let's look very briefly at three of these projects. The three chosen have had a powerful impact on junior high science for many years and have influenced textbook publishers' junior high science series from then until now.

Junior High Reform Projects

Although the following projects were used in middle schools, the dominant organization was the junior high school when they were developed, field-tested, and implemented into school systems in the United States. We will discuss the middle school later in this section.

Introductory Physical Science (IPS)

It was suggested that the typical junior high science course in the 1960s was general science, in which students studied a variety of topics: chemistry, climate, geology, physics. A new course was proposed by developers at Educational Development Services

in Watertown, Massachusetts, which represented a radical departure from the "general science" approach (see Table 4.1). Instead, with funding from NSF, the developers created the Introductory Physical Science (IPS) course around a central theme. The theme was the development of evidence for an atomic model of matter. All the content in the IPS course was selected and organized to help students build the concept and then use the atomic model of matter to predict ideas about heat and molecular motion.

In the IPS course students performed many experiments (over 50). The experiments were integrated into the text, and were designed to involve the students in gathering data and testing hypotheses about the physical world.

The course was also built on the assumption that students would need to utilize the skills of inquiry to develop the main concept. The hands-on experiments developed such inquiry skills as observing, predicting, and analyzing data. Student obtained data was used

TABLE   4.1

COMPARISON OF IPS
WITH GENERAL SCIENCE

General Science Curriculum	Introductory Physical Science (IPS)
Grade 7	**Grade 7, 8, or 9**
Unit	Chapter
1. Water (suggested time six weeks)	1. Introduction
2. Rocks and Soil (six weeks)	2. Quantity of Matter
3. Air (seven weeks)	3. Characteristic Properties
4. Fire (six weeks)	4. Solubility and Solvents
5. Trees (six weeks)	5. The Separation of Substances
6. The Human Machine (five weeks)	6. Compounds and Elements
Grade 8	7. Radioactivity
	8. The Atomic Model of Matter
1. Earth and Its Neighbors (four weeks)	9. Sizes and Masses of Atoms and Molecules
2. Weather and Climate (five weeks)	10. Molecular Motion
3. Water, Its Effects and Its Uses	11. Heat
4. Gardening (seven weeks)	
5. Safety and First Aid (nine weeks)	

[2]K. T. Henson, *Methods and Strategies for Teaching in Secondary and Middle Schools* (New York: Longman, 1988), p. 371.

TABLE 4.2

ESCP ORGANIZING THEMES

Behavioral Themes	Conceptual Themes	Historical Themes
Science as inquiry	Universality of change	Historical development and presentation
Comprehension of scale	Flow of energy in the universe	
Prediction	Adjustment to environmental change	
	Conservation of mass and energy in the universe	
	Earth systems in space and time	
	Uniformity of process: a key to interpreting the past	

to develop ideas about matter and energy. The authors also believed that students would develop a better understanding of inquiry and science if they studied a topic in depth, as opposed to skipping from one topic to another.

Another significant aspect of the IPS course (as well as with the other NSF reform programs) was the development of apparatus and materials kits that accompanied the IPS textbook. The course could not be studied without using the materials. In addition to the laboratory kits, the IPS curriculum also included laboratory notebooks and a teacher's guide.

The Earth Science Curriculum Project

The Earth Science Curriculum Project (ESCP) designed a course that was also a radical departure from the contemporary earth science course, which was offered in either the eighth or ninth grade. At the time (1962), earth science was not a major curriculum offering in the junior high school. States such as New York, Connecticut, and Massachusetts had strong earth science programs and relied on textbooks that divided earth science into geology, astronomy, and meteorology. Oceanography was not a major topic in these early earth science programs.

ESCP was developed by scientists (geologists, oceanographers, astronomers, meteorologists, pale-ontologists), high school teachers, and science educators. The project produced a student textbook, teacher's guide, film series, field guides, and a specialized equipment package.

The ESCP approach to curriculum was designed to give students an understanding of earth processes, of the methods of science, and of what earth scientists do. The ESCP approach emphasized inquiry, discovery, and interpretation of student-obtained data. The laboratory activities, which originally during the field-testing phases appeared in a separate laboratory manual, were integrated into the ESCP textbook, *Investigating the Earth* (Figure 4.3).

ESCP was organized around ten overarching themes (behavioral, conceptual, and historical) as shown in Table 4.2.

A significant contribution of the ESCP course was the organization of the content of the earth sciences. Typically, content is organized into traditional subject matter divisions (geology, meteorology, oceanography, and astronomy). Instead the ESCP course was organized into four interdisciplinary units as follows:

Unit I. The Dynamic Earth
Chapters: The Changing Earth, Earth Materials, Earth Measurement, Earth Motions, Fields and Forces, Energy Flow
Unit II. Earth Cycles
Chapters: Energy and Air Motions, Water in the

Figure 4–1

Figure 4–2 *An imaginary arrow through the same two stars of the Big Dipper always points toward the North Star, Polaris.*

The Many Motions of the Earth

4–1 Investigating motions in the sky–Sky Watch.

When you are outdoors on a clear night and look up at the stars, your first reaction may be that they are beautiful, gleaming brightly in the darkness. But the stars have many things to tell you. For one thing, you know that the stars will have moved if you look up at them again in an hour or two. In what directions do the stars move and how fast? How can you obtain accurate answers to these questions? The men who built Stonehenge found one way to keep track of these motions. Aren't there easier ways than this?

In Sky Watch, which you started several weeks ago, you have learned how to plot the path of the sun. This is much the same thing the observers at Stonehenge did more than three thousand years ago. How can you do exactly the same thing for the stars that you did for the sun?

Procedure

Using the materials shown in Figure 4–1, find the North Star, as indicated in Figure 4–2, and plot its position. Then plot the positions of three other stars, one in the east, one in the south, and one in the west. The brighter the stars you pick, the easier they will be to plot.

An hour later, repeat the procedure, plotting the positions of the same four stars. Try not to move the hemisphere and baseboard between observations. If the baseboard must be moved, make a marker line along one edge of the baseboard and replace it exactly on the marker line for your second set of observations. If possible, plot the positions of these stars a third, and even a fourth time. Then take to class the transparent hemisphere with the positions plotted on it and answer the followng questions.

(1)In which direction did the stars move? (2)Did any of the positions of the stars shift in relation to each other? Explain. (3)How many degrees above the horizon was the North Star?

Place the transparent hemisphere on which you recorded your observations of the star positions on top of the globe and answer the following questions.

(4)Where should you locate the position of the North Star marked on the transparent hemisphere in relation to the globe? (5)What is the relationship of the paths of the stars to latitude lines on the globe?

FIGURE 4.3 A page from the original ESCP text. Investigations were an integral part of the student's experience as shown here in the sky watch activity. From INVESTIGATING THE EARTH, ESCP. Copyright © 1967 by American Geological Institute. Reprinted by permission of Houghton Mifflin Company. Photo of stars courtesy of Lick Observatory.

Air, Waters of the Land, Water in the Sea, Energy, Moisture and Climate, The Land Wears Away, Sediments and the Sea, Mountains from the Sea, Rocks within Mountains, The Interior of the Earth

Unit III. Earth's Biography
Chapters: Time and Its Measurement, The Record in Rocks, Life—Present and Past, Development of a Continent, Evolution of Landscapes

Unit IV. Earth's Environment in Space
Chapters: The Moon: A Natural Satellite, The Solar System, Stars as Other Suns, Stellar Evolution and Galaxies, The Universe and Its Origin

The ESCP had a positive effect on the teaching of earth science in the United States. Interest in the earth sciences increased during the sixties and seventies, and as a result earth science is an integral part of the middle school curriculum today.

The ESCP text is still used, although not as widely as it was in the 1970s. Districts that do use the ESCP text use it in accelerated programs with high-ability students. The most recent edition of the text (1989) has incorporated recent developments in the earth sciences but has reflected a more traditional organization of content. The text has, however, preserved the emphasis on student inquiry and analysis of student-obtained data.

Studies done on new earth science curricula, including ESCP, showed positive results in the effect of these curricula on process and analytic skills. Although not many studies were completed, the new earth science curricula were the only reform projects that did not show a positive achievement result.[3]

The Intermediate Science Curriculum Study (ISCS)

Developed at Florida State University, ISCS was a three-year individualized science curriculum project for grades 7, 8, and 9. The ISCS program was designed to help junior high students develop an understanding of physical and biological principles of science, an understanding of science and the scientific process, and to develop the ability to use the processes of science.

The ISCS program was an individualized curriculum. The text materials were written to guide the student along at his or her own pace. In practice, most teachers organized the students into small teams, and the team moved along at its own pace. The students are led through the materials by being actively involved in hands-on experiments, answering questions posed in the text materials, and solving problems (Figure 4.4).

Student material is organized into core chapters and excursion activities (Figure 4.5). Students are instructed to proceed through the text completing activities and answering questions (in a separate science record book). Excursions are special activities referenced in the core chapters and found in the back of the book. Theoretically, the student was invited to stop and do the excursion if it looked interesting, or if it would help with the current chapter. In practice, teachers usually helped the students by suggesting which excursions were required and which were optional. The ISCS program offered junior high schools a complete three-year science curriculum.

Grade 7: Probing the Natural World (Volume 1) This grade focused on the development of the concept of energy and the development of an operational definition of energy. The focus was on physics.

Grade 8: Probing the Natural World (Volume 2) This grade focused on chemistry, and students developed a particulate model for the nature of matter.

Grade 9: Eight separate books on the following topics:

- Why You're You (genetics)
- Environmental Problems (environmental science)
- Investigating Variation
- In Orbit (space science)
- What's Up? (astronomy)
- Crusty Problems (geology)
- Winds and Weather (meteorology)
- Well-Being (health science)

These programs, and other NSF curriculum projects aimed at the junior high school, influenced the nature of the curriculum for early adolescent students. The NSF reform projects tended to be discipline oriented (e.g., physics, physical science, earth science), whereas the typical junior high science program was three years of general science. By the late seventies, general science was still the typical curriculum sequence in 70 percent of the schools, with life science 21 percent, earth science 20 percent, and physical science 13 percent. By 1986, the pattern was more evenly distributed across the content areas with about 30 percent biology and life science, 22 percent earth science, 22 percent general science, and 21 percent physical science.

[3]For a discussion of the results of the course content projects versus traditional science courses, see J. A. Shymansky, W. C. Kyle, Jr., and J. M. Alport, "The Effects of the New Science Curricula on Student Performance," *Journal of Research in Science Teaching* 20 (5): 387–404.

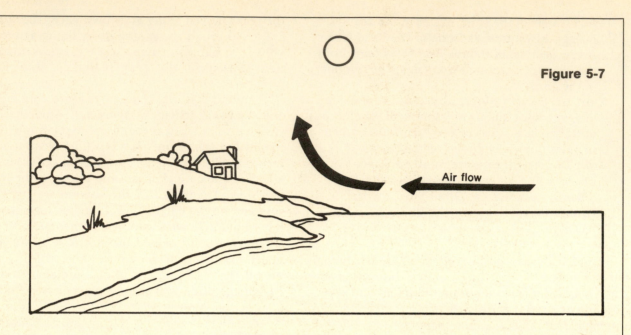

Figure 5-7

Air flow

Suppose the cool area is adding moisture to the air, as in Figure 5-7. If so, the moisture will be lifted as the air moves over the land.

☐ **5-14.** Will the increased moisture content improve chances for cloud formation?

PROBLEM BREAK 5-2

If you live near the ocean or a very large lake, you may have noticed some peculiar things about wind. Wind direction often seems to be related to the time of day. During warm, daylight hours, the wind blows from one direction. Then, during cool, night hours, it blows from the opposite direction.

In Figure 5-8 of your Record Book, indicate the wind direction you predict for the two times of day shown.

Figure 5-8

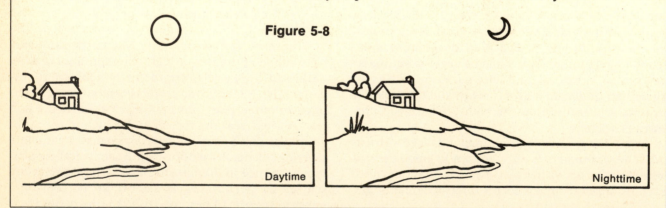

Daytime

Nighttime

FIGURE 4.4 In the ISCS program students moved through the materials at an individualized pace by reading, answering questions, doing hands-on activities, and solving problems. From WINDS AND WEATHER, ISCS. Copyright © 1972 The Florida State University. Reprinted by permission of Silver Burdett.

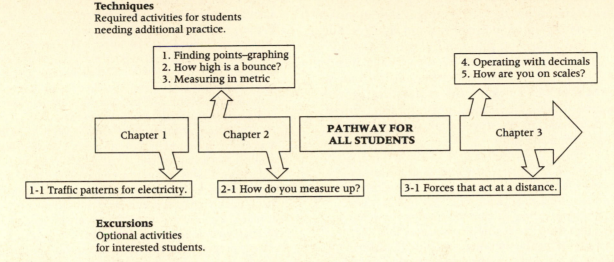

FIGURE 4.5   The Intermediate Science Curriculum Study (ISCS)

The results of studies that compared the effects of new junior high science curricula on student performance versus traditional science programs tended to favor the new science programs.[4] Physical science curricula produced positive effects on student performance in achievement and perceptions (of science). Overall, the junior high programs produced positive effects in achievement (although not in each curriculum case), perceptions, and process skills. These programs did not do as well in analytic and related skills areas, however.

[4]Shymansky, Kyle and Alport, "The Effects of the New Science Curricula," 387–404.

THE MIDDLE SCHOOL MOVEMENT

The odds are that if you are asked to observe or teach early adolescent students in your teacher education program, you will do so in a middle school. The middle school movement began with experimental or reform-minded junior high schools in the 1960s. Some of the early innovations included team teaching, modular and flexible scheduling, and interdisciplinary teaching practices. Wiles and Bondi point out that overenrollment and desegregation had as great an effect on the spread of the middle school movement as did curriculum theory or organizational planning.[5] By the mid seventies a number of national organizations began to focus on the educational problems and issues of the middle school learner. During this period the National Middle School Association was formed, and at annual meetings of organizations such as the National Science Teachers Association, an increasing number of workshops, seminars, and speeches were given on teaching middle school science. The literature on the middle school and the middle school learner exploded, and researchers began to pay attention to the special needs of the preadolescent. The National Science Teachers Association created a journal specific to middle school science called *Science Scope*, and in 1991 the National Science Middle School Teachers Association was formed.

Although the junior high had been developed in the period 1910–1920 to respond to the special needs of the preadolescent student, the original child-centered focus was lost, and junior high schools tended to resemble the high school in curriculum and organization of staff. The focus was on the content rather than on the child.[6]

The middle school educators of the seventies and eighties developed schools that have a philosophy that responds to the needs of the preadolescent learner, which includes physical development, intellectual development, social-emotional development, and moral development. The middle school learner is an emerging adolescent, experiencing changes in

[5]J. Wiles, and J. Bondi, *Making Middle Schools Work.* (Alexandria, Va.: Association for Supervision and Curriculum Development, 1986), p. 1.

[6]Wiles and Bondi, *Making Middle Schools Work*, pp. 3–5.

each of these realms. The middle school philosophy establishes the rationale for creating a school environment that addresses these needs and changes. For example the following philosophy and goal statement, typical of contemporary middle schools, incorporates the needs of these emerging adolescents to create a school that is flexible and exploratory, as well as designed to help students understand the content of the major disciplines, develop personally, and develop basic thinking skills.

Middle School Philosophy and Goal Statement[7]

The middle school offers a balanced, comprehensive, and success-oriented curriculum. The middle school is a sensitive, caring, supportive learning environment that will provide those experiences that will assist in making the transition from late childhood to adolescence, thereby helping each individual to bridge the gap between the self-contained structure of the elementary school and the developmental structure of the high school.

The middle school curriculums are more exploratory in nature than the elementary school and less specialized than the high school. Realizing that the uniqueness of individual subject disciplines must be recognized, an emphasis on interdisciplinary curriculum development will be stressed. Curriculum programs should emphasize the natural relationships among academic disciplines that facilitate cohesive learning experiences for middle school students through integrative themes, topics and units. Interdisciplinary goals should overlap subject area goals and provide for interconnections such as reasoning, logical and critical thought, coping capacities, assuming self-management, promoting positive personal development, and stimulating career awareness.

The academic program of a middle school emphasizes skills development through science, social studies, reading, mathematics, and language arts courses. A well-defined skills continuum is used as the basic guide in all schools in each area including physical development, health, guidance, and other educational activities. Exploratory opportunities are provided through well-defined and structured club programs, activity programs, and special interest courses, thereby creating opportunities for students to interact socially, to experience democratic living, to explore areas not in the required curriculum, to do independent study and research, to develop, and practice responsible behavior, and to experience working with varying age groups.

The middle school curriculum will be a program of planned learning experiences for our students. The three major components for our middle school curriculum are (1) subject content, (2) personal development, and (3) essential skills.

MIDDLE SCHOOL
SCIENCE CURRICULUM PATTERNS

Science curriculum offerings in the middle school can be viewed as traditional, or integrated and applied. These two groupings were used in a study of U.S. curricula.[8] According to the results of the study (which focused on grades 5, 9, and 12), the most emphasized of the traditional sciences in grades 5 and 9 was earth science. The most emphasized topics were the solar system and meteorology. Figure 4.6 shows the mean ratings of traditional science emphasized in grades 5 and 9. Biology, although not emphasized as much as earth science, scored higher ratings than physics and chemistry. Chemistry is the least emphasized traditional science subject in the middle school.

In the survey results for applied and integrated science, environmental science content is the most common in the science offerings in grades 5 and 9 (Figure 4.7). The only subject that is emphasized more is earth science. Note also that health was rated fairly high compared with physics and chemistry.

There are two patterns that typically describe the curriculum of the middle school or junior high school. These depend on whether the school includes grades 6 to 8, or 7 to 9. The typical patterns are as follows:

Grades 6 to 8
Grade 6: general science (Usually based on the sixth-grade edition of an elementary science textbook)
Grade 7: life science
Grade 8: earth science
Grades 7 to 9
Grade 7: life science
Grade 8: earth science or physical science
Grade 9: physical science, earth science or biology

There are two-year sequences of unified or general science available from some publishers. Some districts use these as their seventh and eighth grade

---

[7]Wiles and Bondi, *Making Middle Schools Work*, p. 11. Reprinted with permission of the Association for Supervision and Curriculum Development. Copyright 1986 by ASCD. All rights reserved.

[8]J. K. Miller, "An Analysis of Science Curricula in the United States," unpublished doctoral dissertation, Teachers College, Columbia University, New York, 1985.

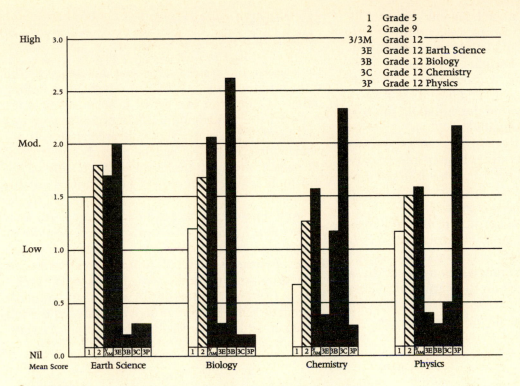

FIGURE 4.6   U.S. Mean Ratings of Traditional Science Content. Jacobson, "The Current Status of Science Curricula," in A. B. Champagne and L. E. Hornig, *The Science Curriculum* (Washington, D.C.: American Association for the Advancement of Science, 1987), p. 159. Used with permission.

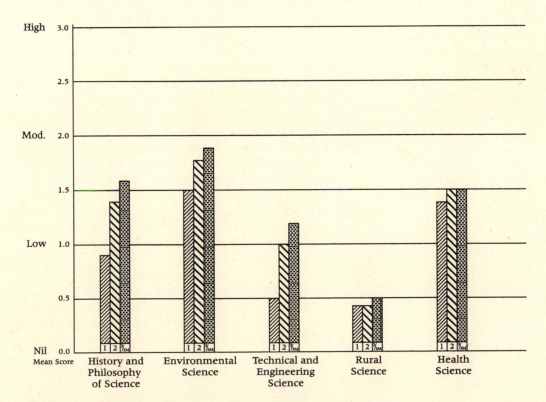

FIGURE 4.7   U.S. Mean Ratings of Applied and Integrated Science Content. W. J. Jacobson, "The Current Status of Science Curricula," in A. B. Champagne and L. E. Hornig, *The Science Curriculum* (Washington, D.C.: American Association for the Advancement of Science, 1987), p. 161. Used with permission.

science options in either of the foregoing patterns.

The middle school science curriculum that is based on the traditional model of the textbook emphasizes coverage of the discipline, as opposed to inquiry or problem solving. Although most of these programs include lists of skills, and include skill activities and investigations within the textbook, the goal statements are generally cognitive.

To see this more clearly, all you need to do is examine the outline of a typical chapter in a science textbook. Figure 4.8 outlines one chapter of the topics and features of an earth science text that is used in many school systems in the United States.

What should science education be like in a middle school? What should be emphasized? What curriculum topics, units, and subjects should be stressed? Let's examine some recommendations made by a group of science educators on the nature of middle school science, and then look at some promising practices in the teaching of middle school science.

Chapter 17: Igneous Rocks

Section A. Igneous Rocks
Goals: Students will
- learn how igneous rocks form
- learn to identify some igneous rocks

17-1 Origin of Igneous Rocks
17-2 Minerals in Igneous Rocks
17-3 Igneous Rock Classification
Skill: Classifying Rocks
Review
Think Section
Career: Technician
Investigation 17-1: Classifying Igneous Rocks

Section B. Igneous Activity
Goals: Students will
- learn about the different types of volcanoes on Earth's surface
- learn about igneous rock structures that result from volcanic activity

17-4 Volcanic Mountains
Investigation 17-2: Volcanic Eruptions
17-5 Igneous Rock Structures
Review
Problem Solving
Chapter Review

FIGURE 4.8   Outline of Chapter in an Earth Science Text

The Desired State for Middle School Science

The concept of "desired state" was developed by the authors of the Project Synthesis reports. The desired state is the opinion of a group of science educators about what the curriculum in a specific content area ought to be. The concept of a desired state was dependent on the intellect, judgment, and experience of the persons involved. Desired states for middle school and high school science will be identified in this chapter.

Although there was not a middle school group in Project Synthesis, an NSTA task force on middle or junior high science developed criteria (using the Project Synthesis model of desired states) for exemplary programs in middle school science. What follows is a summarization of the recommendations of the middle school or junior high task force in the following areas: goals, curriculum, instruction, teachers, and evaluation.[9]

Goals

The middle school science program should emphasize a holistic approach to science teaching. Exemplary middle school science programs contribute to scientific literacy by helping students develop the use of the skills of inquiry and problem solving by paying close attention to investigative and reasoning abilities, gaining independence, practical applications of the methods of science, and learning how to learn. According to the middle school task force, the science program should also help students gain an academic background in science, but the focus should be on developing a foundation in science for them as they function as citizens, rather than as science specialists. The goals of an exemplary program should also include the development of attitudes, the experience of science in interdisciplinary settings and in the world of work, and helping students relate science to the understanding of self, one's relation to the environment, and to the role of science in daily activities.

Curriculum

The curriculum of the middle school is literally caught in the middle between elementary and high school. As such it is a much debated issue. A number of educators think the curriculum should emphasize the in-

---

[9]The discussion of the desired state for middle school science is based on K. E. Reynolds, et al., "Excellence in Middle/Junior High School Science," in J. Penick and J. Krajcik, eds., *Focus on Excellence: Middle School/Junior High Science*, vol. 2, no. 2 (Washington, D.C.: National Science Teachers Association, 1985), pp. 4–9.

terdisciplinary nature of science, and the selection of content should be related to the contemporary needs of preadolescent students, as well the fact that they will be future citizens and in that capacity will need to make decisions and solve problems. Thus, some of the proposals include a curriculum that is heavily oriented toward the science-technology-society theme. Topics such as space exploration, health science, earthquake preparedness, weather phenomena and storms, and animals in the local environment are being included as content topics. In Chapter 6, we will explore the science-technology-society theme, and give examples of curricula and specific lessons that exemplify this approach. You may want to skip ahead to Chapter 6 to explore these ideas.

The desired state of the middle school curriculum is also characterized by the following:

- Science processes focusing on inquiry
- A balance among life, earth science, and physical science
- Practical applications and everyday experiences incorporated into science topics
- Issues in science and society, career awareness, historical aspects of science, as well as the nature of the science community
- Cognitive, affective, and psychomotor learning goals and activities
- An opportunity for student input to curriculum topics, and for independent work
- The interdisciplinary nature of science

Instruction and Learning

Instruction in the middle school should take into account the developmental characteristics of the preadolescent child, realizing that most of them will be

FIGURE 4.9 To what extent should middle school science involve students in hands-on investigations, working in collaborative small groups on topics that have relevance to their lives and societal problems? What do you think? © Daemmrich.

at the concrete and transitional levels in cognitive development. The curriculum should also take into account the varying student learning styles among preadolescents (see Chapter 2). Exemplary science programs should offer a variety of instructional strategies (see Chapters 7 to 8 for specific recommendations). Learning should focus on concrete experiences. Technology and a variety of supplemental teaching aids such as multiple textbooks, microcomputers, hands-on materials, and visual-aids should be central in the learning process. Middle school students should be involved with community resources, including field trip sites, outdoor education programs (see the discussion below of OBIS, Project Wild, and Project Learning Tree), and the opportunity to interact with adult women and men outside the school who can serve as role models.

Instruction should provide opportunities for the students' social growth, for clarifying values (especially on socially and personally relevant topics in the STS realm), and for the development of independence in thought and action. Instruction should involve students in small group learning activities as well as independent study and research projects.

Teachers

Teachers, according to the middle-junior high science task force, are the most significant factor in creating the desired state for middle school science programs. They report that "teachers in exemplary programs are dynamic, thoughtful, young at heart, and eager to learn with their students."[10] Teachers should have a broad background in science and should be seen more as a generalist than as a specialist. They should have a strong grasp of how science relates to individuals and society, and how technology and careers intersect with science teaching. Teachers should have specialized preparation in the teaching of science for the preadolescent and should have deep understanding of the psychosocial development of the student at this age. These teachers also understand how the middle school provides the transition from elementary to high school. Teachers are creative: they are able to make optimal use of classroom space and are able to engineer an interesting classroom given existing facilities, materials, apparatus, time, budget, and community resources.

Evaluation

Exemplary programs should include a provision for continual evaluation of the program. Evaluation does not rest on simply cognitive scores on achievement tests. The evaluation of the science program should

---

[10]Penick and Krajcik, *Focus on Excellence*, p. 2.

include cognitive, affective, and psychomotor measures. Students should be asked to provide feedback on what they like and dislike in the program, and teachers should use this data and their own reflective processes to evaluate the program's effectiveness. Program evaluation should include process and product. Successful evaluation leads to improvements in the quality, efficiency, effectiveness, and functioning of the program.

The desired state for middle school science provides us with a vantage point as we look at middle school science programs. We can ask how the current middle program compares with the criteria identified in the desired state.

EXEMPLARY MIDDLE SCHOOL SCIENCE PROGRAMS

There are a number of exciting programs that you should become familiar with as you formulate your ideas about science education. One of the sources of exemplary science education programs is the National Diffusion Network (NDN). The NDN is a federally funded system that makes exemplary education programs available for adoption by schools, colleges, and other institutions. To be listed, the program must provide evaluation data showing that the program is effective in the school setting in which it was developed or field tested, and that it could be transported and used in other schools. As of 1987, about 450 programs had been approved, and 82 were receiving Federal dissemination funds to help other schools adopt them. The NDN publishes booklets describing programs that work. (You can obtain, for free, awareness materials, usually a brochure or sample activities from the exemplary curriculum. Write to the National Diffusion Network to obtain information on exemplary programs: Office of Educational Research and Improvement, U.S. Department of Education, 555 New Jersey Avenue, N.W., Washington, DC 20208).

We will now look at a number of programs that fulfill some or all of the criteria cited previously. Some of the programs were developed by teachers and science education specialists at the local school or system level; others were developed by science education research and development teams. A few of the programs described are available through the National Diffusion Network.

Unified, Integrated Programs

The traditional approach to the organization of the middle school science curriculum is a plan of separate courses of earth, life, and physical science. An alternative approach is to develop a two- or three-year curriculum sequence around a unifying theme or set of conceptual themes.

PRIS2M

The City School District of Rochester, New York, has developed an integrated K–12 science program called R.I.S.E. (Rochester Integrated Science Education).[11] The elementary program consists of a K–6 general education program that integrates a textbook program with a hands-on activities, writing, and language arts experiences. PRIS2M (Program for Rochester to Interest Students in Science and Math), is a two-year integrated sequence integrating science and math and is followed at the high school level by the U.S.E.-Rochester program (Unified Science Education). The entire Rochester program is based on four underlying beliefs:[12]

- Science is a way of learning about the world
- Science knowledge can be generated through the application of process skills
- Students learn science best by actively "doing" science
- Grade level placement of activities should be consistent with recent research regarding developmental levels of learning

According to Benjamin Richardson and his colleagues, PRIS2M came about as a challenge to increase the number of minority students who would be able to pursue careers in engineering. With the cooperation of the Xerox Corporation and the school system, the goal of PRIS2M was to create an exciting, meaningful, and relevant science and math experience as an alternative to the traditional, and often theoretical, junior high science program.

PRIS2M consists of nine units of instruction for grades 7 and 8. The units, rather than focusing on the

---

[11]This discussion based on B. Richardson, et al., "PRIS2M," in *Focus on Excellence: Middle School/Junior High Science*, vol. 2, no. 2 (Washington, D.C.: National Science Teachers Association, 1985) pp. 37–50.
[12]Richardson, et al., "PRIS2M," p. 37.

traditional disciplines, focus instead on the nature of science, scientific concepts, processes, problems and phenomena, issues and careers in science, and the interrelationships between science, technology, and society.

The program assumes that most students will be operating at the concrete operational level; therefore the curriculum is highly manipulative and materials intensive. The focus is on the development of basic and integrated science process skills, as well as on problem solving. Some of the curriculum highlights of PRIS2M are as follows:[13]

- Students are encouraged to make choices of what they learn; the curriculum provides alternative activities for students
- Each student carries out a long-term science project each year, thereby helping the student understand the nature of science
- Student attitudes toward science and math are improved by involving them in activities that are of interest to middle or junior high students

GEMS (Great Explorations in Math and Science)

The Lawrence Hall of Science, University of California, Berkeley, has produced a series of curriculum units based on successful programs conducted at the Hall. GEMS is packaged as a series of teacher's guides that detail the lessons for each unit.[14]

Each unit focuses on a main topic, such as paper towel testing, and then develops skills and themes around the topic. For example in the paper towel testing project, students are involved in designing controlled experiments, measuring, recording, calculating, and interpreting data, while at the same time finding out about consumer science, absorbency, wet strength, unit pricing, cost-benefit analysis, and decision making.

Some of the units in the GEMS program include

- Paper Towel Testing
- Chemical Reactions
- Discovering Density
- Animals in Action
- Quadice (A mathematical game that encourages students to think and talk about mathematical relationships.)
- Acid Rain
- Global Warming and the Greenhouse Effect

Each of the GEMS units uses unique materials and activities to involve students in discovery activities. For example, in the Chemical Reactions unit, students observe chemical changes (bubbles, changes in color and heat) in a Ziplock bag. Students observe changes and design experiments to explain their observations.

The GEMS units are excellent models for units that can be infused into the ongoing science curriculum.

Science and Technology: Investigating Human Dimensions

This three-year program, developed by BSCS, is a project for middle school students that will integrate science and technology. The program is an extension of the K–6 Science, Technology, and Health program (discussed earlier in the chapter). According to the developers, the program will be used in grades 5 to 9. Here are some of the program characteristics:[15]

- A focus on the development of the early adolescent
- Strategies to encourage the participation of female, minority, and handicapped students
- An emphasis on reasoning and critical thinking
- Cooperative learning as a key instructional strategy
- An instructional model that enhances student learning
- A conceptual approach to science and technology fields
- The inclusion of science-technology-society themes

(Information about this program can be obtained from Biological Sciences Curriculum Study (BSCS), The Colorado College, Colorado Springs, CO 80903.)

Improving Urban Middle School Science

This NSF funded middle school science program targets the problems and needs of preadolescent urban students. Seventh and eighth grade program modules integrate scientific concepts and understandings from the physical, human and health, life, and earth sciences within the science, society, and technology framework. (Information about Improving Urban Middle School Science can be obtained from Educational Development Center, Watertown, MA.)

Earth Science Programs

There are many examples of outstanding earth science programs. Recent thinking by earth science educators has suggested that there is a need to reformulate the

[13]Richardson, et al., "PRIS2M," p. 43.
[14]To Build a House: GEMS and the Thematic Approach to Teaching Science Great Explorations in Math and Science. (Berkeley, Calif.: Lawrence Hall of Science, 1991)

[15]L. Trowbridge and R. Bybee, *Becoming a Secondary School Science Teacher* (Columbus, Ohio: Merrell, 1990), p. 282.

traditional perceptions that teachers hold of the earth sciences. Programs that are exemplary in earth science will design curricula in which students[16]

1. Have gained knowledge of facts, concepts, and principles related to the major unifying themes in the earth sciences
2. Are able to use and understand both holistic and reductive scientific methods as ways to acquire new knowledge
3. Are able to process relevant information and make responsible decisions regarding science and technology issues
4. Are aware of careers in the earth sciences and how the earth sciences affect such non-earth science careers as law, politics, and economics
5. Have developed an interest in, and critical attitudes toward, science and technology in society

The exemplary curriculum in earth science is organized around unifying themes, problem solving, and the relationship between earth science and society. The American Geological Institute, one of the major organizations that advocates a new approach to teaching earth science, suggests that the traditional divisions of rocks, air, ice, and oceans need to be replaced with a holistic view of the earth sciences. A holistic view would focus on the interaction of human beings and the earth and would lead to investigations of real problems and issues. Students would be involved, for instance, in humanity's role in exploiting and potentially destroying resources and the environment.

With these ideas in mind, here are brief descriptions of two exemplary earth science programs.

Geology Is

This program provides geosciences learning activities based on an interdisciplinary approach.[17] It includes a broad range of materials and media-delivery instruments that promote variety in teaching and learning. The program emphasizes the content of geology, as well as the social implications in the wise use of earth resources. The content of the program is designed to help students become more responsible consumers of Earth's resources and make more informed decisions for the future regarding energy, geologic hazards, and land use. The program is organized around five units:

- Introduction to Geology
- Earth Materials
- Observing the Earth
- Internal Processes
- External Processes

Each unit contains multiple chapters, with the five units constituting a full year course. Each unit contains text material, laboratory exercises, and activities, as well as objective and subjective tests. Slide tapes, films, and videotapes are provided as well. Instruction encourages small group and individual exploration. Because of the importance of outdoor applications, off and on campus field experiences are outlined.

The materials are available through the National Diffusion Network. Awareness materials are available at no cost (Write to Geology Is, O'Fallon Township High School, 600 South Smiley, O'Fallon, WA 62269. Geology Is is published by Kendall-Hunt, Dubuque, Iowa.)

Marine Science Project: For Sea

The developers of this program point out that by the year 2000, three out of four Americans will live within an hour's drive of the sea or the Great Lakes.[18] The developers at the Marine Science Center in Poulsbo, Washington, have created curriculum materials for grades 6 to 8, as well as for grades 2 and 4 in the elementary school, and grades 9 to 12 at the high school level. There are curriculum guides for each grade level containing teacher background material for each activity, student activity, and text material, as well as a listing of vocabulary words and a selected bibliography of children's literature of the sea and additional information books.

The curriculum materials can be infused into the existing science curriculum or can be used to replace the existing science program. Awareness materials are available from the Marine Science Center, 17771 Fjord Drive, Poulsbo, WA 98370. The program has also been approved by the National Diffusion Network.

Life Science Programs

Life science for preadolescent students ought to be oriented to help them (a) understand themselves and other human beings as distinct organisms, (b) appreciate the universal need to be in touch with their

---

[16]Rodger Bybee, et al., "Excellence in Earth Science Education," in J. E. Penick, ed., *Focus on Excellence: Earth Science,* vol. 3, no. 3 (Washington, D.C.: National Science Teachers Association, 1984), p. 7.

[17]U.S. Department of Education, *Science Education Programs That Work* (Washington, D.C.: Office of Educational Research and Improvement, 1988), p. 5. You can write for this document: 555 New Jersey Avenue, NW, Washington, DC 20208.

[18]U.S. Department of Education, *Science Education Programs That Work,* p. 9.

own nature, and (c) learn to live in harmony with nature and to minimize the separation between the human and natural environments.[19] A desirable life science program will focus on many if not all of these:[20]

- Human adaptation and alternative futures
- Biosocial problems and issues
- Inquiry processes unique to life science disciplines
- Decision making involving life knowledge in biosocial contexts
- Career awareness
- A problem-centered curriculum
- The use of the natural environment, community resources, and students themselves as foci for study
- Cooperative, small-group work on problems and issues

There are many examples of life science programs for the middle school that meet most of the foregoing criteria. Here are descriptions of two.

Wildlife Inquiry through Zoo Education (WIZE)

This grades 7 to 9 program, developed at the Bronx Zoo, New York, explores issues related to wildlife survival in the twenty-first century.[21] The program uses a nontraditional, multidisciplinary approach to help students understand concepts related to population ecology, wildlife conservation, and species survival. The program consists of two modules: Module I (Diversity of Lifestyles) and Module II (Survival Strategies). The program, which involves about 15 weeks of instruction, is a hands-on approach in which students learn that animals are members of populations that interact with one another, and that the ecological processes affecting animals also affect humans.

The instructional materials include student resource books, photo cards, discovery cards, worksheets, a teacher's manual, audio cassettes, filmstrips, and wildlife games. The program is activity oriented and includes zoo visits. The two modules form a continuum in the study of wildlife ecology, although either could be used separately in a life science or general science program.

The materials have been approved by the National Diffusion Network and are available from Curator of Education, Bronx Zoo, New York Zoological Society, 185th Street and Southern Boulevard, Bronx, NY 10460.

Stones and Bones

This program, developed by the Los Angeles Unified School District under the direction of Sid Sitkoff, is an interdisciplinary program that focuses on active student involvement through laboratory activities. The program has three strands, or pathways, designed to study human beings: a general science pathway, a biology pathway, and a semester course pathway. The general science pathway is designed for the non-college-oriented student. The program consists of 20 laboratory investigations on topics such as geologic time, measuring radioactivity, mapping, the behavior of primates, and replica casts of fossil hominids. The students also simulate an archaeological excavation. The biology pathway is a four to six week introduction to physical anthropology. The program consists of 11 investigations focusing on topics such as primate behavior and distribution, interpreting archaeological records, primate locomotion and morphology, and replica casts of fossil hominids. The semester course involves the students in an in-depth study of the development of humankind. Through a series of laboratory investigations, students study phylogeny through time, continental drift, the locomotion and behavior of primates, and classification and morphology, as well as examine 14 fossil replica casts of *Australopithecus*, Homo erectus, Neanderthal, and Cro-Magnon.

Stones and Bones has been approved by the NDN, and awareness materials are available from the Los Angeles Unified School District, Office of Instruction, 450 N. Grand Ave., Los Angeles, CA 90012.

[19]Based on P. D. Hurd, "Biology Education," in N. C. Harms and R. E. Yager, eds., *What Research Says to the Science Teacher*, vol. 3 (Washington, D.C.: National Science Teachers Association, 1981), p. 13.
[20]Based on Hurd, "Biology Education," pp. 28–29.
[21]U.S. Department of Education, *Science Education Programs That Work*, p. 15.

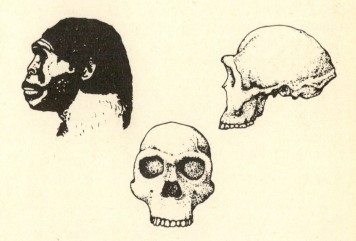

FIGURE 4.10 The Stones and Bones Curriculum involves middle school students in an interdisciplinary study of geology, biology, and physical science.

Physical Science Programs

The characteristics of desirable physical science programs will enable students to pursue individual goals and interests by involving them in projects and activities that relate physical science content to real-world problem situations. Desirable programs will help students overcome their fear of the physical sciences, especially because of the mathematics that it involves.

Sci-Math

This program is designed to help students apply mathematics to introductory science.[22] The course can be used as a physical science course for eighth or ninth grade students. The NDN suggests that it can be used for high achievers at the seventh-grade level, and slow learners at the eighth- or ninth-grade levels.

The program consists of two modules. Module 1 focuses on the mathematics involved in prealgebra, dealing with arithmetic and the logic of operations. Module 2 explores how algebraic equations express proportions and studies the graphical representation of proportions.

There are 23 hands-on activities in the program. The problems and activities deal with variables that are familiar to the student, and materials used in the investigations are readily available and inexpensive, such as rulers, string, pennies, spoons, jars, and masking tape. (Awareness materials are available from Sci-Math Director, Education & Technology Foundation, 4655 25th Street, San Francisco, CA 94114.)

Informal Science Study (ISS)

This program offers an alternative and supplement to the physical science course.[23] The ISS materials present several miniunits based on the students' involvement with popular park rides, sports, and playground experiences to develop physical science concepts. The program topics include motion, acceleration, relativity, forces, gravity, time, graphing, conservation of energy, and frames of reference.

Each miniunit is designed to involve the students in dialogue through an introduction and a review and application of the physical science concepts in low-key, nontechnical language. Terms are kept to a minimum and are introduced only after instruction. Some of the units include laboratory activities using toys and playground equipment.

The developers have shown that students significantly increase their knowledge and comprehension of science concepts after as few as three weeks of instruction. (These National Diffusion Network materials are available from Informal Science Study, University of Houston, Room 450 Farish Hall, Houston, TX 77004.)

---

[22]U.S. Department of Education, *Science Education Programs That Work*, p. 12.
[23]U.S. Department of Education, *Science Education Programs That Work*, p. 7.

Case Study 1.
A National Science Curriculum

The Case. Albert Shubert is a prominent science educator who is the chief education officer of the prestigious American Science Commission on Education. In a recent international study it was concluded that Japan, China and the Soviet Union had a strong national commitment to science and mathematics education. Shubert, an advocate of a rigorous, compulsory, national science curriculum for American youth, said in a major American newspaper that "the U.S. has no comparable commitment to science and mathematics education. To ignore the fact that these countries (Japan, China and the Soviet Union), whose history, resources, politics, economics and culture is so different, have made strong national commitments to science and mathematics, is to be left behind. To assume our ways are better is to be arrogant."

Shubert went on to say that science education in the United States should involve national leadership, national planning, and national policy making.

The Problem. Should the United States have a national science curriculum? Do you agree or disagree with Shubert? Why? Is the American decentralized system of education an ineffective model? Should the United States move toward a more centralized system? □

Case Study 2.
Unified Science

The Case. Sarah Jenkins is the science department head in a large high school in an urban area. During the summer the new principal of the high school holds a two-day retreat for all the department heads to discuss plans for the year and curriculum changes. Sarah Jenkins, like all the other department heads, wonders what the principal means by "curriculum changes." At the meeting the principal, a middle-aged woman with a Ph.D. in curriculum and administration, leads the group in a discussion of how the curriculum could be changed to make it more interesting and relevant to students. When the discussion gets to the science curriculum, a proposal that emerges from the group is the possibility of integrating the separate subjects of biology, chemistry, and physics into a single, unified science program, for example, Science I, Science II, Science III, Science IV. The principal grabs on to this idea and charges Sarah Jenkins to come up with a plan that might be implemented on a small scale starting this year.

The Problem. What should Sarah Jenkins do? Is this a valid approach to high school science? Is the principal justified in making this demand? □

Think Pieces

- What do you think should be the main goals of the science curriculum for middle schools? What differences in goals, if any, should exist between middle and high school science?
- Should the United States adopt a national science curriculum, or should it maintain the emphasis on a decentralized system of science education? Find and locate in the literature information about NSTA's "Scope, Sequence, and Coordination Project."
- What are the most important things a student should learn from (choose one area) earth science, life science, physical science, or general science?
- Do you think requiring students to take a course in science in each year of high school is a good idea? Why? □

Research Matters:
Student Behavior and Motivation in the Middle School
by Anne C. Howe[24]

What motivates junior high school pupils to spend their time productively in science class? Does increased effort really lead to higher achievement? In order to study these and other questions we made systematic observation of the behaviors of boys and girls in junior high school science classes. In all the classes studied teachers used an activity centered approach to teaching with much of the time spent in hands-on or laboratory work. Our sample included an equal number of boys and girls

(Continued on p. 142)

(Continued from p.141)
black and white students. In addition to observational data we obtained scores on standard reading and mathematics tests and used final class grades as the measure of achievement.

We defined three categories of behavior. Active Learning Behavior was defined as any behavior, other than reading or writing, directly related to the assigned activity. This included setting up equipment, carrying out an experiment, observing an experiment or discussing the work with a partner. The second category was Passive Learning Behavior—listening to the teacher, reading or writing. The third category was Non-Learning Behavior, defined as any non-task oriented or non-productive behavior.

Our first finding was that mathematics and reading skills are the most important factors in science achievement at the eighth grade level. Boys and girls who have poor basic skills when they enter eighth grade are not likely to succeed even when the class is based on active involvement. We found that the next most important factor in science achievement was Active Learning Behavior. We found no race or sex differences in behaviors—an interesting finding in view of the assumption held by many people that girls are less active than boys in school and that black students' lower achievement is due to non-attention in class.

We found that reading and math scores and active learning behavior were related to achievement for all groups.

But what motivates pupils to be active in class in productive ways, to engage in Active Learning as opposed to Passive Learning or Non-Learning behaviors? Many researchers who have studied pupil motivation have based their work on the assumption that pupils will direct their efforts toward achieving goals that they value. Therefore, if pupils value success they will work to achieve it. In order to study this question we asked pupils to answer questionnaires to determine whether they valued success in school and particularly in science and whether they attributed success, when they attained it, to their own effort, to their ability, to the ease of the task or just to luck. Then we asked teachers to rate pupils on these variables and pupils responded to questions about their effort, the level of difficulty and the environment of that class.

We found that pupils of both races and sexes do value success in science and that they believe their parents and teachers value it also. There was little to suggest, however, that valuing success is translated into effective effort in the classroom except for one group, the black boys. For that group positive correlations were found between value of success and Active Learning, between teachers' valuing success and Passive Learning, and a negative correlation between teachers valuing success and Non-Attending. Apparently those in this group who value success *themselves* engage in more Active Learning behaviors and those that think *teachers* value success engage in more Passive Learning behavior.

Two sex differences in causal attributions were found; boys attributed success more to ability and girls attributed success more to effort. Attribution of success to ability is a sign of self-confidence, stronger in boys than in girls, that is not strongly related to actual ability and does not lead to more or less effortful classroom behavior. It is interesting that those who attribute success to ability do not attribute failure to lack of ability. Both attribution of success to effort and to ease of task are associated with increased Passive Learning, the traditional mode of classroom behavior. These pupils are "trying hard" by listening to the teacher and working on in-class assignments. Unfortunately for them, we did not find a relationship between Passive Learning and final grade.

Those who attribute failure to lack of ability do, in fact, have lower grades (i.e., they have more failure), lower SRA-Math scores (possibly a sign of lower ability) and engage in Active Learning less often. A similar pattern emerges for those who attribute failure to the environment (i.e., luck or other external forces). They engage in Active Learning less often, Non-Attending behaviors more often, and have lower SRA-Math scores. ☐

[24]Anne C. Howe, "Pupil Behavior and Motivation in Eighth Grade Science," Research Matters...To the Science Teacher Monograph Series (National Association for Research in Science Teaching). Used with permission.

The Middle School Student
by Steven J. Rakow[25]

While the phrase "middle school" is often used to refer to an administrative structure, it also refers to the particular age group of students that middle school serves—10 to 14 year olds. What are the unique characteristics of these learners?

Physically, middle school students are developing at a rate faster than at any time in their lives except infancy. This rapid growth may bring with it clumsiness and poor coordination. Often the young compound their anxiety; growth rates among classmates might differ by several years, leading some children to wonder if they will ever catch up

(Continued on p. 143)

(Continued from p. 142)
with their rapidly maturing class-mates. Early adolescents tend, therefore, at times, to be very self-conscious and may become "behavior problems."

Socially, middle school children seek to establish their own identity and their own self-concept. As they seek to "fit in," their peers become an increasingly important influence on their lives. Along with this newly established identity is the need to make their own decisions. Often, middle school students are reluctant to act in accordance with the demands of their parents or those of other adults. While this stage of development may bring a certain anxiety to parents and teachers, it is an indication that the early adolescent is developing a greater capacity to conceive of alternative approaches to solving personal problems.

Middle school classrooms contain a fascinating array of cognitive abilities. Barbel Inhelder and Jean Piaget provide one model that attempts to describe the intellectual development of the young. They suggest that individuals progress through four states of intellectual development: sensori-motor (ages birth to 2 years), preoperational (ages 2 to 7), concrete operational (ages 7 to 11), and formal operational (ages 11 to 14). In brief, Inhelder and Piaget propose that, at different stages of cognitive development, people have a reper-toire of patterns of reasoning—which individuals may apply to solving problems. The "concrete operational" student will be able to carry out logical thought processes using concrete objects, while the "formal operational" student will be able to engage in the manipula-tion of purely men-tal constructs. Several other differences also dis-tinguish these two categories, such as the ability of "formal operational" students to use proportional and probabilistic logic

patterns, to consider the influence of many variables simultaneously, and to examine many possible combinations of a group of objects. Inhelder and Piaget's conclusions about patterns of reasoning could have implications for the kind of cognitive demands that we place upon students in science classrooms.

Subsequent research, however, has questioned Piaget's formulations, particularly as applied to specific age groups. It has been suggested that an individual's ability to approach a problem may be as much a factor of experience as of his or her gen-eralized cognitive development.

In any case, successful middle/junior high school teach-ers need to be aware of and sen-sitive to the physical, emotional, affective, and cognitive charac-teristics of their middle school students. Because of the hetero-geneity of abilities and physical characteristics, and because of the rapid rate at which these early adolescents are developing, the identification of special ability is particularly difficult. □

[25]Steven J. Rakow, "The Gifted in Middle School Science," in *Gifted Young in Science: Potential through Performance* (Washington, D.C.: National Science Teachers Association, 1988), pp. 143–144. Reproduced with per-mission of the National Science Teachers Association. Copyright 1988 by the National Science Teachers Association, 1742 Connec-ticut Avenue, NW, Washington, DC 20009.

Problems and Extensions

- Suppose that you are a member of a curriculum selection team for your middle school. The task of the team is to make recom-mendations to the district cur-riculum department concerning the goals and curriculum for a middle school program, grades 6 to 8. What would your recom-mended curriculum look like?
- Investigate one of the exem-plary middle school science pro-grams. You should locate the materials of the program, and write a brief report outlining the strengths and weaknesses of the program. What are the goals of the curriculum? How do the goals of the program compare with the recommendations for exemplary middle schools developed by Project Synthesis?
- Demonstrate at least one activ-ity from one of the exemplary middle school science programs with a group of peers or middle school students.

RESOURCES

Adey, Philip, ed. Adolescent Development and School Science. New York: Falmer Press, 1989.

This book is based on the proceedings of an international seminar held at King's College Centre for Educational Studies, London, in 1987. The book examines the impact of cognitive and social development on the teaching of science in middle and secondary school.

Brown, Faith. Five "R's" for Middle School: Strategies for Teaching Affective Education. Columbus, Ohio: National Middle Schools Association, 1980.

This booklet focuses on affective learning in the middle school. It includes a rationale for including affective programs in middle schools, as well as outlining a sample program.

George, Paul S. *The Theory Z School: Beyond Effectiveness*. Columbus, Ohio: National Middle Schools Association, 1984.

Based on the management theory applied in some businesses in the United States and abroad, theory Z proposes a school that is good for both productivity and personal growth. George analyzes the characteristics of successful American and Japanese corporations based on theory Z, and applies the concepts to the middle school.

Merenbloom, Elliot Y. *The Team Process in the Middle School: A Handbook for Teachers*. Columbus, Ohio: National Middle Schools Association, 1986.

This is a very useful book if you want to find out about how middle school teachers successfully work in clusters or teams.

Padilla, Michael J. *Science and the Early Adolescent*. Washington, D.C.: National Science Teachers Association, 1983.

This book contains a collection of writings by science teachers about middle and junior high school science, including developmental characteristics of the early adolescent, teaching strategies, and sample science activities appropriate for the middle school student.

Padilla, Michael J., Dallas Stewart, and Al Evans. *Attributes of Exemplary Middle Grades Science Teachers*. Georgia Science Teachers Association, 1988.

This monograph summarizes the knowledge, competencies, and understandings that a middle grades science teacher should have to be a master of the subject at the middle school level.

CHAPTER 5

THE SCIENCE CURRICULUM: HIGH SCHOOL PATTERNS AND PROGRAMS

In this chapter we turn our attention to the high school science curriculum. How is it different from the middle school curriculum? How is it related to the middle school curriculum? Grade 10 is the year when a large number of students end their science curriculum career. How can we encourage students, especially girls and minority students, to take more sciences in high school, not so that they will become scientists but rather so that they can be more aware and responsible citizens as they continue their lives, and choose careers in science, technology (and mathematics), and related fields. We'll look back at the curriculum, move to the contemporary scene, and then make some predictions about the future by looking at some exemplary high school programs.

Finally, the chapter concludes with an examination of the science curriculum in five other countries, Australia, China, Japan, Nigeria, and Russia. We live in a period of human history in which all the nations of the earth are interdependent in many ways, and connected by the advances in technology and communications. A global perspective on science education is necessary in this period of global connectedness. What is the nature of science education in these countries? How is it different from the curriculum in the United States? What comparisons can be made? What are some contemporary problems and concerns?

The high school science and mathematics curriculum is under a great deal of pressure to produce the most academically talented group of high school students in the world. Most of this pressure is coming from politicians, from governors' houses to the White House. In the 1990 State of the Union Address, the President indicated that the improvement of mathematics and science was of the highest priority in his administration. A few days later, he appointed a commission to advise him on mathematics and science. Yet with all this rhetoric, which is not new to the high school mathematics and science scene, life goes on in American high schools.

PREVIEW QUESTIONS

- What is the nature of the contemporary high school science curriculum?
- What were the characteristics of the high school science curriculum reform projects of the 1960s?
- What are the characteristics of exemplary high school science programs?
- What are the science curriculum patterns in other nations?
- How does the science curriculum of the United States compare with the science curriculum in other nations?

145

HIGH SCHOOL
SCIENCE REFORM PROJECTS

It is valuable to look back at the nature of the curriculum reform projects of the 1950s and 1960s because some of them are still used in American high schools, and many, if not all, had an impact on one or more of the science disciplines. It all started with the improvement of high school physics, so that's where we'll start.

Physics

Prior to 1956 the content and organization of physics at the high school level was highly influenced by the Harvard Descriptive List of Experiments, or by the periodic emphasis on the application of physics (toy physics, household physics, consumer physics, atomic age physics). In 1956 a conference was held at the Massachusetts Institute of Technology by Jerrold R. Zacharias, a professor of physics at MIT. The state of high school physics teaching was discussed, and the seeds were planted for the development and implementation of a new course for physics in the high school.

PSSC Physics

During the period 1956–1960, the PSSC Physics course was developed by several hundred physicists, high school teachers, apparatus designers, writers, and editors. The result was a course that contained a student textbook, a teacher's guide, laboratory experiments, tests, films, and a series of paperback books on selected topics in science.

The PSSC Physics course departed from the traditional course in physics, which emphasized the facts and description of physics concepts; the PSSC course was designed to help students "do physics," by engaging the students in the activities and thought processes of the physicist. The goals of this new program included helping students[1]

- Understand the place of science in society
- Understand physics as a human activity, and a product of human thought and imagination
- Appreciate the intellectual, aesthetic, and historical background of physics
- Appreciate the limitations of knowledge about the physical world

- Understand that knowledge of physics comes about from observation and experimentation
- Appreciate the spirit of inquiry
- Appreciate the logical unity of physics and the way that physicists think about the world
- Understand basic principles of physics that manifest themselves on an astronomical as well as human and atomic scale

In the 1960s and 1970s there were a number of studies done by science educators to evaluate the effectiveness of PSSC Physics and the other course improvement projects, and also to compare their effectiveness with the traditional courses in the respective disciplines. Studies showed that PSSC students did better on higher-level cognitive tasks than their peers in traditional physics courses.[2]

The PSSC course involved the students in a series of laboratory activities that were unique (Figure 5.1). Over 50 experiments were designed to support and help develop the concepts in the textbook. The experiments were not designed to verify a concept that had been introduced by the teacher or the textbook. Instead, the laboratory experiments created a novel situation in which students had to think about a problem, gather relevant data, and analyze results. To accomplish this, the PSSC developers designed special laboratory equipment that was simple, easily assembled, and inexpensive. All other course improvement projects followed this pattern of designing special equipment. The PSSC equipment included roller-skate carts, doorbell timers, and ripple tanks.

Project Physics

Another physics course developed during this period was Project Physics. It departed from the PSSC model, perhaps because the developers were science educators, involving high school teachers from the beginning. Project Physics set out to develop a general education physics course based on good physics but designed for today's citizen.[3]

Project Physics developed a course along humanistic lines, in that the developers were interested in emphasizing human values and meaning, as opposed to PSSC Physics, which focused more on the intrinsic

[1]P. D. Hurd, *New Directions in Teaching Secondary School Science* (Chicago: Rand McNally, 1969), pp. 187–188.

[2]J. A. Shymansky, W. Kyle, and J. Alport, "The Effects of New Science Curricula on Student Performance," *Journal of Research in Science Teaching* 20, no. 5 (1983): 387–404.
[3]Hurd, *New Directions*, p. 194.

FIGURE 5.1   A PSSC Laboratory, "Refraction of Particles." In this laboratory exercise, students use a "rolling ball model" to find out how the particle theory of light explains refraction. Physical Science Study Committee, *Physics Laboratory Guide*, Lexington, Ma.: D.C. Heath, ©1960 by Education Services Incorporated. Reproduced with permission of the publisher.

structure of physics. Project Physics objectives were designed to help students understand and appreciate[4]

- How the basic facts, principles, and ideas of modern physics developed
- Who made the key contributions and something of their lives
- The process of science as illustrated by physics
- How physics relates to the cultural and economic aspects of society
- The effect of physics on other sciences
- The relationship and interaction between physics and contemporary technology

Project Physics produced a vast array of teaching materials including (1) six student texts, called Student Guides (Concepts of Motion, Motion in the Heavens, Energy, Waves and Fields, Models of the Atom, and The Nucleus); (2) Physics Readers (articles and book passages related to the topics in the Student Guides); (3), a Laboratory Guide (student experi-

[4]Hurd, *New Directions*, p. 195.

ments); (4) laboratory equipment; (5) film loops; (6) films, and (7) a teacher's guide.

PSSC Physics and Project Physics were the two major physics curriculum projects developed with funds from the NSF. Physics enrollments continued to decrease during the period of time that these courses were developed and thereafter. Research results showed that the courses were effective in improving students' understanding of physics and their ability to accomplish high cognitive tasks.

Chemistry

During the golden age of science education, three major chemistry curriculum projects were developed with support from the National Science Foundation, namely, the Chemical Education Materials Study (CHEM Study chemistry), the Chemical Bond Approach (CBA), and Interdisciplinary Approaches to Chemistry (IAC). Table 5.1 compares some of the features of these three programs.

These three chemistry programs represent two very different approaches to the subject. The CHEM

TABLE   5.1

GOLDEN AGE CHEMISTRY PROGRAMS

Chemistry Course	Development Center	Approach	Curriculum
CHEM Study	Harvey Mudd College, Claremont, Calif.	Chemical concepts should be developed inductively by active student involvement. The laboratory experiments are designed for students to gather data, not verify concepts. Three commercial versions of CHEM were created after the original curriculum was designed.	Student text: *Chemistry: An Experimental Science* (W. H. Freeman) Laboratory manual Teacher's guide Tests Films Programmed materials
CBA	Reed College, Portland, Ore.	The course is organized around a conceptual theme: the chemical bond. Students are encouraged to use theoretical models to explain data. The lab is integrated with discussion with the text.	Student text: *Chemical Systems* Laboratory Manual Teacher's guide Teacher's laboratory guide
IAC	University of Maryland	IAC is an interdisciplinary approach to chemistry. After an introductory module, unifying themes connect students to various areas of chemistry such as organic, nuclear, environmental.	Student texts (modules): • *Reactions and Reason* • *Diversity and Periodicity* (inorganic) • *Form and Function* (organic) • *Molecules in Living Systems* (biochemistry) • *The Heart of the Matter* (nuclear) • *The Delicate Balance* (environmental) • *Communities of Molecules* (physical)

Study and CBA programs were innovative but were closely aligned to the traditional organization and approach to chemistry. The IAC program, on the other hand, was more socially relevant in that there was a closer connection to the real world of the student, and the materials were interdisciplinary. Furthermore, IAC was designed as a series of modules, a forerunner of the Individualized Science Instructional System (ISIS), which designed over 30 minicourses in general science, earth science, biology, chemistry, and physics.

According to Shymansky, Kyle, and Alport, CHEM Study and CBA did not fare as well as the NSF projects in physics and biology. As they put it, "of the three traditional secondary disciplines (biology, chemistry, and physics), it is probably safe to conclude on the basis of the data...that the new chemistry curricula produced the least impact in terms of enhanced student performances."[5] Student perfor-

[5]Shymansky, Kyle, and Alport, "The Effects of New Science Curricula," 387–404.

mance was low in achievement and process skills. The authors speculate that the small differences in achievement may have been due to the fact that CHEM Study and CBA were not very different from traditional chemistry courses.

Biology

Biology education in the United States has been influenced to a greater degree (aside from the socio-political-cultural issues of the evolution-creation science debate) by the Biological Sciences Curriculum Study (BSCS) than by any other force or group. The BSCS was organized in 1959 at the University of Colorado. It is still active as a curriculum force in science education today, with headquarters at Colorado College, Colorado Springs.

The BSCS developed three approaches and separate curricula for the high school biology program. The BSCS developed two organizational structures that were used to give direction to the three approaches to biology. BSCS, according to William Mayer, identified seven levels of organization as follows:[6]

1. The molecular level
2. The cellular level
3. The organ and tissue level
4. The organismic level
5. The population level
6. The community level
7. The world biome level

The three versions of BSCS that were developed were organized around three of the foregoing levels, and according to Mayer, "in order not to prejudice prospective users for or against any one of the three approaches, there were no descriptive titles: rather, the three texts were designated by color."[7] The versions and their "level of organization" are as follows:

BSCS Blue Version: (molecular) Biological Sciences: Molecules to Man

[6]W.V. Mayer, ed., *Biology Teachers' Handbook*, 3d ed. (New York: John Wiley and Sons, 1978), pp. 8–9.
[7]Mayer, *Biology Teachers' Handbook*, p. 9.

BSCS Yellow Version: (cellular) Biological Science: An Inquiry into Life
BSCS Green Version: (world biome) Biological Science: An Ecological Approach

BSCS materials were also organized around a set of unifying themes, "the characteristics and concepts that provide the most comprehensive and reliable knowledge of living things known to modern science."[8] The unifying themes were identified as follows:[9]

1. The change of living things through time: evolution
2. The diversity of type and unity of pattern in living things
3. The genetic continuity of life
4. The complementarity of organism and environment
5. The biological roots of behavior
6. The complementarity of structure and function
7. Regulation and homeostasis: the preservation of life in the face of change
8. Science as enquiry
9. The history of biological concepts

Hurd points out that the first five unifying themes were used to organize the content in each version, themes 8 and 9 refer to the internal structure of biology, and themes 6 and 7 refer to content and structure.[10]

The three versions of BSCS had an enormous impact on high school biology. Shymansky, Kyle, and Alport reported that there was more research data available on the BSCS programs than on any other new curriculum project. BSCS programs compared very favorably with traditional biology courses. With the exception of physics, the new biology programs "showed the greatest gains across all criteria measured"[11] (achievement, perceptions, process skills, analytic skills).

[8]Mayer, *Biology Teachers' Handbook*, p. 9.
[9]Mayer, *Biology Teachers' Handbook*, pp. 11–14.
[10]Hurd, *New Directions*, p. 154.
[11]Shymansky, Kyle, and Alport, "The Effects of New Science Curricula," p. 395.

CONTEMPORARY HIGH SCHOOL SCIENCE CURRICULUM PATTERNS

Iris Weiss, in a study of the science curriculum found that half of all sciences in grades 10 to 12 are biology and life science, with most of the rest chemistry and physics. Furthermore, over 95 percent of secondary classes use published textbooks and programs, although fewer than one in five science teachers re-

TABLE 5.2

PERCENTAGE OF TEXTBOOK
COVERED IN SCIENCE COURSES

Percentage of Textbook Covered	Percentage of science classes	
	Grades 7–9	Grades 10–12
Less than 25	2	1
25–49	10	11
50–74	27	37
75–90	39	33
More than 90	20	15

I. R. Weiss, "The 1985–86 National Survey of Science and Mathematics Education," in A. B. Champagne, and L. E. Hornig, *The Science Curriculum* (Washington, D.C.: American Association for the Advancement of Science, 1987) p. 230. Used with permission.

ported they cover the entire book (Table 5.2) And only two textbook publishers account for more than half the textbook usage in secondary science.[12]

These facts describe the picture of the traditional science curriculum. Although the United States does not have a national curriculum, the evidence from survey research studies shows that the American secondary science curriculum can be described in terms of a few science textbooks, about five or six different kinds of courses, and a limited variety of instructional techniques.

Weiss reports that lecture and discussion are the mainstays of the high school science curriculum. In 1977, lecture and discussion were considerably more common than laboratory activities, and in a more recent study (1986), the same was found, only the difference was even greater.[13]

When a randomly selected sample of high school science teachers was asked what their most recent lesson was, 75 percent in 1977 said it was a lecture-discussion lesson, and slightly more than 50 percent said it *included* hands-on activities. In 1986 more than 80 percent reported lecturing as their most recent lesson, and only 40 percent said they used a hands-on lesson.[14]

The science curriculum in American schools has been described as a layer cake model (Figure 5.2). In the layer cake model, a different science course is of-

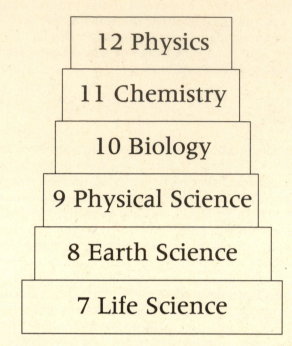

FIGURE 5.2 Layer Cake Model of the American Science Curriculum

fered each year as the students progress through the science curriculum. This stacking of one science course on top of another is the pattern that was established in the early part of the century. However, the model is actually more complicated than that.

The National Science Teachers Association has recommended a radical change in the science curriculum model that is employed in most school systems in the United States. Borrowing on the models in other countries, especially the Soviet Union, the NSTA Scope, Sequence and Coordination proposal calls for offering four strands (biology, chemistry, earth science, and physics) that would run through the science curriculum from grades 7 through 12 (Table 5.3). Each year, students would be exposed to each strand; that is, they would have experiences in biology, chemistry, earth science, and physics starting in grade 7.

Although not all school systems track students, the pattern shown in Table 5.4 is typical. Survey courses are designed for students who are either at risk or have not performed very well academically. General courses are for the large population of secondary students—the average student who is college bound but not necessarily focusing on science. Accelerated and advanced placement courses are designed for high school students who are contemplating a career in science or mathematics in college, and who have been successful in previous mathematics and science courses.

[12]These data reported by I. R. Weiss, "The 1985–86 National Survey of Science and Mathematics Education," in A. B. Champagne, and L. E. Hornig, *The Science Curriculum* (Washington, D.C.: American Association for the Advancement of Science,1987), pp. 225–233.
[13]Weiss, "The 1985–86 National Survey," pp. 230–233.
[14]Weiss, "The 1985–86 National Survey," pp. 230–233.

Table 5.4 shows the science curriculum for one of the largest school systems in the Southeast for the years 1990–1995. Notice that for each subject there are survey, general, and accelerated (or advanced) courses. And for juniors and seniors, many school systems offer advanced placement courses in which the student can earn college credit for a high school science experience. Also note that at the 11th and 12th grade levels, there are elective courses in marine biology, biology II, and anatomy and physiology.

TABLE 5.3

NSTA SCOPE, SEQUENCE AND COORDINATION MODEL

	Grade Level						
	7	**8**	**9**	**10**	**11**	**12**	**Total Time Spent**
	Hours per Week by Subject						
Biology	1	2	2	3	1	1	360
Chemistry	1	1	2	2	3	2	396
Physics	2	2	1	1	2	3	396
Earth / Space Science	3	2	2	1	1	1	360
Total Hours per Week	7	7	7	7	7	7	
Emphasis	Descriptive; phenomenological		Empirical; Semi-quantitative		Theoretical; abstract		

"Scope, Sequence, and Coordination of Secondary Schools: A Rationale," (Washington, D.C.: National Science Teachers Association, 1990), p. 4.

TABLE 5.4

THE TYPICAL SECONDARY SCIENCE CURRICULUM

Grade	Course	Textbook
7	Life Science	*Life Science* (Prentice-Hall, 1988)
8	Earth Science	
	Survey	*Earth Science* (Prentice-Hall, 1988)
	General	*Earth Science* (Prentice-Hall, 1988)
	Accelerated	*Earth Science* (Silver Burdett, 1987)
9	Physical Science	
	Survey	*Physical Science* (D.C. Heath, 1987)
	General	*Physical Science* (Prentice-Hall, 1988)
	Jr./Sr.	*Physical Science* (Addison-Wesley, 1988)
	Biology Accelerated	*Modern Biology* (Holt, 1989)
10	Biology	
	Survey	*Biology: An Everyday Experience* (Merrill, 1988)
	General	*Biology* (HBJ, 1989)
	Accelerated	*Modern Biology* (Holt, 1989)

(Table continued on p. 152)

(Table continued from p. 151)

Grade	Course	Textbook
11	Chemistry	
	Survey	*ChemCom: Chemistry in the Community* (Kendall/Hunt, 1988)
	General	*Chemistry: A Study of Matter* (Prentice-Hall, 1989)
	Accelerated	*Chemistry* (Addison-Wesley, 1987)
	Advanced Placement	*Chemistry* (Random House, 1988)
	Biology	
	Biology II	*Biology* (Addison-Wesley, 1987)
	Marine Biology	*Marine Biology* (Benjamin/Cummings, 1986)
	Anatomy and Physiology	*Essentials of Human Anatomy & Physiology* (William C. Brown, 1989)
12	Physics	
	Physics I	*Modern Physics* (Holt, 1990)
	Conceptual Physics	*Conceptual Physics* (Addison-Wesley, 1987)
	Physics II	*College Physics* (HBJ, 1987)
	Advanced Placement	*College Physics* (HBJ, 1987)

Physical Sciences Course Guide and *Life Science Course Guide* (Decatur, Ga: DeKalb County School System, 1989).

INQUIRY ACTIVITY   5.1

Science Curriculum Patterns

Curriculum patterns in science vary from state to state, and within school districts within the states. In this inquiry activity your task is to compare the curriculum patterns of two school districts in your state. To do this, you will have to obtain copies of the science curriculum guide from two school districts. You can obtain these by contacting the district's administrative office, or the science coordinator.

Level	Course	Textbook
Elementary School		
Middle/Junior High School		
High School		

FIGURE 5.3   Curriculum patterns

Materials

Curriculum guides from two school districts
Curriculum patterns chart (Figure 5.3)

Procedure

1. List the overall goals of the science curriculum of the districts you are investigating.

2. Identify the curriculum for the elementary, middle or junior high, and high school grades by completing the chart in Figure 5.3.

Minds on Strategies

1. How do the patterns compare? Are there similarities between the two districts? What are the differences?

2. Are there any differences in the stated goals that give you insight into each district's philosophy of science education?

3. To what extent is the layer cake model of science curriculum implemented in these districts?

EXEMPLARY HIGH SCHOOL SCIENCE CURRICULA

A number of recommendations have been made regarding the future direction of high school science curricula. Especially relevant here are the recommendations by Project Synthesis, the reports of the NSTA's Focus on Excellence series, and Project 2061.

Biology

Biology is one of the most important courses in the American science curriculum because for many students it is their last experience with a formal course in science. Also, as was noted earlier, more than half of all high school science courses are in the field of biology. This includes survey, general, and accelerated biology courses, biology II courses, marine biology, and human anatomy and physiology.

Biology teaching at the high school level has emphasized the vocabulary and very specific cognitive objectives, as opposed to focusing on more general goals such as understanding biological systems, how biological knowledge is developed, and how biological information can be of benefit to humans and society.

As has been pointed out by biology educators, the rapid changes in the biological sciences in the last 40 years, and the impact of these changes on the individual and society have been profound. Students leaving high school today are living in a "new biological world." One of the major recommendations of the biology group of Project Synthesis was that the high school biology program should center on the study of human beings as a part of nature. A greater effort needs to be made to help students make connections between biological knowledge, society, and technology.[15]

Bybee suggests a reformulating of the goals of biology teaching and suggests the following as the foci for biology education:[16]

- Biology education should lead to a fundamental understanding of biological systems
- Biology education should develop a fundamental understanding of, and the ability to use, the methods of scientific investigation (use science processes to solve problems in biology)
- Biology education should prepare citizens to make responsible decisions concerning science-related social issues
- Biological education should contribute to an understanding and fulfillment of personal needs, thus contributing to the development of the individuals
- Biological education should inform students about careers in biological and health sciences

In a more recent report on the teaching of biology, the National Research Council recommends a biology program at the high school level that is leaner (in the number of terms and concepts introduced), and for the course "the scaffolding should include an understanding of basic concepts in cell and molecular biology, evolution, energy and metabolism, heredity, development and reproduction, and ecology. Concepts should be mastered through inquiry, not memorization of words."[17]

As was pointed out in the report, the formula for change in the current teaching scheme of high school biology should be focused on teaching scientific concepts, reasoning, and learning through inquiry. The committee sees the following as the scope of change that is needed to ensure that high school biology meets these criteria:

1. In designing a course, we must identify the central concepts and principles that every high-school student should know and pare from the curriculum everything that does not explicate and illuminate these relatively few concepts
2. The concepts must be presented in such a manner that they are related to the world that students understand in a language that is familiar

---

[15]For an in-depth discussion see R. W. Bybee, "A New View for Biology Education", in J. E. Penick, and R. J. Bonnstetter, eds., *Focus on Excellence: Biology,* vol. 1, no. 3 (Washington, D.C.: National Science Teachers Association, 1983), pp. 4–11.

[16]Bybee, "A New View for Biology Education," pp. 4–11.

[17]Committee on High School Biology, *Fulfilling the Promise: Biology Education in the Nation's Schools* (Washington, D.C.: National Academy Press, 1990), p. 21.

3. They must be taught by a process that engages all the students in examining why they believe what they believe. That requires building slowly, with ample time for discussion with peers and with the teacher. In science, it also means observation and experiments, not as an exercise in following recipes, but to confront the essence of the material[18]

The following exemplary biology programs below reflect the goals and desired directions of this new view of biology education.

Experiential Biology I and II

The biology I and II sequence is a common model in most American high schools. This exemplary program developed by teachers at Addison High School in Michigan focuses on the active involvement of all students with a view to these broad goals:[19]

- Learn and experience the basic inquiry and investigative processes of science
- Show students that biology is not remote and above the society that supports it
- Learn basic facts, concepts, and principles of biology, and the field of science in general
- Develop favorable attitudes toward biology and science
- Gain a knowledge of science as a career choice
- Develop an awareness of the environment and humanity's role in it

Biology I emphasizes the investigation of local plants, animals, and their ecological interaction, DNA, and genetics. Students work in small teams on activities and experiments. In the biology I course, special socially relevant problems and issues are presented. Students are given some choices with regard to activities and projects in biology I. For example, during each of the three nine-week sessions, the students identify the research problem they will work on.

Biology II is highly individualized. Since the focus of the course is on having the students engage in biological research and on the social implications of biology, individual students are guided along by the teacher toward their areas of interest. According to Ellis, students play a major role in planning lessons, decision making, and classroom management. In the biology II course, students work in teams during the first and third nine-week term and do individual research projects during the second and fourth terms.[20]

Human Ecology

Human ecology, according to many biological educators, is one of the important directions for biology teaching. Human ecology becomes a unifying theme for the biology teacher and the student. Human ecology is based on the relationship between human beings and the total environment. Bybee points out that human ecology as a unifying theme can focus biology teaching toward these ends:[21]

- *Holistic:* Because of the complexity of human problems related to biology, they need to be considered in relation to their interrelatedness and holistic nature
- *Ecological scarcity:* Humans need to understand the limits of the environment's ability to provide resources and degrade wastes
- *The ecosystem concept:* The human ecosystem is where teaching should begin, with departures to the big (the biosphere as a whole) and the small (microbes)
- *Social responsibility:* Students, as future citizens, need to learn how to interact with the environment responsibly

The human ecology program described here was developed in Claymont, Delaware. According to the developers, it is an innovative science elective course with the focus being the development of human beings and the well-being of individuals and families in their social and physical environments.[22]

The curriculum is described by the developers as community based. The staff developed their own teacher and student handbook and use other resources as well, including videotapes and filmstrips. The curriculum includes six units, as follows: introduction, self-inventory, physical development, psychological development, social institutions, and environmental factors. The outline of the curriculum is shown in Table 5.5.

Chemistry

The authors of Project Synthesis pointed out that the physical sciences are not valued in the curriculum of American schools.[23] This is quite evident in our high

[18]Committee on High School Biology, *Fulfilling the Promise*, p. 25.
[19]B. Ellis, "Biology I and II," in J. E. Penick, and R. J. Bonnstetter, eds., *Focus on Excellence: Biology*, vol. 1, no. 3 (Washington, D.C.: National Science Teachers Association, 1983), pp. 65–72.
[20]B. Ellis, "Biology I and II," pp. 65–72.
[21]R. Bybee, "A New View," p. 6.
[22]J. Carney, and F. Castelli, "Human Ecology," in J. E. Penick, and R. J. Bonnstetter, eds., *Focus on Excellence: Biology*, vol. 1, no. 3 (Washington, D.C.: National Science Teachers Association, 1983), pp. 106–116.
[23]R. D. Anderson, "Physical Science Education," in N. C. Harms, and R. E. Yager, eds., *What Research Says to the Science Teacher*, vol. 3 (Washington, D.C.: National Science Teachers Association, 1981), pp. 33–52.

TABLE 5.5

HUMAN ECOLOGY CURRICULUM

Introduction	Self-Inventory	Physical Development	Psychological Development	Social Institutions	Environmental Factors
Rationale/goals	Who am I?	Passages	Values clarification	Family	Social agencies and institutions
Historical development	What do I want to be?	Embryonic development	Effects of drugs	Terms and functions	
	What you may want to be	Developmental handicaps	Effects of alcoholism on the family	Family functions	
	Human relations in the School	Human genetics	Motivation	Family tree and roles	
	What? Why? How?		Abnormal behavior	Women and work	
			Suicide	Family relationships	
			Stress	Marriage	
				Divorce	
				Parent-child relationships and divorce	

Carney and Castelli, "Human Ecology," pp. 106–116.

schools. Not as many students take chemistry (and physics) in American high schools compared with biology enrollments. Perhaps this is due to the fact that physical science courses are viewed as elitist, offered generally for the college bound or science-prone student. One of the goals of physical science educators is to increase the proportion of students taking high school physical science courses. Naturally, this can be done by requiring these courses (as some school districts have), or by reforming the nature of physical science courses in a way that makes them more relevant to the needs of students.

There have been attempts to make courses more relevant; however, these courses, for the most part, were aimed at reaching a limited population, often being designed for the lower-ability students. The desired state for chemistry education starts with this goal: Chemistry will cease being taught only as a college prep course; chemistry instruction will be broad based and designed for all high school students.[24]

The desired state for chemistry education will also include the following:

- The focus will be on observation and description of common chemical reactions
- There will be a reduction in the number of topics covered in high school chemistry courses. The chemistry curriculum would organize content around conceptual themes or socially relevant topics (as in ChemCom). Conceptual development would be achieved by showing students the interrelatedness and interconnectedness of ideas
- Greater emphasis will be placed on common chemical reactions, and on the use of simple models to explain chemical change. Student understanding will be enhanced if these chemical reactions are connected with real problems in the students' community and world
- Social issues will play an increasing role in making chemistry content relevant to students' lives
- A greater connection between careers in chemistry and the teaching of high school chemistry will be made. Students will have the opportunity to see how people use chemistry in their professional work, and how it affects them as citizens

[24]See J. D. Herron, et. al., "Ideals in Teaching Chemistry," in J. E. Penick, and J. Krajcik, eds., *Focus on Excellence: Chemistry*, vol. 3, no. 2 (Washington, D.C.: National Science Teachers Association, 1985), pp. 4–10.

With these notions in mind, let's explore a few exemplary chemistry programs.

The High-Interest Activities Approach

Chemistry lends itself to the EEEP concept (developed more fully in Chapter 8). For now, an EEEP is an Exciting Example of Everyday Phenomena, and we use it to integrate science with the real world. An EEEP is a nontraditional way to involve students in science activities.

The chemistry program described here uses high-interest laboratories to pique the students' curiosity.[25] According to the developers, about 15 percent of class time is spent on nontraditional activities including labs, computers, photography, and science movies.

Some of the nontraditional labs used in their program are shown in Table 5.6.

Chemistry for All

To deal with students coming from varied backgrounds, the faculty at Coral Springs High School in Coral Springs, Florida, have developed a laboratory-oriented approach to chemistry at four levels (Consumer Chemistry, Chemistry I, Advanced Chemistry, and Organic Chemistry).

Perhaps the most important course is the Consumer Chemistry course, for students who have not

[25]S. J. James, "Using High-Interest Activities," in J. E. Penick, and J. Krajcik, eds., *Focus on Excellence: Chemistry* (Washington, D.C: National Science Teachers Association, 1985), pp. 28–30.

TABLE 5.6

NONTRADITIONAL CHEMISTRY LABS

Nontraditional Lab	Concepts Illustrated
Hydrogen balloon	The production of hydrogen gas by a simple displacement reaction.
Root beer titration	The molarity of the base, sodium hydroxide, is found by titrating the base into a known molarity of hydrochloric acid. Then the normality of hydrochloric acid is compared with different substances such as root beer, Coke, or Sprite.
Ice cream	This experiment shows the effect of a solute (salt) in lowering the freezing point of water (mixture of sugar, cream, flavoring).
Hot air balloon	Students find the hover temperature of the balloon
Rock candy	Students make rock candy and study supersaturated solutions and crystal formation.

James, "Using High-Interest Activities," pp. 28–30.

done well previously in mathematics or in abstract quantitative reasoning. According to the developers of the course, the purpose is to give the students a general understanding of chemistry and its role in their lives. The authors point out that this is done in a diversified laboratory hands-on program in a warm and accepting environment.

The content of the course includes foods and food additives, pharmaceuticals and drugs, nuclear energy, plastics and polymers, environmental chemistry, agricultural chemistry, and the chemistry of home care and personal products.

In the teaching of chemistry at this high school, the teacher is perceived as a facilitator of the learning environment. Very few lectures are given. Instead, students are given the syllabus of the course, which contains due dates for all labs and reading assignments. Experiments are posted by the week and can be done any day during the week. Students are involved in planning their research and laboratory activities to fit their needs. The program features individualizing the instruction to fit the needs of the student. Worksheets are used to help students master the material, and computer software for drill and problem solving are available for student use.

The program pairs students as laboratory partners. Laboratory periods include a two-hour period, thereby giving students ample time to complete experiments. The faculty also work together as a team of instructors, thereby giving students access to more than one faculty member.

Physics

Physics has the reputation of being the most elitist of the high school science courses. A very small proportion of high school students enroll in physics. Exemplary physics programs, such as Physics for Everyone described presently, focus in on how physics can be relevant to all students. One of the problems physics teachers have had to come to grips with is the "mathematics" issue. Can students understand physics without a strong background in mathematics? It is conventional wisdom that many students avoided physics because they suffered from math phobia. The recommendations of groups like Project Synthesis and of physics teachers are that a multilevel approach to physics in the high school is in order. The course Conceptual Physics, which is being offered in an increasing number of high schools, is an example of physics teaching in which the quantifiable approach to physics is downplayed in favor of a qualitative approach. Concepts in physics can be understood without a lot of mathematics.

Anderson points out that the scientific and technological world that students live in should have some impact on the nature of high school physics teaching.[26] Many social issues, such as nuclear weapons, energy needs, global warming, national security, world food supplies, health, environmental protection, and mineral resources, are topics that students need to understand as citizens; but more important, these issues offer the opportunity to help students see the interrelatedness of physics concepts and the world around them.

One program that is exemplary in this way is Christensen's Global Science: Energy, Resources, Environment.[27] This program, which we will discuss in greater detail in Chapter 6, involves the student in a study of science (many of the concepts being derived from physics) in the context of real-world problem solving. Topics such as "energy and the future," "making peace with our environment," and "energy and society" form the foundation for his approach.

Here are some other exemplary programs in the area of physics.

Physics for Everyone

In this approach to the physics curriculum, a multi-level approach offers three approaches to physics to the students.[28] The levels include

- Advanced placement physics (for able math students, and those considering careers in science and engineering)
- General level physics (for students comfortable with math but not pursuing careers in science or engineering)
- Qualitative physics (the same concepts and principles of physics but with an emphasis on a qualitative understanding)

According to the developers of this multilevel approach to physics, the students are not ability grouped; students may sign up for any course and move freely from one to the other.

The description that follows pertains to the qualitative physics course. The instructor approaches the course in a qualitative manner. The underlying theme of the course is teaching for transfer: How can physics be relevant in everyday life? Each week, the students work on an assignment in which they must recognize an application of a principle in physics. Students must report on their findings.

The course is taught in an environment of positive reinforcement. Emphasis is on positive grades. Laboratory activities emphasize creativity. Instead of using labs to verify principles, labs in this course are used to help students discover principles, as well to develop problem-solving skills. For example, students are asked to determine a way to transfer heat from a hot object to a cold object. In small teams, they brainstorm and diagram ways to do this. Suggestions are grouped to identify pertinent variables. Later they devise experiments to test their ideas.

The qualitative course emphasizes the application of physics to the student's present and future world. Examples include driving and safety, athletics, music, photography, the human eye, and computers.

The instructor also uses an unorthodox evaluation process. No written exams are given. Instead a system of "up arrows" in the grade book are used to evaluate the students daily class work, laboratory performance, laboratory reports, and weekly outside work. The "up arrow" strategy is simply marking in a positive way a students performance. It is a powerful application of reinforcement theory.

Conceptual Physics

This is a nontraditional approach to traditional physics topics.[29] The content of conceptual physics includes the following units of study:

- Mechanics
- Properties of matter
- Heat
- Sound
- Electricity and magnetism
- Light
- Atomic and nuclear physics
- Relativity

Conceptual physics focuses on comprehension (and understanding) of physics concepts, without the mathematical elegance that characterizes the traditional physics approach. Conceptual physics does not downplay high-level thinking. On the contrary, the course is full of problem-solving exercises and activities that will challenge all students. Through some well-conceived review exercises, home projects, and in-class exercises, students are engaged in a wide range of intellectual activities.

[26]Ronald D. Anderson, "Excellence in Physical Science Education," in *Focus on Excellence: Physical Science*, John E. Penick and Vincent N. Lunetta, eds. (Washington, D.C.: National Science Teachers Association), p. 4.

[27]J. W. Christensen, *Global Science: Energy, Resources, Environment*, 2nd ed. (Dubuque, Iowa: Kendall/Hunt, 1984).

[28]Based on A. V. Farmer, "Physics for Everyone," in J. E. Penick, and V. N. Lunetta, eds., *Focus on Excellence: Physical Science*, vol. 1, no. 4 (Washington, D.C.: National Science Teachers Association, 1984), pp. 61–67.

[29]P. G. Hewitt, *Conceptual Physics*, 6th ed. (Glenview, Ill: Scott, Foresman, 1989).

Think Pieces

- What are the advantages and disadvantages of the present American science curriculum model (layer cake)? What are the advantages and disadvantages of the NSTA Scope, Sequence and Coordination curriculum model?
- What should characterize a high school biology, chemistry, or physics course in light of recent reports, especially Project 2061?
- Find and locate literature to support the development of a science course that is conceptually oriented, and that is designed to develop scientific thinking skills to inquire about natural phenomena and develop scientific explanations.

- What are the science education programs like in some other countries, in addition to the ones described here? You might want to investigate science education in Mexico, France, Great Britain, Hungary, or the Philippines. Prepare a brief report by creating a poster report for the country you choose.
- What are some of the differences and similarities between U.S. science education and science education in Australia, Japan, and Nigeria? Make a chart identifying the variables that you will use to make comparisons and how each country fared on the variable. ☐

The Science Curriculum: A Global Perspective

The science education community extends beyond the boarders of the United States and is an active force throughout the world. We live on a planet that some describe as a global community. Computers, satellites, FAX machines, telephones, and television bring educators together from countries as far apart as Costa Rica and the Soviet Union. What is the education of students in other countries about? When do students begin studying science in other nations? What is the nature of the science curriculum in other

(Continued on p. 159)

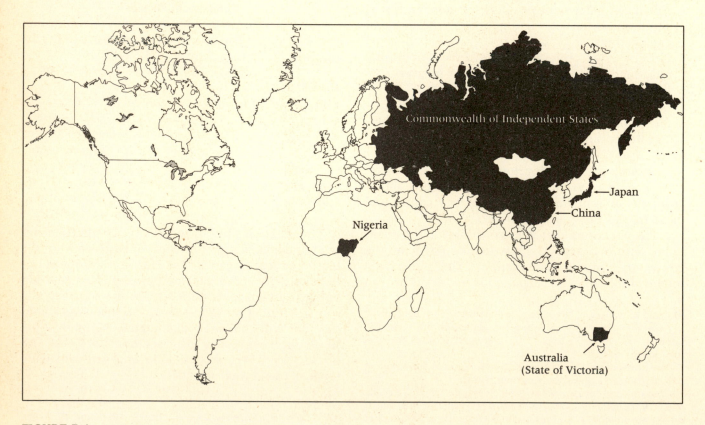

FIGURE 5.4   Science Education: A Global Perspective

(Continued from p. 158)
countries? In this Gazette, the science curriculum of several countries (see Figure 5.4) is highlighted to give you a global perspective of science teaching.

Australia
by Roger T. Cross[30]

The education system in each of the Australian states and territories has distinctive features. This resulted from the autonomy of the states from each other prior to federation in 1901. There are six states and two territories (although both the Northern Territory and the Australian Capital Territory now have parliaments of their own and are no longer administered by the central government). Consequently there are eight separate educational systems which, while they contain many common features, also have particular characteristics of their own, which makes it difficult to generalize except to say the Australian educational system owes much to that of the United Kingdom, both in the structure of primary and secondary schooling and the mix of state schools and private schools.

One interesting feature is the strength of the Catholic segment of private schooling, which stems from the large predominantly Catholic Irish and Italian immigration to Australia, particularly in the state of Victoria. The state educational authorities exercise different control over the curriculum in their schools to different extents, some very prescriptive and others giving control to the schools. It is this latter type that is perhaps the most interesting and is seen to the greatest extent in the state

[30]The section on Australian science education was written by Dr. Roger T. Cross, professor of science education at La Trobe University, Bundoora, Victoria. Used with Dr. Cross's permission.

of Victoria, the second most populous state. It should be noted that the federal government has been attempting to convince the states of the need for a common core curriculum across the states. The rationale for this has been the difficulties children experience when they move from one state to another. This has met with limited success at the moment because of the differing philosophical stances taken by states toward education.

The Example of Victoria. Approximately one-third of all children are in private schools; of these, Catholic religious based schools predominate. Schools have total autonomy to set the curriculum for the first eleven years of schooling (called P–10), however the Ministry of Education has, in the last two years, released framework documents for the key subject areas, which can be viewed as a retreat from this position. It should be noted that the documents are in no way imposed on schools. The political and educational debate surrounding the battle for autonomy was, I believe, lively and initiated by reformists in education in the sixties and early seventies.

The schooling consists of

1. *Kindergarten:* optional for 4 year olds
2. *Primary school:* Seven years of schooling, classes preparatory to grade 6. Children enter at age 5.
3. *Secondary school* (now called postprimary school): Six years of school, of which grades 7 to 9 are compulsory.

The retention rate (percentage of students who graduate) to year 12 is rapidly increasing and is now approaching 70 percent. The retention rate has been steadily increasing from a rate of 45 percent and is expected to rise to 85 percent by the end of the decade

Students can leave school at the end of year 10 (approximately 16 years). Those students who complete postprimary school go on to year 12 and upon graduation receive the Victorian Certificate of Education (VCE). This new policy requires all students to take four units of mathematics/science/technology studies.

Australian Science Education: The Case of the State of Victoria. Primary school science (P–6) is quite variable. Students may or may not receive what can be described as a science education. Many schools run a completely thematic approach to teaching, although mathematics and language are separated out for special attention. The amount of science may vary from grade to grade as much as it varies from school to school. A recent study by the federal government into math and science has recommended an increase into the time devoted to science in preservice training. It is probably valid to say that few children receive what can be identified as science for primary children in a systematic way throughout their primary schooling.

Secondary school science is quite a different story. In the first four years of secondary school (years 7 to 10), science is a core part of the curriculum and is generally compulsory to year 10. The allocation varies but is normally in the order of three to four 50 minute periods per week (150–200 minutes). Although there is autonomy and a wide range of textbooks to choose from, the curriculum doesn't vary a great deal from school to school. Most often it is a topical approach, as shown in Table 5.7. Students are involved in laboratory activities. Students work in small teams to investigate phenomena related to the topic (Table 5.7).

(Continued on p. 160)

TABLE 5.7

AUSTRALIAN SCIENCE CURRICULUM, YEARS 7 TO 10

Year	Core	Additional Units
7	Classification of Living Things	Forces
	Metric Measurement	Things and Places
	Mice and Men	Plants
	Introductory Chemistry	Sky throughout the Year
		Crystals
8	The Changing Earth	Astronomy
	Heat Transfer	Energy and Change
	When Substances are Heated	
	Cells	
9	Electricity	Coordination of the Body
	Reactions	Space Science
	Forces/Mechanics	Water
	Digging up Evidence	Consumer Science
	Human Physiology	Sound
10	Further Chemistry	Geological Mapping
	Geological Processes	Microorganisms
	Light	Human Sexuality
	Heredity	Drugs
	Forest	Insight into Measurement
	Ecology/Conservation	

(Continued from p. 159)

Some schools organize each year of science around a theme, such as "Sun," or "Earth." Refer to Figure 5.6 on page 161 for a Level 10 science course organized around the theme Earth. By the use of the theme Earth as a unifying concept, it is believed that students will better appreciate that compartmentalization of science is artificial and unnecessary, and come to know and understand just a little more about the "spaceship" that hurtles earthlings through space.

The science subjects for the last two years are part of the offerings of the Victorian Certificate of Education administered by the Victorian Curriculum and Assessment Board. The following subjects are available to students: biology, chemistry, geology, science, psychology, and physics. Each subject is divided into four parts and counts 4 units of the 24-unit-value Victorian Certificate of Education. Students are expected to complete at least two of the four units of a course. Some flexibility is built in, allowing students to enter the beginning of unit 1, unit 2, or unit 3 of each course.

One of the innovative attempts of the Victorian Curriculum and Assessment Board is the science syllabus for each subject. There is a strong attempt to give students a picture of the role of science in society, thereby emphasizing an STS approach. Central ideas or concepts are identified in the science syllabus, followed by very explicit contexts in which the ideas can be explored. The contexts involve the students in exploring science in everyday environments, that is, household for electricity, car for electrical systems. □

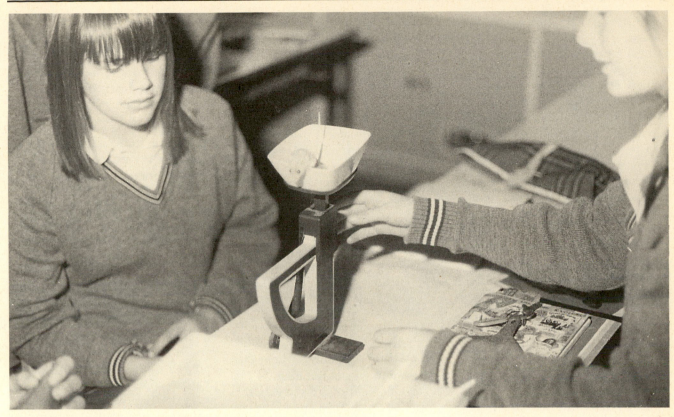

FIGURE 5.5 Australian students doing a small group, hands-on activity.
(Roger T. Cross)

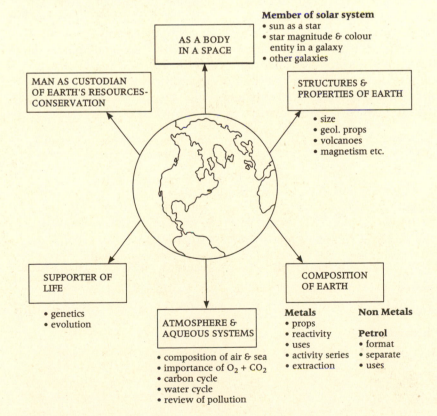

Member of solar system
- sun as a star
- star magnitude & colour entity in a galaxy
- other galaxies

AS A BODY IN A SPACE

MAN AS CUSTODIAN OF EARTH'S RESOURCES-CONSERVATION

STRUCTURES & PROPERTIES OF EARTH
- size
- geol. props
- volcanoes
- magnetism etc.

SUPPORTER OF LIFE
- genetics
- evolution

ATMOSPHERE & AQUEOUS SYSTEMS
- composition of air & sea
- importance of O_2 + CO_2
- carbon cycle
- water cycle
- review of pollution

COMPOSITION OF EARTH

Metals
- props
- reactivity
- uses
- activity series
- extraction

Non Metals

Petrol
- format
- separate
- uses

FIGURE 5.6 Level 10 Australian Science Courses: Note how each science topic (supporter of life, composition of Earth) use "Earth" as a unifying theme.

China[31]

China has the largest educational system in the world, with over 200 million students enrolled in public schools. Estimates, however, indicate that only 20 percent of the students who enter kindergarten complete the ten-year program of formal schooling. About 80 percent of the population of China lives in rural areas, and enrollment in school tends to be lower in the rural areas when compared with Chinese cities.

Science education, like other subjects in the Chinese curriculum, has experienced change. During the Great Proletarian Cultural Revolution in China (1966–1976), schooling was given less emphasis, being shortened from 12 to 10 years, as well as becoming less academic and more vocational. Cross and Price point

[31]Roger Cross and Ronald Price of LaTrobe University, Bundoora, Victoria, provided papers and documents on China's science curriculum.

out, however, that "this period in education in China is of undoubted interest for teachers in the West because of the way in which Chinese curriculum writers appear to have attempted to produce textbooks that stressed the relevance of science rather than its theoretical aspects."[32]

For example, their analysis of Chinese physics texts revealed that there appeared to be a greater emphasis on "useful knowledge rather than understanding laws and principles."[33] Although the present Communist regime in Beijing has rejected the Great Cultural Revolution, Cross and Price point out that what might be lost is the attempt of Chinese educators during this period to combine education with "productive labor in contrast to traditional 'goose-stuffing' methods."

In 1978 two conferences were held to rebuild the science curriculum after the Great Cultural Revolution. These conferences, attended by thousands of scientists and educators, shaped the new emphasis on science in China, which according to Hurd, is as follows: "In contrast with science education policies characteristic of the Cultural Revolution, the new science program was viewed as one emphasizing basic theories and stressing logical and abstract thinking. Problem solving, conceptualizing, and applying knowledge to production were to be stressed."[34]

The new science program in China, which has resulted in new textbooks and curricula, is based on goals formulated by scientists and teachers as follows:

1. Mastery of key concepts and basic information
2. Ability to conceptualize and make inferences
3. Development of systematic and logical methods for analysis and synthesis in solving problems
4. Appreciation of the importance of physical models in thinking
5. Facilitation of the student's ability to apply knowledge to practical problems, especially in agriculture and production
6. Appreciation of the evolution of science concepts to foster a dialectical-materialistic point of view and way of thinking
7. Development of skills in experimental procedures and in the use of scientific instruments[35]

Science education in China is driven by textbooks that are written by committees of scientists and teachers. According to Hurd, the first set of texts produced after the Cultural Revolution were criticized as being too theoretical and impractical for many Chinese students.

Following is a brief examination of the scope and sequence of the Chinese science curriculum, and some comments about science instruction in Chinese schools.

The Scope and Sequence of the Science Curriculum. Primary school science in grades 1 to 3 is not taught as a separate "subject" but is part of the physical education program. The texts use stories about science, inventions, animals, personal hygiene, and community sanitation to present science. In grades 4 and 5, science is taught twice a week throughout the school year. Grade 4 science emphasizes weather and the atmosphere, and the biology of plants and animals. The course also emphasizes applied topics such as water purification, environmental protection, and control of harmful insects. Grade 5 science emphasizes human physiology and diseases, physics topics including simple machines, sound, electricity, heat and light, and topics in earth and space science, including the earth's rocks and soils, the solar system, stars, and the universe.[36]

[32]Roger Cross and Ronald Price, "School Physics as Technology in China During the Great Proletarian Cultural Revolution: Lessons for the West," *Research in Science Education* 17, (1987): 165–174.
[33]Cross and Price, "School Physics as Technology in China," 165–174.
[34]Paul DeHart Hurd, "Precollege Science Education in the People's Republic of China," in Margrete Siebert Klein, and F. James Rutherford, eds., *Science Education in Global Perspective: Lessons from Five Countries* (Washington D.C.: American Association for the Advancement of Science, 1985), p. 70.

[35]Hurd, "Precollege Science Education," pp. 69–70.

[36]Hurd, "Precollege Science Education," p. 94.

TABLE   5.8

TIMETABLE OF THE CHINESE CURRICULUM

Course	Junior Level			Senior Level			Hours	%Curr.
	I	II	III	I	II	III		
Biology	2	2				2	198	3.5
Chemistry			3	3	3	3	372	6.7
Physics		2	3	4	3	4	500	9.0
Math	5	6	6	5	5	5	1,026	18.5
Total Hours per Week	30	31	31	29	26	26	—	—

FIGURE 5.7   Chinese senior level biology classroom (Roger T. Cross)

FIGURE 5.8   Chinese senior level chemistry classroom. (Roger T. Cross)

Secondary level science is divided into three years each at the junior and senior levels. Science courses make up slightly more than 20 percent of the Chinese student's curriculum in most of the nation's schools. The curriculum is arranged as shown in Table 5.8.

Although there are different arrangements of Chinese schools (five-year versus six-year schools, priority versus non-priority), the science curriculum is generally as shown in Table 5.9.

As in many of the other countries discussed in this section, science is given high priority in the curriculum. One question that we might raise given the priority of science in China's schools is, How is science taught?

In an analysis of science textbooks in Chinese schools, Cross, Henze, and Price concluded that the aim, after an exhaustive analysis of topics and questions at the end of each chapter, was to produce specialists, and the textbooks did not appear to provide training skills that would be useful in a variety of situations. They also concluded that the following goals were not emphasized: problem solving, the application of principles to novel situations, and interpreting and predicting.[37]

Although it is difficult to generalize on the actual nature of classroom instruction in Chinese classrooms, Cross, Henze, and Price make these observations:

Class sizes are very large, often ranging from fifty to even seventy students per class. Only in a few schools within each province is equipment comparable to what one finds in schools in the advanced industrial countries, though in the best equipped schools it is excellent, both in quality and quantity. Probably for reasons of class size and availability of equipment class experiment is confined to the few 'Student Experiments' listed in the textbooks or may even be totally absent. But teacher demonstration is common, often involving special apparatus. Lessons are formal, closely following the textbook. Students memorize a great deal of material and work through exercises, regularly being brought to the blackboard to answer questions orally.[38]

TABLE   5.9

....

CHINESE SECONDARY SCIENCE CURRICULUM

....

Junior Level

Physics: The physics course begins in the second year of the junior school with the study of mechanics. Topics include measurement, weight, force, pressure in liquids and gases, buoyancy, force and movement, simple machines, work and energy, change of state, heat energy, and heat engines. In the third year, the course includes current electricity, electrical work, transmission in liquids and gases, electromagnetism, and an introduction to light.

Chemistry: The chemistry course begins in the third year of the junior level with topics including oxygen, hydrogen, solutions, moles and heats of reaction, the structure of matter, nitrogen subgroup, speeds of reaction and equilibrium, the carbon group, and organic compounds.

Biology: The biology course begins in the first year of primary school and continues into the second year. The course begins with a consideration of the structure of living things, including cells, tissues, and organs. This is followed by the organs of flowering plants, seeds, roots, stems and leaves, flowers, and fruit, Students then study the structure and functions of the major groups.

Senior Level

Physics: Three years of physics are offered. Topics include mechanics, including concepts of equilibrium and motion including Newton's laws; mechanical vibration and wave motion and the gas equations; electric and magnetic fields, static electricity, alternating current electricity, the nature of light, and elementary concepts of atomic structure.

Chemistry: Senior level chemistry begins in year 1 with sodium and sulfuric acid, heats of reaction, the periodic table of the elements, the nitrogen group, the carbon group, and colloids. In year 2 topics include electrolysis, magnesium and aluminum, transitional elements, hydrocarbons, sugars and proteins, and high polymer compounds.

Biology: Senior level biology begins with the structure and function of the cell, then considers the origin of life, assimilation and metabolism, reproduction and development, regulation and control, heredity and variation.

[37]Roger T. Cross, Juergen Henze, and Ronald F. Price, "Science Education for China's Elite Secondary Students: The Example of Chemistry," *School Science and Mathematics,* in press for 1992.
[38]Cross, Henze, and Price, "Science Education for China's Elite Secondary Students," in press for 1992, p. 4.

Japan

The Japanese system of science education has been designed after the American model. Its overall organization is a 6-3-3 structure (elementary school, junior high school, high school), paralleling the U.S. model. Japanese students are tracked with 70 percent in a college-university curricula, and the other 30 percent in one of six vocational courses.[39]

The Japanese system is highly centralized, being controlled by a national Ministry of Education, Science and Culture (Mombusho). This body regulates textbooks, accreditation, teachers' salaries, and curriculum.[40]

A very strong teachers union (Nikkyoso) counterbalances the bureaucratic Mombusho by influencing class activities and supplementary curriculum. According to Troost, the Nikkyoso has been a spokesagent for "progressive values, whole person education, science for understanding, the creative use of science, and the impact of science on the world."[41]

In the Japanese system students are not tracked during the compulsory years (1 to 9), but rather follow the same curriculum. Equal mastery of content within classes is expected for all students during these years. The curriculum in grades 10 to 12 provides some flexibility in that students can elect courses of study.

The Japanese primary science curriculum, according to Troost, begins in grades 1 and 2 with the study of plants, animals, objects, and phenomena of everyday life.

The primary science curriculum focuses on the process of science. Professor Shigekazu Takemura, former head of the Mombusho's science curriculum, describes the focus of primary science with phrases such as "science play" and "do-it-yourself fun."[42] The primary science in grades 3 through 6 follows a spiral curriculum in which topics are reintroduced as students progress through the year. Textbooks are based on the national curriculum guide. The primary-level textbooks are written and designed to encourage observation and inquiry; they are colorful, lightweight, and students might receive three or four during a single year.

The junior high science curriculum is organized into two content areas: physics and chemistry, and biology and earth science (Table 5.10). The Japanese junior high curriculum covers a wide range of topics, and in fact, Troost reports that in international comparisons of curriculum, the

[42]Troost, "Science Education," p. 34.

Japanese science curriculum was ranked first in providing students with the opportunity to learn a range of science content.

Students in high school can elect science courses from the general science offerings:

Basic (Integrated) science

Physics I
Physics II

Chemistry I
Chemistry II

Biology I
Biology II

Earth science I
Earth science II

According to Troost, most students take three or four courses in science during their high school career, choosing from integrated science, biology, chemistry, physics and earth science. □

[39]Troost, K. M., "Science Education in Contemporary Japan," in M. S. Klein, and F. J. Rutherford, eds., *Science Education in Global Perspective: Lessons from Five Countries* (Washington, D. C.: American Association for the Advancement of Science, 1985), p. 13.
[40]Troost, "Science Education," p. 16.
[41]Troost, "Science Education," p. 18.

TABLE   5.10
....

MAJOR ITEMS IN THE JAPANESE JUNIOR HIGH SCIENCE CURRICULUM
....

Area I: Physics and Chemistry	Areas II: Biology and Earth Science
Grade 7	**Grade 7**
Substance and reaction	Kinds of living things and their lives
Forces	Earth and the universe
Grade 8	**Grade 8**
Substance and atom	Physical mechanisms of life
Electric current	Weather changes
Grade 9	**Grade 9**
Substance and ion	Relationships between living things
Motion and energy	Changes in Earth's crust
	Human beings and nature

Nigeria[43]

In population (about 100 million), Nigeria is larger than any other nation in Africa or Europe, except for the USSR. It is one of the leading producers of energy (in the form of oil). A former colony of Great Britain, Nigeria gained its independence in October 1960. Of the 48 countries of sub-Saharan Africa, Nigeria has the ninth largest per capita income. Its languages include English (official), Ibo, Hausa, and Yoruba; its ethnic groups include Hausa, Fulani, Yoruba, and Ibo; and 47 percent of the population are Moslem, 34 percent Christian, and 18 percent indigenous beliefs.[44]

Education in Nigeria is compulsory for students ages 6 to 17. The educational system is organized into a 6-3-3 plan. The Federal Ministry of Education develops a core curriculum for primary and secondary education (divided between junior and senior levels).

The organization having the greatest influence on science education is the Science Teacher Association of Nigeria (STAN). At the 1990 annual conference of STAN, the association made the following recommendations for Nigerian science education:[45]

- Nigeria should embark on programs designed to achieve scientific literacy for all by the year 2000
- Computer education should be introduced at all levels of education as soon as possible
- Teachers should adopt a humanistic approach to teaching that will make the learning of science, technology, and mathematics (STM) attractive to the learner
- STM curricula should be reviewed to ensure that they are based on the environment. STM teachers should also draw examples from their local environment. This will help to improve learning and also demystify science. The use of low-cost equipment should also be emphasized
- A well-articulated integrated science program should form the foundation for achievement of scientific and technological literacy for all
- The preservice preparation of science teachers should be approached with a greater degree of seriousness than in the past. Degree programs should be designed in integrated science.
- Educational research in STM should be directed toward strategies for improving learning through problem solving.

Primary Science. Science education at the elementary level is outlined in a national syllabus entitled Integrated Science for Primary Schools. The primary science curriculum can be described as integrated and spiral. Each year, students are introduced to a repeating series of topics at levels appropriate to the age of the students. The topics include air (1 to 5), animals (2 to 6), changes in nonliving things (5), classification and colors (3 and 4), earth and sky (5 and 6), electric circuits (5 and 6), environment (1 and 6), food (1 to 4), force (3 and 4), health and safety, and heat (4 and 5), housing and clothing (1 to 4), light (3 to 5), magnets (6), measurement (3 to 6), modeling and relevant technology, and plants (2 to 5), rocks and minerals (5 and 6), senses (1 and 2), sound (3), and water (1 to 6).[46]

Secondary Science. Secondary science is based on a two-tier system, similar to China and Japan, in which the curriculum is organized into junior-level (three years) and senior-level (three years) courses. The science curriculum at the senior level is an integrated, spiral curriculum consisting of courses in agricultural science, biology, health science, chemistry, and physics.

At the junior level, an integrated curriculum and textbook and laboratory-oriented program named the Nigerian Integrated Science Project was developed by the STAN. This science program was designed to fit the three-year junior-level science curriculum. The curriculum materials have been designed to embody a hands-on approach to science teaching based on principles of discovery learning. The authors of the three-year course describe the materials as follows: "The level of language used in the text is simple and vivid. The illustrations are appropriate and relevant to the day-to-day experience of the Nigerian child. Most of the suggested *pupil activities* can be easily carried out in the average Nigerian learning environment. The teacher will receive adequate guidance from the statement of objectives, the summary, and the exercises associated with each unit."[47]

Sample units (from the second year) around which topics are organized include you as a living thing, you and your home, living components of the environment, nonliving components of the environment, and controlling the environment. The pupil textbooks are designed in a spiral approach. Students are frequently asked to relate the topic of study to work they did the previous year. The texts include many activities integrated into the books.

The curriculum for the senior level is a spiral approach and includes study in agricultural science, biology, chemistry, and physics. According to the National Curriculum for Senior Secondary Schools, the emphasis on teaching science should include "field stud-

(Continued on p. 167)

(Continued from p. 166)
ies, guided-discovery, laboratory techniques and skills coupled with conceptual thinking."[48]

The agricultural science curriculum emphasizes guided discovery and learning by doing in that students are able to produce food and other agricultural products for themselves and their communities. The program is organized around the three concepts of production, protection, and economics; all concepts are organized into six units: basic agricultural concepts, crop production, animal production, agricultural ecology, agricultural engineering, and agricultural economics.

The biology curriculum is organized around seven conceptual structures: life, ecology, plant and animal nutrition, conservation of matter, variations and variability, evolution, and genetics. The three years of biology are intended to help students acquire

1. Adequate laboratory and field skills in biology

2. Meaningful and relevant knowledge in biology

3. The ability to apply scientific knowledge to everyday life in matters of personal and community health and agriculture

4. Reasonable and functional scientific attitudes

The chemistry curriculum is organized around the major concepts of energy, periodicity, and structure. The authors of the chemistry program assumed that most chemistry concepts can be subsumed under these concepts. The curriculum thus is composed of three years of study organized into 16 teaching units based on the following chemical principles: the particulate nature of matter, periodicity, chemical combination, quantitative aspects of chemical reaction, rates of reaction, equilibrium, carbon chemistry, and industrial applications of chemistry.

As with all the other sciences, the guided discovery approach to teaching has been used to design the physics syllabus. Physics teachers are strongly encouraged to use

a student-centered and inquiry-based approach to teaching. The content of the physics program is organized around two major themes: motion and energy. The curriculum for each of the three years is organized around five topics: space, time and motion; conservation principles; waves; fields; and quanta. □

[43]This section on Nigerian science education was based on material provided by and in consultation with, Dr. Babatunde Abayomi, professor of science education at Albany State College, Georgia.
[44]Statistics from Sanford J. Ungar, *Africa: The People and Politics of an Emerging Continent* (New York: Simon & Schuster, 1989), pp. 521–544.
[45]Science Teacher Association of Nigeria, *STAN Bulletin* 7, no. 3 (October 1990), p. 7.
[46]Federal Ministry of Education, *Integrated Science for Primary Schools* (Lagos, Nigeria: Nigeria Educational Research Council, 1982).
[47]*Nigeria Integrated Science Project: Pupils' Textbook Two* (Jericho, Ibadan, Nigeria: Heinemann Educational Books [Nigeria], 1984), p. iii.
[48]*National Curriculum for Senior Secondary Schools*, vol. 3, Science (Lagos, Nigeria: Federal Ministry of Education, 1985), p. 20.

Russia

The Russian educational system provides a general education to all of its citizens. According to a study of the Soviet mathematics and science curriculum, on which the present Russian system is based, practically all students (99.3 percent) complete secondary school, and all participate in a compulsory science and mathematics curriculum.[49] This is in stark contrast with science and mathematics curricula in the United States, in which less than one-sixth of American students take high school physics (whereas all students take five years of physics), and one-third of American students take chemistry (Russians take four years).

Although most American students take high school biology, usually in grade 10, Russian students study biology for six years!

The Russian educational system, like most of the other countries reviewed here, has been a centralized system controlled by the State Committee on Public Education. However, because of the policies of perestroika (restructuring), and glasnost (openness), the August 1991 revolution, and the transformation of the Soviet Union into the Commonwealth of States, education decision making has shifted away from the center and into the various republics and regional school districts. Even prior to these events, the trend was to give local schools more autonomy in curriculum content decisions (as

much as 40 percent), and textbook selection (both resulting from perestroika), the decision on the nature and scope of the curriculum is a national one.

The overall curriculum for Russian science is shown in Figure 5.10. All students are introduced to science at the junior level (classes 1 to 5), and then begin their study of science in earnest in the sixth level. Let's begin by looking at science at the junior level.

Junior Level Science. The science curriculum begins with a two-year experience called "The World Around" in levels 1 and 2. The focus of this experience is the exploration of the natural

(Continued on p. 168)

FIGURE 5.9 Ludmila Bolshakova's 10th level chemistry class, School 157, St. Petersburg. Credit: Debra Ault-Butenko

(Continued from p. 167) environment including gardens, parks, the countryside, changing seasons, and caring for pets and indoor plants.

This is followed by three years of nature study in classes 3 through 5. Nature study involves the students in developing skills in observation, inquisitiveness, and the love of nature. The aims of the three years of nature study include

- Observing the weather, ants, behavior of wild creatures, and consequences for human health and agriculture
- Focusing on natural surroundings, the human body, health and fitness, and the concern for the health of the people
- Development of a holistic picture of the natural world
- Emphasizing the processes of science, including observing natural objects, comparing

and contrasting, deduction, and generalizing

The content of nature study includes Earth and other bodies, air and the water cycle, conservation, rocks and soil, living organisms and their environment, the effect of human activity on the living world, and ecology.

Senior Level Science. The science curriculum from classes 6 to 11 is described as a spiral curriculum, contrasting it with the nonspiral (layer cake) model of the American curriculum. In the Russian's spiral science curriculum each science is taught over a period of several years. For example, in class 6, students begin their study of biology, which continues for the next five years. (Figure 5.10). Notice that in each of the next two years students pick up first physics and then chemistry

and continue studying these subjects each year. Let's look at each subject.

The biology curriculum in the sixth class is botany, and students meet for two periods of instruction per week. Zoology is introduced in the seventh level and continues into the eighth level. Human anatomy and physiology is taught in level 9, and general biology is offered in the 10th and 11th levels.

Students study physics two periods per week during the seventh and eighth levels. The content of physics in the seventh level includes physical phenomena, the structure of matter, the interaction of bodies, pressure, work, power, and energy. In the eighth level students study thermal, electrical, electromagnetic, and light phenomena. The physics curriculum is more intense in the next three years, with students meeting

(Continued on p. 169)

Russian Curriculum: Subjects and Grades

	1	2	3	4	5	6	7	8	9	10	11
The World Around	▬	▬									
Nature Study			▬	▬	▬						
Biology						▬	▬	▬	▬		
Physics							▬	▬	▬	▬	
Chemistry								▬	▬	▬	
Computer Technology										▬	
Astronomy											▬
Mathematics	▬	▬	▬	▬	▬	▬	▬	▬	▬	▬	

FIGURE 5.10   Russian Science Curriculum

(Continued from p. 168)
three years, with students meeting three periods per week in level 9, and four periods each week in the last two years. In level 9, students study kinematics, conservation laws, and waves. In the 10th level they study molecular physics, and electrodynamics, followed in the 11th level with a continuation of electrodynamics, as well as electromagnetic waves and quantum physics.

The chemistry curriculum begins in level 8. The chemistry curriculum is focused on exploring two basic systems of knowledge, namely, substances and processes. The Soviet approach to chemistry includes the development of a conservationist attitude to natural resources. The content in the 8th and 9th levels is the study of inorganic chemistry, while in the 10th and 11th levels the focus is on organic chemistry.

In the last year of school students also take a course in astronomy that introduces a contemporary view of the universe, focusing on the practical use of astronomy. Successes in space and the developments of international cooperation are also part of the course.   □

[49]I. Wirzup, "The Soviet Challenge in Science, Mathematics and Technology Education, " in M. S. Klein, and F. J. Rutherford, eds., *Science Education in Global Perspective: Lessons from Five Countries* (Washington, D.C.: American Association for the Advancement of Science, 1985), p. 177.

Case Study 1.
The Science Proficiency Race

The Case. At a recent conference on science teaching it was reported that

- U.S. fifth graders performed at about the average level of the 15 countries in an international study
- U.S. grade 9 students and advanced science students (second year biology, chemistry, and physics) had lower performance levels than their counterparts in most other countries
- Only 42 percent of the U.S. 13 year olds demonstrated an ability to use scientific procedures and analyze scientific data, compared with more than 70 percent in Korea and British Columbia

A professor from a very prestigious American university reported the results to an audience of about 500 science educators. To make the results more visual, the professor showed a number of graphics showing the results among more than ten countries who had participated in an international test of science proficiency. One graphic that was shown compared ninth-grade students' ability to analyze experiments (Figure 5.11). On this ability the United States ranked ninth out of 12 competing countries. A teacher from the audience disputed the professor's results, claiming that these other countries have different goals and commitments to science and that the comparison wasn't fair.

A number of other teachers seemed to agree because they started nodding their heads in agreement. The professor, who appeared unshaken by the response, went on to describe other areas where U.S. students lagged behind their counterparts.

The Problem. If you were in the audience, how would you react to the professor's test results? Would you side with the teachers? Why?   □

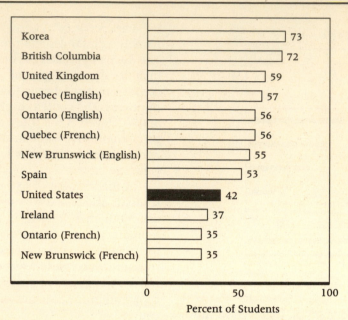

FIGURE 5.11 Percentage of 13 year olds able to analyze experiments (Level 500). Iris R. Weiss et al., *Science and Mathematics Education Briefing Book* (Chapel Hill, NC: Horizon Research, Inc., 1989). Source: *A World of Differences*, 1988. © Horizon Research, Inc., Chapel Hill, NC, 6/89.

Problems and Extensions

- Suppose that you were asked to make recommendations to a school system on the content for the ideal course in biology, chemistry, physics, or earth science for high school students. (a) What are the aims of the course? (b) What are the major curriculum topics?

- Evaluate one of the biology, chemistry, or physics "exemplary" science programs, and provide evidence to support the program's effectiveness with high school students.

- Find and locate one research study that compares the effectiveness of a golden age high school science course, for example, biology, chemistry, or physics with traditional science programs. What variables were used to make the comparisons?

- Visit a high school science teacher, and interview the teacher with regard to his or her science course. What program or textbook does the teacher use? How effective is the course with the teacher's students?

RESOURCES

Champagne, Audrey B. *Science Teaching: Making the System Work.* Washington, D.C.: American Association for the Advancement of Science, 1988.

This is one of a series of books published annually by the AAAS, entitled This Year in School Science. This volume describes the current condition of science teaching and assesses progress toward achievement of a national educational goal: establishing and achieving a sufficient supply of teachers.

Champagne, Audrey B., and Leslie E. Hornig. *The Science Curriculum.* Washington, D.C.: American Association for the Advancement of Science, 1987.

This outstanding volume is designed to provide a source of information and ideas about the future of the school science curriculum. The volume raises critical questions and provides tentative answers about the science curriculum.

Champagne, Audrey B., and Leslie E. Hornig. *Science Teaching.* Washington, D.C.: American Association for the Advancement of Science, 1986.

This volume contains a series of statements and papers on the state of science education by leading educators, politicians, and scientists.

Champagne, Audrey B., Barbara E. Lovitts, and Betty J. Calinger. *Scientific Literacy.* Washington, D.C.: American Association for the Advancement of Science, 1989.

This volume includes papers that describe the elements of the diverse ideas about the characteristics of scientific literacy, written by science teachers, philosophers, science educators, and scientists.

Committee on High School Biology. *Fulfilling the Promise: Biology Education in the Nation's Schools.* Washington, D.C.: National Academy Press, 1990.

This book offers a vision of what biology education could be, along with practical suggestions for putting the recommendations into practice.

Purves, Alan C. *International Comparisons and Educational Reform.* Alexandria, Va.: Association for Supervision and Curriculum Development, 1989.

This excellent book, edited by the chairperson of the International Association for the Evaluation of Education Achievement (IEA), contains a collection of papers on results and programs of this international organization.

6

SCIENCE, TECHNOLOGY, AND SOCIETY IN THE SCIENCE CLASSROOM

Imagine secondary students inspecting bags of garbage, counting bottles and cans that their family uses in a week, or collecting the exhaust gases from the school principal's car. Odd activities for a science class? Not exactly. In fact, a growing number of science teachers are designing lessons similar to the following:

- Organizing a recycling campaign in the school in which all cans, bottles, paper, plastic containers and bags, and newspapers are separated, counted and recycled.
- Debating issues such as, In an environmentally sustainable global economy should most people ride bicycles (or use some energy efficient and environmentally safe form of transportation) rather than drive cars?
- Participating in an electronic international conference by linking classrooms around the globe with computers to discuss and exchange ideas with fellow students in other nations about environmental problems such as global warming, ozone depletion, acid rain, or rainforest degradation.
- Engaging in a brainstorming session to identify environmental problems and issues that impinge on their lives. Teams of students will be organized to select one of the problems and then plan a project to take action to solve the problem.

These and other activities similar to them enable secondary students to explore real issues in the classroom. The activities give a context for science learning: the local community, the school grounds, the water system in a school, an automobile, a local pond or lake, the local atmosphere. Although not limited to local environments, acting locally on prob-

lems that have relevance to students increases the chances for participation and action taking. These activities, and indeed this approach to science teaching, has been called the science-technology-society (STS) approach and represents a major trend in science education for the 1990s. This chapter will be devoted to exploring STS not only from a theoretical point of view but, more important, from the practical aspect of implementing STS in the classroom.

PREVIEW QUESTIONS

- What are the characteristics of environmental education and STS programs?
- What strategies do science teachers use to present STS lessons in the classroom?
- What are some of the significant STS themes, and how do teachers present them in the classroom?

FIGURE 6.1 The science curriculum can engage students in discussions, activities, and projects that involve real issues as advocated by the STS approach. Copyright 1990 Tony Freeman.

- What are some STS curriculum examples used in today's secondary science classrooms?

- How are STS modules evaluated? Are there criteria that science teachers agree on?

THE NATURE OF STS

The central premise of STS teaching is to help students develop the knowledge, skills and effective qualities to take responsible citizenship action on science and technology oriented issues.[1] According to a number of science educators, much can be learned about how to develop successful STS lessons, modules, and courses by examining a closely related movement, the environmental education (EE) movement.

Environmental educators, motivated by the first Earthday celebration in April 1970, recognized the need for educational programs that would develop an awareness and appreciation for the environment. Furthermore, these educators, who represented a large and growing number of scientists, teachers, and ordinary citizens around the world, were concerned about the serious deterioration of the atmosphere, water resources, biosphere, and other Earth systems. Environmental educators began to focus on teaching materials that enabled students to explore the local environment and develop skills and abilities that would enable them to do so. In general, environmental education involved the following characteristics:[2]

- EE programs are typically oriented toward a problem
- EE focuses on realistic situations
- EE helps students develop alternatives that exist for situations and the skill of choosing between them
- EE empowers students to take action on issues and problems
- EE uses the real environment of the school and its surroundings as a context
- EE involves the clarification of values
- EE aspires to increase the ability of students to improve their own environmental situation

Environmental education programs such as Project Learning Tree and Project Wild (see pages 192–193) were developed in the 1970s. They are examples of curriculum projects that are used in the schools today and serve as a model for EE in the classroom.

STS programs parallel the goals of EE programs, and indeed, they are often difficult to differentiate. STS was given impetus for inclusion in science education curricula through the efforts of Project Synthesis. Project Synthesis outlined the following as characteristics of STS programs:[3]

- STS programs should prepare students to utilize science for improving their own lives and coping with an increasingly technological world
- STS programs should prepare students to utilize science to deal responsibly with science and technology oriented issues
- STS programs should identify fundamental knowledge in science and technology so that students can deal with STS issues
- STS programs should provide the student the appropriate expertise and experience to make decisions and take advantage of career options in science and technology

STS and EE programs are oriented in the same direction and place a great deal of emphasis on problem solving, empowering students to take action, the application of scientific knowledge to real issues and problems, and an awareness of careers.

For purposes of discussion in this chapter and in the remainder of this book, we will let the concept of STS subsume EE programs, and when we speak of or make reference to STS, it is understood that EE is included as well. Whether it is an STS or an EE program, there are at least five elements that characterize these programs: they are (1) problem oriented, (2) interdisciplinary, (3) relate science to society, (4) they promote global awareness and (5) incorporate relevance.

---

[1]See Randall L. Wiesenmayer and Peter A. Rubba. "The Effects of STS Issue Investigation and Action Instruction and Traditional Life Science Instruction on Seventh Grade Students' Citizenship Behaviors," paper presented at the National Association for Research in Science Teaching meeting, Atlanta, Georgia, April, 1990.

[2]Based on P. J. Fensham, "Changing to a Science, Society and Technology Approach," in *Science and Technology Education and Future Human Needs*, J. L. Lewis and P. J. Kelly, eds. (Oxford: Pergamon, 1987), p. 72.

[3]E. Joseph Piel, "Interaction of Science, Technology, and Society in Secondary Schools," in *What Research Says to the Science Teacher*, vol. 3, Norris C. Harms and Robert E. Yager, eds. (Washington, D.C.: National Science Teachers Association, 1981), pp. 94–112.

Some Characteristics of STS and EE Programs

Problem and Issue Oriented

STS and environmental education teaching is problem and issue oriented. Students often choose the problem or issue they will investigate, which is contrary to most of the schooling process. Problem- and issue-oriented teaching involves two important dimensions of learning that characterize STS and EE teaching. These dimensions are

1. Anticipation
2. Participation

Anticipation in STS and EE teaching implies that students will develop the capacity to face new situations. Anticipation is the ability to deal with the future, to predict coming events, and to understand the consequences of current and future actions. Anticipation also implies "inventing" future scenarios and developing the philosophy that humans, especially ordinary citizens, can influence future events.

Participation, on the other hand, is the complementary side of anticipation. Students must participate directly in learning. As was pointed out in Chapter 2, cognitive psychologists have developed powerful ideas that suggest that knowledge is constructed by the learner, not transmitted from teacher to student. STS and EE programs underscore participation by designing activities and strategies that involve students in a process or a series of stages, beginning with problem identification and ending with action on the problem or issue.

According to many educators, the decade of the 1990s will be the "environmental education" decade, and educators around the world will be focusing attention on STS and EE teaching, but from a global perspective.[4] Students will learn that the problems and issues they tackle locally will not only be similar to problems and issues people encounter elsewhere but will have global consequences. James Botkin, Mahdi Elmandjra, and Mircea Malitza in their report *No Limits to Learning* make the point very clearly:

> Participation in relation to global issues necessarily implies several simultaneous levels. On the one hand, the battleground of global issues is local. It is in the rice fields and irrigation ditches, in the shortages and overabundances of food, in the school on the corner and the initiation rites to adulthood. It is in the totality of personal and social life-patterns. Thus participation is necessarily anchored in the local setting. Yet it cannot be confined to localities. Preservation of the ecological and cultural heritage of humanity, resolution of energy and food problems, and national and international decisions about other great world issues all necessitate an understanding of the behaviour of large systems whose complexity requires far greater competence than we now possess. The need to develop greater competence and to take new initiatives is pressing. For example, during times of danger or after a natural catastrophe, nearly everyone participates. Can we not learn to participate constructively when animated by a vision of the common good rather than a vision of common danger?[5]

Interdisciplinary Thinking

As we will discuss later in this chapter, a number of individuals as well as groups have proposed lists of topics upon which STS and EE programs should be based. Among others, these topics include health, food and agriculture, energy, land, water and mineral resources, industry and technology, the environment, information transfer, and ethics and social responsibility.[6] These topics fall outside the traditional disciplines of biology, chemistry, physics, and earth science and instead require an interdisciplinary view.

Students working on any one of the foregoing topic areas will be required to gather information from several disciplines, thereby bringing the student into interdisciplinary spaces. The content of biology will be seen as relevant to chemistry if students investigate health problems. Yet as you look at the problems, students will not be confined to the disciplines of science. Philosophy and psychology (especially since most of the STS and EE problems and issues involve ethics, values, and decision making) will be important to students, as well as geography, history, and sociology (Figure 6.2).

Connect Science to Society

Peter Fensham made an important point: "most of the great reforms in science curriculum looked inwards to science and scientific research for their inspiration." On the other hand STS and EE curricula "look outwards from science to society to see how science is, or could be, applied" (Figure 6.3).[7]

---

[4]See Victor J. Mayer, "Teaching from a Global Point of View," *Science Teacher*, January, 1990, pp. 47–51.

[5]James W. Botkin, Mahdi Elmandjra, and Mircea Malitza. *No Limits to Learning*. (Oxford: Pergamon, 1979), p. 31. Used with permission.
[6]Suggested by the Bangalore Conference on Science and Technology Education and Future Human Needs. See J. L. Lewis and P. J. Kelly, *Science and Technology Education and Future Human Needs* (Oxford: Pergamon, 1987).
[7]Fensham, "Changing," p. 69.

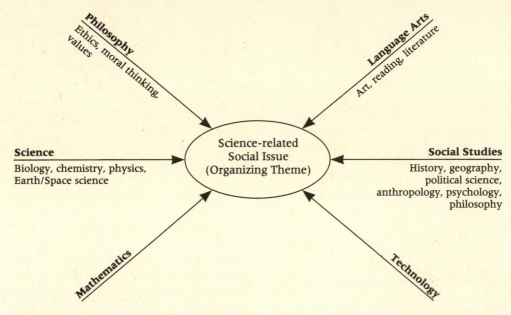

FIGURE 6.2 Interdisciplinary Thinking Model

The STS and EE teacher or curriculum developer looks first to important social issues (where will all the garbage go? earthquake preparedness, human diseases, pesticides and home gardens, to name a few) and then helps students explore the "appropriate" science content that is relevant to the social issue or problem. In this sense, science content is not seen in isolation (as it is typically presented in textbooks and by teachers) but rather in a real context that is related to students, community or personal lives.

Global Awareness and the Gaia Hypothesis

Global awareness was surely manifested when the first pictures were sent back to Earth by Apollo astronauts giving the single-celled picture of Earth. Yet global awareness is more than a visual picture of the earth, it implies something more powerful. Global awareness implies that all things are connected, that the atmosphere over Toledo can affect trees in

Boston, that removing trees from the forests of Brazil could change the temperature of Moscow, and recycling newspapers could reduce the chances of oil spills. The environmental motto Act Locally, Think Globally has relevance here.

And just as the space age has given us new visual images of Earth, it has led to new questions and theories. One of the scientists to work on the Martian project that looked for signs of life on the "red planet" was James Lovelock. Lovelock and his colleagues on the Martian project devised a number of "life-detection" experiments. One of their suggestions was that a planet bearing life might have an unexpected mix of gases in its atmosphere if life's chemistry were at work. Dr. Norman Myers, editor of *GAIA: An Atlas of Planetary Management*, described Lovelock's breakthrough this way:

> When they looked at Earth in this light [having an unexpected mix of gases], their predictions were borne out with a vengeance. Earth's mix of gases, and temperature, were hugely different from what they predicted for a "nonliving" Earth, as well as from neighboring planets. The fact that these conditions appeared to have arisen and persisted alongside life led to the Gaia hypothesis—the proposal that the biosphere, like a living organism, operates its own "life-support" systems through natural mechanisms.[8]

What Lovelock and microbiologist Lynn Margolis, coauthor of the Gaia hypothesis, suggested was that

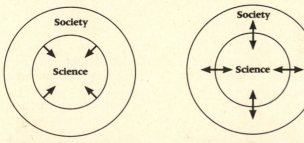

FIGURE 6.3 Traditional vs STS Science Curricula: STS Curricula look outward from science to society to show how science can be applied.

[8]Norman Myers, ed., *GAIA: An Atlas of Planetary Management* (Garden City, N.Y.: Anchor, Doubleday, 1984), p. 13.

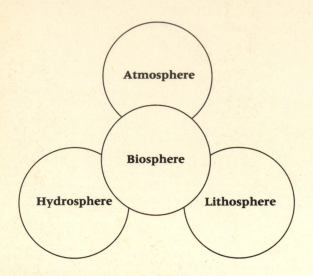

FIGURE 6.4   Interacting/Interdependent Elements of Gaia.

the earth's atmosphere was not simply a product of the biosphere but was "a biological construction—like a cat's fur, or a bird's feathers; an extension of a living system, designed to maintain a chosen environment."[9]

The Gaia hypothesis is a useful tool to help students think about the interrelationship of Earth's basic resources—energy, water, air, and climate. According to the Gaia hypothesis these elemental resources can be radically affected by changes in any one of them. Many of the STS themes that are identified later in this chapter are evident in one of these elemental resources. Students will discover that management of these elemental resources is what many environmental action groups advocate.

Global awareness and the philosophy of the Gaia hypothesis lead to a new way of reasoning about the earth, its environment, and inhabitants, namely global thinking. Global thinking is systemic thinking in which whole systems are perceived, as well as the consequences of changes in any aspect of the system. Global thinking is not a foreign concept to students. Here are remarks made by middle school students about global thinking. They responded to the question, What, in your opinion, is the meaning of global thinking?[10]

• Global thinking is thinking how our actions and reactions affect all the world

[9]James E. Lovelock. "Elements," in *GAIA: An Atlas of Planetary Management,* Norman Myers, ed. (Garden City, N.Y.: Anchor, Doubleday, 1984), p. 100.
[10]These student comments were provided by Phil Gang, The Institute for Educational Studies, Atlanta, Ga. The comments were made by junior high students in a workshop on global thinking conducted by Dr. Gang.

• Thinking globally means the world must come together and solve the problems of today so that the world of tomorrow can be a better place
• To me, global thinking means being in tune with the different cultures of the world, and everybody having its environmental problems and social issues strongly at heart
• Thinking of the world as a whole with differences but not divided

Relevance

A useful way to understand STS is to contrast STS programs with traditional or standard science education programs (Table 6.1). STS programs tend to be problem oriented, whereas traditional science programs are based on concepts found in the textbook or curriculum guide. STS programs depend on a facilitative teacher who organizes the learning environment so that students engage in identifying problems, seek resources and information to learn about the problem, and develop action plans to solve the problem. STS teaching deals squarely with the problem of relevance by involving the students in problems that relate to the students' world.

TABLE   6.1

TRADITIONAL VERSUS STS SCIENCE PROGRAMS

Traditional or Standard Science Programs	STS Programs
Survey of major concepts found in standard textbooks	Identification of problems with local interest or impact
Use of labs and activities suggested in textbook and accompanying lab manual	Use of local resources (human and material) to locate information that can be used in problem resolution
Passive involvement of students assimilating information provided by teacher and textbook	Active involvement of students in seeking information that can be used
Science being contained in the science classroom for a series of periods over the school year	Science teaching going beyond a given series of class sessions, meeting room, or educational structure
A focus on information proclaimed important for students to master	A focus on personal impact that makes use of students' own natural curiosity and concerns

(Table continued on p. 177)

(Table continued from p 176)

Traditional or Standard Science Programs	STS Programs	Traditional or Standard Science Programs	STS Programs
A view that science is the information included and explained in textbooks and teacher lectures	A view that science content is not something that merely exists for student mastery simply because it is recorded in print	Science occurring only in the science classroom as a part of the school's science department	Students learning what role science can play in a given institution and in a specific community
Practice of basic process skills—but little attention to them in terms of evaluation	A deemphasis of process skills that can be seen as the glamorized tools of practicing scientists	Science being a study of information in which teachers discern the degree students acquire it	Science being an experience students are encouraged to enjoy
No attention to career awareness, other than an occasional reference to a scientist and his or her discoveries (most of whom are dead)	A focus upon career awareness—emphasizing careers in science and technology that students might expect to pursue, especially those in areas other than scientific research, medicine, and engineering	Science focusing on current explanations and understandings; little or no concern for the use of information beyond classroom and test performance	Science with a focus on the future and what it may be like
Students concentrating on problems provided by teachers and text	Students becoming aware of their responsiblilties as citizens as they attempt to resolve issues they have identified		

Robert E. Yager. "STS: Thinking over the Years," *The Science Teacher*, March, 1990, pp. 52–55. Reproduced with permission of *The Science Teacher*. Copyright 1990 by the National Science Teachers Association, 1742 Connecticut Avenue, NW, Washington, D.C. 20009

STRATEGIES FOR TEACHING STS IN THE CLASSROOM

An important aspect of the STS movement is that science teaching must go beyond teaching information and help students clarify their values, develop the skills to take action on issues, and learn how to discuss the moral and ethical implications of science. These foci (clarifying values, action skills, and moral-ethical implications) underlie three strategies that STS science educators advocate.

Clarifying Values

- Chemicals are harmful to humans
- Nuclear power plants should be banned
- Fast foods are not nutritious and should be avoided
- Smoking should be banned in public settings

STS teaching creates an environment in which students are confronted with situations in which they must clarify their values on a variety of issues. The nature of today's value-laden issues and problems is contrary to the neutral or value-free connotation of science. How can students be helped to clarify their values about science issues?

The question that always arises in discussions of values teaching is what values should be taught, if they should be taught at all. In a class discussion of abortion, what is the role of the teacher, and whose value beliefs should guide the direction of a decision? Most educators would claim that class activities based on value-laden issues should involve a process of valuing as opposed to a particular value direction.

A process model of values clarification was developed by Raths, Harmin, and Simon and is known as the values clarification approach.[11] They define a value in terms of three key valuing processes: choosing, prizing, and acting (Figure 6.5).

According to Raths, Harmin, and Simon, the valuing process, whether the subject be abortion, drugs, birth control, health, population control, or the social responsibility of scientists, collectively involves seven criteria. In the spirit of science, Raths, Harmin, and Simon claim that these criteria must emerge in a climate of free inquiry independent of authoritarian persuasion. Table 6.2 outlines the major aspects of the valuing process as it relates to science teaching.

[11]Louis Raths, Merril Harmin, and Sidney Simon, *Values and Teaching* (Columbus, Ohio: Charles E. Merrill Publishng Company, 1966).

TABLE 6.2

VALUES CLARIFICATION PROCESS

Key Valuing Process	Student Actions	Criteria	Key Teacher Questions
Acting	Repeating	At this level, the action will reappear a number of times	How did you decide which had priority?
	Acting on choices	Demonstrating choices by integrating the choice in our life	How has it already affected your life?
Prizing	Affirming	Willing to affirm the choice publicly	Would you explain to others how you feel about this?
	Prizing and cherishing	The individual esteems, respects, and holds the value dear	Are you glad you feel that way?
Choosing	After thoughtful consideration	Reflective thought based on research and investigation	What is the basis for your choice?
	Choosing from alternatives	Examining alternatives and then making a choice	What ideas did you reject before making this choice?
	Choosing freely	Choosing without coercion	Where did the idea for your choice come from?

After Raths, Harmin, and Simon, *Values in Teaching.*

One of the keys to implementing values clarification in the science classroom is the utilization of methods that enable students to choose, prize, and act on STS issues. Five methods are presented that are designed to help students develop clearer values about STS issues. These include the value dilemma sheet, STS action dramas, action voting, case studies, and STS action projects.

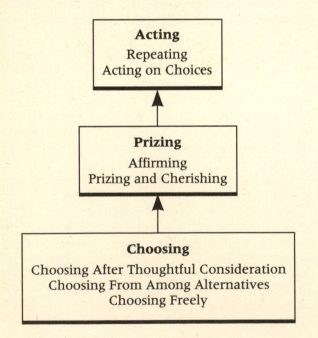

FIGURE 6.5 A Values Clarification Model. After Raths, Harmin, and Simon, *Values in Teaching.*

The STS Value Dilemma Sheet

The STS value dilemma sheet consists of a provocative statement (or illustration) and a series of questions. The purpose of the provocative statement (or illustration) is to raise an STS issue that has value implications for students. The questions are designed to take the student through the value clarifying process (choosing, prizing, acting). The STS value dilemma sheet should be used individually with students. They should be given the chance to reflect on the dilemma, write their responses, and then perhaps participate in a small group discussion. Large group discussions do not necessarily help students clarify their values. Here is an example.

STS Value Dilemma Sheet 1: Nuclear Reactors

In the Spring of 1979 a reactor at the Three Mile Island nuclear power plant in Pennsylvania suffered a partial core meltdown, releasing radioactive material into the surrounding environment. Thousands of people in the vicinity claim to suffer from cancer or thyroid damage as a result of the accident. Thus far cleanup has cost close to $1 billion, not including the cost of the reactor itself (about $2 billion) or the more than 2500 lawsuits filed by nearby residents.

On April 26, 1986, in what was the world's largest nuclear disaster ever, a reactor at the Chernobyl nuclear power plant in the Ukraine exploded, releasing vast quantities of radioactive material into the atmosphere. Clouds of fallout covered large areas of Europe, contaminating food supplies and increasing the rate of cancer in human beings. The ongoing cleanup has cost $14 billion so far, and over 250 people have died.

Following the Chernobyl disaster, one NRC member estimated the chance of an accident in the United States as big or bigger occurring by 2005 to be as high as 45 percent. In 1989, citizens of Sacramento, California, voted to shut down the publicly owned Rancho Seco nuclear power plant because it was unsafe and uneconomical.

There are 560 commercial nuclear power plants in operation worldwide, 112 of which are in the United States.

1. Is there justification for building new nuclear power plants?
2. Do you think nuclear power plants should continue to be part of the world's energy sources?
3. Some people think that nuclear power plants are unsafe not because an accident might happen but because the nuclear industry has not figured out a safe method for discarding the radioactive waste products produced by nuclear plants. Discuss your feelings on this matter.
4. Would you buy a house or take a job that was within a mile of a nuclear power plant?
5. What would you do to show your position on this issue?

STS Action Dramas

STS action dramas, or role playing, sociodramas, or dramatic improvisation, as they are variously called, give students the opportunity to get out of their role and into a new one to explore an STS issue. STS action dramas do not have to be elaborately directed. They work best if students are left to devise their own approaches. The important element is for the teacher to establish a condition in which the students can temporarily take on a new character, express ideas, and then reflect (with the whole class participating).

Sources of ideas for STS action dramas include newspapers (especially papers that regularly feature debates), congressional decisions, actions and new laws, environmental incidents (accidents), and STS issues featured in textbooks.

No More Bikes!

The city council has decided to ban bicycles on main roads from 7 A.M. to 9 A.M. and 3 P.M. to 6 P.M. on weekdays because of the number of bicycle-related accidents. Students at the city's only middle school have protested and will present their petition and arguments at the next council meeting.

Voting

Voting is a form of survey research in which students are asked to vote on questions related to a particular topic. Voting is an action that citizens can take regarding issues, and their vote affirms the choice they have made on an issue.

To use the voting method in the classroom, prepare a ballot similar to the one shown in Figure 6.6. Have each student complete the ballot independently, and then summarize the results on chart paper or on the chalkboard.

Case Studies

The case study is a popular teaching strategy to help students learn STS principles by analyzing an STS issue. Agne points out that the case study can be used to study science-technology interactions, can be used quite readily throughout the year, and simply requires the identification of a relevant case for student analysis.[12] Case studies are readily available in most contemporary science textbooks. Most textbooks contain one or more case studies for each instructional

---

[12]R. M. Agne, "Teaching Strategies for Presenting Ethical Dilemmas," in *Ethics and Social Responsibility in Science Education*, M. J. Frazer and A. Kornhauser, eds. (Oxford: Pergamon, 1986), pp. 165–174.

Put a check mark (✓) if you agree with the statement. Put a double check (✓✓) if you agree emphatically. Put a cross (✗) if you disagree with the statement.

1. Birth control is a legitimate health measure.
2. Birth control is necessary for women who must help earn a living.
3. The practice of birth control may be injurious physically, mentally, or morally.
4. We simply must have birth control.
5. The practice of birth control is equivalent to murder.
6. Birth control has both advantages and disadvantages.
7. Only a fool can oppose birth control.

FIGURE 6.6 Voting Ballot: Birth Control. Biological Sciences Curriculum Study 1980, Human Reproduction: Social and Technological Aspects (Dubuque, Iowa: Kendall/Hunt, 1984) p. 57. Copyright BSCS.

unit in the course. The following list identifies the case studies used to introduce STS issues in a contemporary middle school physical science text.[13]

- Do the benefits of industrial robots outweigh their disadvantages?
- Should all forest fires in national parks be extinguished?
- Is information stored in computers a threat to people's privacy?
- How useful are instant replays in helping football officials judge plays?
- How can people be protected from lead in the environment?
- To protect the ozone layer, should chlorofluorocarbons be banned?
- What can be done about the buildup of greenhouse gases in the atmosphere?
- Should highly radioactive waste be stored in Yucca Mountain?

Action Projects

Action projects are designed to get small teams of students involved in researching and investigating a local STS or environmental issue. Issues students might investigate are limitless. Some could include

- Waste disposal
- Pollution problems and control
- Recycling efforts and conservation
- The zoning of land
- Energy issues
- Environmental protection
- Local endangered wildlife concerns

Action projects begin with the identification of a local problem and result in taking some action on the problem. What follows is a six-step procedure designed to help students carry out a project.[14]

1. *Problem identification:* Students brainstorm possible problems or situations to improve the local community.
2. *Fact-finding:* Students find out information to understand why the problem exists. They are informed that they can obtain information from (1) community resources, (2) national and international organizations, (3) opinionnaires, (4) resource people, and (5) independent and group study.
3. *Problem selection and definition:* Students choose one or two problems that if solved could make the biggest difference. They must decide if they have the tools and resources available to successfully solve the problem.
4. *Brainstorming solutions:* Students generate as many solutions to the problem as possible.
5. *Evaluating solutions:* Students rank-order the effectiveness of their solutions and choose the top two to three ideas. They list criteria to help them decide what might be the best solution. Criteria include, Will the solution make a long term difference? Are there adequate resources to help them in acting on their solution?
6. *Taking action:* In this step, students decide the different ways to carry out their solution. They decide which strategies will be most effective. They create a time line and plot a course of action.

The STS Module

Except for the action project, most of the values clarification strategies are single-lesson approaches to teaching STS. An alternative strategy is to develop an STS or EE module that would be taught as part of an existing science course. This approach seems to be more powerful. Rather than adding a day on acid rain or global warming or endangered species, the module approach advocates that teachers identify a significant science-based social issue and use the learners' interest in the topic as the basis for a two-to-four-week module within the existing science program.[15]

Although there are many commercially available STS modules on the market, the purpose of this section is to outline criteria that you might use to develop your own STS module. According to Wiesenmayer and Rubba, environmental educators developed a model of instruction in which students develop the ability and desire to take responsible environmental actions. The model, Investigating and Evaluating Environmental Issues and Actions Skill Development Modules, was developed by Harold Hungerford and his colleagues at the University of Southern Illinois.[16] The model went well beyond the environmental awareness level and

[13]Mark A. Carle, Mickey Sarquis, and Louise Mary Nolan, *Physical Science: The Challenge of Discovery* (Lexington, Mass.: D. C. Heath, 1991).

[14]Excerpted from Micki McKisson and Linda MacRae-Campbell, *No Time to Waste! The Greenpeace East-West Educational Project, Teacher Guide* (Washington, D.C.: Greenpeace International, 1989), p. 23.

[15]See Herbert D. Their, "The Science Curriculum in the Future: Some Suggestions for Experience-Centered Instruction in the Fifth through Ninth Grades," in *Redesigning Science and Technology Education,* Rodger W. Bybee, Janet Carlson, and Alan J. McCormack (Washington, D.C.: National Science Teachers Association, 1984), pp. 162–169.

[16]Harold R. Hungerford et al., *Investigating and Evaluating Environmental Issues and Actions Skill Development Modules* (Champaign, Ill.: Stipes, 1988).

achieved success in getting students involved in action taking. Details of the model are described in the STS Curriculum Examples section of this chapter. Rubba and Wiesenmayer adapted the environmental model to STS and proposed four levels or stages of activities in developing an STS module. The goal of any STS module would be to "aid citizens in developing the knowledge, skills and effective qualities needed to make responsible decisions on STS issues, and to take actions on those decisions toward issue resolution."[17] The levels of activities include level I: STS foundations; level II: STS issue awareness; level III; STS Issue Investigation; and Level IV; STS action skills development.

The starting point for an STS module is a list of possible social issues that might interest the students. Here are some suggestions:

Oil spills
Acid rain
AIDS
Solid waste disposal
Recycled paper
Garbage

[17]Wiesenmayer and Rubba, "The Effects of STS Issue Investigation and Action Instruction," p. 4.

Drinking water
Tornado preparedness

An STS module on any one of these topics should normally take between one and four weeks. The time spent on the unit should be divided among the four levels of instruction. To help you in planning an STS module, let's look briefly at each level.[18]

STS Foundations

At the foundation level, you must identify the concepts that are related to the issue in question, for example, acid rain, oil spills, or AIDS. However, in the STS approach the concepts that should be identified are those that cut across the traditional disciplines of earth, life, and physical science. Candidates for this would include unifying themes such as change, field, interaction, model, system, or energy transfer, evolution, stability, patterns of change, systems and interaction, and scale and structure. For example, in a module on oil spills, STS foundation concepts might include system, change, energy, and interaction. If students were investigating the 1988 Alaska oil spill they might explore the interaction effects of oil on the ocean-beach system.

Selecting unifying concepts and themes leads students to new ways of thinking. Students learn to con-

[18]Based on Wiesenmayer and Rubba. "The Effects of STS Issue Investigation and Action Instruction," pp. 5–6.

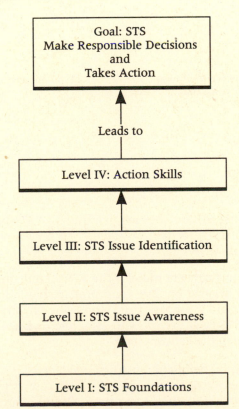

FIGURE 6.7 STS Action Model. After Hungerford, *Investigating and Evaluating Environmental Issues.*

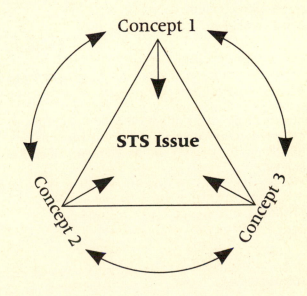

FIGURE 6.8 Concepts in Relationship to an STS Issue

nect ideas—to think holistically. Facts and concepts are not taught in a linear fashion— concept 1 → connect ideas—to think holistically. Facts and concepts are not taught in a linear fashion— concept 1 → concept 2 → concept 3—but are taught in the context of an issue. In this light concepts are seen as being interdependent and interdisciplinary.

Finally students are engaged in exploring major ideas not for their own sake but as useful constructs in helping them solve relevant social and global problems. The foundation level should also show the students how science and technology interact, and their interdependence.

STS Issue Awareness

A series of lessons should be planned to help students realize how STS interactions result in controversy surrounding an issue. The lessons would help students examine (a) all sides of the issue, (b) beliefs and attitudes affecting the issue, and (c) alternative solutions to the issue. For example, in a module on the risks from smoking, students might explore

- The negative physiological effects of cigarette smoking
- The nature of addiction and the difficulty experienced by many smokers who attempt to quit but can't
- The efforts made by the cigarette industry to attract young people between the ages of 12 and 18 years to smoking
- The attempts of the cigarette industry to keep important information about the health hazards of smoking secret

Values clarification and awareness activities such as value dilemma sheets, STS action dramas, case studies, and voting are useful tools at this level.

Issues Investigation

The issues investigation level is similar to the action project discussed in the Clarifying Values section. At this level students are taught how to investigate an issue. It would involve training with problem identification skills, including problem identification and statement, the use of secondary sources, data collection via primary sources, data analysis, and drawing conclusions. The newly acquired skills would be applied to an investigation of a real local issue.

Action Skills Development

Taking action on an issue involves an understanding of new skills that many students are unfamiliar with. Wiesenmayer and Rubba suggest that lessons should be designed to help students with consumerism, legal action, persuasion, physical action, and political action. They also point out that students should have the opportunity to apply these skills.

Wiesenmayer and Rubba found in a study of seventh graders that an STS module that employed all four levels promoted greater understanding of STS and responsible action on an STS issue (trash disposal).[19]

[19]Wiesenmayer and Rubba, "The Effects of STS Issue Investigation and Action Instruction."

INQUIRY ACTIVITY   6.1

....

STS Module Design

....

What are the elements of an STS module? In this inquiry activity you will use the four instructional levels of STS modules to sketch the elements of an STS module.

Materials

None

Procedure

1. Select an STS issue that you think would be interesting and suitable for secondary students. You might want to refer to the next section in this

chapter for some suggestions of STS themes, including issues such as tropical rainforests, nuclear energy, population growth, and acid rain.

2. Create a concept map for the STS module that you have decided on. The concept map should identify the main concepts (unifying themes) related to the STS issue. Work with someone else while you develop the concept map.

3. Using the following instructional levels, briefly describe two or three lessons for each category in the STS module. The descriptions should identify the purpose of the lesson and what the students would do.

STS Levels:
   Level I: STS foundations
   Level II: STS issue awareness
   Level III: Issues investigation
   Level IV: STS action skills development
4. Identify an evaluation plan for the STS module. How are going to find out what the students have learned?

Minds on Strategies

1. What unifying concepts did you identify? How did these unifying concepts compare with those identified by other students in your class?
2. What social action (on the part of the students) did you suggest in your module?

STS THEMES AND HOW TO TEACH THEM

The content of STS lessons, modules, and courses of study are derived from students and teachers constructing problems, and investigations from real-world issues and concerns. In this section we will explore a number of issues and concerns that we will refer to as STS themes. A number of individuals and groups have identified themes that ought to be addressed in STS lessons, modules, and courses (Table 6.3). Project Synthesis identified eight areas of concern and admitted that these may not be the "most significant" but exemplified the kinds of issues that students should explore (list 1, Table 6.3). At the international conference on Science and Technology Education and Future Human Needs held at Bangalore, India, in 1985, the conference coordinators identified eight issues that they referred to as interdisciplinary topics (list 2). They made the point that STS requires that teachers and students work outside the normal boundaries of science. The third list of topics is one that was proposed by Rodger Bybee. Bybee referred to these as global problems and surveyed scientists and engineers to find out how they would rank these global problems, which are listed in Table 6.3 according to the ranking reported in Bybee's study.

The purpose of this section is to examine some of the potential STS themes and make suggestions for actions (classroom activities) that you might take with

TABLE   6.3

STS THEMES

List 1: Project Synthesis Areas of Concern[a]	List 2: Bangalore, India, Conference Interdisciplinary Topics[b]	List 3: Global Problems proposed by Bybee[c]
Energy	Health	Population growth
Population	Food and agriculture	War technology
Human engineering	Energy resources	World hunger and food resources
Environmental quality	Use of land, water, and mineral resources	Air quality and atmosphere
Utilization of natural resources	Environment	Water resources
National defense and space	Industry and technology	Land use
Sociology of science	Information transfer and technology	Energy shortages
Effects of technological development	Ethics and social responsibility	Hazardous substances
		Human health and disease
		Extinction of plants and animals
		Mineral resources
		Nuclear reactors

[a]See Piel, "Interaction of Science," pp. 94–112.
[b]See Lewis and Kelly, *Science and Technology Education*, p. 11.
[c]Rodger Bybee, "Global Problems and Science Education Policy," in *Redesigning Science and Technology Education*, by Rodger W. Bybee, Janet Carlson, and Alan J. McCormack (Washington, D.C.: National Science Teachers Association, pp. 60–75.

students. The information and ideas proposed can be the seeds for lessons or modules that you might develop as part of your science teacher education program. The following themes will be presented:

Population growth
Air quality and atmosphere
The effects of technological development
The utilization of natural resources

Population Growth

The size of the human population effects virtually every environmental condition facing our planet. As our population grows, demands for resources increase, leading to pollution and waste. More energy is used, escalating the problems of global warming, acid rain, oil spills, and nuclear waste. More land is required for agriculture, contributing to deforestation and soil erosion. More homes, factories, and roads must be built, occupying habitat lost by other species that share the planet, often leading to their extinction. Simply put, the more people inhabiting our finite planet, the greater the stress on its resources.

Why Population Growth Is an Environmental and STS Issue[20]

It took from the beginning of time to about the year 1810 for the human population to reach 1 billion people. Just more than 100 years passed before the next billion were added, and the population doubled again to 4 billion people by 1974. By 1987 Earth was home to 5 billion human beings, and this number is

[20]Ron Anastasia and Susan Weber, "Population Growth Fact Sheet," EarthNet, A Forum on the Connect Business Telecommunications System. Used with permission of Earth Day Resources. Copyright 1990. Earth Day Resources, 116 New Montgomery St., San Francisco, Calif. 94105

growing. If the global population continues to increase at the current rate of 1.8 percent annually, it will double again in just 39 years.

A society is not sustainable when it consumes renewable resources faster than they can be replenished. In other words, an overpopulated society clears forests and uses water supplies faster than they are renewed, or pollutes faster than the environment can adapt to sustain life. By these measures, the United States and most other nations of the world are overpopulated.

Contrary to some people's impression, the population explosion has not stopped. In 1990 another 95 million people were added to the earth, more than in any previous year. At this rate, the world's population would easily surpass 10 billion and could exceed 14 billion people late in the next century. No realizable amount of improvement in agriculture, pollution control, energy efficiency, or other areas would be able to keep up with this pace of growth. Today's 5.2 billion humans is already more than our planet can handle.

The major consumers of the earth's resources are the developed countries, such as the United States. While these countries contain less than 20 percent of the world's people, they consume 80 percent of its resources. Although the United States is home to just 5 percent of the human population, we use 25 percent of the total energy. The current population of the United States is about 250 million people. At the current rate of growth we are expected to add 60 million more people in the next 50 years—110 times as many people as now live in Boston.

Vast areas of land in the United States have been cleared to support our population. Over 3 billion tons of topsoil are lost annually as a result of intensive farming and overgrazing. Large stretches of forest have been cut to provide wood and paper, leaving only 5 percent of our ancient (uncut) forests standing. In water-poor areas, high rates of growth are leading to water diversion and depleted groundwater reserves. As urban areas expand, air and water pollution are amplified.

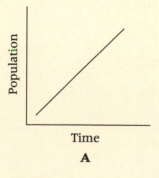
A

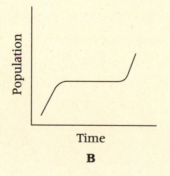
B

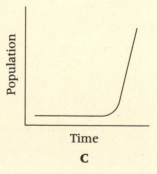
C

FIGURE 6.9 Population Graphs

As Susan Weber, executive director of Zero Population Growth, put it, "overpopulation does not happen only in the Third World. Each year, the U.S. adds the equivalent of another Los Angeles to its population. In just 35 years, the industrialized nations together will add another U.S. for the earth to support. The future depends on our putting the brakes on now."[21]

STS Actions

- Ask students to identify which graph describes the growth pattern of the human population over the past 2000 years. Then ask them to identify the impact of the human population growth pattern on

  1. The earth's atmosphere
  2. The availability of mineral resources
  3. Water resources
  4. Trees
  5. The temperature of the earth

  Use the results of this exercise to identify student misconceptions.

- Get students involved in writing elected officials to support legislation to fund family planning, develop better contraceptives, and promote equality for women, and break the cycle of poverty.
- Have students do research on the size of families in the United States and other "developed" countries, and compare with family sizes in "underdeveloped" countries. Create a values dilemma sheet based on this idea: we should encourage small families by example and by educating others about the need to make environmentally responsible reproductive choices.
- Have students find out what efforts are being made in their own community to limit the impact of growth on the environment.
- Have students make graphs showing how the population has changed in the last 20 years in their school, their community, and their state.

Air Quality and Atmosphere

The quality of the earth's atmosphere is a significant social issue, not only in the United States but in every industrialized nation on the planet. One issue that has captured the attention of governments and ordinary citizens is the effects of acid rain on the environment. Acid rain has also been the topic of some commercially produced STS units, and physical science and earth science teachers have easily incorporated an STS acid rain module into their ongoing curriculum.

[21]Anastasia & Weber.

The Acid Rain Issue[22]

Acid rain, more accurately called acid precipitation, is rain, snow, or fog that contains a significant amount of sulfuric acid or nitric acid. The primary cause of acid rain is the combustion of coal and oil, processes in which large quantities of sulfur dioxide and nitrogen oxides are released into the atmosphere. Once in the atmosphere, sulfates and nitrates combine with moisture to form sulfuric acid and nitric acid. A small percentage of acid precipitation results from natural causes, such as volcanic activity. However, approximately 90 percent is caused by humans.

The Effects of Acid Rain

Acid rain is harmful to human health, increasing the rate of infant mortality and reducing proper lung functioning in many Americans. Research has implicated airborne sulfates in at least 50,000 premature deaths each year, as many lives as are lost in automobile accidents. For people with asthma, even low levels of sulfur pollution can cause serious lung damage in a matter of minutes.

Acid rain also causes acidification of rivers and lakes, a process disruptive to aquatic ecosystems. When acidity levels become too high, fish can no longer reproduce and soon die out. High acidity levels leach heavy metals, such as mercury and lead, from lake and river beds. These metals often end up in the tissues of fish, making them toxic to humans. Acid levels in rivers and lakes are highest in the spring when snowpacks melt and release high concentrations of pollutants. Unfortunately, this happens at a crucial time in the lives of fish, during their spawning season.

Damage to tree foliage and degraded soil quality are other effects of acid rain. About half of West Germany's forests show signs of damage caused by high levels of acidity in rain and fog. In Canada and northern New England acid rain has been linked to large-scale damage to maple trees, threatening the maple syrup industry. Since factories are often equipped with tall smokestacks that send sulfur dioxide and nitrogen oxides high into the atmosphere, they may be carried by wind streams for hundreds of miles. Thus, one country will often receive another's acid rain. Half of Canada's acid rain originates in the United States.

[22]Ron Anastasia and Michael Fischer, "Acid Rain Fact Sheet," EarthNet, A Forum on the Connect Business Telecommunications System. Used with permission of Earth Day Resources. Copyright 1990. Earth Day Resources, 116 New Montgomery St., San Francisco, Calif. 94105

Numerous buildings and construction materials are also subject to damage caused by acid rain, making the problem directly economic as well as environmental. A draft EPA study estimated the damage to building materials in 17 northeastern and midwestern states to run as high as $6 billion annually. Not even our cultural history is safe, as thousands of statues and monuments are crumbling and corroding. Some of the hardest hit include the Statue of Liberty and the Gettysburg Battlefield.

STS Actions

- Ask students to make a diagram showing how they think acid rain is formed and how it affects the environment. Give the students the following list of terms that they may use to aid in the creation of the diagram: coal, power plant, smokestacks, water vapor, rain, acid, combustion products, trees, rivers. Use the results to identify student misconceptions.

- Collect rain samples over a sustainable period and determine the pH using cabbage juice (boil four or five red cabbage leaves in two quarts of water) as an indicator. The purple juice turns pink in acid solutions, and green in basic solutions. Students can also collaborate with counterparts in other schools to gather data on acid rain over a larger area.

- Participate in the National Geographic KidsNet System, which enables students in different states to share acid rain data. Write to National Geographic, Washington, D.C.

The Effects of Technological Development

The Nuclear Energy Issue[23]

Created only 45 years ago, the first nuclear reactors were used only to produce plutonium for nuclear weapons. Most reactors today make primary use of the heat energy from the nuclear reaction to produce electricity, propel navy ships, or power space satellites. To create power, radioactive uranium atoms within the nuclear reactor are split, shooting particles in every direction with great force. The particles slam into other atoms, shearing off their electrons or, in the case of a direct hit, splitting their nuclei. The process releases tremendous amounts of heat, which is

[23]Ron Anastasia and Joan Claybrook, "Nuclear Energy Fact Sheet," EarthNet, A Forum on the Connect Business Telecommunications System. Used with permission of Earth Day Resources. Copyright 1990. Earth Day Resources, 116 New Montgomery St., San Francisco, Calif. 94105

used to convert water to steam. The steam spins a large turbine, producing electricity. There are 560 commercial nuclear power plants in operation worldwide, 112 of which are in the United States.

Problems with Nuclear Power

All nuclear reactors produce plutonium and a variety of other radioactive wastes. Radiation changes the way cells divide, causing mutations that prevent organs from functioning properly. Even a small dose of radiation can cause cancer in living organisms, so radioactive waste must be kept out of the biosphere for tens of thousands of years. There are currently 18,000 tons of high-level radioactive waste in the United States alone, yet there is no permanent waste site for this material.

In the process of producing electricity, radiation permeates machinery, equipment, and the entire reactor building. After about 20 years, materials and equipment become so degraded that the plant must shut down. More than half of all U.S. reactors will have to be decommissioned by 2010, at an estimated cost of $400 million to $3 billion each. Dismantling one nuclear power plant would yield enough low-level radioactive waste to cover a football field 13 feet deep. Decommissioning all U.S. reactors would yield a wall of radioactive material 3 feet wide and 10 feet high stretching from Washington, D.C. to New York City.

Chernobyl

On April 26, 1986, in what was the world's largest nuclear disaster ever, a reactor at the Chernobyl nuclear power plant in the Ukraine exploded, releasing vast quantities of radioactive material into the atmosphere. Clouds of fallout covered large areas of Europe, contaminating food supplies and increasing the rate of cancer in human beings. The ongoing cleanup has cost $14 billion so far, and over 250 people have died. Hundreds of thousands of others will die prematurely, because radiation concentrates in body tissues where it decays slowly.

Following the Chernobyl disaster, one NRC member estimated the chance of an accident in the U.S. as big or bigger occurring by 2005 to be as high as 45 percent. In 1989, citizens of Sacramento, California, voted to shut down the publicly owned Rancho Seco nuclear power plant because it was unsafe and uneconomical to operate.

Nuclear advocates are mounting a $40 million-a-year campaign to promote nuclear power as a solution to the greenhouse effect. However, the Rocky Mountain Institute estimates that to reduce the emissions that cause global warming 20 to 30 percent by the middle of the twenty-first century (substituting nuclear power for coal power) would mean building one new plant every 3 days for the next 40 years at a cost of $9 trillion.

Beyond speculation, over the past 15 years conservation has saved about six times more energy than nuclear power. Furthermore, improving the energy efficiency of buildings, appliances, and lighting costs one-seventh the price of building a new nuclear power plant. According to Public Citizen, during the 1980s the average construction costs for nuclear reactors increased fourfold, operating and maintenance costs grew by 70 percent, and major repair costs rose 30 percent. Coupled with increased reliance on renewable energy, such as solar and wind, energy efficiency is the most promising solution to our energy needs.

According to Joan Claybrook, president of Public Citizen, "the United States is in the midst of an energy revolution.[24] Renewable sources are already making a bigger contribution to our energy needs than nuclear power. In fact, while nuclear power now provides less than 7 percent of our energy needs and its share is likely to diminish in the future, renewables already provide almost 8 percent and are growing rapidly."

STS Actions

- Give students a drawing of a building that contains a nuclear power plant. Have the students explain how they think a nuclear power plant produces electricity. Tell them to illustrate their ideas on the drawing of the plant. Use the results to identify student misconceptions.
- If there is a nuclear power plant in your community, arrange for a field trip to the plant. In advance of the trip distribute information on nuclear power available from the power company. Also provide students with information from environmental groups such as Nuclear Information and Resource Service (1424 16th St., NW Suite 601, Washington, D.C. 20036). Have the students prepare questions that will help them in their inquiry about nuclear power plants. (Note: if there isn't a nuclear plant, take the students to a power plant that uses some other fuel.)
- Have students investigate power generating plants that use nonrenewable energy sources (coal, oil, nuclear material) and renewable energy sources (wind, water, volcanic steam).
- Organize a debate on the issue, Should nuclear power be eliminated from power companies' future construction plans?

[24]Anastasia & Claybrook.

THE UTILIZATION OF NATURAL RESOURCES

The Tropical Rainforests Issue[25]

Tropical rainforests are broad-leafed evergreen woodlands that receive at least 100 inches of rain annually. Rainforests once covered about 5 billion acres in the tropics (Figure 6.10). As a result of human interference, only half of the original rainforests exist today. Nevertheless, they are home to at least 5 to 10 million species of plants and animals, approximately one-half of Earth's life forms. Remaining rain forests are disappearing at a rate of 100 acres per minute, an area the size of Kansas every year.

Many natural resources and much of our food come from tropical rainforests. Rainforests serve as a genetic pool for many fruits and vegetables, and new varieties continue to be discovered. Only 1 percent of the tropical rainforest plants that have been identified have been scientifically analyzed, yet they are the source of more than 25 percent of the medical compounds sold on the market today.

Nearly the entire acreage of tropical rainforests lies within the borders of developing countries. Often the governments of these countries are encouraged to exploit the resources of their forests to pay off foreign loans. External financial pressures have forced them to sacrifice long-term sustainability to service short-term national debt. Population growth and inequitable distribution of land have further contributed to the problem.

Each year millions of acres of tropical rainforests are burned to make way for agriculture, much of it for export. The nutrients of the rainforest are stored in its multi-layered canopy. When forests are burned,

[25]Ron Anastasia and Randy Hayes, "Tropical Rainforests Fact Sheet," EarthNet, A Forum on the Connect Business Telecommunications System. Used with permission of Earth Day Resources. Copyright 1990. Earth Day Resources, 116 New Montgomery St., San Francisco, Calif. 94105

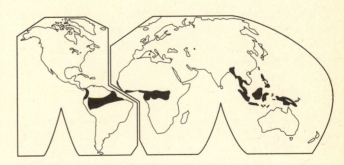

FIGURE 6.10 Location of Tropical Rainforests

these nutrients mix with the barren topsoil, where they are quickly eroded by rain. When the land is depleted of nutrients, the farmer moves on and clears more rainforest.

In Central America the primary motive for clearing rainforest is to make way for cattle ranching. Most of the beef, however, is produced for export to developed countries to be used by fast food restaurants. Over 120 million pounds of beef are imported by the United States from Central America annually.

According to the World Bank and the United Nations Development Programme, at least 12.5 million acres of tropical rainforest are logged every year. Much of this lumber is exported for use in furniture and other hardwood products. Teak, mahogany, rosewood, purpleheart, and ramin are some of the more common tropical hardwoods imported by developed countries. The United States imports about 15 percent of the world's hardwood products.

Many acres of rain forest are flooded each year as a result of large hydroelectric projects built to provide energy for large metropolitan areas and for multinational industrial projects. As a result of these hydroelectric projects, thousands of indigenous peoples who have relied on the sustenance of the rain forests for thousands of years have been relocated, and their cultures destroyed.

Clearing tropical rainforests means destruction of habitat for the millions of species of plants and animals that live in these regions. Furthermore, forests act as a natural sink for carbon dioxide, the major "greenhouse" gas responsible for global warming. As rainforests are destroyed, carbon dioxide is released into the atmosphere, leading to higher global temperatures. Some scientists predict that as global temperatures rise we will face an increase in crop failure, oceans will rise and flood coastal areas, and many species of plants and animals will become extinct.

Two-thirds of the world's fresh water, excluding that which is locked in the polar ice caps, is cycled within tropical rainforest systems. Rainforests absorb this large amount of water, releasing it slowly and evenly through the process of evapotranspiration. But as rainforests are cleared, soils become exposed to heavy rain, leading to flooding and erosion. It is often impossible to reestablish a rainforest once it has been cleared.

STS Actions

- Provide students with a map of the world and ask them to identify the location of at least one tropical rainforest. Tell them that about 50 acres of rainforest are destroyed each minute (almost 27 million acres per year, equal in size to the state of Pennsylvania). What impact could this deforestation have on

  1. The earth's temperature
  2. Animal and plant extinction
  3. The amount of carbon dioxide in the atmosphere
  4. The quality of life for people in the tropics

- Engage students in a debate regarding this statement made by Randy Hayes, Executive director of Rainforest Action Network: "We believe that tropical rainforests are one of the most important global ecological issues of our time. These forests are a vital part of the life support systems of the planet. To ensure their survival is to ensure our own survival. If we don't act now they could be gone by the year 2050. We may be the last generation that will have a chance to save the rainforests."

INQUIRY ACTIVITY 6.2

STS Issues in Science Textbooks

In this inquiry activity you are going to find out what issues are emphasized in secondary science textbooks. To do this you will analyze one middle school-junior high or high school science textbook using a classification system of STS issues.

Materials

A secondary science textbook

Procedure

Look over the categories of STS issues in Figure 6.11. The categories represent a comprehensive listing of STS issues. Analyze the text by identifying all the STS issues that are presented by studying the table of contents and the pages of the text. Indicate the number of pages devoted to the issue when identified. Illustrations count as much as the verbal text. Estimate the amount of space devoted to the issue on each page and record as 1 page, 1/2 page or 1/4 page.

STS Issue Category*	STS Issue in Textbook (Use key words to identify the issue)	Number of Pages Devoted to Topic	Percentage of Total Pages in the Book. Total Pages in Book
Nature of Science definitions, assumptions, aims of science			
Social System of Science history, lives of scientists, science and government, industry, public			
Human Behavior genetic, environmental factors, human intelligence, behavior modification, theories of development, ethical issues related to control of behavior			
Population growth, overpopulation, problems in developing countries, control, projections, population ethics			
Food Supply and Agriculture history, food production, scarcity, crops, economics of food, ethics of food			
Human Reproduction aspects of human reproduction, sex education, venereal diseases, reproductive control and engineering, fetal rights			
Human Genetics traits and disorders, mutations, genetic diagnosis, counseling and interventions, ethical issues			
Human Health health and disease, causes of disease, health care, aging, death and dying, health habits, world health problems			
Evolution origin of life, evidence for evolution, theories of, application to social theories, evolution vs. creationism, population genetics, human evolution			
Energy Resources sources, energy conservation, new sources, economics of, energy policies, quality of life, international politics and energy			

FIGURE 6.11   Textbook Analysis Chart of STS Issues

(Figure continued on p. 190)

(Figure continued from p. 190)

STS Issue Category*	STS Issue in Textbook (Use key words to identify the issue)	Number of Pages Devoted to Topic	Percentage of Total Pages in the Book. Total Pages in Book
Environment human view of nature, effect of human activities on biosphere, natural resources, pollution, endangered species, biological and nuclear weapons			
Space Research and Exploration extraterrestrial life, space stations, colonization, planetary exploration, militarization of space, effect of space exploration of view of human species and the planet			
Totals			

*These categories are based on a study reported by Dorothy B. Rosenthal, "Social Issues in High School Biology Textbooks: 1963–1983," *Journal of Research in Science Teaching* 21, no. 8 (1984): 819–831. Copyright© 1984, John Wiley & Sons, Inc. Reprinted by permission of John Wiley & Sons, Inc.

Minds on Strategies

1. In Rosenthal's study the social system of science and evolution were the most commonly presented STS issues in biology textbooks. You might want to read her study, and compare her results with yours. How do your results compare with Rosenthal's?[26]

2. What categories are least presented in secondary science textbooks? (You'll need to collaborate with others in your class to make this conclusion.)

3. In examining biology, physical science, and earth science texts, what differences in the relative emphases of STS issues did you find?

[26]Dorothy B. Rosenthal, "Social Issues in High School Biology Textbooks: 1963–1983," *Journal of Research in Science Teaching* 21, No. 8 (1984): 819–831.

STS CURRICULUM EXAMPLES

STS curriculum materials are becoming a prominent feature in the science curriculum landscape. Since the early 1980s the number of projects devoted to developing STS curriculum materials has increased, and they are readily available. In this section several STS curriculum modules, courses of study, and teacher resource books will be reviewed.

STS Evaluation Criteria

The curriculum materials presented here have a number of characteristics in common. These characteristics are the criteria for evaluating STS materials established by a task force of the SSTS (Science through Science Technology Society). According to the SSTS task force, the following criteria constitute the characteristics of STS modules and courses of study:

1. *Societal relevance:* The relations of technological or scientific developments to societally relevant issues are made clear to capture attention.

2. *Relate STS:* The influence and effect of technology, science, and society on each other are presented.

3. *Self and society:* The material develops learners' understanding of themselves as interdependent members of society and of society as a responsible agent within the ecosystem of nature.

4. *Balance of views:* A balanced treatment of differing viewpoints about the issues are presented without author bias.
5. *Broaden interest:* The material helps students go beyond specific subject matter to broaden considerations of STS, which include a treatment of personal and societal values and ethics.
6. *Decision skills:* The material engages students in problem-solving and decision-making skills.
7. *Suggest action:* The material encourages learners to become involved in a societal or personal course of action.
8. *Introduce some science or qualification:* The STS material helps students develop an understanding of at least one area of science, and of using qualification as a basis for making judgments.[27]

The STS materials should help students make connections or linkages with other disciplines, as shown in the STS connection diagram (Figure 6.12) The STS connection diagram can be used to help you highlight linkages as you examine curriculum materials, and as you develop your own modules.

The Chemical Education for Public Understanding Program

The Chemical Education for Public Understanding Program (CEPUP) is a project that has developed STS modules for middle school or junior and high school students. The modules (see the following list) are designed to highlight chemicals and their uses in the context of societal issues, so that the that learners interact with real science.

[27]Catalog of STS Instructional Materials, *SSTS Reporter: Science Through Science Technology and Society.* University Park, Pa: The Pennsylvania State University, March 1988, pp. 3–4.

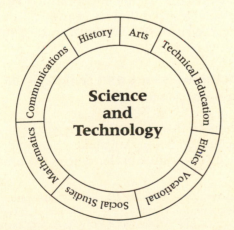

FIGURE 6.12   STS Connection Diagram. Catalogue of STS Instructional Materials used with permission.

Chemical Survey and Solutions and Pollution
Risk Comparison
Determining Threshold Limits
Investigating Groundwater: The Fruitvale Story
Toxic Waste: A Teaching Simulation
Plastics in Our Lives
Investigating Chemical Processes: Your Island
  Factory
Experiencing Scientific Research: An Air
  Pollution Study

(CEPUP is available from CEPUP, Lawrence Hall of Science, University of California, Berkeley, Calif. 94720; or Addison-Wesley, 2725 Sand Hill Road, Menlo Park, Calif. 94025, (800) 447-2226.)

The goals of CEPUP are[28]

• To provide educational experiences focusing on chemicals and their interaction with people and the environment
• To promote the use of scientific principles, processes, and evidence in public decision making
• To contribute to improving the quality of science education in America
• To enhance the role of science teachers as educational leaders

The CEPUP materials are experience based and allow adolescent students to develop "evidence-based decision-making skills and capabilities necessary to become effective, contributing members of society."[29] The materials emphasize concrete experiences as well as inference and decision making to help students in their transition from concrete to abstract thinking.

Each module in the CEPUP series focuses on an issue such as toxic waste, groundwater, risk comparison, solutions, and pollution. The module consists of a teacher's manual and modular kits which contain equipment and teaching materials. The teacher's manual provides an overview of the activities, a conceptual overview, a description of the six to eight activities for each CEPUP module, evaluation items, and a glossary of terms.

The activities include traditional science experiments as well as STS-type activities. For example, in the module Investigating Groundwater: The Fruitvale Story, students role-play different members of Fruitvale at a town meeting. In another unit, Determining Threshold Limits, students investigate threshold by determining the threshold of taste for salt in a solution. In the Risk Comparison module

[28]CEPUP, "Issue-Oriented Science Using CEPUP" (Berkeley, Calif.: Lawrence Hall of Science, April 1990), p. 3.
[29]CEPUP, "Issue-Oriented Science Using CEPUP," p. 3.

students are introduced to chance and probability by flipping coins and rolling dice. The case study is introduced in this module. Students read and react to a story of smallpox inoculation and immunization, and in another activity they investigate John Snow's research on cholera.

Chemistry in the Community

Chemistry in the Community (ChemCom), a one-year chemistry course developed by the American Chemical Society (ACS) with support from the National Science Foundation and other sources, was designed for students who intend to pursue non-science careers.

ChemCom is designed to help students[30]

- Realize the important role that chemistry will play in their personal and professional lives
- Use principles of chemistry to think more intelligently about current issues they will encounter that involve science and technology
- Develop a lifelong awareness of the potential and limitations of science and technology

The ChemCom curriculum consists of eight units, each of which focuses on a chemistry-related technological issue now confronting society and the world. The issue is the vehicle used to introduce the chemistry needed to understand and analyze it. Each topic is set in a community. Communities include the school, the town or city in which the students reside, or the whole earth.

This highly structured yet unique chemistry curriculum involves the students in real issues facing society. Students are engaged in hands-on laboratory activities, as well as in reading about environmental problems, solving problems, and answering questions. The units in the ChemCom curriculum include

- Supplying Our Water Needs
- Conserving Chemical Resources
- Petroleum: To Build or to Burn?
- Understanding Food
- Nuclear Chemistry in Our World
- Chemistry, Air and Climate
- Chemistry and Health
- The Chemical Industry: Promise and Challenge

The curriculum involves the students in a variety of decision-making activities. For example, one type

of decision-making activity is called Chem Quandary and is designed to get the students to think about chemical applications and societal issues that are often open-ended. Often these activities generate additional questions beyond specific answers.
(ChemCom is available from ChemCom, Chemistry in the Community, Kendall/Hunt Publishing Company, 2460 Kerper Boulevard, P.O. Box 539, Dubuque, Iowa 52001.)

Project Learning Tree

Project Learning Tree (PLT) is an environmental education project that is jointly sponsored by the Western Regional Environmental Education Council and the American Forest Foundation.[31] The program consists of two teacher's activity guides, one for K–6 and the other for 7–12.[32] The activity guides contain hands-on experiential activities organized around seven key unifying principles:

- Environmental awareness
- The diversity of forest roles
- Cultural contexts
- Societal perspectives on issues
- Management and interdependence of natural resources
- Life support
- Lifestyles

The activities in PLT are interdisciplinary, and as a result they can be used in a variety of curricula areas. This makes them especially valuable to middle schools that are planning to develop interdisciplinary units for their students.

The PLT program involves students in a variety of indoor and outdoor activities. The curriculum consists of ready-made lessons on trees and forest ecology that can be used to supplement existing curricula. (PLT is available from Project Learning Tree, 1250 Connecticut Avenue, NW, Washington, D.C. 20036.)

Environmental Issues and Action Skills Development Modules

The authors of this curriculum project have developed teaching materials designed to help middle and junior high school students become environmentally

---

[30]American Chemical Society, *ChemCom: Chemistry in the Community* (Dubuque, Iowa: Kendall/Hunt, 1988), p. ix.

[31]*Project Learning Tree Workshop Handbook* (Washington, D. C.: American Forest Institute, 1986).
[32]*Project Learning Tree Activity Guide K-6* and *Project Learning Tree Activity Guide 7–12* (Washington, D. C.: American Forest Institute, 1988).

literate. Their concept of literacy involves the development of environmental knowledge for the sake of helping students take action on environmental problems. The project has developed a series of modules (packaged as a single book, with a teacher's edition) based on an environmental philosophy and four literacy components. The environmental philosophy is expressed in this "key statement":

> The environmentally literate citizen is able and willing to attempt to make environmental decisions which are consistent with both a substantial quality of human life and an equally substantial quality of the environment. Further, this individual is motivated to act on these decisions either individually or collectively.[33]

The four literacy components include ecological foundations, awareness of issues and human values, investigation and evaluation of issues and solutions, and citizenship action.

The six modules that make up the program taken as a whole teach the student the concepts, issues, and skills needed to take environmental action. The themes for each of the modules are as follows:

Module I: Environmental problem solving
Module II: Getting started on issue investigation
Module III: Using surveys, questionnaires, and opinionnaires in environmental investigations
Module IV: Interpreting data from investigations
Module V: Investigating an environmental issue
Module VI: Environmental action strategies

(Materials are available from Stipes Publishing Company, 10–12 Chester Street, Champaign, Ill. 61820.)

Other STS Materials

EEE: Interdisciplinary Student/Teacher Materials in Energy, the Environment and the Economy. Modules on STS topics including Two Energy Gulfs, Transportation and the City, Energy in the Global Marketplace, Western Coal: Boom or Bust? How a Bill Becomes a Law to Conserve Energy.

U.S. Department of Energy
Technical Information Office
P.O. Box 62
Oak Ridge, TN 37830

[33]Harold R. Hungerford et al., *Investigating and Evaluating Environmental Issues and Actions: Skill Development Modules* (Champaign, Ill.: Stipes, 1988), p. 1.

BSCS Modules. Separate modules on STS, human reproduction, television, computers, biomedical technology, the human environment, and genetics.

Kendall/Hunt Publishing Company
2460 Kerper Boulevard
PO Box 539
Dubuque, IA 52001

Investigating and Evaluating Environmental Issues and Skills Development. Workbook containing six modules that develop skills (problem solving, issue investigation, questionnaires, interpreting data, taking action strategies) in students to investigate environmental issues.

Stipes Publishing Company
10-12 Chester Street
Champaign, IL 61820

Project Wild. Interdisciplinary, supplementary environmental and conservation education program emphasizing wildlife, published in two volumes: an elementary and a secondary activity guide.

Salina Star Route
Boulder, CO 80302

Chemicals, Health, Environment, and Me (CHEM). Ten units on societal issue-oriented science topics.

CHEM
Lawrence Hall of Science
University of California
Berkeley, CA 94720

Science in Society. Modules on disease, population and health, food, agriculture, mineral resources, industry, science and social development, and engineering.

The Association for Science Education
Heinemann Educational Books
70 Court Street
Portsmouth, NH 03801

Educators for Social Responsibility STS Materials. Modules on topics such as Perspectives: A Teaching Guide to Concepts of Peace; Why Nuclear Education; Investigations: Toxic Water.

ESR
23 Garden Street
Cambridge, MA 02138

INQUIRY ACTIVITY   6.3

Evaluating an STS Module

The purpose of this inquiry activity is for you or a team of peers to evaluate an STS module using criteria established by the STS Task Force.

Materials

An STS module of your choice, including the equipment module if one is provided

Procedure

1. Examine the STS module in light of the following criteria. Complete the chart (Figure 6.13) for each criteria.
2. Determine the curriculum linkages for the module and shade in the appropriate STS connections in the STS Connections Diagram (Figure 6.14).

Minds on Strategies

1. Would you recommend this STS module for use in the science classroom? In what contexts?
2. What evaluation materials were provided?

STS Criteria	Excellent	Good	Fair	Poor
Societal relevance of the issue				
Relate STS				
Self and society				
Balance of views				
Broadens interest				
Decision skills				
Suggested action				
Introduces some science or quantification				

FIGURE 6.13   STS Evaluation Chart

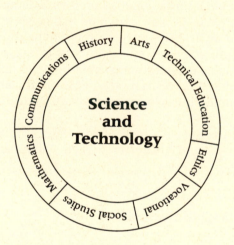

FIGURE 6.14   STS Connections Diagram. Catalogue of STS Instructional Materials used with permission.

Case Study 1.
STS as the Entire Science Program: Some Questions

The Case At a seminar to examine the future directions of the science curriculum, a high school chemistry teacher, who has been a proponent of environmental chemistry, makes a brief presentation regarding her views on STS and the curriculum. The focus of her presentation is as follows:

When heroes of mine, such as David Suzuki, point out that there is such a glut of scientific knowledge now available that we cannot expose students to all of it with any realistic expectation of retention, then it does seem to indicate that there must be a change in what we're doing. Many of my colleagues would argue that the content is not really being taught for content's sake alone, but used as a vehicle to develop a number of extremely important tools including process skills and, believe it or not, to actually develop discussion skills the students have, and to make them aware of the impact of science on our society. There seems to be a real concern that students be able to discuss the interface between science and society, and be able to look at societal problems in a scientific manner. While I think this is good and certainly should be a part of courses, to dedicate the entire program to it is, I feel,

(Continued on p. 197)

Case Study 2.
A Controversial Student Project

The Case. In your first-year biology course, you have decided to integrate STS into the course by having teams of students do a project. The results of the project will be presented to the whole class. Project teams must do three projects throughout the year. The projects must be related to one of the nine biology units that make up your course. One of the teams decides to do an STS project related to the unit on human reproduction. They submit their proposal (as all teams must do). The student project is entitled Birth Control Centers on the High School Campus. In their proposal, the students write, "It is evident that teenage pregnancy is a problem at our high school. Last month, twelve girls had to leave school because of pregnancy. We plan to investigate this problem and make recommendations for action to the school administration. One idea that we have in

(Continued on p. 196)

Think Pieces

- Why should STS be part of all science courses?
- In your opinion, what are the ten most significant global STS issues facing the planet today? How are they related to each other?
- What is global thinking?
- What qualities should be included in STS modules?

April 22, 1990: Threshold of the Green Decade by Denis Hayes[36]

(Note: Denis Hayes, chairman of Earth Day 1990, practices law in San Francisco and teaches engineering at Stanford University. Hayes also organized the first Earth Day, April 22, 1970.)

For many of my generation, involvement with serious issues—adult issues—began with some form of unconventional politics. Passive disobedience and freedom rides in support of civil rights. The endless town meetings of Vietnam Summer. Wearing gas masks down Fifth Avenue on Earth Day. Picketing a state legislature in support of the Equal Rights Amendment. Breaching the exclusion zone around the Seabrook or Diablo Canyon nuclear plants. Blocking a train carrying fissionable material to the Rocky Flats bomb factory in Colorado.

We were impatient and idealistic. The first generation with strontium-90 in our bones (from atmospheric nuclear testing), we trusted no one over thirty. Outraged over the state of the

(Continued on p. 196)

(Continued from p. 195)
mind is to recommend a birth control information center on our campus. We will, however, investigate many alternatives and recommend what we find to be the most effective in reducing pregnancies among high school students." As with all your projects, the student team has two weeks to do their research and prepare for their presentation to the class. During the two-week research period, the principal of the school finds out this team's research. The principal puts a note in your mailbox, demanding a meeting with you to discuss the students' project.

The Problem. How will you prepare for the conference? What evidence will you present to the principal to defend the work that the students' are doing? What will you say to the principal? ☐

(Continued from p. 195)
world we were inheriting, we vowed that we would pass on to our children a world that was peaceful, just, and ecologically sustainable.

Twenty years after Earth Day, those of us who set out to change the world are poised on the threshold of utter failure. Measured on virtually any scale, the world is in worse shape today than it was twenty years ago. How could we have fought so hard, and won so many battles, only to find ourselves now on the verge of losing the war? The answers are complex. But if we can understand the mistakes that led to our current dilemma, we may yet be able to redeem our youthful promises to the next generation.

The American conservation movement has a long, distinguished tradition, traceable back to such giants as Henry David Thoreau, John James Audubon, John Muir, and Aldo Leopold. However, the environmental movement is of much more recent origin. Individuals like Rachel Carson and David Brower sounded the environmental alarm in the 1960s, and events such as the Storm King fight against a power project on the Hudson River and the oil spill off Santa Barbara, California, in 1969 gave rise to local waves of concerned activists. But a full-blown national movement emerged only in 1970.

The modern environmental movement has enjoyed a string of spectacular successes—on Capitol Hill, in the courts, and in the streets. [The first] Earth Day's 25 million participants could not be ignored. Within months, the federal Environmental Protection Agency was created. Congress then swiftly passed the Clear Air Act, the Clean Water Act, and a host of other laws that fundamentally changed the rules under which American enterprise operates.

Yet despite all these accomplishments, we are in serious trouble, and the problems are being compounded with every passing year. Environmental threats now vie with nuclear war as the preeminent peril to our species because our leaders have displayed neither the intelligence nor the integrity nor the guts to lead us into the Green Decade. Those of us who care about the earth must provide the direction and the energy for change if the world is to avoid calamity.

(Continued on p. 197)

Case Study 3. Biased Teaching?

The Case. You are teaching chemistry to college-bound students in a suburban community. Since this is your second year of teaching, you've decided to get students involved in some "environmental chemistry" activities. During the first marking period you've decided to have students read high-interest articles that you have copied (after securing permission, of course) from popular magazines such as *Time, Newsweek, Natural History,* and *National Geographic.* Your plan is to give the students one article every other week, followed by small and large group discussions. The first article you have selected is entitled "Nuclear Energy: Its Time Has Passed." To conserve paper, each small team of four students gets one article that they must share among themselves. The day after you distribute the article, you receive a phone call from one of your student's parents. He explains that he is quite concerned that students are reading such a highly biased article on nuclear energy. He accuses you of indoctrinating the students with environmental sentimentality and says that if you continue having students read these "outside" articles, he will complain to the school board.

The Problem. What do you do in this situation? Did you expect to get this kind of reaction from any of your students' parents? What information can you provide the parent that might remove his objection that your teaching is indoctrination? Who would you see before responding to the parent, and what would you say? ☐

(Continued from p. 196)

What went wrong during the last twenty years? Occasionally we were blindsided. Problems snuck up on us before anyone recognized the threat they posed. We possess only a rudimentary understanding of the complex interactions of life in the biosphere and of the myriad subtle effects of human action upon long-established processes. If at the time of the first Earth Day a poll had been taken of industrial chemists, asking each to name ten triumphs of modern chemistry, most would probably have listed chlorofluorocarbons (CFCs). Not until 1974—four years after Earth Day—did Professor Sherwood Rowland and his colleagues at the University of California at Irvine discover that CFCs posed a theoretical danger to the stratospheric ozone layer that protects the earth from ultraviolet radiation. And it was not until 1985 that a British team discovered the huge seasonal thinning of the ozone layer over the Antarctic.

A common feature of all the problems we have been discussing is that none are the result of forces beyond human control. None are caused by sunspots or the gravity pull of the moon or volcanic activity. All are the result of conscious human choices. All can be cured by making other choices.

First, we need to make our lives congruent with our values. For most of us, there is room for improvement in virtually all spheres. We should conserve energy with easy things, such as replacing incandescent light bulbs with folded fluorescents, which are five times as efficient, insulating our water heaters, and doing laundry in cold water.

Integrating your values into your job and your other activities is another important step, but it still does not discharge your responsibilities.

On Earth Day—April 22, 1990—more than 100 million people around the world will make a personal affirmation of their environmental commitments. At the same time, we will send a message to our leaders that talk is no longer sufficient. Time is running out. We have, at most, ten years to embark on some undertakings if we are to avoid crossing some dire environmental thresholds. Individually, each of us can do only little.

Together, we can save the world. □

[36]Denis Hayes, "Earth Day 1990: Threshold of the Green Decade," *Natural History*, April, 1990. With permission from Natural History, April 1990; Copyright the American Museum of Natural History, 1990.

(Continued from p. 195)

highly questionable. Where do students get the knowledge base to discuss these at anything other than an emotional or gut level if they don't have some content that they can fall back on? I also think when you get into an area such as the values aspect of science, that you're starting to place teachers, who have been trained in a completely different manner, in a position that's a little suspect.[34]

The Problem. How would you react to this teacher? Is it highly questionable to base the science program on STS? Will students only react on an emotional or gut level without formal instruction in science content? And by the way, who is David Suzuki?[35] □

[34]Excerpted from Douglas A. Roberts, "What Counts as Science Education?" in *Development and Dilemmas in Science Education*, Peter Fensham, ed. (London: Falmer, 1988), p. 41.
[35]See David Suzuki, *Inventing the Future* (Toronto: Stoddart, 1989).

Environmental Education for the 1990s and Beyond by Louis A. Iozzi[37]

"It was the best of times, it was the worst of times," commented Charles Dickens in his *Tale of Two Cities*. Surely, Dickens' comment is at least as appropriate today as it was when he first stated it many years ago, for while we live in a magnificently exciting world, we also live in a world of great uncertainty and almost overwhelming challenges.

Recent advances in science and technology are providing us with unprecedented powers—capabilities heretofore only experienced in dreams or in science-fiction stories. Unfortunately, as those of us interested in and concerned about the environment are painfully aware, many of these achievements have not been attained without paying costly penalties. As Barry Commoner remarked, "There is no such thing as a free lunch." Future applications of new technologies will, undoubtedly, present greater challenges and place still greater demands on society.

The responsibility for determining if and where modern technologies are to be utilized is being thrust upon society at a phenomenal rate. For me, this raises at least two critical questions: (1) Is society adequately prepared to make wise and responsible decisions about how new technologies are used? and (2) Will the future leaders of society be any better prepared than we are now to deal with the decision-making challenges of a rapidly changing and increasingly complex world of "tomorrow"!?

Clearly if we are to have a better prepared society and informed decision makers, we need to edu-

(Continued on p. 198)

Problems and Extensions

- What research skills will students need to develop to carry out STS action projects?
- Select one of the following areas—earth science, physical science, life science—and develop a list of STS issues that could be integrated into a one-year course in the field you select.
- Cartoons (especially those found in the editorial pages of newspapers) could be used as value dilemmas for STS activities. Collect at least five and identify the STS issue that each reflects, and explain how you would use them in the science classroom.
- Design either a case study, a voting activity, or a value dilemma for any one of the following STS issue areas: health care, the prolongation of life, euthanasia, artificial organs, genetic counseling, abortion, human experimentation, suicide, and biological weapons. Field-test your activity with a group of peers, friends, or secondary students.
- Do a survey of a group of secondary students to find out how much emphasis is given to STS teaching in their science courses. To begin, ask the students;[38] In your opinion, how much emphasis has been given in the science classes you have taken until now to the following ideas? (You could give them a sheet of paper with this list, and the direction to check write 1 for a lot, 2 for some, or 3 for not much for each.)

- Society controls science and technology
- Science and technology influence society
- Science and technology have limitations
- Science and technology are useful in advancing human welfare
- There is a difference between scientific knowledge and personal opinion
- People must make careful decisions about the use of technology
- There are many good, reliable sources of scientific information that a person can find and use in making decisions

[38]Excerpted from Faith Hickman, "An Instrument for Assessing STS Courses," *Global Science: Energy, Resources, Environment Newsletter* 2, no. 1 (Fall, 1987); 3.

(Continued from p. 197)
cate our young people today. EE can develop those skills to prepare today's youth to be effective decision-makers in the future.

EE for the 1990s should

- Consider the Total environment
- Be infused into existing curricula at all levels
- Emphasize the development of positive environmental values; and more mature moral reasoning/judgement
- Emphasize the development of problem-solving and thinking skills
- Develop decision-making skills that consider both long and short term futures; local-international concerns and problems
- Emphasize inquiry methods that directly involve students in investigating real problems
- Utilize a variety of media in the classroom
- Utilize high quality-low cost/free materials and resources
- Be considered a life-long process
- Be provided to all children—regardless of age, SES, place of residence, and gender
- Not neglect information transfer but also emphasize the development of information access/retrieval skills as well
- Seek to develop greater industry-education cooperation and alliances   □

[37]Louis A. Iozzi, "Environmental Education for the 1990s and Beyond," *Branch: Project Learning Tree*, Spring/Summer, 1987, pp. 1–6. With permission of Project Learning Tree

Science Teachers Talk

"How do you deal with STS issues in the classroom?"

Jerry Pelletier. It is not my style to hide from STS issues, so these issues are part of my curriculum. This year we dealt with the problem of nuclear energy. In order to make the students more aware of this issue we read *Hiroshima* by John Hershey and discussed the ramifications of the release of radioactive particles within the earth's atmosphere. We analyzed the effects of that event with nuclear accidents such as Three

Mile Island and Chernobyl. I handled the issue of evolution in the same manner. We read the play *Inherit the Wind* and discussed the controversy which is still brewing today between creationists and evolutionists. I find that by reading and sharing in the same literary experience students find it easier to discuss and understand various STS issues.

Mary Wilde. The STS issues I deal with in the classroom are the origin of our universe and the

(Continued on p. 199)

(Continued from p. 197)
geological time span of the earth. When I introduce the big bang theory, I spend a lot of time discussing the definition of a theory, and we identify theories developed in the past that have been proven false. We discuss how portions of theories can be correct or inconclusive, and how one theory can lead to the development of another theory. If questions continue to rise concerning creationism versus evolution, I explain that scientific theories are based on scientific evidence, while the creation theory is based on the Bible, which was not meant to be a scientific document.

Dale Rosene. There are a number of guidelines that I follow when dealing with STS issues. These include the following: (1) Be open—allow all students to voice their opinions and views. Encourage them to examine the basis for these views. (2) Try to provide balance when appropriate. (3) Invite experts into the classroom to provide their point of view. (4) Use writing exercises, because these cause the students to more carefully examine their beliefs. (5) Role-playing activities provide an excellent forum for STS issues. (6) Don't try to infuse your views in the class—unless appropriate. (7) Involve community groups when integrating new curricula such as sex education.

John Ricciardi. My curriculum content is full of controversy and speculation. The entire knowledge base of "quantum" and "astro" phenomena is built on human subjectivity. To object, dispute, and oppose is also to be thinking scientifically. There are many "pictures," perceptions, and schools of thought concerning "what is." Science's controversy is science's excitement, strength, and vitality.

In my classes, popular STS issues, such as creationism versus evolution, high tech mechanical/biological creationism versus environmental preservation are presented. However, within the context of the entire curriculum, their significance becomes deaccentuated. It seems that the issues are realized for what they are—only small pieces of the whole of nature's puzzle, only a "fuzziness" of parts to a grander, unseen clearness of "what is."

Ginny Almeder. An important goal of education is to develop the students' ability to deal with societal issues. Many of these issues result from theoretical research and scientific technology, and are controversial by their very nature. STS issues such as creation science, AIDS, in vitro fertilization, genetic testing, and environmental hazards should not be ignored. Students need opportunities to develop critical thinking skills and well-informed opinions.

In our biology classes, we deal with such topics as creation science and evolution, the ethical and legal implications of genetic counseling, animals in research, the use of steroids, and environmental issues such as the "greenhouse effect" and ozone depletion. The students are encouraged to discuss their political and ethical positions regarding the various topics. With this approach, we are able to have nonthreatening discussions which promote both an increased understanding of the issues and a greater acceptance of other viewpoints.

For example, a discussion of creation science and evolution provides a fertile setting for distinguishing between a scientific theory and a religious belief. If students are able to understand the difference, they are more likely to appreciate that science and religion are not mutually exclusive. In addition, such a discussion can be used to develop arguments and counterarguments for various issues and thus improve critical thinking skills as well as an appreciation for the differences of opinion that characterize our pluralistic society. ☐

RESOURCES

Botkin, Daniel B. et al. *Changing the Global Environment.* Boston: Academic Press, 1989.

The book examines a wide variety of environmental issues as well as global environmental changes and the extent to which technology can be employed to improve our global environment.

Bybee, Rodger W. ed. *Science, Technology and Society.* Washington, D.C.: National Science Teachers Association, 1985.

The articles in this book provide background, rationale, and goals for technology's and society's inclusion in science programs; it presents various curriculum materials, new approaches, and instructional strategies; and suggests methods for the theme implementation in traditional programs.

Chalk, Rosemary. *Science, Technology, and Society: Emerging Relationships.* Washington, D.C.: American Association for the Advancement of Science, 1988.

This is a collection of articles that originally appeared in *Science.* Topics range from science and responsibility to science and ethics.

Frazer, M. Y., and A. Kornhauser. *Ethics and Social Responsibility in Science Education.* Oxford, England: Pergamon Press, 1986.

Essays presented at the ICSU Conference on Science and Technology in Bangalore, India, 1985. The book deals with questions of ethics and social responsibility in science education and provides information on the efforts being made to deal with these problems in the practice of science education.

Gibbons, Michael, and Philip Gummett. *Science, Technology and Society Today.* Manchester, U.K.: Manchester University Press, 1989.

Four different themes concerning STS are presented in the various chapters of the book: the nature of and origin of scientific knowledge; scientific knowledge and political authority; the economic impact of science and technology; and the control of science and technology. This book aims at providing teachers with basic information and some sense of how to approach each of the foregoing topics.

Head, John. *The Personal Response to Science.* Cambridge, England: Cambridge University Press, 1985.

This book aims at showing that the relation between science education and the social context of beliefs and practices concerning science is closer than generally recognized. It sketches an intricate interaction between adolescent development, curriculum, and the biases in the ideology and the personality of the scientists.

Jacobs, Heidi Hayes, ed. *Interdisciplinary Curriculum: Design and Implementation.* Alexandria, Va.: Association for Supervison and Curriculum Development, 1989.

The authors of this book describe ways to develop interdisciplinary curriculum units focusing on high-order thinking skills.

Peterson, Rita, et al. *Science and Society: A Source Book for Elementary and Junior High School Teachers.* Columbus, Ohio: Merrill, 1984.

This book contains methods, activities, and strategies designed to help the classroom teacher translate STS into practice.

Robinson, Barabara, and Evelyn Waltson. *Environmental Education.* New York: Teachers College Press, 1982.

The book provides basic information, activities, and teaching tips on environmental education topics as well as a course evaluation procedure.

7

MODELS OF
SCIENCE TEACHING

A chemistry teacher, and chair of the science department in a large school district in the southeastern part of the United States, believes that excitement, enthusiasm, and inquisitiveness should reign in science class. She uses an "inquiry-oriented" approach to teach chemistry. To drop in and visit her classroom is to observe not only an exemplary teacher but one who puts into practice what science educators claim should characterize high school science teaching. Students are involved in watching minidemonstrations and then trying to figure out what happened, testing the acidity of rain (with cabbage juice as the indicator) in the Atlanta area over a long period of time and then drawing conclusions based on their own data, conducting microchemistry experiments designed to help them learn chemistry concepts inductively. Furthermore, students in her classes are linked with students in the Soviet Union by means of a computer-based telecommunications system to explore cooperatively environmental chemistry issues and problems from a global perspective. In short, her method of teaching gives the students the opportunity to inquire, to question, and to explore.

The approach to teaching that this teacher uses in her classes is an inquiry approach, one of many models that science teachers employ in their classes. A model (of teaching) is a plan or pattern that organizes teaching in the classroom and fashions the way instructional materials (books, videos, computers, science materials)

are used and the curriculum is planned.[1] We will investigate several models of teaching in this chapter that will be important to you as you begin your career. The models that have been chosen are based on the learning theories described in Chapter 2. In addition to the inquiry model of teaching, we will explore the following models of teaching: the direct-interactive teaching model, the learning cycle model of teaching, and cooperative learning models of teaching, as well as several additional models including synectics, imagineering, person-centered learning, and the integrative model of learning.

We know from research and experience that practice makes perfect. A model of teaching, to be learned, must be practiced, and practiced and practiced. Unfortunately, some teachers will try a new idea, technique, or model once, not obtain very good results, and consequently abandon the notion. Some researchers report that teachers need to practice new approaches many times (perhaps as many as 20) before the new model is integrated and part of their style of teaching. Thus, in this chapter and the next you will be introduced to two laboratory strategies that are designed to help you "practice" new ideas about teaching. Reflective teaching, which you will learn about in this chapter, will be used to help you implement the models of science teaching. By using another laboratory strategy called microteaching you will learn how to implement specific teaching strategies and skills. These laboratory strategies have been developed to help you learn about teaching through teaching.

[1] Bruce Joyce and Marsha Weil, *Models of Teaching* (Englewood Cliffs, N.J.: Prentice-Hall, 1986), p. 2.

FIGURE 7.1   Inquiry teaching is one of the several approaches that effective science teachers use. Copyright Daemmrich.

PREVIEW QUESTIONS

- What is a model of teaching?
- When and under what conditions should different models of teaching be used?
- What is the relationship between models of teaching and theories of learning?
- What are the direct-interactive teaching functions?
- What are some effective ways to organize content for direct-interactive teaching?
- How is inquiry teaching different than direct-interactive teaching?
- How do the models of inductive inquiry, deductive inquiry, discovery learning, and problem solving compare?
- What is the learning cycle? On what learning paradigm is the learning cycle based?
- What is conceptual change teaching?
- What is the difference between peer tutoring and the conceptual and problem-solving models of cooperative or collaborative learning?
- What characterizes the following models of teaching: synectics, person-centered learning, integrative learning, and imagineering? How can they be used to help students understand science?

MODELS OF TEACHING: HOW CAN THEY BE OF HELP TO THE SCIENCE TEACHER?

Why is it important to know about models of teaching? Scientists use models to help them understand natural systems like rivers, atoms, and cells. They are used to describe a pattern or a phenomenon. Models are also like the scaffolding of a building. It holds the building up, and gives the building its shape and integrity. A model of teaching lays the foundation for the actions and interactions between students and teachers. For example, a teacher-centered model of teaching would imply a set of actions and interactions different from those of a student-centered model of teaching. In the teacher-centered model, teachers would make most of the decisions about curriculum and learning, whereas in a student-centered model, students would be more involved in these decisions. What are some other differences you could name?

Models of teaching are designed to help students learn, and as you will see they are prescriptive. Each model of teaching has its set of propositions and directions, enabling you to implement them in classrooms or in tutorial situations. Bruce Joyce and Marsha Weil describe 20 models of teaching in their book *Models of Teaching*, and they point out that the many models of teaching that are used in the schools are designed to give students the tools to grow. The models that are described in this chapter are designed to help the beginning science teacher get started in the classroom, as well as provide the tools to help secondary students learn science.

It is not the purpose of this book to show that one model is better than another. Rather, each model has its inherent qualities and purposes for helping students learn. Many teachers use a combination of models, integrating them into a personal model of teaching, while other teachers focus on one of these models and build their teaching repertoire around this favored approach.

There is a substantial body of research supporting the models of teaching selected for inclusion in this chapter. Naturally, science educators make a strong claim on the inquiry approach to teaching, and rightfully so. Inquiry certainly is an integral aspect of the nature of science, as was discussed in Chapter 1. But as science teachers, we must go beyond a singular view of teaching and incorporate a variety of models of teaching. Recent research and trends in practice support an integrative view for formulating instructional plans. For example, most of the recent curriculum development projects at the elementary and middle school level have described models of teaching that include not only inquiry (with hands-on activities) but cooperative learning as well. A further examination of these projects also reveals that direct-interactive teaching strategies (especially teacher-directed activities and heavy reliance on teacher questioning) are an integral aspect of the approach.

What models should we explore? Using the learning theories presented in Chapter 2 as a conceptual rationale, eight models of teaching are described in this chapter (Table 7.1), thereby presenting a kaleidoscope for the science teacher. The first model—direct-interactive teaching—is based on behavioral psychology. It is a teacher-centered model emphasizing the teaching of specific information in as direct a manner as possible. Cognitive psychology claims several models of teaching, including inquiry teaching, conceptual change teaching (or constructivism), and synectics. Finally, two models are derived from social and humanistic psychology, namely cooperative learning and person-centered learning.

TABLE 7.1

ORGANIZATION OF MODELS OF TEACHING

Learning Theory Category	Model of Teaching
Behavioral psychology	Direct-interactive teaching model
Cognitive psychology	Inquiry teaching model
	Constructivist models
	Synectics
	Imagineering
	Integrative
Social and humanistic psychology	Cooperative learning
	Person-centered

TEACHING TO LEARN

Before reading and engaging in a study of the various models of teaching presented in this chapter, we will begin by participating as teachers and learners in a series of reflective teaching lessons. In this way you will have a common experience, and can examine the models in light of this laboratory encounter with teaching. One of the assumptions made here is that one way to learn about teaching is to teach.

INQUIRY ACTIVITY 7.1

Reflective Teaching

In this activity you will teach a science lesson to a small group of peers using any model of teaching you desire to accomplish a stated objective. This experience will provide you with the opportunity to find out why you were successful or unsuccessful in a teaching situation. In this activity you will have the opportunity to share teaching methods and become reflective about science teaching as well.[2]

Several members of your class will be given the same instructional objective. They will prepare a lesson based on the objective and then teach it to a group of peers concurrently. At the conclusion of the lesson they will assess the learners to find out how many of them can perform the objective. The "teachers" will conduct brief reflective teaching sessions in the small groups, and then all will assemble for a reflective discussion of the lessons.

Materials

Reflective teaching lesson plans (see the Science Teaching Gazette at the end of this chapter)
Teaching materials as required by each lesson

Procedure

1. *Prepare.* Your class will be divided into small teams (of about five); one person from each will be selected as the first designated teacher (DT). Each of

[2]The concept of reflective teaching is based on Donald R. Cruickshank et al., *Reflective Teaching*, 2nd Ed. (Bloomington, Ind.: Phi Delta Kappa, 1991).

these designated teachers will be assigned a reflective teaching plan by the professor. The DTs will have until the next class to prepare the lesson, which will be taught to a small group of peers. In preparing the lessons, any materials and visual aids can be used to teach the lesson.

2. *Teach.* On the day that you are to be assigned, teach your lesson to your small group. You will have 15 minutes to teach the lesson. Administer any test required for your lesson. Collect the tests, and then give each of your learners the learner satisfaction form to complete (Figure 7.2). Collect the learner satisfaction forms (give students about three minutes to complete them).

3. *Reflect.* Facilitate a reflective teaching session (about five minutes) with your learners using the following questions as a guide. Select only a few of these questions for your reflective teaching session.

- What knowledge, skills, or attitudes were you hoping to develop in your learners? To what extent do you believe you were successful in your effort? What do your learners think?
- In planning your lesson, which model of teaching influenced you most (e.g., direct-interactive teaching, inquiry, cooperative learning, the learning cycle)?
- What did you learn about teaching?
- How did your learners react to the model of teaching you used?
- What influenced you most in planning your lesson (your knowledge about the subject, your

attitudes about teaching, materials available to you)?

4. *Whole group reflection.* When your reflective discussion is complete, join the whole class to discuss each of the reflective teaching lessons.

Minds on Strategies

Use these questions to write a critique of your teaching session and the corresponding reflective discussion session.

1. What model of teaching was used by each designated teacher? In what ways did the models vary?
2. How successful were learners in achieving the stated objectives for the lesson?
3. What was learned about effective teaching?

Name_____
1. During the lesson how satisfied were you as a learner?
_____ very satisfied
_____ satisfied
_____ unsatisfied
_____ very unsatisfied
2. What could your teacher have done to increase your satisfaction?

FIGURE 7.2 Learner Satisfaction Form

THE DIRECT–INTERACTIVE TEACHING MODEL

An eighth-grade earth science teacher begins a lesson by showing students several slides depicting the moon at successive one-hour intervals. Then the teacher says, "Today we are going to discuss the sightings you made of the moon and stars in the western sky last night for homework, and then I will do a demonstration on how to measure the altitude and compass location of a star. You will use this information to practice making measurements of fictional stars in the classroom in small groups. Finally, I'll give you some problems to solve, and then you'll use what you learned today to measure changes in star motion in tonight's sky for homework."

This opening statement by a junior high science teacher conveys some of the elements of the direct-interactive teaching (DIT) model. The DIT model is a good place to begin because it flourishes in many classrooms and calls on the teacher to direct students

FIGURE 7.3 Direct-Interactive teaching is a teacher-centered model designed to teach science information in a direct manner. Teachers use the direct-interactive model to engage students by using questioning strategies, to introduce lab, to carry out post-lab discussion, or review for a test. Copyright Rhoda Sidney.

by assigning specific tasks that must be completed under direct teacher supervision. It is, however, a dynamic model in that the most effective form of direct teaching implies interaction between the teacher and the students.

The teaching of science information and skills can be accomplished quite effectively through the direct-interactive teaching model. In the science classroom, the material to be learned is subdivided into smaller chunks of information and is presented directly to the students. In this teacher-centered model, the teacher's role is very clear: to teach science information and skills in the most direct manner possible.

Research by a number of science educators shows that when direct instruction strategies are used a notable increase in achievement occurs. The science classroom that is based on the direct-interactive teaching model appears to be characterized by a number of factors, as follows:[3]

1. Instructional objectives are formulated and communicated to the students prior to the start of a unit of teaching

2. Teachers gain attention at the start of each lesson by using focusing behaviors and strategies such as advanced organizers and set induction. These typically include asking questions, performing a short demonstration, or the use of an EEEP (see Chapter 8)

3. Students handle, operate on, or practice with science teaching materials. This includes the full range of manipulative materials, from familiar science objects, such as rocks, fossils, and plant specimens, to pictorial stimuli, as well as cardboard cutouts depicting science concepts such as crystal form, as well as paper products such as cards with the names and pictures of atoms, organisms, chemical equations, and the like

4. Teachers alter instructional materials or classroom procedures to facilitate student learning. Rewriting activity or experiment procedures from a textbook, making audio tapes of the science textbook, and giving students directions in writing are examples of how science teachers alter instructional procedures

5. The science teacher focuses attention on the type and placement of questions asked during lessons

6. In effective science classrooms, teachers provide immediate as well as explanatory feedback during the instructional process, rather than waiting until a quiz or major test

[3]Clifford A. Hofwold, "Instructional Strategies in the Science Classroom," in *Research with Reach: Science Education,* David Holdzkom and Pamela B. Lutz, eds. (Washington, D.C.: National Science Teachers Association, 1985), pp. 43–57.

The DIT model fosters a learning environment characterized by teacher-directed learning and high levels of teacher-student interaction. Rosenshine has identified six teaching functions that taken together constitute the essential principles of direct-interactive teaching. These functions (Table 7.2) include checking the previous day's homework, presenting and demonstrating new content and skills, leading the initial student practice session, providing feedback and correctives, providing independent practice, and doing weekly and monthly reviews.

TABLE 7.2

DIRECT-INTERACTIVE TEACHING FUNCTIONS

Teaching Function	Specific Behaviors
Checking previous day's work and reteaching	Check homework.
	Reteach areas where there are student errors.
Presenting or demonstrating new content and skills	Provide overview.
	Proceed in small steps but at a rapid pace.
	If necessary, give detailed or redundant instructions and explanations.
Leading initial student practice	Provide a high frequency of questions and overt student practice.
	Provide prompts during initial learning.
	Give all students a chance to respond and receive feedback.
	Check for understanding by evaluating student responses.
	Continue practice until students are firm.
	Ensure a success rate of 80% or higher during initial practice.
Providing feedback and correctives (and recycling of instruction if necessary)	Give specific feedback to students, particularly when they are correct but hesitant.
	Student errors provide feedback to the teacher that corrections and reteaching is necessary.

(Table continued on p. 206)

Teaching Function
(Table continued from p. 205)

Specific Behaviors

Teaching Function	Specific Behaviors
	Offer corrections by simplifying question, giving clues, explaining or reviewing steps, or reteaching last steps.
	When necessary, reteach using smaller steps.
Providing independent practice so that students are firm and automatic	Assign seatwork or homework.
	Ensure unitization and automaticity (practice to overlearning).
	Ensure student engagement during seatwork (i.e., monitor students).
Providing weekly and monthly reviews	Reteach if necessary.

Based on Barak Rosenshine, "Teaching Functions in Instructional Programs," *The Elementary School Journal,* March 1983, vol. 83, no. 4, pp. 335–351.

The DIT model can be represented as a cycle of teaching (Figure 7.4). As you implement this model of teaching it is important to note that four important aspects of the model stand out:

- You will need to develop and implement a variety of learning tasks
- The learning tasks you develop should engage the learner at high levels
- You should strive for high levels of teacher-student and student-student interaction. You can achieve this by the use of teacher questions, hands-on activities, and small group work
- Your students should perform at moderate-to-high rates of success

Structuring Content for Direct-Interactive Teaching

Another important aspect of the DIT model and is the presentation and structuring of science content. One of the key ingredients is to break content into manageable, teachable and learnable chunks. Borich points out that most teachers "divide" content based on the content divisions in science textbooks. This organization is often arranged to help students read the text and therefore may not be the best way to present content. There are a number of ways to structure new science content. Following are four suggestions that you should find helpful in dividing science content for the DIT model. They include whole-part, sequential, combinatorial, and comparative methods of content structuring.[4]

Whole-Part

Organizing content in a whole-to-part format is useful in introducing science content in its most general form. For instance, if you were presenting information on rock types, you might start with the question, What are the types of rocks? This would lead to natural subdivisions (igneous, metamorphic, and sedimentary) that can be easily learned by students. (Figure 7.5)

[4]Gary D. Borich, *Effective Teaching Methods* (Columbus, Ohio: Merrill, 1988), pp. 148–152.

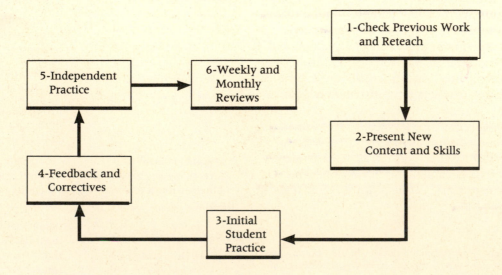

FIGURE 7.4 Direct-Interactive Model of Teaching

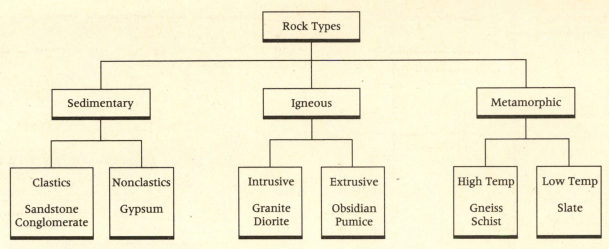

FIGURE 7.5 Structuring Content: Whole to Part

Whole-part structuring is a powerful way to organize information. Recent research using the technique of concept mapping (see Chapter 2) is based on the organization of knowledge from whole to part. Whole to part thinking gives students an organizational framework from which to operate. Big ideas can be used to "hook" subconcepts and subideas, rather than as isolated bits of information.

Sequential Structuring

Sequential structuring is organizing content and skills by ordering. Typically the content or skills are presented from simplest to most complex. Sequential structuring is based on a hierarchical arrangement of science content or science skills. In a way, sequential structuring is an alternative to the whole-part organization. Typically in the sequential structuring of science content or skills, students would be introduced to prerequisite content or skills first and then be introduced to content or skills that are dependent on the previously learned material or skills.

For example, science skills or processes can be broken down into a number of simpler skills that can be mastered in sequence. Simpler or more basic process skills would be introduced first. Examples would include

1. Observing
2. Using space-time relationships
3. Using numbers
4. Measuring
5. Classifying
6. Communicating
7. Predicting
8. Inferring

More complex science skills would be introduced later after students had mastered the more simple or basic skills. Complex science skills often involve the integration of two or more process skills and therefore are usually refered to as integrated process skills. Some include

1. Formulating hypotheses
2. Controlling variables
3. Interpreting data
4. Defining operationally
5. Experimenting

Science content can similarly be arranged in a hierarchy. Scientists classify matter according to size and function. This classification leads to levels of organization. The levels of organization can be useful as a mechanism for sequencing content. For example, if we were to organize a unit of instruction on ecology, we could use the levels of organization of matter shown in Figure 7.6 as the sequence of presentation of content from the microworld, which would contain nonlife, subatomic particles, atom, molecules, protoplasm, and cells to the supermacro world consisting of stars, galaxies, and the universe.

Combinatorial Organization

Science content can be presented by highlighting connections between the various elements of content to be presented. One very effective means of doing this is to present the elements of the content in a cycle. There are many cycles in the various disciplines of science. In earth science content could be organized, for example, by means of the rock cycle or the water cycle instead of learning simply about rivers. In biology, photosynthesis and the Krebs cycle could be organizational cycles to show relationships and combinations.

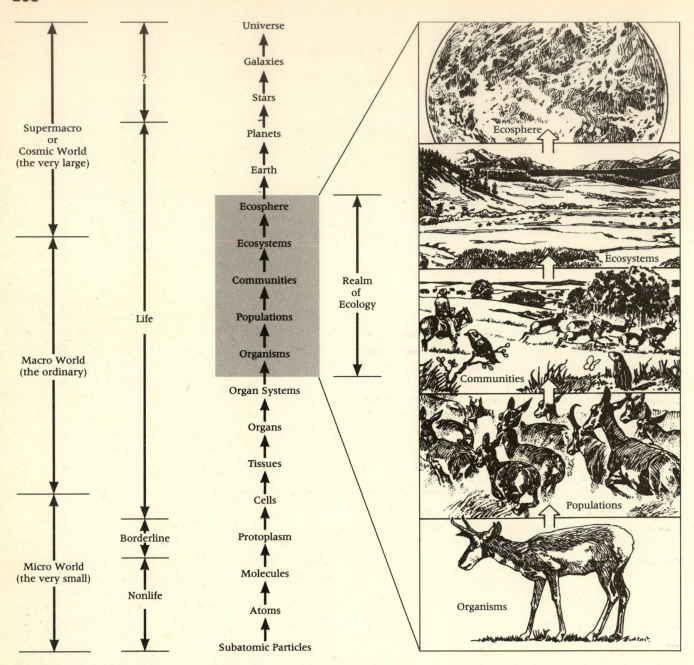

FIGURE 7.6 Levels of Organization of Matter as a way to sequentially structure content. From *Living in the Environment, 2nd ed.*, by G. Tyler Miller, Jr. Copyright © 1979 by Wadworth Publishing Co., Inc. Belmont, CA 94002. Reprinted with permission of the publisher.

Figure 7.7 shows a summary of the major components of an ecosystem and how they are interconnected through chemical and energy cycles. Nonliving chemicals, producers, macroconsumers, and microconsumers are presented in relationship to each other. The presentation of "new content" would include the relationships as well as the specific details of the elements (e.g., macroconsumers). The cycle itself is another powerful organizing element to help the student learn and understand the material being presented.

Comparative Relationships

Another effective way to present content is comparing categories of content in order to heighten similarities and differences. Table 7.3 contrasts the similarities and differences between four types of drugs. The advantage of this approach is that the organization of the content into a table or chart is a useful learning device for the student, and individual elements (e.g., type of drug) are learned in relationship to others. Students can graphically compare, for instance, the relative dependence and effects of the drugs.

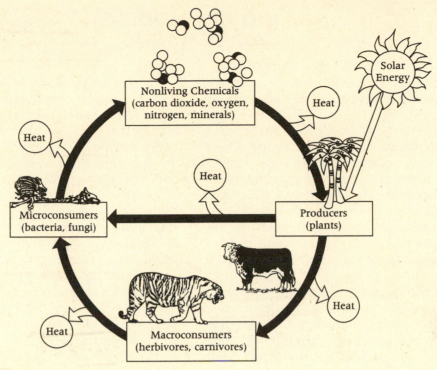

FIGURE 7.7 The Ecosystem Cycle can be used as a combinatorial organization to present science content.

TABLE **7.3**

STRUCTURING CONTENT:
COMPARATIVE RELATIONSHIPS–DRUGS

Drug	Dependence Potential	Short-Term Effects	Long-Term Effects
Stimulants			
Caffeine	Probable	Increased heart rate, increased blood pressure	Irregular heartbeat, high blood pressure, stomach disorders
Amphetamines	High		
Cocaine	High		
Depressants			
Barbiturates	High	Drowsiness, decreased coordination	Depression, emotional instability, hallucinations, death
Tranquilizers	High		
Psychoactive Drugs			
Lysergic acid diethylamide (LSD)	Probable	Hallucinations	Psychosis
Cannabis sativa (marijuana)	Probable	Short-term memory loss, disorientation	Lung damage, loss of motivation
Narcotics			
Heroin	High	Drowsiness, respiratory depression, nausea, constricted pupils	Convulsions, coma, death
Codeine	High		
Morphine	High		

Albert Towle, *Modern Biology, Teacher's Edition*. (Austin: Holt, Rinehart and Winston, 1989), p. 753. Used with permission of the publisher.

INQUIRY MODELS
OF TEACHING

- A *physics* teacher asks students, "Is it a good idea to continue to develop and build new nuclear power plants?"
- An *earth science* teacher asks students to interpret a set of dinosaur footprints, and generate several alternative hypotheses to explain the pattern of the prints
- A *biology* teacher takes students on a field trip to collect leaves from different trees. Students are asked to create a classification system using the leaves
- A *chemistry* teacher gives students an unknown substance, and asks them to use scientific tests to determine the composition of the material

In each of the foregoing situations, the science teacher has created a situation in the classroom in which students are asked to formulate their own ideas, state their opinion on an important issue, or find things out for themselves. It is a radical departure from the direct-interactive teaching model in which the teacher engages students to learn science information or skills. In each of the foregoing scenarios, the student is encouraged to ask questions, analyze specimens or data, draw conclusions, make inferences, or generate hypotheses. In short the student is viewed as an inquirer—a seeker of information—and a problem solver. This is the heart of the inquiry model of teaching.

What is Inquiry?

J. Richard Suchman, the originator of an inquiry teaching program that was widely used throughout the United States, once said that "inquiry is the way people learn when they're left alone."[5] To Suchman, inquiry is a natural way that human beings learn about their environment. Think for a moment about a very young child left in a play yard with objects free to explore. The child, without any coaxing, will begin to explore the objects by throwing, touching, pulling, banging them, and trying to take them apart. The child learns about the objects and how they interact by exploring them, by developing his or her own ideas about them—in short learning about them by inquiry.

Many authors have discussed the nature of inquiry, referring to it as inductive thinking, creative thinking, discovery learning, and the scientific method. Many trace the essence of inquiry to John Dewey.

Dewey proposed that inquiry is the "active, persistent, and careful consideration of any belief or supposed form of knowledge in the light of the grounds that support it and the further conclusions to which it tends."[6] To Dewey the grounding of any belief occurs through inquiry processes: reason, evidence, inference, and generalization. Recently, science educators have proposed various lists of inquiry processes. One such list includes observing, measuring, predicting, inferring, using numbers, using space-time relationships, defining operationally, formulating hypotheses, interpreting data, controlling variables, experimenting, and communicating. Another list by Aikenhead proposes the following as inquiry processes:[7]

1. Observing objects, materials, and phenomena
2. Creating hypotheses
3. Deciding on pertinent variables
4. Setting up an experiment (either according to precise directions or according to self-composed directions)

[6]John Dewey, *How We Think: A Restatement of the Relation of Reflective Thinking to the Education Process* (Boston: D.C. Heath, 1933), p. 9.
[7]Glen S. Aikenhead, "Decision-Making Theories as Tools for Interpreting Student Behavior during a Scientific Inquiry Simulation," *Journal of Research in Science Teaching* 26, no. 3 (1989): 189–203.

FIGURE 7.8 The inquiry model of teaching has been shown in research studies to enhance student academic performance, attitudes towards science, as well as scientific skill development. Copyright Daemmrich.

[5]J. Richard Suchman, *Developing Inquiry* (Chicago: Science Research Associates, 1966).

5. Measuring variables
6. Collecting, organizing, graphing, and interpreting data
7. Cooperating as a research team
8. Participating in scientific consensus making

In the context of learning, students will engage in inquiry when faced with a "forked-road situation," or a problem that is perplexing and causes some discomfort to them.[8] In the model of inquiry presented in this chapter, the creation of forked-road situations, or discomforting problems, will be the essence of science inquiry activities.

What about inquiry teaching in school situations? If you recall from Chapter 1, it was shown that the predominant method of teaching in science is recitation, not inquiry. In fact, the evidence is that very little actual time is spent by students doing inquiry activities. Holdzdom and Lutz report that direct teaching strategies have greater impact than indirect ones.[9] *However, they also report that when inquiry models of teaching were implemented, they were very effective in enhancing student performance, attitudes, and skill development.* They reported that student achievement scores, attitudes, and process and analytic skills were either raised or greatly enhanced by participating in inquiry programs.

While the research supports the inclusion of inquiry models of teaching in secondary science classrooms, there appears to be a reluctance on the part of the science teachers to implement inquiry in the classroom. Several problems need to be recognized to overcome the reluctance to implement inquiry in the classroom.

Why is it that science teachers express the importance of inquiry yet pay little attention to it in the classroom? One reason may have to do with teacher education. It is possible that many teachers have not been exposed to inquiry teaching models in their preparation and therefore lack the skills and strategies to implement inquiry. Some teachers report that inquiry teaching models are difficult to manage, and some report that they don't have the equipment and materials to implement inquiry teaching. Another concern expressed by teachers is that inquiry doesn't work for some students. These teachers claimed that inquiry was only effective with bright students and caused too many problems with lower ability students.

In spite of these problems the evidence is that inquiry models of teaching are viable approaches to teaching and should be part of the science teacher's repertoire. Science teachers have had a love affair with inquiry, and feel strongly that it should be a fundamental part of science teaching. Read about what teachers think about inquiry in the Science Teachers Talk section in the Science Teaching Gazette of this chapter. Note how the teachers interviewed link inquiry with discovery and indicate that the reason they liked science was because of the excitement of finding out about things, probing, exploring—in short, inquiring.

Inductive Inquiry

We will explore four models of inquiry teaching: inductive inquiry, deductive inquiry, discovery learning, and problem solving. We begin with inductive inquiry.

Perhaps the best example of inductive inquiry is the Inquiry Development Program developed a number of years ago by J. Richard Suchman.[10] Suchman produced a number of inquiry programs designed to help students find out about science phenomena through inquiry. Suchman's views on inquiry are quite applicable today, and this statement by him is worth pondering:

> Inquiry is the active pursuit of meaning involving thought processes that change experience to bits of knowledge. When we see a strange object, for example, we may be puzzled about what it is, what it is made of, what it is used for, how it came into being, and so forth. *To find answers to questions* (emphasis mine) such as these we might examine the object closely, subject it to certain tests, compare it with other, more familiar objects, or ask people about it, and for a time our searching would be aimed at finding out whether any of these theories made sense. Or we might simply cast about for information that would suggest new theories for us to test. All these activities—observing, theorizing, experimenting, theory testing—are part of inquiry. The purpose of the activity is to gather enough information to put together theories that will make new experiences less strange and more meaningful.[11]

The key to the inquiry model proposed by Suchman is providing "problem-focus events." Suchman's program provided films of such events, but he also advocated demonstrations and developed a series of idea books for the purpose of helping students organize concepts. It is the inquiry demonstrations that we use to help you develop inquiry lessons.

[8]Proposed by Byron G. Massialas and Jack Zevin, *Creative Encounters in the Classroom* (New York: John Wiley & Sons, 1967), p. 1.

[9]David Holdzdom and Pamela B. Lutz, *Research Within Reach: Science Education* (Washington, D.C.: National Science Teachers Association, 1985).

[10]J. Richard Suchman, *Developing Inquiry* (Chicago: Science Research Associates, 1966). Also consult Suchman, *Inquiry Development Program in Physical Science* and *Developing Inquiry in Earth Science*, each published by Science Research Associates, Inc.

[11]J. Richard Suchman, *Developing Inquiry in Earth Science* (Chicago: Science Research Associates, 1968), p. 1.

The Inquiry Session

The inquiry demonstration is a method to present a problem to your students. The demonstration is *not* designed to illustrate a concept or principle of science. It is instead designed to present a discrepancy or a problem for the students to explore. In fact, we refer to inquiry demonstrations as discrepant events. An inquiry session is designed to engage the class in an exploration of a problem staged by means of the discrepant event. An inquiry session should begin with the presentation of a problem through a demonstration (the discrepant event), a description of an intriguing phenomena, or a problem posed by the use of prepared materials.

Suchman proposed six rules or procedures that teachers have found helpful in conducting inquiry sessions (Table 7.4). According to the inquiry model, students learn that to obtain information they must ask questions. Questioning becomes the students' initial method of gathering data. Thus, the climate of the inquiry classroom must foster the axiom: "There are no dumb questions." Students must come to believe that you will accept their questions—no holds barred. Once an event is presented, the teacher must be sure the students understand the real problem. Once the problem is established, the students engage in the inquiry session to construct a theory to account for the focus event. The major portion of the inquiry session is devoted to the students asking questions to gather data, which are then used to formulate one or more theo-ries. You should refer to Table 7.4 and Figure 7.9 before conducting an inquiry session yourself.

Inquiry Activities

There are numerous sources of inquiry activities (including discrepant events). Tik L. Liem has put together hundreds of discrepant events and inquiry activities that you can use in all areas of science.[12]

Following are some examples of inquiry activities and discrepant events.

The Inquiry Box

Of all the approaches to help students learn about inquiry, the inquiry box might be considered the universal strategy. The inquiry box can be made with a shoe box and should be painted black. For a classroom of students, you could prepare several inquiry boxes. Students are given the box, and asked to determine what the inside of the box is like. An inquiry box contains a marble, which is the main probe that the students can use to determine the pattern that exists within the box. You can prepare different patterns by taping pieces of cardboard in interesting and perplexing patterns as shown in Figure 7.10.

The inquiry activity consists of having teams of students explore each inquiry box that you have pre-

[12]For sources of inquiry activities refer to Tik L. Liem, *Invitations to Science Inquiry*, 2nd ed. (Chino Hills, Calif.: Science Inquiry Enterprises, 1990).

TABLE   7.4

PROCEDURES FOR AN INQUIRY SESSION

Rule	Procedure
Rule 1: Questions	The questions by the students should be phrased in such a way that they can be answered yes or no. This shifts the burden of thinking onto the students.
Rule 2: Freedom to ask questions	Students may ask as many questions as desired once they begin. This encourages students to use their previous questions to formulate new ones to pursue a reasonable theory.
Rule 3: Teacher response to statements of theory	When students suggest a theory, the teacher should refrain from evaluating it. The teacher might simply record the theory, or ask a question about the students' theory.
Rule 4: Testing theories	Students should be allowed to test their theories at any time.
Rule 5: Cooperation	Students should be encouraged to work in teams to confer and discuss their theories.
Rule 6: Experimenting	The teacher should provide materials, texts, and reference books so that the students can explore their ideas.

Based on J. Richard Suchman, *Developing Inquiry* (Chicago: Science Research Associates, 1966).

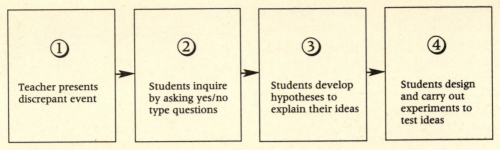

As an example:

① Teacher conducting a demonstration in front of the class. Teacher holding a pendulum.

② Students raising hands as if to ask questions.

③ Teacher writing students ideas on board:
Write on board these sentences:
--length of string
--weight of the pendulum

④ Students working in small groups at tables of four. String, and various weights on the table, as well as a watch.

FIGURE 7.9   Suchman Inquiry Model

pared. The students' theory consists of a diagram of the possible pattern in each box.

The Coin Drop and Throw

The teacher places one coin (a quarter) on the edge of a table and holds another in the air next to it (Figure 7.11). At the same instant, he or she flicks the quarter on the table so that it flies horizontally off the table and drops the other quarter straight down. Both coins strike the floor at the same time. An inquiry about why the coins strike the floor at the same time ensues. (Hint: practice this demonstration before you perform it with a group of students.) Science principles that will emerge from this inquiry include vectors, universal gravitation, and Newton's second law of motion.

The Double Pendulum

The teacher places a long rod (meter stick) across the backs of two chairs. From the rod two simple pendulums of the same length are hung. One of the pendulums is started swinging. The other is allowed to

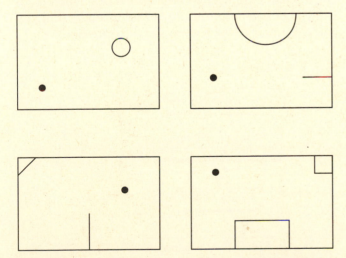

FIGURE 7.10   Inquiry Boxes: A box with an internal pattern designed so the student must use imagination and inquiry to probe its structure: a universal tool for focusing on scientific thinking skills.

FIGURE 7.11   The Coin Drop and Throw

hang straight down. In a few minutes the stationary pendulum begins swinging as the arc of the swinging pendulum decreases. The inquiry focuses on why the second pendulum begins to swing and why the arc of the first pendulum decreases. The science principles in this inquiry include periodic motion and conservation of energy.

The Balloon in Water

A balloon is partially inflated, tied shut, and tied to a heavy object (Figure 7.12). It is dropped into the bottom of a tall cylinder filled almost to the top with water. A rubber sheet is placed over the top of the cylinder and sealed with a rubber band. The teacher pushes on the rubber cover, and the balloon becomes slightly smaller. When the rubber cover is released, the balloon returns to its original size. The inquiry focuses on why the balloon becomes smaller and then larger again. Principles of science in this inquiry include pressure, gases, liquids and solids, and Newton's first law of motion.

Deductive Inquiry

Another form of inquiry teaching is deductive inquiry, which we can contrast with inductive inquiry (Figure 7.13). In this approach to inquiry, the teacher presents a generalization, principle, or concept and then engages students in one or more inquiry activities to help understand the concept. For example, suppose the teacher's lesson plan calls for the introduction to the differences between physical and chemical weathering. The lesson begins with an explanation of physical weathering. Next the teacher discusses the attributes of chemical weathering. During the discussion of physical weathering the teacher would discuss various types of mechanical weathering including frost action, drying, and cracking. Chemical weathering processes such as

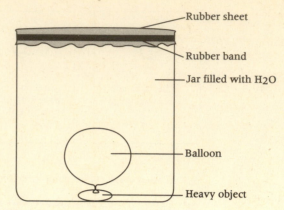

FIGURE 7.12 Submarine

carbonation, oxidation, hydration and leaching would be introduced.

After the development of the major concepts through presentation and questioning, the teacher then engages the students in an inquiry activity in which they explore the concepts of physical and chemical weathering. One approach that has been successful with these concepts is the following: The teacher places in individual trays examples of earth materials that have been affected by either physical or chemical weathering (e.g., rocks with cracks, soil, mud cracks, plants growing in cracks in rocks, and staining evident in the minerals of rocks). Working in teams, students study each tray and hypothesize what caused the change they observe. After all teams have investigated each tray, students write their hypotheses on a large chart on the chalkboard. The teacher leads a postactivity discussion in which the students defend their hypotheses.

Most of the science textbooks written for middle and secondary science courses contain hands-on activities that reinforce the deductive inquiry approach. If the activities are used in the context of deductive inquiry, they can be extremely helpful in aiding students' understanding of the concepts in the course.

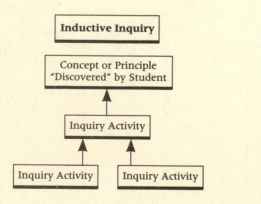

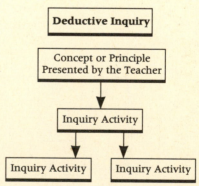

FIGURE 7.13 Comparison of Inductive and Deductive Inquiry

INQUIRY ACTIVITY   **7.2**

Inductive Versus Deductive Inquiry

In this inquiry activity you will examine some secondary science textbooks and make decisions about how inductive and deductive inquiry learning could be implemented.

Materials

A secondary science textbook

Procedure

1. Examine the textbook and identify a chapter that interests you. Identify in the chapter the flow of concepts and how they are presented in the text.

2. Design the chapter inductively. That is, organize a series of lessons (describe them very briefly) in which the students will be introduced to the main concepts inductively (Figure 7.13).

3. Now design the chapter deductively. Organize a series of lessons in which the students will be introduced to the main concepts deductively (Figure 7.13).

Minds on Strategies

1. What did you discover about inductive versus deductive teaching?

2. In which manner was the textbook organized, inductively or deductively?

3. Which model do you think is better for student learning? Why?

Discovery Learning

Discovery learning, a concept advocated by Jerome Bruner, is at the heart of how students learn concepts and ideas. Bruner talked about the "act of discovery" as if it were a performance on the part of the student. To Bruner, discovery "is in its essence a matter of rearranging or transforming evidence in such a way that one is enabled to go beyond the evidence so reassembled to new insights." Bruner believed that discovery learning could only take place if the teacher and student worked together in a cooperative mode. He called this type of teaching "hypothetical teaching" and differentiated it from "expository teaching."[13] In Chapter 1 these forms of teaching were referred to as engagement versus delivery.

Discovery learning in the science classroom engages students in science activities designed to help them assimilate new concepts and principles. Discovery activities help guide the students to assimilate new information. In such activities students will be engaged in observing, measuring, inferring, predicting, and classifying.

There are a number of practical suggestions that you can implement to foster discovery learning in the classroom.

1. *Encourage curiosity.* Since the student in discovery learning is the active agent in learning, the science teacher should foster an atmosphere of curiosity. Discrepant events and inquiry activities are excellent ways to foster curiosity. Having interesting and thought-provoking bulletin boards is another way to arouse curiosity.

2. *Help students understand the structure of the new information.* Bruner stressed that students should understand the structure of the information to be learned. He felt that teachers needed to organize the information in a way that would be most easily grasped by the student. Bruner suggested that knowledge could be structured by a set of actions, by means of graphics, or by means of symbols or logical statements. Demonstrating the behavior of objects is a more powerful way for some students to grasp Newton's laws of motion, than the three classic verbal statements.

3. *Design inductive science labs or activities.* The use of inductive science activities is based on the assumption that the teacher is aware of the generalization, principle, or concept that the students are to discover. An inductive lab or activity is designed so that the student is actively engaged in observing, measuring, classifying, predicting, and inferring. Generally speaking, the teacher provides the specific cases, situations or examples that students will investigate as they are guided to make conceptual discoveries. An example of an inductive science activity is the footprint puzzle (Figure 7.14), in which students are guided to make discoveries about the behavior and environment of dinosaurs that made these prints.

[13]Jerome Bruner, *Toward a Theory of Instruction* (New York: Norton, 1966).

FIGURE 7.14   The Dinosaur Footprint Activity: Give cooperative teams the footprint puzzle. Ask students to observe the footprint and then to generate at least three hypotheses to explain the events in the puzzle. Earth Science Curriculum Project, *Investigating the Earth, Teacher's Guide, Part II*, Boston, MA: Houghton-Mifflin Company. © 1967 American Geological Institute. This publication is not endorsed by the copyright holder.

After the students have explored the footprint puzzle and have written at least three alternative hypotheses to explain the tracks, the teacher leads a discussion to help the students discover some concepts about the dinosaurs.

One of the points Bruner made about discovery was that it is the result of making things simpler for the student. In science, making the science concepts simpler through inductive activities is a far more powerful approach than presenting vast amounts of information about the concept.

4. *Encourage students to develop coding systems.* Coding systems help students make connections between objects and phenomena. Bruner felt that students could learn the method of discovery—the heuristics of discovery—if provided with many puzzling situations. For example, giving students a box of rocks and asking them to invent a classification system of their own would help them understand the principles of coding and classification. The computer is a powerful learning tool in this regard. Programs are available to enable students to practice working with puzzling situations and develop expertise in coding.

5. *Design activities that are problem oriented.* Students need to be engaged in problem solving on a regular basis if they are to learn about the heuristics of discovery. Bruner said "It is my hunch that it is only through the exercise of problem solving and the effort of discovery that one learns the working heuristics of discovery." In short, he said that students need practice in problem solving or inquiry to understand discovery. Activities that are problem oriented often have a simplistic ring to them. For example, here are two problems, any one of which could be a learning activity for students:

- Find a million of something and prove it.
- Go outside and find evidence for change.

6. *Foster intuitive thinking in the classroom.* Intuitive thinking to Bruner implied grasping the meaning, significance, or structure of a problem without explicit analytical evidence or action. Here is where Bruner thought that playfulness in learning was important. Students in a classroom whose teacher values intuition know that it is acceptable to play with all sorts of combinations, extrapolations, and guesses and still be wrong. Including some science activities that encourage guessing and estimating will foster intuitive thinking. Qualitative activities in which students are not encouraged to find a specific answer to a problem will encourage intuitive thought. Skolnick, Langbort, and Day, in their book *How to Encourage Girls in Math and Science*, suggest a number of intuitive strategies, including estimating and engaging in activities with many right answers and multiple solutions.[14]

Problem Solving

Problem solving as a method of inquiry can be used to teach problem-solving skills and to engage students in the investigation of real problems. Don't be fooled by questions and problems that appear at the end of the chapters in science textbooks. For the most part, these "problems" are merely questions that require students to look up the answer in the text or plug in the numbers of a formula.

Problem solving in the context of inquiry engages students in problems that are real and relevant to them. The problems do not have to be ones that students generate (although this approach is probably more powerful). They can be problems that the teacher has presented to the students for investigation.

[14]Joan Skolnick, Carol Langbort, and Lucille Day, *How to Encourage Girls in Math and Science* (Englewood Cliffs, N.J.: Prentice Hall, 1982).

Science, unfortunately, is often presented in textbooks as "problem free." That is, the content of science is arranged in a very neat and tidy way. The truth of the matter is that science is often messy and cluttered, and full of problems.

There are many approaches to dealing with the question of problem solving as a model of inquiry teaching. In the publication *What Research Says to the Science Teacher: Problem Solving,* Dorothy Gabel points out that some science educators prefer an approach that focuses on process skills, others concentrate on helping students solve global qualitative problems, and still others focus on mathematical problem solving.

For example, Charles Ault makes the following suggestions for teaching problem solving in earth science classes.[15] These suggestions are quite applicable to physical and life science, as well.

- Identify the conceptual models needed to reason in specific domains. For example, accounting for falling raindrops is not simple. How do droplets form in the sky? How do they get bigger? What keeps small droplets suspended? Do ice and water coexist in clouds? Do static charges make drops coalesce or not? When rain stops, why are there often still clouds in the sky? Good conceptual models have the raw materials for constructing answers to unusual questions as well as standard ones.

- Solve problems about phenomena familiar to students' experiences. Include plenty of usable content that can resolve dilemmas, such as those dealing with condensation nuclei and the vapor pressure of ice crystals versus water droplets in the preceding example.

- Use props to assist visualization and abstract reasoning. If there are distortions of scale, make them explicit.

- Ask for oral and written restatements of problems, emphasizing precise meanings of terms and relationships in models.

- Connect abstractions to everyday experience by analogy: for example, compare escalators and merry-go-rounds to relative motion and orbit. Be certain that important relationships are well understood in the context of the analogies.

- Use imagination and imagery to express scale: contrast ancient toeholds in the Betatakin ruins of the Anasazi people with the even older excavation of the canyon and cavern. Try body language to convey patterns in earth forms or motion in celestial bodies.

- Remember that the complexity of teaching and learning earth science vastly exceeds the ability of research to offer prescriptive advice.

CONSTRUCTIVIST MODELS

Suppose that you are teaching a general science course in a middle school and the first unit of the year is ecosystems. Suppose further that your college professor introduced you to a method called concept mapping, and you decide to use it to identify the concepts that students might need to understand a unit on ecosystems. Figure 7.15 shows the subconcepts and their relationship to the general concept of ecosystem.

What is the best way to teach students concepts as depicted on this concept map? Recent research (see Chapter 2) has suggested that students construct their own ideas about their world—including concepts—and any attempt to teach "new" concepts to students must take this into consideration. Furthermore, researchers have recommended that linking new concepts to a students' prior knowledge is an integral part of learning. Researchers who have supported this view have been labeled constructivists because of their belief that students construct their own knowledge structures—concepts, beliefs, theories. More often than not, the concept base that students bring to your class is naive and full of misconceptions. To understand science from the constructivist view means that students will have to change their concept—hence the term conceptual change.

The Learning Cycle: A Conceptual-Change Model

A number of conceptual change models have emerged over the last 30 years that suggest that teachers should sequence instruction into a series of teaching and learning phases.[16] The sequences have been described

---

[15]Charles R. Ault, "Problem Solving in Earth Science Education," in *What Research Says to the Science Teacher: Problem Solving,* vol. 5, Dorothy Gabel, ed. (Washington, D.C.: National Science Teachers Association, 1989), pp. 46–47.

[16]There are many learning cycle models. For an excellent discussion of four different models, please refer to Roger Osborne and Peter Freyberg, *Learning in Science* (Auckland: Heinemann, 1985), pp. 101–111.

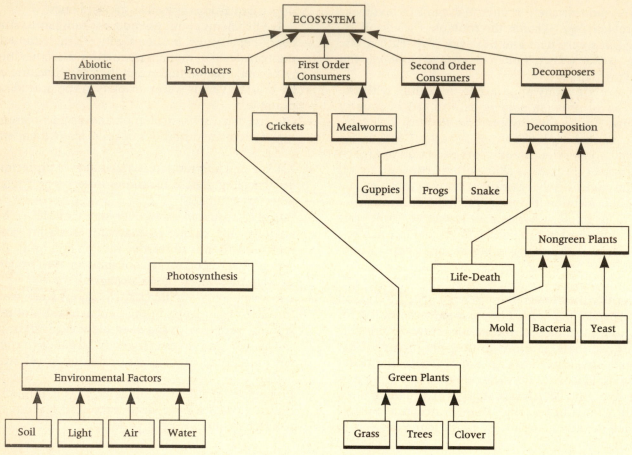

FIGURE 7.15 Ecosystem Concept Map. Anton E. Lawson, Michael R. Abraham, and John W. Renner, *A Theory of Instruction: Using the Learning Cycle to Teach Science Concepts and Thinking Skills*, NARST Monograph, No. 1, (National Association for Research in Science Teaching, 1989), p. 14. Used with permission of NARST.

as learning cycles. For example, the earliest example of a learning cycle was suggested by Chester Lawson, in which he described scientific invention as "belief-expectation-test."[17] But the first direct application of a learning cycle to science teaching was proposed by Robert Karplus, director of the Science Curriculum Improvement Study (SCIS) in 1970. He proposed a three-phase cycle consisting of *preliminary exploration, invention, and discovery*. In essence, Karplus believed that students need to first explore the concept to be learned using concrete materials. The initial introduction of the concept was called invention. In this phase the teacher assumed an active role in helping the students use their exploration experiences to invent the concept. To Karplus, the discovery phase provided the student with the opportunity to verify, apply, or further extend knowledge of the "invented" concept.

Charles R. Barman and the team of Lawson, Abraham, and Renner have proposed a learning cycle based on the work of Karplus but have changed the terminology.[18] We will use their terminology in this book when we refer to the learning cycle model.

The learning cycle has three phases that form the foundation for sequencing science lessons (Figure 7.16): exploration, concept introduction, and concept application. Normally a sequence would take at least three sessions (see the sample lesson that follows).

The Exploration Phase

During this phase the students explore a new concept or phenomenon with minimal guidance. Students might make observations of and classify objects. They might be involved in "messing about" with batteries, bulbs, and wires to find out how a light bulb works.

[17]Anton E. Lawson, Michael R. Abraham and John W. Renner, *A Theory of Instruction: Using the Learning Cycle to Teach Science Concepts and Thinking Skills*, NARST Monograph, No. 1, (National Association for Research in Science Teaching, 1989), p. 9.

[18]Charles R. Barman, *An Expanded View of the Learning Cycle: New Ideas About an Effective Teaching Strategy*, Monograph and Occasional Paper Series, No. 4, (Council for Elementary Science International, 1990).

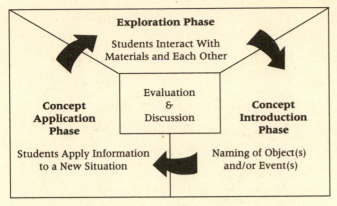

FIGURE 7.16   Learning Cycle Model. Charles R. Barman, *An Expanded View of the Learning Cycle: New Ideas About an Effective Teaching Strategy*, Monograph and Occasional Paper Series, No. 4, (Council of Elementary Science International, 1990). Used with permission.

The Concept Introduction Phase

During this phase, also called the term introduction phase, the teacher assumes a more direct, active role and uses the students' exploratory activities as a means of introducing the scientist's view of the concept or theory that was investigated in the exploratory phase. During this phase students express their ideas about the concepts and ideas, and the teacher presents in very succinct ways the meaning of the concepts and ideas from a scientific point of view. The teacher assumes the direct-interactive mode during this phase, planning lessons along the guidelines presented in the direct-interactive section. The concept introduction phase is an intermediary step, and the teacher should move quickly to the next phase.

The Concept Application Phase

The concept application phase is a student-centered phase in which small teams of students engage in activities designed to apply and extend their knowledge of scientific concepts. The teacher should design activities that challenge the students to debate and defend their ideas. Activities in the concept application phase should be problem oriented. The teacher resumes the facilitative role in the concept application phase.

Students might also perform experiments to gather data to test a hypothesis. In short, the exploration phase allows the students to examine new ideas and test them against their own ideas. Students are actively engaged in interacting with ideas as well as with their peers during the exploration phase. During this phase the teacher should facilitate the work of the students by establishing reasons for exploring new ideas. The use of discrepant events, followed by interesting science activities, is a way to get into the exploration phase. The teacher plays a facilitative role during this phase.

Learning Cycle Lesson:
What Caused the Water to Rise?

Overview.   Students invert a cylinder over a candle burning in a pan of water. They notice that the flame soon goes out and water rises into the cylinder. They engage in discussions to explain their observations. They then test their explanations, which leads to new explanations and understanding of combustion, air pressure, and scientific inquiry.[19]

Materials: Aluminum pie tins, birthday candles, matches, modeling clay, cylinders (open at one end), jars (of various shapes and sizes), syringes, rubber tubing.

Exploration: Begin the lesson by giving each team a student handout describing the inquiry procedure (Figure 7.17), as well as the foregoing materials. Students should then be given the opportunity to explore the phenomenon by following the inquiry procedure.

[19]Charles R. Barman, *An Expanded View of the Learning Cycle: New Ideas about an Effective Teaching Strategy*, Monograph and Occasional Paper Series, No. 4 (Council for Elementary Science International, 1990), p. 19. Used with permission of CESI.

1. Pour some water into the pan. Stand a candle in the pan using the clay for support.
2. Light the candle and put a cylinder, jar, or beaker over the candle so that it covers the candle and sits in the water.
3. What happened?
4. What questions are raised?
5. What possible reasons can you suggest for what happened?
6. Repeat your experiment in a variety of ways to see if you obtain similar or different results. Do your results support or contradict your ideas in Step 5? Explain.

FIGURE 7.17

After 30 minutes of experimenting, stop the students for a discussion of their results. Focus the students on the questions *Why did the flame go out?* and *Why did the water rise?* The most likely explanation (misconception) to the second question is that since the oxygen was "burned up," the water rose to replace the oxygen that was lost.

Lead the students to realize that this hypothesis predicts that varying the number of burning candles will not effect the level of water rise. Four candles, for instance, would burn up the available oxygen faster and go out sooner than one candle, but they would not burn up more oxygen; hence the water should rise to the same level.

Have students do the experiment. The results will show that the water level is affected by the number of candles (the more candles, the higher the water level). Their ideas has been contradicted. Explain that an "alternative explanation" is needed and ask the students to propose one.

As students propose alternative ideas, do not tell them if they are correct. For example, the "correct" explanation (the heated air escaped out the bottom) should not be revealed even if students suggest it. Ask students to think of ways to test their hypotheses. If they propose the heated-air hypothesis, this should lead to the prediction that bubbles should be seen escaping from the bottom of the cylinder. As alternative hypotheses are suggested, have the students test the hypotheses and look for evidence to support predictions. If students do not suggest the "correct" explanation, suggest it yourself. You might say, "What do you think about this idea? The heat from the flame heats the air and forces it out the bottom of the cylinder." Encourage students to test your explanation rather than accepting it as is.

Concept Introduction: After students have collected data testing various hypotheses, you should introduce the "correct" explanation again and introduce the term "air pressure" and a molecular model of gases, which assumes air to be composed of moving particles that have weight and can bounce into objects (such as water) and push them out of the way.

Concept Application: Provide a number of problem-solving situations in which students must apply air pressure and the molecular model of matter.

Application Problem 1. Give students rubber tubing, a syringe, a beaker, and a pan of water. Tell them to invert the beaker of water in the pan of water. Challenge them to find a way to fill the beaker with water in that position (Figure 7.18). (The students will try forcing water in, before discovering they must extract air from the beaker.

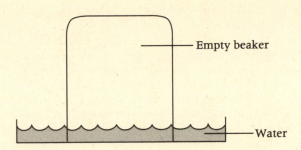

FIGURE 7.18   Fill the Beaker Challenge. How can you fill the beaker in this position with water?

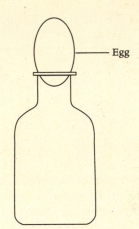

How can you get the peeled, hard-boiled egg in the bottle without touching the egg?

FIGURE 7.19   Egg-in-the-Bottle Inquiry.

Application Problem 2. Challenge the students to find a way to insert a peeled, hard-boiled egg into a bottle with an opening that is smaller in diameter than the egg (Figure 7.19). They cannot touch the egg after it is placed on the mouth of the jar. (After a small amount of water in the bottle has been heated, it is only necessary to place the smaller end of the egg over the opening of the bottle to form a seal. The egg will be forced into the bottle by the greater air pressure outside as the air cools inside.)

Application Problem 3. Pour a small amount of very hot water into a large (2 liter) plastic soda bottle. Then screw the cap on tightly to form a seal. Place the bottle on a desk so that students can view it. The plastic bottle will begin to be crushed. Challenge the students to explain the result using the molecular model of gases and air pressure.

INQUIRY ACTIVITY   7.3

Designing Learning Cycle Activities

Most of the concepts that you will teach will be determined by your textbook and the curriculum scope and sequence for your school district. In this activity you will select a section from the textbook and design lessons based on the learning cycle to teach the content in that section.

Materials

Science textbook, learning cycle check list chart

Procedure

1. Select a chapter from the science textbook you have chosen. Within the chapter, identify a section of content (content chunk) that you want to teach.

2. Design a lesson or a sequence of lessons using the three phases of the learning cycle, and similar in design to the examples previous learning cycle lesson example.
3. Use the learning cycle check list (Figure 7.20 on p. 222) to make sure that you develop complete learning cycle lessons.
4. Meet with a small team of peers. Use the learning cycle check list to evaluate the lessons of members of your team.
5. If you have an opportunity, teach the lessons to a group of peers or secondary students.

Minds on Strategies

1. What thinking skills did students have to use in your lessons? Differentiate between the three

phases of the learning cycle. How do they compare?

Observation
Classifying
Predicting
Inferring
Analyzing data
Controlling variables
Testing hypotheses
Designing experiments

2. How would you classify your lessons based on the types of thinking skills in your lessons (e.g., descriptive, empirical, hypothetical-deductive)? Explain.

The Generative Learning Model

The generative learning model is a teaching sequence based on the view that knowledge is constructed by the learner. It is, therefore, a constructivist model. As James Minstrell, a high school physics teacher and cognitive researcher says, "restructuring students' existing knowledge has become the principal goal of instruction."[20] Minstrell shows how he begins a unit of teaching based on the generative model with this "preliminary phase" lesson:[21]

[20]J. A. Minstrell, "Teaching Science for Understanding," in L.B. Resnick and L.E. Klopfer, eds., *Toward the Thinking Curriculum: Current Cognitive Research* (Alexandria, Va.: Association for Supervision and Curriculum Development, 1989), pp. 129–149.

	Yes	No
I. Exploration Phase		
A. The lesson contains an exploration phase that is activity oriented.	_____	_____
B. Ample time is provided for the exploration phase.	_____	_____
C. The exploration activity provides student-student and student-teacher interaction.	_____	_____
II. Concept Introduction Phase		
A. The concept(s) is/are named or appropriate vocabulary is developed after an exploration activity.	_____	_____
B. The concept(s) and term(s) is/are an outgrowth of the exploration phase.	_____	_____
III. Concept Application Phase		
A. The students extend the concept(s) to a new situation.	_____	_____
B. Appropriate activities are used to apply the concept(s).	_____	_____

FIGURE 7.20 Learning Cycle Check List

Teacher: Today we are going to explain some rather ordinary events that you might see almost any day. You will find that you already have many good ideas that will help explain those events. We will find that some of our ideas are similar to those of scientists, but in other cases our ideas might be different. When we are finished with this unit, I expect that we will have a much clearer idea of how scientists explain those events, and I know that you will feel more comfortable about your explanation.

Teacher: A key idea that we are going to use is the idea of force. What does the idea of force mean to you? (A discussion follows. My experience suggests that the teacher should allow this initial sharing of ideas to be very open.)

Teacher: You've mentioned words that represent many ideas. Most of them are closely related to the scientist's idea of force, but they also have meanings different from the scientist's ideas. Of the ones mentioned, probably the one that comes closest to the meaning the physicist has is the idea of push or pull, so we'll start with that. We'll probably find out that even that has a slightly different meaning to the physicist. (The teacher should allow the class to begin with this meaning for force rather than present an elaborate operational definition.)

Teacher: OK, let's begin (drops a rock): Here's a fairly ordinary event. We see something like this happening every day. How would you explain this event, using your present ideas about force? Instead of speaking right now, make drawings of the situation and show the major forces acting on the rock when it falls. Use arrows to represent the forces, and label each as to what exerts the force.

Students (naming the forces they have represented):
Gravity by the earth
Weight of the rock
Both gravity and the weight
Air
Friction
The spin of the earth
Nuclear forces

Teacher: Which of these is the major force, or which are the major forces acting on the rock while it is falling?

Students: Its weight. Gravity.

[21]This teaching sequence was slightly modified from Minstrell, "Teaching Science for Understanding," pp. 131–137. Reprinted with permission of the Association for Supervision and Curriculum Development. Copyright 1989 by ASCD. All rights reserved.

Teacher: Is the falling rock moving at a constant speed, or is it speeding up or slowing down? How do you know? (Teacher waits three to five seconds.)

Students: The same speed.
(More wait time.)
No wait, if two things fall, they both fall equally fast.
I don't know.
I think the rocks speeds up.
(Students continue making suggestions. The teacher encourages as many students as possible to comment.)

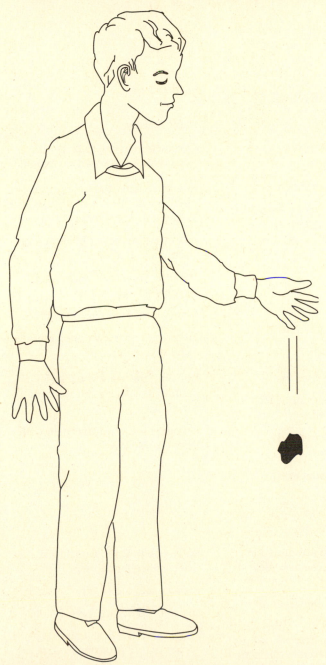

FIGURE 7.21   The Falling Rock Event: How would you explain the falling rock event using your present ideas about force? Use arrows to represent the forces, and label each as to what exerts the force.

Teacher: During the next several days, we will look more closely at the idea we call force. Many of the ideas you've suggested today will be useful, but we may also find that we will want to change some of our notions about force to make them more consistent with the phenomena.

The "preliminary phase" of the generative model of learning is designed to identify existing student ideas. In the preceding example, the teacher, performing a simple demonstration, asked students to make a drawing of the event and then engaged the whole class in a discussion designed to identify the students' existing ideas. No attempt was made by the teacher to correct the student responses or label them wrong or give the scientific meaning of the concepts. Students' existing ideas can also be determined by giving a diagnostic test at the beginning of a course, or a short preinstruction quiz at the beginning of a unit.

Osborne and Freyberg, advocates of the generative learning model, have identified three distinct phases to the model in addition to the preliminary phase, namely focus, challenge, and application.[22] The generative learning sequence is shown in Figure 7.22 and Table 7.5.

The focus phase is designed to help the teacher and student clarify the students' initial ideas.[23] Osborne and Freyberg suggest that the focusing phase is the time to involve the students in activities that focus on phenomena related to the concepts, to get the students thinking about these phenomena in their own words. The teacher's role is a motivational one as well during this phase. As was mentioned earlier, motivation to learn, in the cognitive view, is related to the intrinsic nature of learning. Providing interesting activities that focus attention on getting the students involved is suggested.

The challenge phase focuses attention on challenging student ideas. The teacher, through small group discussion or with the class as a whole, creates an environment in which students can articulate their ideas and hear other students' points of view. The students are challenged to compare their ideas to the scientists' view. The discussion during the challenge phase centers on the experiences students encountered during

[22]Osborne and Freyberg, *Learning in Science*, pp. 108–111.
[23]The discussion that follows is based on Osborne and Freyberg, *Learning in Science*, pp. 108–111.
Based on R. Osborne and P. Freyberg, *Learning in Science: The Implications of Children's Science*, (Auckland: Heinemann, 1985), pp. 109– 110. Used with permission.

Preliminary → Focus → Challenge → Application

FIGURE 7.22   Generative Learning Model Teaching Sequence

the focusing phase. During this phase, some degree of conceptual conflict will occur as students accommodate new ideas. This is in the words of the cognitive psychologists, the tension or struggle that occurs mentally to accommodate new structures or modify existing ones.

The application phase is the instructional period in which students can practice using the new idea in differing situations. The teacher's role is one of creating problem situations for student application of the new ideas. Designing small group activities and indepen-

dent investigations that challenge students to apply the new concepts to different phenomena will facilitate the accommodation of the new idea and provide the time students need to reflect or think about their new learnings.

By now you should have noted the high degree of similarity between the generative model of learning described here and the learning cycle model. Note that the major difference is the identification of a preliminary stage in the generative model; otherwise the phases correlate.

TABLE 7.5

GENERATIVE LEARNING MODEL

Phase	Teacher Activity	Pupil Activity
Preliminary	Ascertain students' views	Complete surveys, quizzes, or activities to ascertain existing ideas.
Focus	Provide motivation experiences.	Engage in activities in order to become familiar with phenomena related to "new concepts."
	Ask open ended questions. Interpret student responses. Interpret and elucidate students' views.	Ask questions about phenomena and activities.
		Describe what they know about events and phenomena.
		Clarify own views on concepts.
		Present own view to small groups and whole class.
Challenge	Facilitate exchange of views.	Consider the view of another student, as well as all students in class.
	Create an environment in which all views are considered.	
	Demonstrate procedures and phenomena, if necessary.	Compare the scientists' view with the class's view.
	Present evidence to support scientists' ideas.	
	Explore the tentative nature of students' reaction to new view.	
Application	Design problems and activities that can be solved with the new idea or concept.	Solve practical problems using the new concept as a basis.
	Help students clarify views on the new ideas.	Present solution to other students.
		Discuss and debate the solutions.
	Encourage an atmosphere whereby students verbally describe solutions to problems.	Suggest further problems arising from the solutions presented.

Based on R. Osborne and P. Freyberg, *Learning in Science: The Implications of Children's Science* (Auckland: Heinemann, 1985), pp. 109–110.

COOPERATIVE OR COLLABORATIVE LEARNING MODELS

A number of studies, as well as numerous practical applications, have led to a mountain of evidence that supports the use of small, mixed-ability cooperative groups in science classes. We are using the term collaborative in the heading of this section to emphasize the importance of verbal communication among the students within the small teams. Students need to talk about their observations, their ideas, and their theories in order to understand science. Students need to get along with each other.

As we said in Chapter 2, cooperative learning is a model of teaching in which students work together to achieve a particular goal or complete a task. A variety of cooperative learning models have been developed, field-tested, and evaluated. Some delineate how tasks are structured and how groups are evaluated. In some models, students work together on a single task; in others, group members work independently on one aspect of a task, pooling their work when they finish.

Regardless of the model, there appear to be essential components of cooperative learning that are integral to any model of cooperative learning. We will examine these essential components first, then move on and look at several cooperative learning models.

Essential Aspects of Cooperative Learning

Cooperative learning brings students together to work on tasks—solving problems, reviewing for a quiz, doing a lab activity, completing a worksheet. In all cases it is the working together that is important. And this has presented a challenge to teachers, as well. When cooperative learning was introduced (in the early 1970s), American schools were faced with two important social changes. One was the mainstreaming of handicapped and disabled students into the regular classroom; the other was the integration of school in

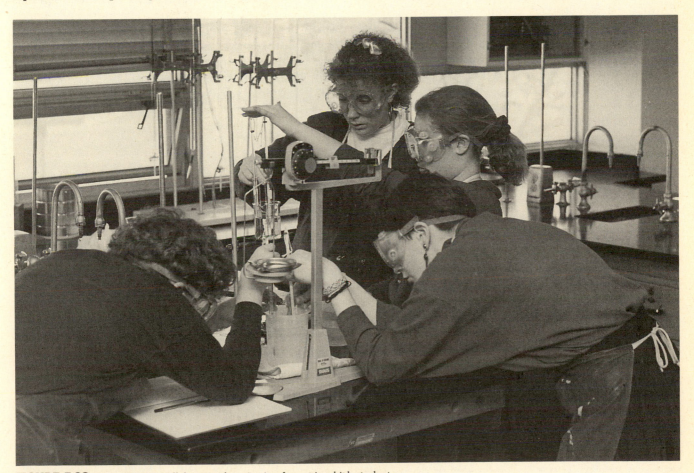

FIGURE 7.23 Cooperative/collaborative learning is a format in which students work together interdependently to achieve a particular goal or complete a task. Copyright James L. Shaffer.

which students from different cultural backgrounds were brought together in desegregated settings.[24] One of the solutions proposed to help students from different racial groups work together in the same school setting, and for many handicapped students who lacked many "social skills," was cooperative learning. David and Roger Johnson outlined the essential components of a teaching and learning strategy that would focus on bringing small groups of students together in teams to work cooperatively as opposed to individually or competitively.[25]

The Johnsons proposed a model in which five elements were essential for cooperative learning groups to be successful. That is, the lessons that teachers taught needed to be characterized by the following elements:

Face-to-Face Interaction

The physical arrangement of students in small, heterogeneous groups encourages students to help, share, and support each other's learning.

Positive Interdependence

The teacher must structure the lesson either through a common goal, group reward, or differentiated role assignments to achieve interdependence between students in a learning team. One way to achieve this is instead of having each member of a lab team write up his or her own report, each team has *one* lab sheet and all students might sign off, agreeing with its contents. A single grade is given based on the *team's* lab sheet. Assigning each person a role in the group is another way to achieve positive interdependence; for example, members are assigned one of the following roles for an activity: communicator, materials handler, checker, tracker.

Individual Accountability

Each student in a learning team must be held accountable for learning and collaborating with other team members. Teachers can achieve individual accountability by focusing on (a) individual contributions of students (using roles, dividing the task, using experts, giving feedback) and (b) individual outcomes of students (using tests, quizzes, grading homework, giving group rewards for individual behavior, using random calling-on procedures).

Cooperative Social Skills

The Johnsons found that students needed to learn interpersonal skills such as active listening, staying on task, asking questions, making sure everyone contributed, and using agreement. Just as science teachers focus on scientific thinking skills (observing, inferring, hypothesizing, experimenting) and assume that students need to be taught these skills, cooperative learning experts have discovered that students need to be taught cooperative social skills.

Group Processing

The fifth essential element of cooperative learning is group processing. Students need to reflect on how well they worked together as a team to complete a task such as a laboratory activity. The teacher can structure this by simply asking the students to rate how well they did in the activity, or how well they "practiced" the social skill that was central in the activity.

The models that are presented here fall into two broad categories: tutorial methods, and problem-solving and conceptual methods. The tutorial methods tend to be structured and teacher centered, whereas problem-solving methods tend to be more open-ended and student centered. The tutorial methods possess many of the characteristics of direct instruction, while the problem-solving methods facilitate inquiry learning. Elements of both methods will be seen to be useful in learning cycle models.

Tutorial Models

In tutorial models students work in small teams to rehearse and learn science information that has been identified by the teacher. Often the material is based on the science textbook, and because of this, tutorial models are easily applied to the secondary science classroom. These methods tend to be motivational because they often involve teams competing against each other for reward structures (e.g., points, prizes, free time). Three models are presented: student teams-achievement divisions, jigsaw II, and two-level content study groups.

Student Teams-Achievement Divisions

Student teams-achievement divisions (STAD) was originated by Robert Slavin and his colleagues at Johns Hopkins University.[26] The STAD model underscores many of the attributes of direct instruction, and it is a very easy model to implement in the science class-

[24]James Bellanca and Robin Fogarty, *Blueprints for Thinking in the Cooperative Classroom* (Palatine, Ill.: Skylight, 1991), p. 241.
[25]See David Johnson and Roger Johnson's classic books on cooperative learning: *Learning Together and Alone* (Englewood Cliffs, N.J.: Prentice-Hall, 1976); and *Circles of Learning* (Alexandria, Va.: Association for Supervision and Curriculum Development, 1986).

room. As in all the cooperative learning models to follow, STAD operates on the principle that students work together to learn and are responsible for their teammates' learning as well as their own.

There are four phases to the STAD model: teaching (class presentation), team study, testing and team recognition. We will illustrate how STAD works by using an example from life science, food making (photosynthesis).

Phase I: Teaching (Class Presentation)

The class presentation is a teacher-directed presentation of the material (concepts, skills, and processes) that the students are to learn. Carefully written and planned objectives should be stated and used to determine the nature of the class presentation, and the team study to follow. Examples from a unit on food making would be

- Students will identify the steps in the food-making process
- Students will compare the light and dark phases of photosynthesis

Key concepts should be identified as well. In this case the following concepts would be presented: ATP, chlorophyll, dark phase, energy, glucose, light phase, and photosynthesis.

The presentation can be a lecture, a lecture-demonstration, or an audiovisual presentation. You also could follow the lesson plans in your science textbook, including the laboratory activities in this phase of STAD. Several lessons would be devoted to class presentations.

Phase II: Team Study

In STAD, teams are composed of four students who represent a balance in terms of academic ability, gender, and ethnicity. The team is the most important feature of STAD, and it is important for the teacher to take the lead in identifying the members of each team. Slavin recommends rank-ordering your students in terms of performance. Each team would be composed of a high- and a low-ranking student and two near the average. The goal is to attempt to achieve parity among the teams in the class. Teams should also be formed with sex and ethnicity in mind. Each team should be more or less an average composite of the class.

Team study consists of one or two periods in which each team masters material that you provide. Team members work together with prepared worksheets and make sure that each member of the team can answer all questions on the worksheet. Students should move their desks so that they face each other in each small team. Give each team two worksheets and two answer sheets (not one for each student). For example, in the case of the food-making unit, the teacher would provide the diagram shown in Figure 7.24, summarizing photosynthesis, and construct a worksheet consisting of about thirty questions related to food making (see Figure 7.25 on p. 229).

In the STAD model the following team rules are explained and posted on the bulletin board:

1. Students have the responsibility to make sure that their teammates have learned the material
2. No one is finished studying until all teammates have mastered the subject
3. Ask all teammates for help before asking the teacher
4. Teammates may talk to each other *softly*

It is important to encourage team members to work together. They work in pairs within the teams (sharing one worksheet), and then the pairs can share their work. A principle that is integral, not only to STAD, but to all cooperative learning models, is that students must talk with each other in team learning sessions. It is during these small group sessions that students will teach each other, and learn from each other. One of the ways to encourage deeper understanding is for students to explain to each other their answers to the questions. One way to facilitate this process is for the teacher to circulate from group to group asking questions and encouraging students to explain their answers.

Phase III: Testing

After the team study is completed, the teacher administers a test to measure the knowledge that students have gained. Students take the tests individually and are not permitted to help each other. To encourage students to work harder, STAD uses an "individual improvement score." Each student is assessed a base score—based on his or her previous performance on similar quizzes and tests. Improvement points, which are reported for each team on a team recognition chart on the bulletin board, are determined based on the percentage of improvement from the previous base score. Generally speaking, if the student gets more than 10 points below the base score, the improvement score is 0, 10 points below to 1 point

[26]Robert E. Slavin, *Using Team Learning*. 3d ed. (Baltimore, Md.: Center for Research on Elementary and Middle Schools, The Johns Hopkins University, 1986).

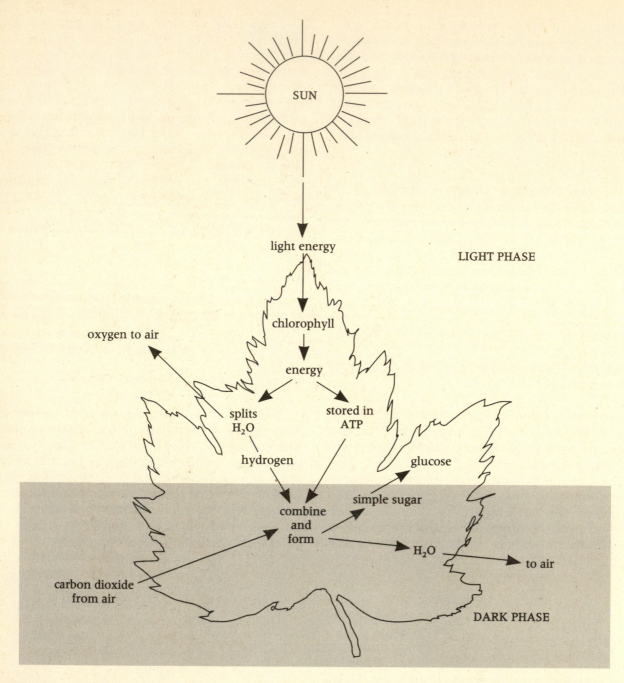

FIGURE 7.24   Summary of Photosynthesis. Slavin, *Using Team Learning*, used with permission.

below results in 10 improvement points, base score to 10 points above gives a score of 20, and more than 10 points above is worth 30 improvement points. (A perfect score, regardless of base score, earns 30 improvement points.)

Phase IV: Team Recognition

Team averages are reported in the weekly recognition chart. Teachers can use special words to describe the teams' performance such as *science stars, science geniuses,* or *Einsteins.*

Recognition of the work of each team can occur by means of a newsletter, handout, or bulletin board that reports the ranking of each team within the class. Report outstanding individual performances, too. Sensitivity is required here. It is important to realize that praising students academically from low-status groups is an integral part of the effectiveness of cooper-

| 1. The organ of the plant in which photosynthesis most often takes place is the _____.

a. stem

b. root

c. leaf | 2. Plants need which of the following to carry on photosynthesis? _____.

a. O_2, CO_2, chlorophyll

b. H_2O, CO_2, light energy, chlorophyll

c. H_2O, O_2, light energy, sugar | 3. The energy stored in plants comes from _____.

a. soil

b. air

c. sunlight | 4. The first phase of photosynthesis is sometimes called the _____ _____.

a. light phase

b. dark phase

c. chlorophyll phase |
| 5. The oxygen released during photosynthesis comes from the _____.

a. chlorophyll

b. carbon dioxide

c. water | 6. Photosynthesis takes place in the _____. of a plant cell.

a. cell wall

b. cytoplasm

c. chloroplast | 7. The energy from the sun is stored in a chemical compound called _____.

a. ATP

b. CO_2

c. H_2O | 8. The second stage of photosynthesis is called the _____.

a. light phase

b. dark phase

c. chlorophyll phase |

FIGURE 7.25 Sample Worksheet Questions (STAD). Slavin, *Using Team Learning*, used with permission.

ative learning. Elizabeth Cohen has found that it is important to be aware of students who you suspect have consistently low expectations for competence.[27] When such a student performs well (not just on the quiz), give immediate, specific, and public recognition for this competence.

One final note about STAD. There is another model developed by Slavin called teams-games-tournaments (TGT) that is also recommended and is very similar to STAD. The same materials used in the team study are used in a series of games in a tournament style class session.[28]

Jigsaw II

Jigsaw II, developed by Eliot Aronson, is a cooperative learning model in which students become experts on part of the instructional material about which they are learning.[29] By becoming an expert, and then teaching other members of their team, students become responsible for their own learning. The jigsaw II model has the advantage of encouraging students of all abilities to be responsible to the same degree, although the depth and quality of their reports will vary

In jigsaw II students are grouped into teams of four members to study a chapter in a textbook. However, the content is broken into chunks, letting each team member become an expert on one of the chunks and then responsible for teaching his or her fellow team members that chunk. The phases of jigsaw II are as follows:[30]

Phase I: Text

The first step in the use of jigsaw II is to select a chapter that contains material for two or three days. Divide the content in the chapter into chunks based on the number of team members. For example, if you have four members in a team, a chapter would be divided into four chunks (as shown for a chapter on electronics in aphysical science course):[31]

- Chunk 1: How do early electronic devices work?
- Chunk 2: How are electronic devices used?
- Chunk 3: How is information processed in a computer?
- Chunk 4: What are some ethical issues concerning the use of electronic devices?

Each member of the team is assigned one of the topics and must read the chapter to find information about his or her assigned "chunk." In the next phase,

[27]Elizabeth G. Cohen, *Designing Groupwork: Strategies for the Heterogeneous Classroom* (New York: Teachers College Press, 1986)
[28]See Slavin, *Cooperative Learning*, or Slavin, *Using Student Team Learning*, 3rd Edition.
[29]Eliot Aronson, *The Jigsaw Classroom* (Beverly Hills, Calif.: Sage, 1978).
[30]Slavin, *Using Student Team Learning*. pp. 36–39.
[31]Questions based on Mark A. Carle, Mickey Sarquis, and Louise Mary Nolan, *Physical Science: The Challenge of Discovery* (Lexington, Mass.: D. C. Heath, 1991), chap. 10, pp. 245–267.

each team member will meet with experts from other groups in the class.

Phase II: Expert Group Discussion

Expert groups should meet for about one class period to discuss their assigned topic. Each expert group should receive an expert sheet. The expert sheets should contain questions and activities (optional) to direct the group's discussion. Encourage diversity in learning methods. Groups might do hands-on activities, read from other source books, or use a computer for a game or simulation. The groups' goal is to learn about the subtopic and prepare a brief presentation that group members will use to teach the material to members of their respective learning teams.

Expert Sheet for Igneous Rock Group

Provide the group with an assortment of igneous rocks, dilute HCl acid, and hand lenses.

1. How were these rocks formed?
2. Under what environmental conditions are these rocks formed? Try to illustrate your theory.
3. Observe each rock and collect observations on a chart that should include the name of the rock or mineral, the color (light or dark), the shape and color of crystals, the arrangement of crystals (even or banded), the effect of acid, the presence of fossils, and the use of the rock.
4. What are some minerals in these rocks?
5. Where would you find these rocks in (your state)?
6. Would you find igneous rocks in, on, or under your school grounds?
7. What would you predict about the size of crystals in igneous rocks if magma (a) cooled slowly, (b) cooled rapidly, or (c) cooled in water? Explain your prediction.
8. What is the difference between an intrusive and an extrusive igneous rock? Give some examples.
9. If you wanted to demonstrate how an igneous rock is made, what would you do?
10. Do you think there are igneous rocks on Mars?

Expert Sheet for Sedimentary Group

Provide the group with an assortment of sedimentary rocks, mixed-sized sand, glass jar, water, HCl acid, and hand lenses.

1. How were these rocks formed?
2. Under what environmental conditions are these rocks formed? Try to illustrate your theory.
3. Observe each rock and collect observations on a chart that should include the name of the rock or

mineral, the color (light or dark), the shape and color of crystals, arrangement of crystals (even or banded), effect of acid, presence of fossils, and use of the rock.
4. What are some minerals in these rocks?
5. Where would you find these rocks in (your state)?
6. Would you find sedimentary rocks in, on, or under your school grounds?
7. Why might you expect to find fossils in sedimentary rocks?
8. Does shaking a jar of mixed-sized sand and water and observing the results demonstrate part of the sedimentary process? Explain by using a diagram and words.
9. Do you think there are sedimentary rocks on Mars?

Expert Sheet for Metamorphic Group

Provide the group with an assortment of metamorphic rocks, HCl acid, and hand lenses.

1. How were these rocks formed?
2. Under what environmental conditions are these rocks formed? Try to illustrate your theory.
3. Observe each rock and collect observations on a chart that should include the name of the rock or mineral, the color (light or dark), the shape and color of crystals, the arrangement of crystals (even or banded), the effect of acid, the presence of fossils, and the use of the rock.
4. What are some minerals in these rocks?
5. Where would you find these rocks in (your state)?
6. Would you find metamorphic rocks in, on, or under your school grounds?
7. If you wanted to demonstrate how a metamorphic rock is made, what would you do? Illustrate and explain.

Home Sheet

Note: This sheet is to be used in home groups after expert groups have met. Give this study sheet to each home team in preparation for a quiz on rocks.

1. Consider the following rocks: sandstone, granite and marble.
   a. Under what environmental conditions is each formed?
   b. Would they be found on Mars?
   c. Where would you find these in (your state)?
2. What are the differences between igneous, sedimentary, and metamorphic rocks?
3. What are some rocks that can be identified by an acid test? What does this tell you about the composition of the rocks?
4. Complete these sentences:
   a. Metamorphic rocks from when

b. The southern part of (your state) contains rocks such as

c. In (a county next to yours), you'll find these rocks

d. Fossils can be found in

Phase III: Reports and Test

In the next phase of jigsaw II, each expert returns to his or her learning team and teaches the topic to the other members. Encourage members to use a variety of teaching methods. They can demonstrate an idea, read a report, use the computer, or illustrate their ideas with photographs or diagrams. Encourage team members to discuss the reports and ask questions, as each team member is responsible for learning about all of the subtopics.

After the experts are finished reporting, conduct a brief class discussion or a question and answer session. The test, which covers all the subtopics, should be administered immediately and should not take more than 15 minutes. Test design should include at least two questions per subtopic.

Team recognition should follow the same procedures used in STAD.

Two-Level Content Study Groups

Jones and Steinbrink have made changes in jigsaw II and developed a model specific to science in what they refer to as the 'two-level content study groups' model of cooperative learning.[32] Their method involves the identification of two levels of cooperative groups, namely a "home team" and an "expert group." They also have developed job descriptions to clarify roles in these two levels of groups. Another change they have made is the size of the group. Home teams consist of from three to five members, depending on the way the content can be divided in the text. They point out that if a chapter has three equal divisions, each home team would consist of three members.

Problem-Solving and Conceptual Models

The cooperative learning models presented to this point have stressed rehearsal for a quiz, or placed emphasis on having students master a body of information. In each of the models, students tutor each other and then compete against other teams either through tests, tournaments, or games.

What cooperative learning models can be implemented if the science teacher's goals include problem solving and the development of higher-order cognitive thinking skills? The cooperative learning model that follows emphasizes a structure in which students share ideas, solve problems, discover new information, learn abstract concepts, and seek answers to their own questions.

Group Investigation

Developed by Shlomo Sharan and Yael Sharan and their colleagues, the group investigation (GI) method is one of the most complex forms of cooperative learning.[33] Its philosophy is to cultivate democratic participation and an equitable distribution of speaking privileges. It also encourages students to study different topics within a group and to share what they learn with group members and with the whole class.

This method places maximum responsibility on the students, who identify what and how to learn, gather information, analyze and interpret knowledge, and share in each other's work. It is very similar to another method of cooperative learning developed by Spencer Kagan called co-op co-op,[34] and would have the greatest chance of success if students have experience with other forms of cooperative learning.

There are five phases to the group investigation method:

Phase I: Topic and Problem Selection

Students organize into groups of five (or fewer) and choose specific topics or problems in a general subject area of science. Group investigation lends itself to a wide range of problem-solving strategies. Students can ask questions that require empirical research, surveys or questionnaires, or historical reporting. Each group plans its own topic or subject of investigation, and strategy for exploration. Then individuals or pairs within the group select subtopics or specific investigatory tasks and decide how they will carry them out.

Phase II: Cooperative Planning

You and the students in each learning team plan specific learning procedures, tasks, and goals consistent with the subtopics of the problem selected.

Phase III: Implementation

Students carry out the plans formulated in the second step. Learning should involve a wide range of ac-

[32]Robert M. Jones, "Cooperative Learning in the Elementary Science Methods Course," *Journal of Science Teacher Education* 1, no. 1 (Spring 1989): 1–3.

[33]Shlomo Sharan et al., *Cooperative Learning in the Classroom: Research in Desegregated Schools* (Hillsdale, N.J.: Lawrence Erlbaum Associates, 1984).

[34]Spencer Kagan, *Cooperative Learning: Resources for Teachers* (Riverside, Calif.: University of California, 1985).

tivities and skills, and should lead students to different kinds of sources, both inside and outside of school. Students might work in small groups or individually to gather data and information.

Phase IV: Analysis and Synthesis

Students meet to discuss the results of their subgroup or individual work. Meetings of this nature will naturally take place more than once during the implementation of GI. The information that has been gathered on the topic or problem is analyzed, and each group member's piece is synthesized to prepare a report for the whole class.

Phase V: Class Presentation

One of the attractive features of GI is that each team makes a presentation to the whole class. Students have to cooperate to prepare a presentation. Teams should be encouraged to prepare presentations that involve the audience, such as debates, demonstrations, hands-on activities, plays, or computer simulations.

During the presentation, the presenting team is responsible for setting up the room, gathering any equipment and materials, and preparing necessary handouts. Encourage the teams to allow at least five minutes for questions and comments from the class. In some cases, you should facilitate this aspect of the presentation.

Although GI places more responsibility on each group to make decisions about content and process, the teacher must stay in contact with the groups to facilitate their flow through the phases. The best topics for group investigations are ones that require problem solving on the part of each team. Although topics can be fairly descriptive, encouraging an investigative approach will reinforce inquiry learning in the science classroom. Group investigation might also be used in place of individual science fair projects (see Chapter 8).

OTHER MODELS

So far we have presented three types of models based on behavioral, cognitive and social-humanistic learning theory. Several additional science teaching models are described that extend the previous models.

Synectics

Synectics is a process in which metaphors are used to make the strange familiar and the familiar strange. Synectics can be used to help students understand concepts and solve problems. Synectics was developed by William Gordon for use in business and industry, but it has also been used as an innovative model in education.

According to Gordon, "the basic tools of learning are analogies that serve as connectors between the new and the familiar. They enable students to connect facts and feelings of their experience with the facts that they are just learning." Gordon goes on to say that "good teaching traditionally makes ingenious use of analogies and metaphors to help students visualize content. For example, the subject of electricity typically is introduced through the analogue of the flow of water in pipes."[35] Synectics can be used in the concept introduction phase of the conceptual change teaching model.

The synectics procedure for developing students' connection-making skills goes beyond merely presenting helpful comparisons and actually evokes metaphors and analogies from the students themselves. Students learn how to learn by developing the skills to produce their own connective metaphors.

Gordon and his colleagues, know as SES Associates, have developed texts and reference materials and provide training to help teachers implement synectics into the classroom.[36] Here is an example of a synectics activity that you could do with students. In this example students learn to examine simple analogies and discuss how they relate to each other.

- Give students analogies and then ask them to explain how the content (the heart) and the analogue (water pump) are alike. Here are some examples:
  - The heart and a water pump
  - Orbits of electrons and orbits of planets
  - The nucleus of an atom and a billiard ball
  - The location of electrons in an atom and droplets of water in a cloud
  - Small blood vessels and river tributaries
  - The human brain and a computer
  - The human eye and a camera

After students feel skillful linking the strange with the familiar, challenge them to create analogies for concepts they are studying.

[35]W. J. J. Gordon and Tony Poze, "SES Synectics and Gifted Education Today," *Gifted Child Quarterly* 24, no. 4 (Fall 1980), 147–151.

[36]For training and resource materials on synectics, write: SES Associates, 121 Brattle Street, Cambridge, Mass. 02138.

The Person-Centered Learning Model

The person-centered model of teaching focuses on the facilitation of learning and is based on the work of Carl Rogers and other humanistic educators and psychologists.[37] The model is based on giving students freedom to not only choose the methods of learning, but to engage in the discussion of the content as well. In practical terms, the person-centered model can be implemented within limits. Rogers believed, as do other psychologists, that making choices is an integral aspect of being a human being, and at the heart of learning. Secondly, Rogers advocated trusting the individual to make choices, and that it was the only way to help people understand the consequences of their choices.

There are several aspects of the person-centered model that appeal to the science teacher, namely, the role of the teacher in the learning process and the creation of a learning environment conducive to inquiry learning.

The Teacher as a Facilitator

To implement a person-centered approach, the teacher must take on the role of a facilitator of student learning rather than a dispenser of knowledge or information. Three elements seem to characterize the teacher who assumes the role of learning facilitator: realness, acceptance, and empathy. In the person-centered model, to show realness the teacher must be genuine and willing to express feelings of all sorts: from anger and sadness to joy and exhilaration. In the person-centered model, the teacher acts as counselor, guide, and coach, and to be effective must be real with his or her students.

Rogers also advocated and stressed the importance of accepting the other person—indeed prizing the person and acknowledging that they are trustworthy and can be held responsible for their behavior.

Finally, to Rogers at least, the most important element in this triad was empathy. Empathy is a form of understanding without judgment or evaluation. Empathy in the science classroom is especially important in developing positive attitudes, and helping students who have been turned off to science to begin to move toward it.

Naturally there is more than these three elements to being a learning facilitator. Technical aspects such as setting up a classroom environment conducive to learning, providing learning materials, and structuring lessons that encourage person-centered learning are involved as well.

The Person-Centered Environment and Inquiry

In the person-centered classroom, students are encouraged to ask questions, choose content, decide on methods and resources, explore concepts and theories, and find out things on their own and in small teams. Clearly these are elements that foster inquiry. Teachers who truly implement inquiry will find themselves fostering the attitude advocated by person-centered educators. Here is a checklist of elements that signal the existence of a person-centered environment:[38]

- [ ] A climate of trust is established in the classroom, in which curiosity and the natural desire to learn can be nourished and enhanced.
- [ ] A participatory mode of decision making is applied to all aspects of learning, and students, teachers, and administrators each have a part in it.
- [ ] Students are encouraged to prize themselves, to build their confidence and self-esteem.
- [ ] Excitement in intellectual and emotional discovery, which leads students to become life-long learners, is fostered.

The Integrative Learning Model

Imagine for a moment a physics classroom. After the students have come in and seem ready to begin class, the teacher says that they are going to begin a new unit on mechanics. He begins the lesson by getting students (and himself) to stand in a circle and begin passing a tennis ball around. At first the teacher tells the students to pass the ball at a constant speed or velocity. Then he says, "Accelerate the ball!" After a few moments, "Now, decelerate it!" The teacher now turns the activity into a game, students may take turns calling out "constant velocity," "accelerate," and "decelerate." As simple and as unusual as this activity is, this teacher has a reputation in the school for doing such activities in his physics class.

In another school, an earth science teacher is playing classical music while she reads a story to the class about how the giant continent of Pangea broke up and drifted apart, creating new ocean basins, pushing rocks together to form huge mountain chains, and causing earthquakes and volcanoes. After the story is read, students get into small groups to collaborate and create a metaphor of the story, for example, a drawing, a clay model, or a diagram.

The head of the biology department is seen taking her students outside (once again). This time, the

[37]Carl R. Rogers, *Freedom to Learn* (Columbus, Ohio: Charles E. Merrill, 1983).

[38]Paul S. George, *Theory Z School: Beyond Effectiveness* (Columbus, Ohio: National Middle School Association, 1983).

teacher explains that the students are going on a "still hunt." Once outside, the students are assigned to sit in an area of the school grounds (this school has a wooded area to which the teacher takes the class quite often). For five minutes the students sit in their assigned area watching for the presence of organisms—ants, spiders, earthworms, birds, mammals—anything that they can see. They are asked to observe and to record their observations in a naturalist's notebook. The students are assembled at the edge of the wood and report their findings from the still hunt. Then the teacher gives the students modeling clay, string, paper, yarn, buttons, cloth, and toothpicks and says, "Design an animal that will fit into this environment but will be difficult to be seen by other animals." When the creatures are completed students place them in "their habitat." The teacher then has the whole class walk through the area looking for the creatures to find out how well they were designed to survive unnoticed in their environment.

Each of these teachers is implementing a model of learning that some refer to as integrative learning. It is a model of learning that suggests that all students can learn with a limitless capacity, and that students can learn by interacting with their "environment freely, responding to any and all aspects of it without erecting barriers between them."[39]

The learning cycle used in the integrative learning model consists of three phases: input, synthesis, and output. Equal emphasis is placed on all three of these phases. The origin of integrative learning can be traced to Georgi Lozanov, of the University of Sofia (Bulgaria). Lozanov had discovered a method of learning that involved the use of music to help relax the learner, the creation of an atmosphere in which the mind is not limited, and the presentation of new material to be learned with what is called an "active concert," followed by a period of relaxation, and ending with a series of games and activities to apply the new material that was learned.

The Lozanov method made its way to the United States and can be found in such work as superlearning, accelerated learning, and whole brain learning. Peter Kline, the developer of integrative learning, has stressed the importance of student synthesis and output in learning. In the integrative classroom, the teacher encourages students to use their personal styles of learning and thus provides auditory, kinesthetic, visual, print-oriented, and interactive learning activities. Music, movement, color, minifieldtrips,

painting, the use of clay, and pair and small group discussion are an integral part of the integrated learning model.

Imagineering

Imagineering is a model of teaching developed by Alan J. McCormack and is designed to encourage visual and spatial thinking.[40] McCormack explains that imagineering is the result of fusing parts of the words "imagination" and "engineering." Imagineering is designed to apply existing scientific knowledge to visualizing solutions to challenging problems.

McCormack, who is well known for his presentations at local, state, and national science teacher meetings, advocates the use of teacher demonstrations to set the stage for imagineering. He has designed an array of imagineering activities that can be used in any science curriculum.[41] Here is one example, called the water-expanding machine (Figure 7.26), and this is how McCormack explains the use of this device in an imagineering lesson:

> This device is a cardboard box with an input funnel located on its top and an output tube that extends through one of its sides. The teacher states that this great new invention can expand by three times any volume of water that is poured into the funnel to be "processed" by the device. An actual demonstration follows: 500 mL of water is poured into the machine and 1500 mL flows from the output tube. Teacher asks the key question that already puzzles everyone: "Has water actually been expanded?" Few students believe that it has, so they are challenged to draw an "imagineering blueprint" of what might be inside the machine that could account for the apparent volume expansion of the water. Later ideas are presented, compared, and criticized. Invariably, a number of good, plausible explanations have been invented. Meanwhile, youngsters have honed both visualization and creative problem-solving skills.[42]

[39]Peter Kline, *The Everyday Genius.* (Arlington, Va.: Great Ocean, 1988), p. 65.

[40]Alan J. McCormack, *Visual/Spatial Thinking: An Essential Element of Elementary School Science* (Washington, D.C.: Council for Elementary Science International, Monograph and Occasional Paper Series, 1988).
[41]For further examples of imagineering, see Alan J. McCormack, *Inventors Workshop* (Belmont, Calif.: David Lake, 1981).
[42]Alan J. McCormack, *Visual/Spatial Thinking: An Essential Element of Elementary School Science* (Washington, D.C.: Council for Elementary Science International, Monograph and Occasional Paper Series, 1988), pp. 23–24.

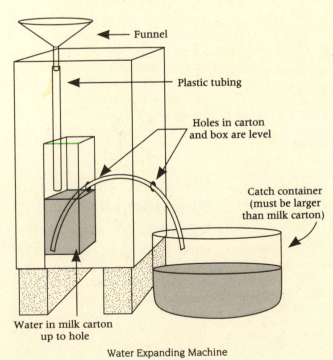

Completed Water Expanding Machine

Water Expanding Machine

FIGURE 7.26 The McCormack Water Expanding Machine. Alan J. McCormack, *Inventors Workshop*, Eden Prairie, MN: Fearon Teacher Aids, 1981, pp. 14–15. Used with permission of the publisher.

Case Study 1. Descent from Innocence[43]

The Case. Richard A. Miller, who grew up in an upper middle class community in upstate New York and graduated (cum laude) from the State University of New York at Buffalo, accepted a teaching position with the Los Angeles Unified School District. He was assigned to Orange High School, which has an enrollment of over 3000 students—approximately 1500 Hispanics, 1000 Blacks, 450 Asians, and about 300 other students. Miller reports his experience teaching with nontraditional activities as follows:

"It wasn't until I had been teaching for a few months that I fully understood the odd smile that appeared on the faces of other teachers when I told them of my class assignment: in addition to two periods of biology, I would be teaching three life science classes. Life science appears to be the dumping grounds for non-college bound, nonacademic, potential dropouts. Now I realize that the smile meant, 'Poor, naive, innocent soul—like a lamb to the slaughter and he doesn't even know it.'

"My descent from innocence was swift and brutal. I was given a temporary roll sheet, assigned a room—actually three different rooms—and with little other preparation was thrust into the world of teaching....I launched into my lessons.

"One of the first units I covered was the metric system....I assumed that this unit would be a brief review for the students. Little did I suspect that, not only did the students have no knowledge of the metric system, they were also ignorant of measuring using the standard English system.

"In order to teach this unit, I planned to conduct a brief lecture on metric prefixes and then have a laboratory exercise in which the students measured various objects and converted from one measurement unit to another. I typed up lab sheets explaining in detail what should be measured and in what units I wanted the measurements. The students were then assigned to lab tables, paired off, and provided with meter sticks. What next ensued can only be described as pandemonium.

"My intention had been to visit with each group of students and answer any questions that they might have....The first group I visited with was made up of four girls. They were having a grand time chatting about local current events and had given up the lab as a futile exercise. It just wasn't possible to measure heights with a meter stick, since they were all taller than the stick. Unable to argue with such logic, I proceeded to the next lab group.

"I arrived just in time to witness the finishing touches that a student was adding to his self-inspired metric project. He had beautifully carved his gang symbol into the meter stick with an eight inch knife he had been carrying. He also asked me if, perchance, I

(Continued on p. 237)

Think Pieces

- Create a poster report that conveys the meaning of any one of the models of teaching presented in this chapter. Assume that you will present the poster at a local science teacher conference whose theme is "Effective Models and Strategies of Science Teaching for the Nineties."
- Design a graphic representation of your model of teaching. Don't be afraid to be inventive, and realize that creative thinking involves combining elements of other ideas. Present your model in class.
- Which of the following models of teaching do you prefer?
  - Direct-interactive teaching
  - Direct inquiry
  - Cooperative learning
  - Integrative learning
  - Person-centered learning

  Prepare a brief speech (three to five minutes) in which you identify the characteristics of the model and why you prefer it over other models. (And of course, you can use visual aids if you are called on to give the speech.) □

Encouraging Student-to-Student Interaction by Roger T. Johnson and David W. Johnson[44]

How should students interact with one another in science class? This question has been neglected by those studying teaching. While science teachers are encouraged to plan carefully the interactions between students and material (specific curriculum, specific

(Continued on p. 237)

(Continued from p. 236)
would like to buy some ludes from him. I declined his offer, asked him to put away the knife, complimented him on his artwork, and proceeded to the third group.

"At this lab table, two young men were having a dispute over a question on the lab handout which directed them to provide the width of their little finger in both centimeters and millimeters. They couldn't decide whether this measurement should be the long or short dimension of a finger. Peter was strongly emphasizing his point of view with well-placed punches on David's arm and chest region. I managed to separate them and clarify the meaning of the lab question. As I left them to visit with group number four, I overheard David say to Peter, 'This side of the stick is meters,' to which Peter replied, 'No it isn't. That's the inches side.' Two or three dull thuds punctuated Peter's response.

"At table four, one student, who had just been released from jail the day before, was sharing with his lab partners the economics lesson he learned while incarcerated. Mercifully, the bell rang. 'Well,' I sighed hopefully to myself, "Only four more periods to go!"

"This was just one of many labs conducted during my first year of teaching that didn't go quite as planned. Although none were as disastrous as the metric lab, each was as much an experiment for myself as it was for the students. Often, I half wished that I had taken my student up on his offer and purchased a healthy supply of central nervous system depressants. As they say, even the best laid plans sometimes go awry.

"Now I am in my second year of teaching..."

The Problem. Do you think Richard had too-high expectations for his students in doing lab activities? If you had watched this lab activity, what specific suggestions would you make to Richard to change the lesson to eliminate the "pandemonium"? □

[43]Judith H. Shulman and Joel A. Colbert, *The Intern Teacher Casebook,* San Francisco, CA: Far West Laboratory for Education Research and Development, ERIC Clearinghouse on Educational Management and ERIC Clearinghouse on Teacher Education, 1988, pp. 22–23. Used with permission.

(Continued from p. 238)
content) and there is a growing concern about the teacher-student interaction, the peer culture of the classroom remains relatively unexplored. Perhaps because we tend to overestimate our own influence on learning as teachers, we have grossly underestimated the power of appropriate student-student interaction on a range of learning outcomes.

There are three basic ways that students can interact with each other. Students can compete with each other to see who are the best students in the class; students can work individually on their own toward an established criterion; or students can work together, cooperatively, taking responsibility for each other's learning as well as their own. Many students in the United States tend to see school as a competitive place where it is important that you do better than the other students. Over the last fifteen years, teachers have been encouraged to structure individual learning in which students work alone at their own pace.

Reports on over 600 research studies, dating back to the late 1800s, which compare learning in cooperative, competitive and individualistic goal structures, have been collected at the Cooperative Learning Center at the University of Minnesota. From these studies it has been concluded that having students work together is much more powerful than having students work alone, competitively or individually. Some of the findings include:

- More students learn more material when they work together, cooperatively, talking through the material with each other and making sure that all group members understand, than when students compete with one another or work alone, individualistically.

- More students are motivated to learn the material when they work together, cooperatively, than when students compete or work alone, individualistically (and the motivation tends to be more intrinsic).

- Students have more positive attitudes when they work together cooperatively than when they compete or work alone, individualistically. Students are more positive about the subject being studied, the teacher, themselves as learners in that class, and are more accepting of each other (male or female, handicapped or not, bright or struggling, or from different ethnic backgrounds) when they work together cooperatively.

The positive effects of cooperative learning in science go beyond the immediate gains in achievement, motivation, self-esteem and acceptance of differences. Students learning in a cooperative goal structure also develop skills in communication, leadership and conflict resolution that are basic to productive, working teams. □

[44]Roger T. Johnson and David W. Johnson, "Encouraging Student/Student Interaction," *Research Matters...To the Science Teacher* Monograph Series (Cincinnati, Ohio: National Association for Research in Science Teaching, 1989). Used with permission.

Case Study 2. Hugging a Tree[45]

The Case. David Brown, an undergraduate student, was granted permission to student teach in the high school he graduated from because it is the type of school in which he would like to teach. David's school, however, is located 200 miles from the university. His college supervisor, Dr. Ahrens, associate professor of biology, was not enthusiastic about having to drive the distance to supervise him. Ahrens was under pressure to conduct research and publish the results, and he was having difficulty finding the time to do so. Furthermore, some of his younger colleagues were already full professors, and were receiving federal grants for their research.

David was accepted with open arms at his "old" high school, and the department head assigned Bob Smith, one of David's favorite high school teachers, as his cooperating teacher. On his first biweekly visit to supervise David, Dr. Ahrens saw performance unlike anything he had witnessed in 15 years of supervising student teachers. The lesson topic focused on trees. David had gotten up at daybreak and gone into a wooded area to collect leaves. But what Dr. Ahrens saw was much more than a leaf collection. David had brought tree limbs—dozens of them—and the classroom looked like an aboretum.

Dr. Ahrens entered through the door at the back of the room and quietly sat down. As usual, he prepared to take notes and complete an assessment form, and he quickly became so surprised by David's activities that he forgot to complete the form. David was running from one part of the room, to another, and then another, taking limbs and small trees with leaves and giving

them to students to examine. First he had the students taste the sassafras leaves. Some comments about root beer were heard. Then David took a double handful of sweet shrubs and crushed them. As he walked down each row of students letting every student smell them, the students "oohed" and "aahed." Next David gave each student a leaf from a cherry tree. He asked them to break the leaves in half and sniff them to see who could tell him the type of tree these leaves came from. Some said that they smelled like a milkshake. Others said the smell reminded them of the chocolate-covered cherry candy that they get at Christmas. After a wisecrack guess that it was a Christmas tree, someone screamed "cherry."

Next, David had some yellow roots. He asked the students to take a small hair of the root and taste it. They were as bitter as quinine. Students gasped in exaggerated disapproval of the bitter taste; some ran to the door and spat outside. By the end of the period every student was grappling with a piece of sugar cane, twisting it and swallowing the sweet juice.

When the bell rang the students applauded and commented about the lesson. Some said that this is the way school ought to be: fun. David was pleased that the students received the lesson so well.

Dr. Ahrens realized that he had become so engaged with observing the activities and so bewildered and upset at having seen something that didn't even resemble a lesson that he failed to complete the rating instrument and was, therefore, unprepared for the assessment conference that was to follow. Never-

theless, he felt he must give David some badly needed feed-back, however general. He would spare no words for this young maverick.

Dr. Ahren's first step toward resolving this perceived disaster was to meet with Mr. Bob Smith, David's cooperating teacher. After sharing a few brief pleasantries he fired the following questions at Mr. Smith.

"What do you think about this lesson?"

"The kids were really excited over it."

"Would you say that this was typical of David's lessons?"

"Yes, he always gets the students fired up."

"Do you think the lesson was well structured, well planned?"

"He obviously kept things moving at a good pace, and he didn't run out of material."

"Does David usually give you a well-prepared lesson plan when he teaches?"

"David has the type of personality that enables him to move through the lesson well without a written plan. I think lesson plans are good if you need them, but they can handicap a natural teacher. David knows that as long as his lessons work, I really don't insist on seeing a written lesson plan."

From this brief conversation Dr. Ahrens concluded that he would get little help from this lackadaisical teacher. In fact, he assumed Mr. Smith was influencing David negatively by providing a loose, unstructured role model. Clearly, it was time to talk to David.

Dr. Ahrens began the session by asking David what he thought he was doing. He continued, "I drove two hundred miles to watch you teach, and instead you provided a circus. My job is to help you raise the achievement scores in this class. From our prestudent teaching seminar, you learned that effective teachers give clear goals,

(Continued on p. 239)

(Continued from p. 238)
hold high expectations, use direct instruction, and closely supervise all assignments. Instead of following this instructional model, you arranged for a disorganized, student-centered picnic, complete with refreshments. I am very disappointed. I will have to record these activities and place the report in your permanent records."

David was shocked. What could he do to salvage his student-teaching grade and his teaching career?

The Problem. Put yourself in David's place. What would you do? ☐

[45]Based on Theodore J. Kowalski, Roy A. Weaver, and Kenneth T. Henson, *Case Studies on Teaching* (New York: Longman, 1990), pp. 28–32. Used with permission of the publisher.

The Reflective Teaching Lessons

These lessons have been designed to be used with Inquiry Activity 7.1, Reflective Teaching on page 203.

Reflective Teaching Lesson 1: Creatures

Description: You are one of several members of the class chosen to teach this short science lesson to a small group of your classmates. Plan to teach it in such a way that you believe both student learning and satisfaction will result.

Objective: Your goal is to get as many learners as possible to classify a set of "creatures"

based on their similarities and differences.

Materials: The creatures (copy them from Figure 7.27), learner satisfaction forms

Reflective Teaching Lesson 2: Shark's Teeth

Description: You are one of several members of the class chosen to teach this short science lesson to a small group of your classmates. Plan to teach it in such a way that you believe both student learning and satisfaction will result.

Objective: Your goal is to get as many learners as possible

to make observations and inferences about shark's teeth.

Materials: Shark's teeth (if you cannot find any, use an interesting substitute), learner satisfaction forms

Special Conditions: Learners can hold shark's teeth but cannot look at them during the lesson.

Reflective Teaching Lesson 3: The Balloon Blower Upper

Description: You are one of several members of the class chosen to teach this short science lesson to a small group of your classmates. Plan to teach it in such a way that

(Continued on p. 240)

All of these are Mellinarks.

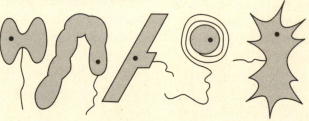

None of these is a Mellinark.

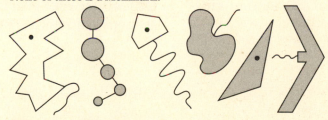

Which of these are Mellinarks?

FIGURE 7.27 Creatures; Lawson, Abraham, and Renner, A *Theory of Instruction*, used with permission.

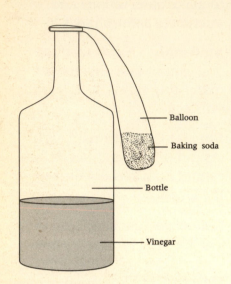

FIGURE 7.28 Balloon-Blower Upper

(*Continued from p. 239*)
you believe both student learning and satisfaction will result.

Objective: Your goal is get as many learners as possible to propose at least two hypotheses to explain the "balloon blower upper" phenomenon.

Materials: Soda pop bottle, balloon, vinegar, baking soda, aluminum foil, learner satisfaction forms

Special Conditions: Make sure you have the demonstration set up so that the students cannot see the baking soda, or that the liquid is vinegar. To make the demonstration work, wrap the bottle with foil. Fill the bottle about ¼ full with vinegar. Pour a small amount of baking soda in a balloon. Carefully cover the mouth of the bottle with the balloon. The powder falls in and reacts with the vinegar making the balloon blower upper work. Up goes the balloon!

**Reflective Teaching
Lesson 4: Mission to Mars**

Description: You are one of several members of the class chosen to teach this short science lesson to a small group of your classmates. Plan to teach it in such a way that you believe both student learning and satisfaction will result.

Objective: Your goal is to get as many learners as possible

to disagree with NASA's desire to jointly plan and execute a manned mission to the planet Mars in the early part of the twenty-first century.

Materials: Pretest and post-test, learner satisfaction forms

Special Conditions: Discussions have been held among scientists and administrators from NASA and the Russian space program about the possibility of sending a joint mission to Mars. Although this seems like a beneficial scientific undertaking, your job is to get your learners to disagree with NASA and the Russians. You may want to consider the following: cost; there are problems on Earth to solve first; the moon might be a better place to go; a station on the moon would be less expensive and more beneficial.

Test: Use the test shown in Figure 7.29 as both a pretest and a postest.

Directions: What is your opinion about each of these statements? Indicate by placing a check mark in the space to the right of each statement.

	Strongly Disagree	Disagree	Agree	Strongly Agree
1. The United States and Russia should combine efforts and send a manned mission to Mars.				
2. The cost of the mission is well worth it in benefits to humankind.				
3. Before such a mission takes place, we should take care of all problems facing us on the earth.				

FIGURE 7.29 Mission to Mars Pre/Post Test

Science Teachers Talk

"Is the discovery or inquiry model of teaching important in your approach to teaching? Why?"

Ginny Almeder. Since science is both a body of knowledge and a process, I value the inquiry learning approach. Knowledge of facts is necessary to develop scientific literacy but having opportunities to apply the knowledge in new contexts in order to develop problem-solving skills is essential. For the teacher, there is a delicate balance between presentation of factual information and the discovery process. However, with careful planning, both approaches can be successfully integrated. For example, after reading introductory material from the text or lab manual, students can be asked to describe the problem of a planned lab activity, devise hypotheses, and make predictions. During the post-lab session, students can discuss alternative hypotheses and experimental design as well as various interpretations of the data and suggestions for further experiments. Projects provide another vehicle for developing problem-solving skills. Individual students can present their proposals or projects and benefit from a class discussion of variables and controls.

John Ricciardi. The essence of sciencing is discovery; the primary discovery of knowing oneself in relation to one's surroundings. For me, the inquiry learning model is the most important if our focus is first to know oneself—through the process of knowing the envelope of nature around us. As a teacher of astronomy and quantum physics, I must help focus my students' attention to aspects of physical reality that normally are not perceived by our primary senses. The size and distance extremes of an electron and a galaxy super cluster are awesome. We can only know these objects by "blindly touching them in the dark" with our instruments. One prime task in these sciences is to visualize, extrapolate, imagine, and wonder about these things we can't naturally see. This kind of thinking is a real, vital part of science; it unfolds and reveals nature by discovering our own "inner tools" and identity—our own potentiality.

Jerry Pelletier. This learning model is the essence of my science teaching approach. In order for students to truly understand scientific ideas and concepts, they must experience them for themselves and question what they observe, hear, and manipulate. Every one of my students is required to participate in a science fair exhibit at school. I feel that this is the most important project in their middle school science education. They are required to use all of the skills that have been developed in our science program. They must develop their own topic, form their own hypothesis, develop an experiment that tests their hypothesis, collect and analyze data, and draw a logical conclusion. This to me is learning. Students questioning, inquiring, observing, solving and analyzing.

Mary Wilde. When I first started teaching eighth-grade science, I studied very hard to make sure I knew all the answers to any questions that might arise in class. I felt it was my duty to present information, not even considering that the students may not have the same interest in the science I was teaching that I did. Then I asked myself, "Why do I enjoy science?" I realized that the excitement of learning science came from discovering answers to proposed questions, solving problems, and reaching a goal through my own investigation. Thus, teaching science through the inquiry method has become very important in my approach to science teaching. I feel that this method provides students with opportunities that allow them to utilize problem-solving skills. They will not only develop higher thinking skills, but will also retain the information learned longer because it has a purpose. The student has now become an active learner and a part of the learning process. Teaching through the inquiry approach also allows for the student to express creativity and individuality. I feel that allowing students to discover, to imagine, and to create, establishes positive attitudes toward science.   □

The Psychology and Philosophy of Inquiry
by J. Richard Suchman[46]

In an inquiry program, the teacher presents the students with a series of objects or events. These events are puzzling. They run counter to the student's expectations; they are discrepant with his beliefs about reality. The student faced with a discrepant event sets out to find some explanation, some theory, that will close the gap between his beliefs and his observations. He begins to gather more data about the event and about similar events, and attempts to find some theory that will account for his new data. If the teacher *now* suggests an appropriate organizer, the student immediately realizes that the organizer helps him to make sense of what he is seeing. It adds more meaning to the data he is collecting. Thus, in an inquiry program, the teacher first lets the student discover a problem and begin the search for a solution,

(Continued on p. 242)

(Continued from p. 241)
and begin the search for a solution, and then at an appropriate moment, introduces a useful organizer. The student has an intrinsic motivation for learning the concept—it assists him in his attempt to increase the meaningfulness of the data he is gathering.

Let us call the first kind of motivation—the attempt to gain good grades or praise from others—social-ego motivation. This is the motivation that is inherent in the desired for survival—physical, social, and psychological survival. "Am I going to be approved? Am I going to make the grade?" It is an ego-centered motivation. "How am I doing? Does the teacher like what I'm doing? Am I going to succeed?" The theories, the questions, the data, the class assignments, the class discussion—all are regarded from the point of view of "How does this affect *my* status?" This kind of motivation is present in all human beings at all times. We are always concerned about our own status in one way or another. In fact, this motivation seems to take priority over other kinds. It is a kind of survival response—we all want to survive, and our survival is ensured by the approval and support of the people around us.

But there is another kind of motivation. If a person is not too pressed by the need for survival—that is, if he is feeling quite safe and secure, unconcerned about approval—his actions are motivated by another question: "Does what I see match what I know?" This is the *motivation for closure,* for closing the gap, for being sure that one's ideas are in accord with what one sees. This is one kind of *cognitive motivation*—motivation that leads a

person to attempt to find an explanation for a discrepant event. He isn't only concerned with what he sees; he is concerned with how his own mental formulations relate to what he sees. If there is a gap between the two, he gets concerned because he wants to have as true a mental picture of the world as possible.

A second kind of cognitive motivation is the *motivation of curiosity.* This is what impels a child to fool around with an old alarm clock. The clock is not a discrepant event. The child is not particularly concerned because he has no complete theory that explains how the clock works. He is tinkering around simply because he is curious. "What's inside this thing? What happens if I do this, or if I do that?" There is no strong pressure to find answers to these questions. It is a toying with ideas, with data, with objects and events, for the sheer pleasure of being engaged in the activity.

The motivation of closure, on the other hand, provides a strong pressure. Something happens that is puzzling, and we feel a definite need to discover an explanation for the discrepant event. The motivation of curiosity doesn't provide the same kind of pressure. It is merely a wondering, not an urgent need for an answer. This kind of motivation prompts people to raise new problems rather than solve old ones. As a matter of fact, the end product of curiosity-motivated inquiry is usually a new question rather than an answer.

The motivation of closure can be used to start almost any student inquiring. Faced with a discrepant event, the student will have a *need* to find an explanation. He will keep

going at the problem, and he will want someone to help him find an answer. But as he becomes more familiar with the inquiry process, more aware of his own ability to formulate and test theories, he will begin to be motivated by curiosity. He will no longer need a discrepant event to start his inquiry. Anyone would start inquiring if he saw red grass, but the student motivated by curiosity may begin to wonder *why grass is green.* He begins to raise questions about familiar events and objects that are not discrepant at all.

The most creative thinking and inquiry seem to occur under the conditions of least pressure, when there is little need to produce an answer. This kind of inquiry is open-ended, exploratory. It can take many different directions; there is no internal or external pressure on the student to find a final answer. He is free to be creative and daring in trying out new ideas. Therefore, the most creative learning, the most productive inquiry, is the result of the motivation of curiosity.

The inquirer motivated by curiosity is *autonomous.* He plans and executes his own strategy; he generates and tests his own ideas and theories. He never achieves complete closure, because phenomena are infinitely varied and variable and thus impossible to sample and catalog in their entirety. In other words, one can always find more data. No matter how powerful a theory or encompassing a generalization may be, a new counterinstance waits just around the corner. Thus inquiry remains a continuously open and moving process. □

[46] J. Richard Suchman, *Developing Inquiry* (Chicago: Science Research Associates, 1966), pp. 70–72.

Problems and Extensions

- Present a discrepant event related to a specific concept drawn from earth science, life science, physical science, or environmental science. Your presentation should be brief and should be designed to engage the class in inquiry.
- Present a five-to-ten-minute lesson based on one of the following models of teaching. Write a one page report on the strengths and weaknesses of the lesson, and how you would change the lesson if you were to teach it again.

  Direct-interactive instruction
  Inductive inquiry
  Deductive inquiry
  Discovery learning
  Constructivist model
  Cooperative learning
- Create a chart in which you analyze at least five models of teach

ing. Your chart should provide insight into the purpose of the model, the essential characteristics of the model, and under what conditions the model should be used.
- One criticism of inquiry and discovery methods of science teaching is that this approach takes too much time, and that students can learn concepts and skills if presented more directly. Debate this criticism by first taking the side of inquiry, and then the side of the criticism. In which were you more convincing? Is there a solution to this problem?
- Observe a video of a science teacher teaching a lesson. Make anecdotal comments on the chart shown in Figure 7.35, indicating examples of the various models of teaching observed during the lesson. What generalizations can you make about this teacher's approach to teaching? View a video of another teacher and compare the two teachers

approaches to teaching.
- Is there a relationship between metaphors of teaching and the models of teaching presented in this chapter? Three metaphors for the nature of teaching are [47]

  Entertainer: Teaching is like acting; you are on a stage and you perform.
  Captain of the ship: Teaching is like directing and being in charge.
  Resource: Teaching is making your self available, assisting, and facilitating.

  Either observe two teachers in their classrooms, or observe videotapes of two teachers. Analyze their teaching in terms of these metaphors. What relationship exists between the perceived metaphor and models of science teaching?

[47]Based on Kenneth Tobin, Jane Butler Kahle, and Barry J. Fraser, *Windows into Science Classrooms* (London: Falmer, 1990), pp. 51–57.

Model of Science Teaching	Anecdotal Comments: Examples of How the Model Was Implemented; Examples of Teacher
Direct-Interactive teaching	
Inductive inquiry	
Deductive inquiry	
Discovery learning	
Conceptual change model	
Generative learning model	
Cooperative learning	
Synectics	
Person-centered learning	
Integrative learning	
Imagineering	

FIGURE 7.30

RESOURCES

Borich, Gary D. *Effective Teaching Methods*. Columbus, Ohio: Merrill, 1988.

This book, better than any other one, presents the elements (among other concepts) of direct-interactive teaching in a way that beginning teachers can immediately translate into lesson plans.

Bellanca, James, and Robin Fogarty. *Blueprints for Thinking in the Cooperative Classroom*. Palatine, Ill.: Skylight, 1991.

This is the best book around on cooperative learning and high-order thinking skills. The book provides practical ideas and model lessons for encouraging cooperative learning, K–12. High-order thinking strengthens cooperative learning through the cycle of small group process.

Graves, Nancy, and Ted Graves. *What Is Cooperative Learning? Tips for Teachers and Trainers*. Santa Cruz, Calif.: Cooperative College of California, 1989.

This is the essential handbook for those interested in implementing cooperative learning in the classroom. It provides specific strategies for achieving positive group interdependence, individual accountability, face-to-face communication, teaching communication skills, and how to process lessons.

Hassard, Jack. *Science Experiences: Cooperative Learning and the Teaching of Science*. Menlo Park, Calif.: Addison-Wesley, 1990.

This book contains theory on science education and cooperative learning, as well as eight science units, with an STS and interdisciplinary flair on topics from space science to human evolution.

Joyce, Bruce, and Marsha Weil. *Models of Teaching*. Englewood Cliffs, N.J.: Prentice-Hall, 1986.

This book presents most of the theoretical views of teaching and shows how these theories are applied in the classroom. The authors describe and analyze each model in terms of the focus of the model, give a description of the model in action, and how the teacher can implement the model in any content area.

McCormack, Alan J. *Magic and Showmanship for Teachers*. Riverview, Fla.: Idea Factory, 1990.

If you want to implement the "imagineering" model into the science class, this book will show you the way. It contains basic techniques of setting up the classroom to arouse student curiosity, then presents chapter after chapter of "imagineering" activities.

Osborne, Roger, and Peter Freyberg. *Learning in Science: The Implications of Children's Science*. Auckland: Heinemann, 1985.

This is the essential book to find out about conceptual change teaching, and how students initial ideas or frameworks affect their learning. Based on the Learning in Science Project at the University of Walkato, this book provides practical suggestions in a theory-rich context for the middle and high school teacher.

Rakow, Steven J. *Teaching Science as Inquiry*. Bloomington, Ind.: Phi Delta Kappa Educational Foundation, 1986.

This is one of PDK's "fastback" series of booklets on education. This booklet describes what inquiry is, relates inquiry to the learning cycle, and discusses the status of inquiry in science teaching.

Resnick, Lauren B., and Leopold E. Klopfer. *Toward the Thinking Curriculum: Current Cognitive Research*. Alexandria, Va.: Association for Supervision and Curriculum Development, 1989.

You should look at chapter 7 of this book, "Teaching Science for Understanding," by James A. Minstrell, a high school physics teacher. Minstrell, in addition to being a science teacher, is also a researcher specializing in conceptual change and cognitive science research. This chapter, in a practical way, integrates the research on cognitive processes by weaving examples from his classroom research with the literature in science education.

Rogers, Carl. *Freedom to Learn for the 1980s*. Columbus, Ohio: Merrill, 1983.

This classic book on the person-centered model of teaching describes the philosophy and gives practical examples of how to develop a person-centered style. The book is essential reading for teachers who wish to understand the roots of the teacher as a facilitator.

Rose, Colin. *Accelerated Learning*. New York: Dell, 1985.

This book presents the results of research on the integrative or accelerated model of teaching. The methods have been shown to be effective with slow and at-risk learners.

Slavin, Robert E. *Cooperative Learning: Theory, Research and Practice*. Englewood Cliffs, N.J.: Prentice-Hall, 1990.

This book shows how teachers in all disciplines can take advantage of the benefits of cooperative learning by introducing such cooperative team arrangements as learning teams, discussion groups, project groups, lab groups, and peer tutoring. Slavin is one of the leading researchers and developers of cooperative learning strategies.

CHAPTER
8

STRATEGIES FOSTERING THINKING IN THE SCIENCE CLASSROOM

Would anyone believe that thinking is not a central part of the science classroom? Who would not emphasize thinking in the science classroom? Unfortunately research studies indicate that the predominant strategy used in science classes is recitation, with the teacher in control. What impact will this strategy have on student thinking? In most cases this form of teaching will reinforce memorization and rote learning. An example of this is reported by Anderson and Smith: they have noted that students can pass a chapter quiz on photosynthesis and still not understand that plants make their own food.[1]

In this chapter we will explore teaching strategies in the context of how they influence and facilitate student thinking. The first set of strategies will be explored in terms of their impact on students' ability to think critically and creatively. We will explore strategies such as questioning, structured discussions and debates, field trips, and role playing.

Thinking in science can also be facilitated by reading and writing strategies. However, many students have trouble comprehending contemporary secondary science textbooks. What strategies can be utilized to solve this problem? We'll examine language abilities and skills strategies that aid student learning in science.

We will also explore strategies that will foster independent thinking among secondary students. How can students be empowered to be thinkers in their own right? We'll examine science process skills more carefully, as well as the nature of problem solving, and make some connections to science projects and science fairs.

Finally, we'll consider how the computer can be instrumental in enhancing student thinking in the science classroom. We'll examine the computer as a medium to enhance students' scientific skills, along with its power to help students write, communicate, and do research.

PREVIEW QUESTIONS

- What teaching strategies can be used to foster critical and creative thinking among students?

- How can the communications skills of reading and writing in the science classroom be improved?

- What strategies aid student independent and collaborative thinking?

- How can computer technologies be used to enhance thinking in science?

[1]C. W. Anderson and E. L. Smith, "Teaching Science," in *Educators' Handbook: A Research Perspective*, Virginia Richardson-Koehler, ed. (New York: Longman, 1987), pp. 84–111.

STRATEGIES FOSTERING CRITICAL AND CREATIVE THINKING

Critical and creative thinking are counterparts of a holistic view of student thinking. They are not opposites; they are indeed complementary (Figure 8.1). Critical thinking is "reasonable, reflective thinking that is focused on deciding what to believe or do."[2] Critical thinkers in a science class have learned how to look at phenomena aware of their own biases, and approach the situation objectively and logically. Creative thinking, on the other hand, is the ability to form new combinations of ideas to fulfill a need.[3] Examples of creative thinking in science classes are brainstorming, creating alternative hypotheses, synthesizing information, and thinking laterally.

Critical thinking and creative thinking can be contrasted (Figure 8.1) to indicate the differences in these forms of thinking. Look over the following list of science teaching tasks and decide which ones are examples of critical thinking and which are examples of creative thinking:

- Summarizing the main ideas in a chapter on forces and momentum
- Building models of atoms using clay and toothpicks
- Inventing a system that will measure the mass of an elephant
- Writing review questions based on specific pages in the science textbook
- Observing an event and then writing at least three alternative explanations
- Estimating the volume of the classroom

[2]Robert J. Marzano et al., *Dimensions of Thinking: A Framework for Curriculum and Instruction* (Alexandria, Va.: Association for Supervision and Curriculum Development, 1988), p. 146.
[3]Marzano, et al., *Dimensions of Thinking*, p. 146.

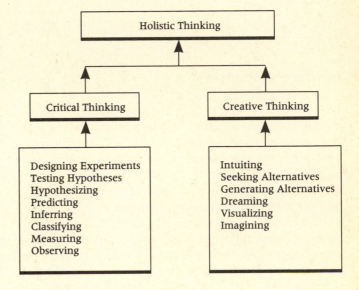

FIGURE 8.1   Holistic Model of Thinking

- Listing as many observations of a burning candle as possible during a five-minute observation session

There are many strategies to foster creative and critical thinking in the science classroom. Three assertions will guide our approach to these strategies that foster this development. Critical and creative thinking are fostered in

- Interactive classrooms that focus on inquiry teaching
- Classrooms that deal with controversies, thereby encouraging discussion, debate, and discourse
- Classrooms that bring students in contact with real-world problem solving

INTERACTIVE TEACHING STRATEGIES

Critical and creative thinking require that students be actively engaged in learning science, as opposed to the more traditional yet typical approach in which the student is on the receiving end of a lecture. The interactive classroom is one in which communication patterns involve students to teacher, teacher to students, and students to students. The interactive classroom is a stimulating place in which students have been motivated to learn and are given the freedom to explore, discover, and inquire. In the interactive classroom you will find teacher-centered as well as student-centered activities. Regardless of the type of activity that the teacher selects, there appear to be at least six specific strategies that teachers use to create an interactive science classroom: interactive teachers

- Use advance organizers to establish interest and instructional goals

- Create a stimulating classroom environment
- Understand the art of questioning
- Use examples to help students understand concepts
- Create a positive learning environment
- Use closure and transitional skills

Let's look at each of these strategies in some detail.

Advance Organizers

Wouldn't you agree that helping students make meaningful connections between what they know and what is to be learned would facilitate learning new ideas? An advance organizer is a device that teachers use to help students make these connections. Advance organizers are frameworks for helping students understand what is to be learned.

An advance organizer is not an overview but rather a presentation of information (either verbal or visual) that is an "umbrella" for the new material to be learned.[4] The concept of advance organizer was proposed by the psychologist David Ausubel. To Ausubel, effective advance organizers are presented by teachers at a "higher level of abstraction, generality, and incisiveness" than the science material that is to follow.[5]

Advance organizers can be useful devices at the start of a unit, before a discussion, before a question-answer period, before giving a homework assignment, before student reports, before a video, before students read from their science textbook, before a hands-on activity, and before a discussion of science concepts based on students' laboratory experiences. Let's look at some examples of advance organizers, and then have you consider designing some of your own.

- A teacher shows a picture of Mendeleyev in his laboratory in St. Petersburg and discusses his contribution to the development of the periodic table before introducing any of the details of the table of the elements.
- A teacher has students bring in pictures that show the destruction caused by the 1989 San Francisco area earthquake (or any earthquake) before introducing earthquake waves and how they are measured.
- A teacher discusses the origin of life on the earth before introducing Darwin's theory of evolution.

- A teacher shows a poster depicting many forms of energy and asks students to discuss and identify the examples of energy before introducing the students to a new unit on heat energy.

Advance organizers help the students organize the conceptual knowledge they are to learn. The teacher can refer back to the organizer and use it to link a sequence of lessons. For example, the science textbook can provide clues and examples of advance organizers. Look at a secondary science textbook and identify advance organizers for one of the chapters in the book. Discuss the advance organizers with a peer in your class. Do they meet the criteria of being "at a higher level of abstraction, generality and inclusiveness," than the materials in the chapter?

TABLE 8.1

INTERACTIVE TEACHING SKILLS

Classroom Teaching	Teacher Actions
Movement	Moving into the classroom and not hovering at the front of the classroom, especially behind the demonstration desk is desirable here.
Gestures	Complimenting verbal messages with body language is an important aspect of communication. Teachers should use hand, head and body to convey meaning.
Focusing	Teachers who focus students use verbal statements (look at this chart of vertebrates), and use gestures (pointing to a fault line on a map projected on the overhead), or use a combination of both.
Interaction styles	Stimulating environments occur where there are a variety of interaction styles among teacher and students. Whole class, small group and individual interaction styles should be utilized.
Multiple sensory channels	Student learning styles research suggests the creation of multiple sensory classrooms. Teachers should provide verbal, tactile, and kinesthetic experiences for students.

[4]Based on Margaret E. Bell-Gredler, *Learning and Instruction* (New York: Macmillan, 1986), p. 172.
[5]David Ausubel, *Educational Psychology: A Cognitive View* (New York: Holt, Rinehart and Winston, 1968), p. 148.

Creating a Stimulating Classroom Environment

Imagine walking down the hall of a school and being able to peer through the door windows enabling you to compare the classroom environments of a variety of teachers. One teacher is sitting on the desk talking with the students; half an hour later you pass by the classroom and the teacher is still sitting there. In another classroom, you have to strain your head to find the teacher; the teacher is in the corner of the room pointing to the aquarium during a discussion.

In a third classroom you are struck by the hand and body movements of the teacher explaining what the students are to do in lab. In the fourth class you observe the teacher walking among six groups of students who appear to be wrapped up in intense discussions.

There are a number of specific teaching skills that impinge on creating a stimulating environment in the classroom and that will have positive effects on the critical and creative thinking of students. The use of movement, gestures, focusing, different interaction styles, and multiple sensory channels (Table 8.1) appear to affect the environment that the teacher establishes in the classroom.

INQUIRY ACTIVITY   8.1

Microteaching: Practicing Science Teaching Skills

In this section we have discussed several teaching behaviors that are related to the establishment of an interactive classroom. In this inquiry activity you will prepare a brief lesson, teach it to a group of peers, view a video of the lesson during a reflective teaching conference, and then reteach the lesson incorporating changes suggested by your reflective teaching coach.

Materials

VCR, camera, videotape

Procedure

1. Prepare a five-minute science lesson on a science topic of your choice. Prepare the lesson in such a way that you focus on the teaching skills listed in Table 8.1. Please note that in five minutes you will not be able to teach a "complete" lesson.
2. Meet with your observer-coach (a peer) prior to the lesson to explain your objectives and the skills you plan to focus on.
3. Teach the lesson to four to six peers. The lesson should be videotaped.

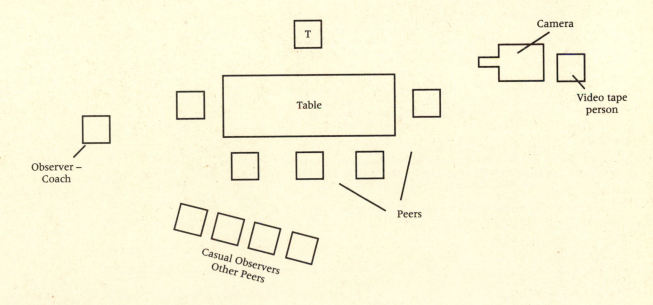

FIGURE 8.2   Microteaching Class Arrangement

4. The observer-coach should use the microteaching evaluation form (Figure 8.3) and use the results in a reflective teaching conference.

5. The observer-coach should conduct a reflective teaching conference immediately following the five-minute lesson during which the video should be viewed and suggestions be made on how the lesson might be changed for the reteaching session.

6. Reteach the lesson to another group of peers. Following the lesson, view the video with your coach and evaluate the effectiveness of the changes that you tried to incorporate into the second lesson.

Minds on Strategies

1. How successful were you in integrating the skills into the five-minute lessons?

2. Use the microteaching format to practice other critical and creative thinking teaching strategies. Your five-minute lessons can focus on the following:

a. Using advance organizers in science teaching

b. Using low- and high-inquiry questions in science lessons

c. Using examples to help students understand science concepts

TEACHER _____ TOPIC _____

OBSERVER/COACH _____ 1st LESSON _____

2nd LESSON _____

INTERACTIVE TEACHING SKILLS	Weak	Below Average	Average	Strong	Superior	Outstanding	Truly Exceptional
TEACHER MOVEMENTS 1. At various times during the lesson, the teacher was noted in the left, right, forward, and back of the teaching space.	1	2	3	4	5	6	7
TEACHER GESTURES 2. The teacher used gestures (hand, body, etc.) to help convey extra meaning in the presentation of the lesson.	1	2	3	4	5	6	7
FOCUSING 3. When the teacher wanted to emphasize a point, it was clearly stressed through the use of gestures (pointing, banging on the board, etc.) or through the use of verbal expressions ("Listen closely," "Watch this," etc.) or by combining both gestural and verbal acts.	1	2	3	4	5	6	7
INTERACTIONS 4. The teacher varied the kind of participation required of the students. That is, students could be directly called on, group questions were asked, student-student interchange could occur. Students could role-play, go to the board, etc. The teacher is to mix these various techniques.	1	2	3	4	5	6	7
PAUSING 5. The teacher gave the students time to think or get ready for new ideas by using silence. That is, all teacher activity ceased for short time periods.	1	2	3	4	5	6	7
ORAL-VISUAL SWITCHING 6. The teacher used visual material (words on blackboard, objects, pictures, etc.) in such a way that the student must look to get the information. That is, the teacher does not say what the word or object is but refers to it in the lesson, making the student look, not listen, to what is going on.	1	2	3	4	5	6	7

COMMENTS: _____

_____

FIGURE 8.3 Microteaching Evaluation Form. Based on Dwight Allen and Kevin Ryan, *Microteaching* (Reading, Mass.: Addison-Wesley Publishing Company, Inc., 1969), pp. 15–18.

The Art of Questioning

Of all the skills discussed in this section, questioning, according to many science educators, is one of the most important. Teachers sometimes ask over a hundred questions in a class session to encourage student thinking. Do science teachers' questions facilitate critical and creative thinking? Are some questioning strategies more effective than others? Let's examine some aspects of the art of questioning, including types of questions, wait time, and questioning and creativity.

Categories of Questions

Examine the following list of questions. Can you assign each question to one of two categories? Please identify the criteria you used to name the categories.

1. Are all the fruit flies alike for each feature?
2. What is weathering?
3. What do you predict will happen if a jar is put over a candle?
4. Using evidence that you choose, do you think scientists should be limited in the areas they want to research?
5. How many elements are in the periodic table?
6. Which planet is largest: Mars, Venus, or Mercury?

There are many systems that teachers use to classify questions. Upon close observation, in most systems questions are typically classified into two categories. Various terms are used to describe these two categories (Table 8.2). The binary approach is useful because two categories are more manageable for a beginning teacher to learn to implement than the typical approach of using systems with six categories.[6]

[6]You will, however, be introduced to Bloom's Taxonomy of Cognitive Objectives in Chapter 11. Bloom's taxonomy contains six categories, each of which can be used as a category of questions.

TABLE 8.2

CATEGORIES OF QUESTIONS

Category 1	Category 2
Factual	Higher cognitive
Closed	Open
Convergent	Divergent
Lower level	Higher level
Low order	High order
Low inquiry	High inquiry

What kinds of questions do teachers ask in the classroom? Gall reports that 60 percent of teachers' questions require students to recall facts, about 20 percent require students to use higher cognitive processes, and the remaining 20 percent are procedural.[7] If teachers want to foster critical and creative thinking in the classroom, this pattern of questioning should be changed. Let's examine more closely how questioning strategies might be used to enhance critical and creative thinking.

One way to classify questions is to determine whether they are low inquiry (closed or convergent) or high inquiry (open or divergent).[8]

Low-Inquiry Questions

These questions focus on previously learned knowledge in order to answer questions posed by the teacher that require the students to perform one of the following tasks:

1. Elicit the meaning of a term
2. Represent something by a word or a phrase
3. Supply an example of something
4. Make statements of issues, steps in a procedure, rules, conclusions, ideas, and beliefs that have previously been made
5. Supply a summary or a review that was previously provided
6. Provide a specific, predictable answer to a question

High-Inquiry Questions

These questions focus on previously learned knowledge in order to answer questions posed by the teacher that require the students to perform one of the following tasks:

1. Perform an abstract operation, usually of a mathematical nature, such as multiplying, substituting, or simplifying
2. Rate some entity as to its value, dependability, importance, or sufficiency with a defense of the rating
3. Find similarities or differences in the qualities of two or more entities utilizing criteria defined by the student
4. Make a prediction that is the result of some stated condition, state, operation, object, or substance
5. Make inferences to account for the occurrence of something (how or why it occurred)

[7]Meredith Gall, "Synthesis of Research on Teachers' Questioning," *Educational Leadership* 42 (November 1984): 40–47.
[8]George T. Ladd, "Determining the Level of Inquiry In Teachers' Questions," *National Association of Research in Science Teaching Abstracts*, Minneapolis, Minn.: National Association of Research in Science Teaching, 1970), p. 84.

TABLE 8.3

DIFFERENCES BETWEEN LOW- AND HIGH-INQUIRY QUESTIONS

Type of Question	Student Responses	Type of Response	Examples
Low Inquiry (convergent)	• Recall, memorize	Closed	How many...
	• Describe in own words		Define...
	• Summarize		In your own words... state similarities and differences...
	• Classify on basis of known criteria		What is the evidence...?
	• Give an example of something		What is an example...?
High Inquiry (divergent)	• Create unique or original design, report, inference, prediction	Open	Design an experiment...
			What do you predict...?
	• Judge scientific credibility		What do you think about...?
	• Give an opinion or state an attitude		Design a plan that would solve...?
	• Make value judgments about issues		What evidence can you cite to support...?

Low-inquiry questions tend to reinforce "correct" answers, or focus on specific acceptable answers, whereas high-inquiry questions stimulate a broader range of responses and tend to stimulate high levels of thinking. There is evidence to support the use of both types of questions. Low-inquiry questions will help sharpen students ability to recall experiences and events of science teaching. Low-inquiry questions are useful if you are interested in having students focus on the details of the content of a chapter in their textbook or of a laboratory experiment.

High-inquiry questions encourage a range of responses from students and tend to stimulate divergent thinking. Table 8.3 summarizes the differences between low- and high-inquiry questions.

Wait Time

Knowledge of the types of questions and their predicted effect on student thinking is important to have. However, researchers have found that there are other factors associated with questioning that can enhance critical and creative thinking. One of the purposes of questioning is to enhance and increase the verbal behavior of students in the science classroom. Mary Budd Rowe has discovered that the following factors affect student verbal behavior:[9]

9Mary Budd Rowe, "Science, Silence and Sanctions," *Science and Children* vol. 36 (October 1969): 22–25.

1. Increasing the period of time that a teacher waits for students to construct a response to a question
2. Increasing the amount of time that a teacher waits before replying to a student response
3. Decreasing the pattern of reward and punishment delivered to students

She has found that if teachers increase the time they wait after asking a question to five seconds or longer, the length of response increases. In the science classroom in which the teacher is trying to encourage inquiry thinking, wait time becomes an important skill, as well as a symbol of the teacher's attitude toward student thinking. Teachers who are willing to wait recognize that inquiry thinking requires thoughtful consideration on the part of the students. Rowe points out that teachers who extend their wait times to five seconds or longer increase "speculative" thinking. The use of silence in the classroom can become a powerful tool to enhance critical and creative thinking.

Rowe also believes that teacher sanctions (positive and negative rewards), if used indiscriminately, can reduce student inquiry. At first glance this doesn't make sense. However, Rowe has found that when rewards are high, students tend to stop experimenting sooner than if rewards are low. When students begin attending to rewards rather than the task, the spirit of inquiry tends to decrease.

Another factor related to questioning is the attitude of the teacher. Have you ever been in a class situation in which you wanted to ask a question but feared the teacher's reaction to your question—it might be a dumb question? One classroom rule that we think is important is "There are no dumb questions." A corollary to this rule is "There are no dumb answers." Students need to believe that their responses will be accepted by the teacher; anything short of this will tend to reduce the probability of student participation.

If your class is engaged in microteaching you might want to focus your lessons on questioning. Prepare the lesson and use the evaluation form in Figure 8.4 to analyze your effectiveness.

Using Examples to Help Students Understand Concepts

The word *example* comes from the word *sample*, which means a portion of the whole that shows the quality and character of the whole. The use of examples is fundamental to helping students understand science concepts. The teacher that identifies examples of concepts or asks students to cite examples of concepts is acknowledging that learning must be tied to students' prior knowledge. Using examples is a way to tie science teaching to the students' world.

Examples should be exciting and should be recognizable as everyday phenomena. We use the acronym EEEP to mean exciting examples of everyday phenomena. EEEPs should be used to help students understand science concepts. EEEPs can exist in the form of an object (a rubber door wedge to represent an inclined plane), an artifact (a piece of pottery to help students understand soil), a machine, a photograph, a piece of technology, or a toy.

One way to help you think about examples is to examine a list of categories and use the list as a stimulus for connecting these categories with science concepts. What kinds of examples can you generate from the following list?

Detergents
Household chemicals
Fertilizer
Nails
Muddy water
Vegetables
Flowers
Building materials (bricks, mortar, wood)
Paper products

TEACHER _____ TOPIC _____

OBSERVER/COACH _____

1st LESSON _____

2nd LESSON _____

ASKING QUESTIONS		Weak	Below Average	Average	Strong	Superior	Outstanding	Truly Exceptional
A.	*Increasing Probability for Student Response* At various times during the lesson the teacher paused after asking questions and called on both volunteers and non-volunteers.	1	2	3	4	5	6	7
B.	*Decreasing Teacher Verbal Participation* The teacher was observed redirecting the same question to several pupils and asked more high-inquiry questions than low.	1	2	3	4	5	6	7
C.	*Probing* The teacher prompted students and asked students to clarify their responses.	1	2	3	4	5	6	7
D.	*Reducing Interference* The teacher made an attempt not to repeat or answer his or her own questions and not to answer student questions.	1	2	3	4	5	6	7

FIGURE 8.4 Asking Questions Evaluation Form. Based on Allen and Ryan, *Microteaching*, modified by S. Smith and J. Hassard.

Beach stones
Playground rocks
Food items
Slinkies
Dishware
Eye glasses and lenses
Recycled newspaper
Biodegradable plastic bags
Balloons
Oil and other viscous fluids
Toy cars

Examples can be used deductively or inductively in helping students understand concepts. However, it is important to keep in mind what we have discussed about the learning cycle and the generative model of learning. As you recall, cognitive psychologists theorize that students construct their knowledge (of concepts) and must do so through their interaction with ideas, phenomena, and people. In the deductive approach of using examples, learning begins with the idea, principle, or concept, and is then followed by an exploration of examples and phenomena. The process culminates in relating the concept to the examples (Figure 8.5)

In the inductive approach, which is closer to the constructivist notion of how students learn concepts, students begin with an exploration of ideas and phenomena, followed by a teacher-directed activity to facilitate the "invention" of the science concepts. (Figure 8.6)

In either approach it is important to start with the simplest examples, and ones that are relevant to your students' experience and knowledge. Examples can be thought of as metaphors and analogies of the science concepts in the curriculum. The more familiar the examples are to the students, the greater the probability of hooking the students' understanding of the concept.

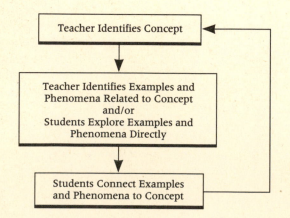

FIGURE 8.5 Deductive Use of Examples

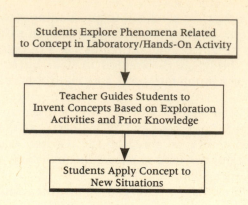

FIGURE 8.6 Inductive Use of Examples

Research by Treagust and his colleagues has shown that examples, especially those in the form of analogies, are rarely used to help students understand science concepts.[10] Typically, the teacher introduces the new science concept by defining it, or by using an example that is not familiar to the students.

An interesting finding by Treagust and his colleagues was that teachers who were not familiar with the content being taught typically use definitions to explain concepts; teachers well versed in the content use definitions the least.

A Positive Learning Environment

The psychologist Carl Rogers has shown that student attitudes (toward themselves, their peers, their teachers, and the subject) are an integral aspect of student learning.[11] He has suggested that a climate of inquiry, which is essential for critical and creative thinking, is fostered when students perceive the learning environment in terms of realness or genuineness, acceptance, and empathy. Throughout this book we have emphasized the importance of inquiry learning and have substantiated the claims of cognitive psychologists that students must explore, invent, and apply their knowledge in real situations to develop and construct science concepts. The science teacher must foster a classroom climate that projects and supports this cognitive perspective. Such an environment is characterized as follows:[12]

[10]David F. Treagust et al., "Science Teachers' Use and Understanding of Analogies as Part of Their Regular Instruction," paper presented at the annual meeting of the National Association for Research in Science Teaching, Atlanta, April 1990.
[11]Carl R. Rogers, *Freedom to Learn for the 1980s* (Columbus, Ohio: Merrill, 1983).
[12]Nancy Battista, *Effective Management for Positive Achievement in the Classroom* (Phoenix, Ariz.: Universal Dimensions, 1984), p. 26.

FIGURE 8.7 Inquiry learning requires a positive classroom environment. According to cognitive psychologists, inquiry takes place in an environment in which students construct ideas about science by exploring, inventing, and applying this knowledge. Copyright Rhoda Sidney.

- The teacher projects an image to the students that tells them, "I am here to help you build your character and your intellect"
- The teacher conveys the notion that each and every student is unique that he or she and is interested in each student as a unique person
- The teacher conveys the idea that all students can accomplish work, can learn, and are competent
- The teacher expects high standards of values, competence, and problem-solving ability
- The teacher conveys, through his or her own behavior, a character of authenticity
- The teacher conveys high ethical standards by establishing a high degree of private or semiprivate communication with students

Rogers claims that the most important aspect of the teacher-facilitator role is that of empathy. In the context of science education, this makes perfectly good sense. The student who has "science anxiety" can only be helped in an environment in which the teacher empathizes with this state. Too often, the student who was interested in science in the early grades gets turned off to science in the middle grades. Perhaps one of the reasons for this low interest in science is the lack of an empathic classroom climate.

Closure and Making Transitions

Closure is the complement of advance organizers. Closure acts as a cognitive link between past knowledge and the new knowledge (experiences). Closure can also function to help give the students a feeling of accomplishment or achievement.

Closure is not limited to the end of a lesson. There are many instances in a lesson in which you will help the students make a transition, for example, from a prelab session to the laboratory activity itself. Closure in this instance functions as a transition from one activity to another.

There are a number of ways to integrate closure and transitions into your lesson plans. Three are as follows:

1. *Drawing attention to the completion of a lesson or a part of the lesson.* The teacher can provide a consolidation of concepts and elements that were covered before moving to a subsequent activity. It is extremely helpful to relate the lesson back to the original organizing principle (advance organizer). Some teachers review the main ideas of the lesson by means of an outline or a concept map. Another closure technique is to stop throughout a teacher-directed lesson and ask student pairs to explain the ideas that were developed.

2. *Making connections between previous knowledge and the new science concepts.* Teachers find it helpful to review the sequence that was followed in moving from previous knowledge to the new ideas. The learning cycle or generative model emphasizes this general sequence. Using examples (EEEPs) can facilitate student transition from a misconception state to one of understanding the concept.

3. *Allowing students the opportunity to demonstrate what they have learned.* It is a much more powerful technique if students can suggest ways that demonstrate closure. One technique that researchers and teachers have found effective is concept mapping. A concept map drawn after a lesson, chapter, or unit of study is a visual mechanism in which students describe their understanding.

CONTROVERSIES IN THE
SCIENCE CLASSROOM: REAL—WORLD
PROBLEM SOLVING

A strategy that appears to encourage critical and creative thinking is to involve students in the exploration of science controversies—a kind of academic conflict that arises when one student's opinions and ideas are incompatible with those of another student.[13] Johnson and Johnson have experimented and field-tested a series of environmental issues structured for controversies in the science classroom. The recent emphasis on STS (Chapter 6) provides the science teacher with a wide range of topics for a structured controversy strategy. You should also refer to Robert Barkman's book, *Coaching Science Stars: Pep Talk and Play Book for Real-World Problem Solving*.[14] Some potential topics include

> Global warming: Is the earth really heating up?
> A hungry earth: Can the earth feed its human population?
> Crisis in the ocean: How polluted is the ocean?
> The garbage problem: What is the best way to manage waste?
> Chemicals on the highways: How can hazardous waste be managed?
> Extinction: How endangered is life on the planet?

The strategy of structured controversies consists of four procedures: selecting a topic, preparing learning materials, structuring the controversy, and conducting the controversy in the classroom.

Selecting a Topic

Teacher and student interest are critical in the selection of a topic. It is important that two positions on the issue can be identified, and that the students are able to deal with the content of the topic. The teacher can either present the class with the topic or have the students choose from a list of topics.

Preparing Teaching Materials

According to the developers of this strategy, the following materials are needed for each position on the issue:

[13]David W. Johnson and Roger T. Johnson, "Critical Thinking Through Structured Controversy," *Educational Leadership* 45 (May 1988): 58–64.
[14]Robert C. Barkman, *Coaching Science Stars: Pep Talk and Play Book for Real-World Problem Solving*, (Tucson, Ariz.: Zephyr, 1991).

- A description of the group's task
- A description of the phases of the process (see Conducting the Controversy in the Classroom)
- A definition of the position to be advocated
- Resource materials including a bibliography, pamphlets, magazines, newspaper articles

Structuring the Controversy

The controversy should be investigated in a cooperative learning context. Students should be grouped by fours heterogeneously; pairs of students will work together on one side of the issue. (Refer to the section on cooperative learning models in Chapter 7).

Conducting the Controversy in the Classroom

To manage the process in the classroom, students should be led through a series of phases, as follows:

- *Learning the position*. Each partner team should become thoroughly familiar with their position on the issue by reading the materials and preparing a persuasive presentation. Additional reading may be required to master the position.
- *Presenting the position*. Each partner team presents their position to the other team in their group. It is important to listen carefully as well as ask questions to clarify points on the issue.
- *Discussing the issue*. During this phase each team should argue their position forcefully by presenting facts to support points on the issue. Students should be encouraged to ask their opposing teammates to support their arguments with information and facts.
- *Reversing positions*. In this phase each pair presents the opposing pair's position in as sincere and forceful a manner as possible.
- *Reaching a decision*. In this final phase each group must prepare a report on the issue that summarizes and synthesizes the best arguments for both points of view. However, as a group, they must reach consensus on a position that is supported by the evidence. Each group should prepare a single report and be prepared to engage in a large group discussion.

FIGURE 8.8   Structured Controversy in Science Teaching: A cooperative learning strategy enabling students to explore science related social issues.

Throughout the process, students should be held accountable by acknowledging a set of discussion rules:[15]

1. I am critical of ideas, not people
2. I focus on making the best decision possible, not on "winning"
3. I encourage everyone to participate and master all the relevant information
4. I listen to everyone's ideas, even if I do not agree
5. I restate (paraphrase) what someone has said if it is not clear

6. I first bring out *all* the ideas and facts supporting both sides and then try to put them together in a way that makes sense
7. I try to understand both sides of the issue
8. I change my mind when the evidence clearly indicates that I should do so

[15]Johnson and Johnson, "Critical Thinking," 58–64. Reprinted with permission of the Association for Supervision and Curriculum Development. Copyright 1988 by ASCD. All rights reserved

Throughout the process, critical thinking emerges as students analyze points of view and search for evidence to support their arguments. Facts and information are analyzed in terms of a position or argument. Creative thinking manifests itself in a number of ways. Students have to reverse positions on the issue requiring them to move from one side of the issue to the other. This encourages flexibility in thinking, an important component of creative thinking.

STRATEGIES THAT AID STUDENT LEARNING

Picture this: The bell is ringing and the teacher shouts, "For tonight's homework, read chapter 8 and do prob-lems 1 to 11 on page 243!" Secretly the teacher knows that very few of the students will "read" the chapter, and maybe half of them will turn in the answers to the questions.

In another classroom the teacher is explaining an assignment in which the students are to identify the main ideas in the first half of a chapter in their text and then write supporting details that explain, prove, or tell something about the main idea.

In a third classroom a teacher has prepared on large sheets of paper the main terms in the current biology unit. The words are written in English and Spanish, and the teacher has pictures pasted next to each word. The teacher is reviewing the words with the class.

To a growing number of science educators, the teaching of language abilities and skills should be an integral aspect of science teaching. The three examples cited here are only the tip of the iceberg with regard to the actual language arts teaching that occurs in the science classroom. To some, the teaching of language or study skills is seen as remedial action necessary for students with low reading scores or abilities. For these students, this is beneficial if they do indeed have teachers that provide "special" instruction to help them comprehend their science textbook, or give them pointers for writing reports.

However, there is a more powerful argument to support the integration of language skills in the science classroom. Three of the four goal clusters recommended by Project Synthesis involve personal, societal, and career awareness objectives. These goals take the purpose of science teaching beyond the laboratory or the academic context and into everyday life: newspapers, magazines, television, and the local as well as global community. Students need to develop the skills (in the science classroom) to evaluate and find information, make decisions about issues, and communicate this information effectively. Language skills are at the heart of this goal. In this section we will explore strategies that will help students in the science classroom to improve their ability to read science information, and to communicate their ideas in writing.

Reading Strategies for the Science Classroom

First, let's start off with the notion that reading is as much a scientific activity as observing, classifying, measuring, and hypothesizing. If we want science to be more accessible to our students, we must recognize reading as a new science process skill. Leslie Bulman in *Teaching Language and Study Skills in Secondary Science* writes:

> If pupils can read about science it makes it potentially more accessible both in terms of present understanding and, even more important, of future interest. Most learning after the exposition of school and lecture hall is over, is via books....Reading is more important as a scientific technique than many practical skills. It is clear from the aims of science education...that they cannot be achieved without advanced reading skills.[16]

[16]Leslie Bulman, *Teaching Language and Study Skills in Secondary Science* (London: Heinemann, 1985), p. 20.

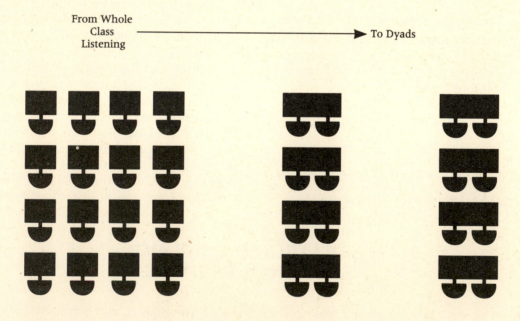

From Whole Class Listening ⟶ To Dyads

FIGURE 8.9 Student learning can be enhanced by encouraging active listening and interaction. This can be accomplished by making a shift from whole class to small group learning.

The most recent edition of *Modern Biology*, the most widely used science textbook in the United States, has over 800 pages of text, and lists over 1100 entries in the glossary.[17] Quite a formidable task for a 14 year old student to undertake! Regardless of the students' intellectual development, student success in science courses could be enhanced if they were taught in a manner that we might call learning how to learn. Recent research in metacognition supports efforts in which students are taught strategies of learning that will enable them to learn science but also understand their own thinking processes. With this in mind, we will explore several thinking strategies that will enhance students' reading abilities.

[17]Albert Towle, *Modern Biology* (Austin, Tex: Holt, Rinehart and Winston, Inc., 1990)

Listening

Teachers' talk about two-thirds of the time in an average classroom; therefore, listening is an important survival skill that students need to develop. A good listener is an active listener, who concentrates and

---

Listening Skill Activity 1

In this activity students practice being an active listener by concentrating and participating. Tell the students that you are going to read a passage to them, and that they should participate in the following ways:

1. When you listen, ask yourself questions about what you are hearing; answer the questions if you can.
2. Try to connect what you hear with what you already know.
3. Try to "picture" in your mind what is being said, and draw a picture if it is appropriate.
4. When the passage has been read, write the main idea in your notebook.

Passage to Be Read: In this science class you will be doing a number of activities in the laboratory and it is important to use the right tools and instruments. You wouldn't use a thermometer to measure the height of a building. You wouldn't use a microscope to observe the moon. You wouldn't collect rocks in a glass beaker. You would choose the equipment that best fits the task you are doing. In this course, you will be measuring and observing objects, living things, and phenomena, and you will learn what tools and instruments you should use.

Listening Skill Activity 2

This activity is a modification of the "Think-Pair-Share" strategy in which students are presented with some information, a problem, or a question, and are asked to think about their response or comprehension, and then share their idea with a partner. In the science classroom this technique is very powerful, especially if it is repeated over the course of the year. Present information to students in the form of a minilecture. At a convenient point in the presentation, stop, and either ask a question or ask the students to identify the main ideas or concepts presented. Give the students a minute to think about their response, then have them share their thinking with their partner. The process enables all students in the class to become active listeners and participants in small group interaction.

participates in the process of communication. Listening is one-half of the act of communication; however in the classroom, students do not always have the opportunity to respond and interact directly with the speaker. The listener must develop skills to "communicate" alone, and the teacher must implement strategies that provide students (in pairs or small groups) the opportunity to interact (Figure 8.9).

Coming to Terms: The Vocabulary Problem

Thelen suggests that science teachers not preteach new vocabulary words.[18] If indeed the words are new, students should be engaged in learning activities to construct knowledge about these new words, most of which are science concepts. Instead, she suggests that science teachers reduce the number of vocabulary words that students need to learn, and use vocabulary reinforcement activities to help enhance vocabulary development.

One technique that science teachers can use to reduce vocabulary is to use a structured overview—a sort of visual overview showing the relationships between science concepts. In this approach the teacher would identify the vocabulary of the learning activity, chapter, or unit and arrange the words in a scheme that depicts the relationships between the concepts. The structured overview is then presented to the students and used to find out what they already know about the concepts and terms. The structured overview or organizer can be used as students explore the topic.

Most science textbooks identify by means of bold print, underline, or italics the key vocabulary words in a chapter. Students need to learn how to interpret

meanings by focusing on context clues. Context clues are the sentences and phrases in the text surrounding the vocabulary word. Reading guides might suggest an activity after the student has read a passage in a text, or might ask the student to read a passage and then write the meaning of several vocabulary words.

For example, you might have students read a passage such as the one that follows and then write the meaning of the words shown beneath the passage. The meaning of the words will be based on the students' prior knowledge as well as on the contextual clues in the passage.

All organisms are composed of and develop from cells. Some organisms are composed of only one cell. These organisms are called *unicellular* organisms. Most of the living things you see around you are composed of more than one cell. Such organisms are called multicellular organisms. *Multicellular* organisms usually arise from a zygote. The process by which a zygote becomes a mature individual is called *development*.[19]

unicellular_____

multicellular_____

development_____

Figure 8.11 shows how a reading guide can be useful in helping students make use of some of the context aids in the text to enhance vocabulary and concept development. In this case the teacher is selecting and identifying what is important and telling the students to focus on these aspects of the chapter in order to develop conceptually.

---

[18]Judith N. Thelen, *Improving Reading in Science* (Newark, Del.: International Reading Association, 1984).

[19]Albert Towle, *Modern Biology* (Austin, Tex.: Holt, Rinehart and Winston, 1989), p. 7.

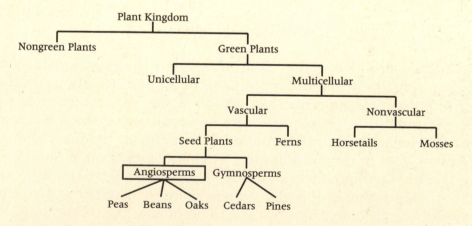

FIGURE 8.10   Structured Overview of the Concept *Angiosperms*. Judith N. Thelen, *Improving Reading in Science* (Newark: Del.: International Reading Association, 1984), p. 16. Reprinted with permission of the International Reading Association.

Paragraph 10 (Yes, that's right, paragraph 10). Read paragraph 10 and keep it in mind as you follow the exercises on this sheet. Now proceed to:

Paragraph 1.   Draw a picture of a light spectrum.

        a. Does it look like a rainbow?

        b. Why do they call it a rainbow pattern?

        c. If you take the orange out what color will replace it—black or white?

Paragraph 3.   a. Read sentence 1.

        b. Look at Fig. 5.8 and read sentence 1.

        c. 1. Does it look like a rainbow?

           2. List 3 things that are the same or different between Fig. 5.8 and a rainbow.

              a.

              b.

              c.

FIGURE 8.11   Science Text Reading Guide. Judith N. Thelan, Improving Reading in Science (Newark, Del.: International Reading Association, 1984), p.33.

Another technique that Thelen recommends is vocabulary reinforcement activities in the context of cooperative small groups.[20] The process apparently enhances vocabulary development, and comprehension of science concepts. The specific method that she suggests also employs recent research on metacognition. Students are informed of the importance and value of thinking and discuss how science words and thinking are related. Students are given information about analytical and intuitive thinking (see box on page 262) and then are told that the vocabulary activities in the science class are designed to help them not only build their science vocabulary but improve their thinking abilities as well.

A number of types of vocabulary activities can be used to reinforce and extend students' vocabulary development. These include categorizing, word puzzles, matching, magic squares, crossword puzzles, and various writing forms (look ahead to the section on science writing). Here is an example:

Reading for Meaning and Understanding

How can textbooks be used so that students will read them for meaning and understanding? A partial answer to this question lies in how we think students learn. Let's accept the cognitive psychologist's assertion that human learning is dependent on the individual's active interaction with the environment, and that new learning must be tied to what the student already knows. If the textbook is to be one of the "environments" with which the learner will interact, we must find a way that involves the student actively with the textbook. Simply saying "Read pages 20 to 30 for homework" will not work. Critics of textbooks charge that "the textbook is weak in that it offers little opportunity for any mental activity except remembering. If there is an inference to be drawn, the author draws it, and if there is a significant relationship to be noted, the author points it out. There are no loose ends or incomplete analyses."[21]

Although textbooks have become more visual (Figure 8.13), there is still the problem that "few users of textbooks, instructors and students alike, are

[20]Thelen, *Improving Reading in Science*, p. 36.

[21]Norris M. Sanders, *Classroom Questions* (New York: Harper & Row, 1966), p. 158.

Word Categories

There are five words in each section below. Cross out the two words in each that you feel are not related to the others. Explain the relationship by titling each group.

1._____   2._____

    amino acids   waste
    energy   storage
    water   food
    enzyme   pocket
    protein   contractile

FIGURE 8.12   Word categories. Judith N. Thelen, *Improving Reading in Science* (Newark, Del.: International Reading Association, 1984), p. 37. Used with permission of the International Reading Association.

Today we would like you to read about something that is very important. It is at the very heart of everything you will ever try to learn. We are talking about *thinking*.

Ever since you first came to school, your teachers have tried to encourage you to think. Too often, however, teachers simply *tell* you to think. During the next few weeks we are going to try to *teach* you some ways of thinking.

First, it is important that you know something about thinking. What is it? How do people think? As you read on, we will discuss two kinds of thinking. Then we will try to show how words and thinking go together. Finally, we will relate what has been said to some activities we shall undertake for the remainder of the school year.

The people who study thinking tell us that it can be broken down into two broad types: analytic thinking and intuitive thinking.

Analytic thinking is a very careful kind of thought. It usually proceeds one step at a time. These steps usually are very clear and each step usually can be reported by the thinker.

Remember when your teachers tried to teach you how to solve word problems in arithmetic. They gave you a series of steps like these:

1. Find out what the problem is asking.
2. Determine what information you have been given.
3. Decide how you should solve the problem—will you add, subtract, multiply, divide, or use a combination of operations?
4. Solve the problem.
5. Check your answer.

When you figure out a problem by following steps like these, you are thinking analytically.

Intuitive thinking, on the other hand, does not proceed in careful, well-defined steps. Thinkers arrive at answers with little, if any, awareness of the processes by which they read them. They can rarely provide an explanation of how they get their answers.

Let us use mathematics again for an example. Did you ever look at a problem and, all of a sudden, seem to know the correct answer? Then, as you tried to tell someone how you got the answer, you found that you could not do so? If this has happened to you, then you have experienced intuitive thinking.

Both types of thinking are important. Intuitive thinking has led to some of the world's greatest discoveries. However, intuitive thinking becomes nothing more than guessing unless one is able to go back and verify what has been found. In other words, one should always attempt to confirm intuitive thinking by a more careful, analytic method.

Now let us consider some relationships between *words* and *thinking*.

What is a word? Whole books have been written about this question. However, for our purpose, let's define a word as a spoken, or written symbol that "stands for a thing, experience, or idea." When we say "apple," all we have done is made a series of noises or sounds (ap'l). Due to the fact that we have experience with apple, we *think* of something round, red, and edible. We all may not think the exact same thought. However, our thoughts usually are along similar lines if we have had similar experiences with the idea represented by the word.

At your present age and grade level, almost all the thinking you do is performed with words. *You think with words*. Would it be possible to think about and learn about any of school subjects without knowing the important words of those subjects? Possibly, but you certainly would have a very difficult time.

It is not enough to know or agree upon the meanings of words. We must also know how words (or rather, the ideas represented by words) are related. For example, how are the following words related: animal, vegetable, matter, mineral? Does one word seem to be more important than the others?

During the next few weeks we will provide you with different kinds of vocabulary activities which will

1. help you learn the meanings of important terms,
2. help you see relationships among these terms and discover which words are most important, and
3. provide opportunities for you to practice intuitive and analytic thinking.

We believe that these activities will cut down on some of your study time as well as help you discover more effective ways to go about learning in other subjects.

Student handout explaining the importance of analytic and intuitive thinking. Judith N. Thelan, *Improving Reading in Science* (Newark, Del.: International Reading Association, 1989), pp. 53–54. Used with permission of the International Reading Association.

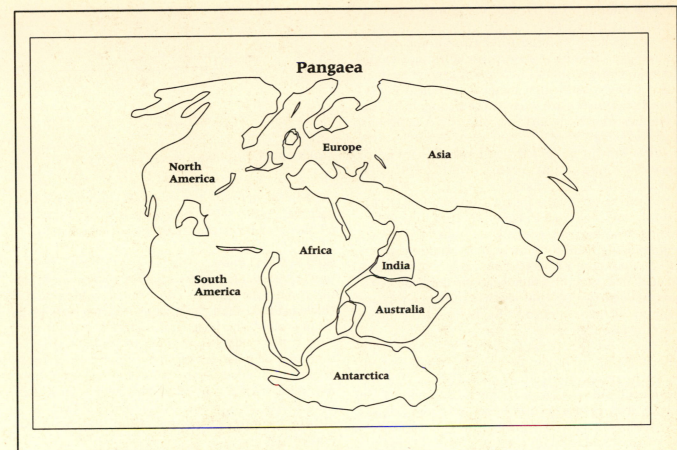

Pangaea

North America

Europe

Asia

Africa

India

South America

Australia

Antarctica

Fossils of the animal shown to the right were discovered in both South America and Africa. These findings convinced Wegener that the continents were once joined. Second, explorers found rocks made of glacial sediments at the equator where no glaciers could exist. How did Wegener explain this finding? He believed that the land mass drifted to a warmer region of the earth.

Wegener's evidence was interesting, but it did not prove that continents moved. Scientists rejected the continental drift theory because Wegener could not explain how or why continents moved. His imaginative theory is not entirely correct, but it set the stage for other bold ideas.

Review it

1. List the continents that were part of Pangaea.
2. How would Wegener account for fossil ferns found in the rocks of Antarctica?

FIGURE 8.13 Illustrations in textbooks can enhance student comprehension if students are asked to analyze and make inferences about the illustrations. Jay M. Pasachoff, Naomi Pasachoff, and Timothy M. Cooney, *Earth Science* (Glenview, IL.: Scott Foresman and Company, 1983), p. 191. Used with permission of Scott Foresman.

literate enough to derive full value from the text-book's illustrations."[22]

For students to read for meaning and understanding, a teaching strategy must be employed that actively engages them in comparing what they already know with the new material, as well as involving them in processes such as predicting, inferring, hypothesizing, summarizing, drawing conclusions, and discussing.

The strategy presented here has been used in science classrooms and has been shown to be effective. It is called the K-W-L procedure.

According to Donna Ogle, the originator of K-W-L, prior knowledge is an integral aspect of how we interpret what we read, and of what students will learn from reading.[23] Unfortunately, most science teachers fail to make use of what their students bring to a topic. The K-W-L procedure supports the main assertion of cognitive psychology that students' preconceptions of science need to be determined prior to learning new concepts.

The procedure consists of three cognitive steps: assessing what we know, determining what we want to learn, and recalling what we did learn. Ogle has developed a K-W-L strategy sheet that students can use as they read a section of the science textbook (Figure 8.14).

---

[22]Robert V. Blystone and Beverly C. Dettling, "Visual Literacy in Science Textbooks," in *What Research Says to the Science Teacher: The Process of Knowing*, vol. 6, Mary Budd Rowe, ed. (Washington, D.C.: National Science Teachers Association, 1990), p. 19.

[23]Donna M. Ogle, "K-W-L: A Teaching Model that Develops Active Reading of Expository Text," *The Reading Teacher* (February, 1986): 564–570. Reprinted with permission of Donna M. Ogle and the International Reading Association.

1. K–What we know	W–What we want to find out	L–What we learned and still need to learn

2. Categories of information we expect to use

A.   E.

B.   F.

C.   G.

D.

FIGURE 8.14 K-W-L Strategy Sheet. Donna M. Ogle, "K-W-L: A Teaching Model that Develops Active Reading of Expository Text," *The Reading Teacher* (February 1986): 564–570. Used with permission.

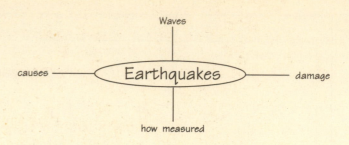

FIGURE 8.15 Map of Main Ideas on Earthquakes

Briefly, here are the essential characteristics of each step in the K-W-L procedure and an example of a lesson plan on earthquakes (see box on pg. 265).

Step K: What I know. This is a brainstorming session in which students express what they know about the topic. What the students know can be written on the chalkboard, on chart paper, or by students working in small groups. The focus at this stage should be specific. If the students are going to read a section in their text on earthquakes, ask "What do you know about earthquakes?" not "What do you know about natural disasters?" or "Have you ever been to San Francisco?" Focusing on the content will help bring out the cognitive structures of the students' prior knowledge.

A second part of the K-step is to have the students categorize the information they have generated during the brainstorming session. For example, in the lesson plan on earthquakes, the teacher might suggest that students group their information in the following categories: causes of earthquakes, how earthquakes are measured, and damage caused by earthquakes.

Step W: What do I want to learn? This step helps the students anticipate the reading that is to come, and helps the students focus on what they want to learn from the reading. This step should be done as a group activity. The teacher should ask the students to write down on the K-W-L worksheet questions that they are most interested in having answered as a result of the prior discussion and brainstorming session. Once the questions are written, the teacher might have the students share their questions in small groups prior to actually reading.

Step L: What I learned. Students can write down what they learned on the K-W-L strategy sheet. They can also check to see if their questions were answered, and if some of their prior knowledge was confirmed. Students should work in small groups and discuss their questions to determine if their questions were answered.

K-W-L Lesson Plan on Earthquakes

Objectives

1. To describe how earthquakes are caused
2. To predict the effects of an earthquake

Reading: Students will read Chapter 19, "Earthquakes," in *Focus on Earth Science* (Merrill Publishing Company, 1989), pp. 411–428.

K: What do students already know about earthquakes?

1. Pairs of students brainstorm and record what they know about earthquakes.
2. Pairs of pairs (groups of four) share lists and prepare a composite list.
3. Groups of four share with the whole class by taping lists on the walls of the classroom. Teacher focuses on the lists, and asks groups of four to categorize the information: topics include—at least—causes of earthquakes, damage caused by earthquakes, earthquake waves, and how earthquakes are measured.
4. Teacher asks students how they got their information about earthquakes. This last procedure personalizes student knowledge and acknowledges the sources of students' prior knowledge.

W: What do students want to know about the topic?

1. Each group of four develops four questions about the topic.
2. Teacher records these questions on the board; teacher can elaborate on the questions, perhaps selecting two or more that seem interesting to the students.

L: What did students learn about the topic?

1. Group circles information on the master list that the text confirmed.
2. Information is crossed off that text refuted.
3. Students contribute to new list: What we learned!
4. Teacher goes around the class and asks each student to indicate (by thumbs-up or thumbs-down) if their question(s) were answered.
5. Teacher asks each group to make a map (Figure 8.15 on pg. 264) of the main ideas and supporting secondary categories of what they learned from the reading.

Semantic Mapping

Although mapping was used in the preceding reading process procedures, the concept of mapping deserves additional discussion. Semantic mapping is the structuring of information in graphic form. It is not a new process, and has been known as concept mapping, webbing, networking, and plot maps.[24] Semantic mapping is a tool that teachers can use to help students connect prior knowledge with new science concepts to be learned in terms of a schema or a holistic conceptual system. In many cases new science words are introduced and defined in isolation from a more general system of ideas.

As Heimlich and Pittelman point out, semantic maps are diagrams that help students see how words or concepts are related to one another. In most cases

semantic mapping begins with a brainstorming session in which students are encouraged to make associations to the main topic or concept presented. Students are actively engaged in using their prior knowledge, as well as new science concepts and experiences that the teacher has provided to develop a semantic map. Semantic maps can be developed individually or in small cooperative groups or with the whole class.

Semantic maps (Figures 8.16 and 8.17) can be used in many different contexts in the science classroom, including the following:

1. As a science vocabulary (concept) building strategy
2. As a pre- and postreading strategy
3. As a science study skill strategy

Heimlich and Pittelman have developed a basic strategy for developing semantic maps; this strategy can be used for any of the three purposes just listed. The following is an adaptation of their strategy:

[24]Joan E. Heimlich and Susan D. Pittelman, *Semantic Mapping: Classroom Applications* (Newark, Del.: International Reading Association, 1986), pp. 1–3.

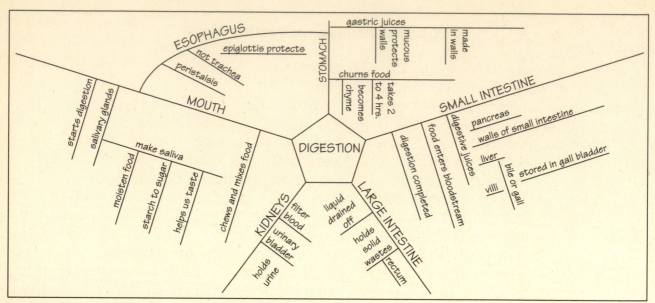

FIGURE 8.16 Semantic Map for Digestion. Joan E. Heimlick and Susan D. Pittelman, *Semantic Mapping: Classroom Applications* (Newark, Del.: International Reading Association, 1986), p. 21. Reprinted with permission of Joan E. Heimlick and the International Reading Association.

1. Choose a science concept related to the chapter or unit students are studying.
2. List the concept on large chart paper or on the chalkboard.
3. Encourage the students to think of as many words or concepts as they can that are related to the main concept (see the example on sharks in Figure 8.17) and to list them in categories on the paper. This step of the process can be done individually, in a small team, or with the whole class. Encouraging interaction among the students is an important part of the process; therefore, most teachers choose to do this as a small group or whole class activity.
4. Students can share their results with each other.
5. Students should discuss by comparing and contrasting their semantic maps to develop deeper understanding of science concepts.

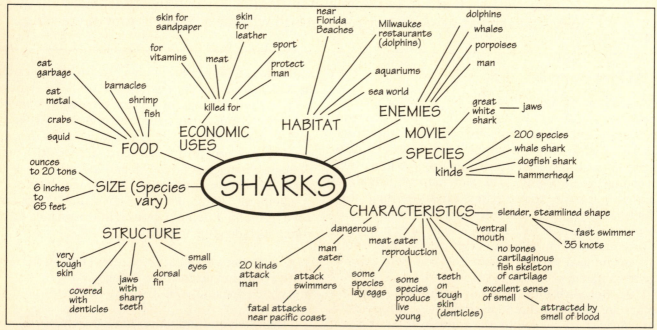

FIGURE 8.17 Completed Concept Map for Sharks. Joan E. Heimlick and Susan D. Pittelman, *Semantic Mapping: Classroom Applications* (Newark, Del.: International Reading Association, 1986), p. 33. Reprinted with permission of Joan E. Heimlick and the International Reading Association.

Title: Crusty (Rock) Writing

Objectives

1. Collect data on rocks using observational skills.
2. Record notes about a natural object, a rock.
3. Reconstruct notes in poetry form.

Description: As part of a science unit on local geology, students select a rock that they observe carefully, using all five senses. They then write words and phrases based on their observations. After reading and thinking about their notes, they write an ode to their individual rock, beginning, "O rock..."

Procedure

Have students gather rocks as part of field trip, or bring in enough rocks from the local scene so that each student will have one to observe.

1. Stimulus: All students have a rock on their desk. Discuss what the students can observe about a rock based on each of the senses. Have a student recorder write key words on the board or on chart paper, such as

Sight—size, shape, color

Hearing—rattle, scraping

Taste—mineral content, dirt

Touch—shape, roughness, smoothness, unevenness, bumps

Smell—sweet fragrance, earthiness

Have students fold a sheet of composition paper in thirds, labeling one section for each sense, and the sixth one entitled "Other Ideas." Ask students to observe their rocks and to jot down notes about what they observe.

2. Activity: After they have had time to observe and write notes, tell them that they can use their observations to write an ode to their rocks. Tell them that an ode is a song that begins "O..." and usually praises a person. They can begin their poem with "O rock..." and speak to their rock as a person, using personification.

3. Follow-up: After students have written for a while, have them read their poems to a partner. Partners can assist each other in adding ideas or revising the poem, as needed. Their poems might be something like this:

Ode to a Rock

O wonderful little gray rock,

Bumpety, lumpety, and tough.

You have tumbled down from the high mountain,

You have survived the trampling of many rough feet,

The crush of an automobile's wheels.

I will give you an easier life now

Perched on my bedroom windowsill.[25]

4. Evaluation: Circulate around the room to observe student participation as they observe, write, and share. Have students determine criteria (scientific and poetic) for an especially good poem after they have shared, answering the question, What made some poems stand out as especially effective? Students can revise their poems based on the established criteria. Have students display the rock writing with rocks laid on a table or shelf.

[25]Iris McClellan Tiedt et al., *Reading/Thinking/Writing: A Holistic Language and Literacy Program for the K–8 Classroom* (Needham Heights, Mass.: Allyn & Bacon, 1989), p. 22. Reprinted with permission of Allyn & Bacon.

Strategies for the Science Classroom

Writing activities in the science classroom have great potential. Consider the science writing lesson used in a middle school earth science course.[26] (see box on pg. 268.)

This lesson, which enables students to make connections between science process skills and writing, is not the typical way in which students write in science classrooms. Unfortunately, creative writing, which can help students become successful science learners, is low on the priority list of science writing assignments. Bulman reports that students spend between 11 percent and 20 percent of their time in science classrooms involved in writing activities. However, Bulman also reports that over half of the writing time is devoted to copying or taking dictated notes (Table 8.4).

What is the purpose of writing in the science curriculum? To improve the writing abilities of students in all subjects, many school districts have implemented a concept known as writing across the curriculum in which subject matter teachers are given training and teaching materials to integrate writing into their subject. Any approach to integrate writing into the science classroom must take into consideration the goals of student writing in science. Bulman has suggested four goals:

1. To help the growth of understanding of science concepts
2. To provide a record of concepts and activities that can be used for revision later
3. To provide feedback to the teacher on the growth of the students
4. To develop students' ability to communicate[27]

How can these goals be achieved? What are some strategies that might help achieve these goals? In general, writing should be viewed as an integral part of the learning cycle that was presented in Chapter 7. Students need the opportunity to reflect on their thinking, and writing about their observations, inferences, hypotheses, and conclusions is a powerful strategy. The following strategies are designed to help

TABLE 8.4

TYPES OF WRITING IN HIGH SCHOOL SCIENCE CLASSES

Types of Writing	% Time Spent	
	1st Year in High School	4th Year
Copying: copying or dictated note-taking	46	56
Reference: making notes from printed material	0	19
Personal: essay writing, writing in own way, diary, some project work, reports of experiments	29	8
Answering: answering worksheets, exercises, test and exam questions	25	17

Lesley Bulman, *Teaching Language and Study Skills in Secondary Science* (London: Heinemann, 1985), p. 54.

students clarify and extend their thinking by having them write in a variety of forms:

Science logs
Letter writing
Science newspapers
Story writing

Science Logs: Visual and Verbal Journals of Students' Ideas

Writing is a way of expressing ideas and concepts that are within us. These concepts and ideas are based on prior experiences, imagination, and the willingness to let these ideas emerge. We have already shown that reading and writing are integrated processes. In real-life experiences of reading and writing, this is known to be true. For example, if you ask a writer what is essential to writing, the answer invariably is read, read, read. Writing and reading are inseparable processes, and in the science classroom the teacher can help students by providing integrated language experiences.

The science log is one such experience. It can become a place—a creative space—in which students organize their ideas from their reading of the science text, from science experiments and activities, and from day-to-day activities that are part of your teaching process. Here are some specific suggestions for using the science log.

[26]Iris McClellan Tiedt et al., *Reading/Thinking/Writing: A Holistic Language and Literacy Program for the K–8 Classroom* (Needham Heights, Mass.: Allyn & Bacon, 1989), pp. 21–22.
[27]Lesley Bulman, *Teaching Language and Study Skills in Secondary Science* (London: Heinemann, 1985), p. 54.

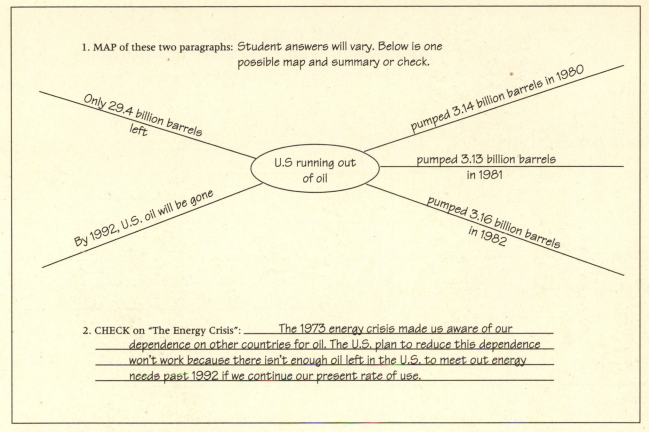

1. MAP of these two paragraphs: Student answers will vary. Below is one
possible map and summary or check.

Only 29.4 billion barrels left

pumped 3.14 billion barrels in 1980

U.S running out of oil

pumped 3.13 billion barrels in 1981

By 1992, U.S. oil will be gone

pumped 3.16 billion barrels in 1982

2. CHECK on "The Energy Crisis": _____ The 1973 energy crisis made us aware of our dependence on other countries for oil. The U.S. plan to reduce this dependence won't work because there isn't enough oil left in the U.S. to meet out energy needs past 1992 if we continue our present rate of use.

FIGURE 8.18   Note-taking for the Science Log

1. *As a note-taking device.* Students should follow the K-W-L procedures and use the log as the place to write their results. Figure 8.19 shows one student's entry in a science log after reading about the "energy crisis." Note that students are encouraged to map and to write brief statements that imply their understanding of what they read.

2. *As a record of experiments and activities.* Visual and verbal thinking should be encouraged as students record the results of experiments and hands-on activities. Figure 8.20 shows a form that is used to encourage students to record the results of a science inquiry activity. Note that the students are involved in making predictions, collecting data, drawing a diagram or picture of the experience, and writing an explanation for the event or phenomenon they observed.

3. *In preparing a daily log.* Some teachers have students use their logs as a postlesson review of the lesson. However, students are asked to make a map of the lesson by writing the main idea in the center of a page in the log and then identifying the supporting ideas and concepts by connecting them to the main idea.

4. *As a learning tool.* Iris McClellan Tiedt suggests that the science log be conceptualized as a learning log.[29] As a learning log, students express their findings or

their questions, and it can be used to clarify concepts, ask questions of the teacher, or set goals for learning. Logs can be kept in spiral-bound notebooks, loose-leaf notebooks, in a folder, or on the computer. As was outlined previously, logs should help students summarize and clarify what they learned in a class session, a laboratory exercise, a class discussion, or a reading. Teachers should read logs to evaluate the effectiveness of lessons, as well as to identify students who need specific help. Refer to the sample lesson on pg. 271 to give you an idea of how to make the science log a learning tool.

In the foregoing lesson the log is used in an active way. Students not only write in their own logs but exchange their logs with other students in the class and have an opportunity to write in their classmates' logs. Logs are also viewed as feedback mechanisms for the teacher. Logs can be an integral part of the science curriculum and can foster critical and creative thinking.

Now let's examine some other types of writing formats appropriate for the science classroom.

Letter Writing

There are at least two aspects of the letter writing format that have been shown to be effective: writing to

[29]Tiedt et al., *Reading/Thinking/Writing*, pp. 248–251.

EEEP Data Sheet
Exciting Examples of Everyday Phenomena

Your prediction _____
_____
_____
_____

Data _____
_____
_____
_____

Drawing

Explanation _____
_____
_____
_____

FIGURE 8.19 EEEP Data Sheet

**Lesson:
Using the Learning Log with a Hands-On Activity**

Objectives
1. To write in order to clarify what they observed in a hands-on activity
2. To ask questions immediately after the hands-on activity
3. To write in science class

Description: This lesson shows you how the science log can and should be used immediately following a hands-on activity. It provides an opportunity for you to have the students express what they have learned from the activity.

Procedure

1. Stimulus: Have students observe a discrepant event or inquiry demonstration. Students should be involved in the activity, not simply passive observers of the event.

2. Activity: Following the discrepant event or activity, the students record their findings in their science logs. Encourage the students to connect new learning with old, express discoveries, ask questions of the teacher, and express any frustration related to the activity.

3. Follow-up: Have students exchange logs with their partner, or read the logs yourself.

4. Evaluation: Have the partner respond to the log in writing to the individual. A brief comment is all that is needed. Peers can answer questions, and comment on the log.

FIGURE 8.20   Using the learning log with a hands-on activity

others for information, and writing "letters to the editor." Letters can be the beginnings of potential research, especially of the survey type. For instance, students might want to write to various organizations related to a topic they are studying, for example, the Sierra Club, the Audubon Society, the National Aeronautics and Space Administration, the National Transportation Board, the Environmental Protection Agency, and the United Nations. Each of these organizations is a source of good information. Letters to the editor are especially significant if students are working on an STS project or are debating an issue that has personal and societal implications.

Students can be given a format to follow to write letters of inquiry and letters to the editor and can use the computer to prepare the documents.

Science Newspapers

Students in your class might be interested in applying desktop publishing to produce a science newspaper that could be distributed to other students in the school, as well as to citizens in the community. There are a number of software packages available that will allow students to design newspapers of a quality that compares with commercially produced newspapers.

The newspaper is especially useful in that it provides a creative mode for students to write about famous scientists, discoveries made by scientists, and ecological and environmental concerns, as well as science and society issues.

Story Writing: Science Fiction

Imagination is as important to science as it is to art forms such as pottery, painting, and writing novels. Reading as well as writing science fiction can be a powerful medium for many students in your science class. Books by Robert Heinlein, Carl Sagan, Isaac Asimov, and Ray Bradbury can provide the stimulus needed for students to write their own science fiction. I recommend that you read passages from your favorite authors and use it as a vehicle for brainstorming about imagination and science.

An excellent resource for the science classroom in this regard is *Fantastic Reading* by Isaac Asimov, Martin Greenberg, and David Clark Yeager (published by Scott, Foresman). It is an activity book providing high-interest science fiction and fantasy stories along with vocabulary, writing, reading comprehension, and study skill activities.

Writing science fiction brings students into the world of invention and creativity. Students can explore the limits of their own creativity by writing imaginative stories about the future, or the past by attempting to integrate science concepts into the story line. What kind of thinking might be activated if students were to write stories that involved the following?

Black holes
The big bang

Antimatter
Speeds approaching that of light
Life on Io, and other moons of Jupiter

Two other resources that are helpful in the science fiction and creative writing area are

Alan McCormack, *Inventors Workshop* (David S. Lake, 1981).

Joe Abruscato and Jack Hassard, *The Whole Cosmos Catalog of Science Activities* (Scott, Foresman, 1991).

STRATEGIES THAT FOSTER INDEPENDENT THINKING

School science needs to help students become independent thinkers. In *Project 2061: Science for All Americans*, the authors pointed out that science teaching related to scientific literacy needs to be consistent with the spirit and character of scientific inquiry and with scientific values. Students must be engaged actively "in the use of hypotheses, the collection and use of evidence, and the design of investigations and processes, and placing a premium on the students' curiosity and creativity."[29] The authors of the report refer to this kind of thinking as the "scientific habit of mind" and view its implementation as an important goal. They put it this way:

> Scientific habits of mind can help people in every walk of life to deal sensibly with problems that often involve evidence, quantitative considerations, logical arguments, and uncertainty; without the ability to think critically and independently, citizens are easy prey to dogmatists, flimflam artists, and purveyors of simple solutions to complex problems.[30]

The habit of mind that is being suggested here is problem solving, and as Stanley L. Helgeson points out, "problem solving has been a concern of science education for at least three quarters of a century."[31] Helgeson goes on to point out that whether we use terms such as *scientific method, scientific thinking, critical*

thinking, inquiry skills, or *science processes,* these are in essence expressions of a more general concept: problem solving.

Thus, in this section we will explore problem solving and pay particular attention to strategies that secondary science teachers can use to emphasize problem solving in their curriculum. We will first begin with the notion of problem solving and then show the value of emphasizing the "processes of science" in secondary science lessons. Finally, we will conclude this section by relating problem solving and science processes to school science fair projects and research investigations.

Problem Solving

One of the problems with problem solving is defining it. There are a variety of definitions of problem solving. Some define problem solving in terms of the skills needed to solve problems, for example, testing hypotheses, and analyzing data. Others define problem solving as a series of steps that people use to find a solution or answer to a question.

Helgeson reports that in the literature of science education there is a strong linkage between problem solving and science process skills. That is, many science teachers teach students science process skills in the context of a subject—earth science, biology, chemistry, physics—because they accept the notion that these process skills are indeed the elements of problem solving. There are two aspects of this that should be pointed out. First, what the generally accepted steps in problem solving are, and second, what the science processes associated with problem solving are.

[29]American Association of the Advancement of Science, *Project 2061: Science for All Americans* (Washington, D.C.: American Association for the Advancement of Science, 1989), p.5.
[30]*Project 2061: Science for All Americans,* p. 13.
[31]Stanley L. Helgeson, "Problem Solving in Middle Level Science," in *What Research Says to the Science Teacher: Problem Solving,* Dorothy Gabel, ed. (Washington, D.C.: National Science Teachers Association, 1989), p. 13.

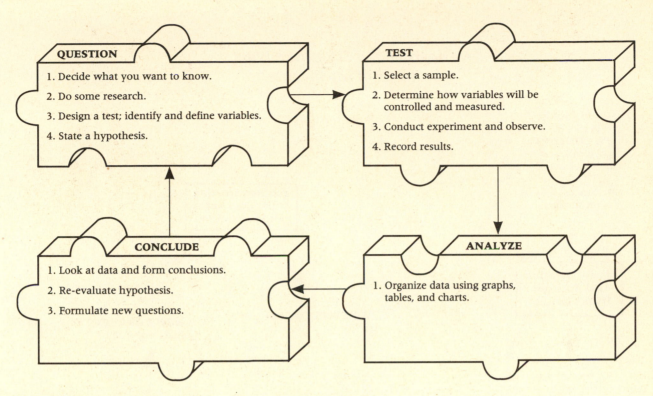

FIGURE 8.21   Problem solving steps described in a modern science textbook. Dale T. Hesser and Susan S. Leach, *Focus on Earth Science* (Columbus, Ohio: Merrill, 1989), p. 10.

A Problem Solving Schema

No doubt you have come across the steps usually associated with problem solving. Are these familiar?[32]

- Problem orientation
- Problem identification
- Problem solution
- Data analysis
- Problem verification

A more elaborate form of the problem-solving strategy is one described in a middle school or junior high science text (Figure 8.21). Four general steps are presented: question, test, conclude, and analyze. Students are shown that there are four steps involved in the scientific method to generate information and questions.

Another way to look at problem solving is from the vantage point of the scientist. Paul Brandwein depicts the scientist's approach to problem solving as the "scientist's methods of intelligence." Notice that the scientist's way begins with a discrepancy—a situation that does not fit the scientist's present "con-

cept"(Figure 8.22). Although this theoretical model is often not applied directly to the classroom, there are a number of elements that are applicable.

Note that the model involves a number of processes: observing, hypothesizing, and designing an investigation. These are examples of scientific thinking skills that have become one of the organizing structures for teaching problem solving. Let's take a closer look at these scientific thinking skills.

Scientific Thinking Skills and Problem Solving

The curriculum projects of the 1960s and 1970s placed emphasis on problem solving and developed as part of the organization of the curriculum a series of problem-solving skills that became known as the processes of science. They are sometimes called the skills of science. Today, these skills are referred to as scientific thinking skills. It is important to note that problem solving as perceived in science classrooms is intimately related to these thinking skills.

The thinking skills of science are conceptualized as belonging to two distinct groups, basic thinking skills and integrated thinking skills (Table 8.5). As you examine different science curriculum project ma-

[32]Helgeson, "Problem Solving in Middle Level Science," p. 15.

A Diagram of a Scientist's Way: methods of intelligence

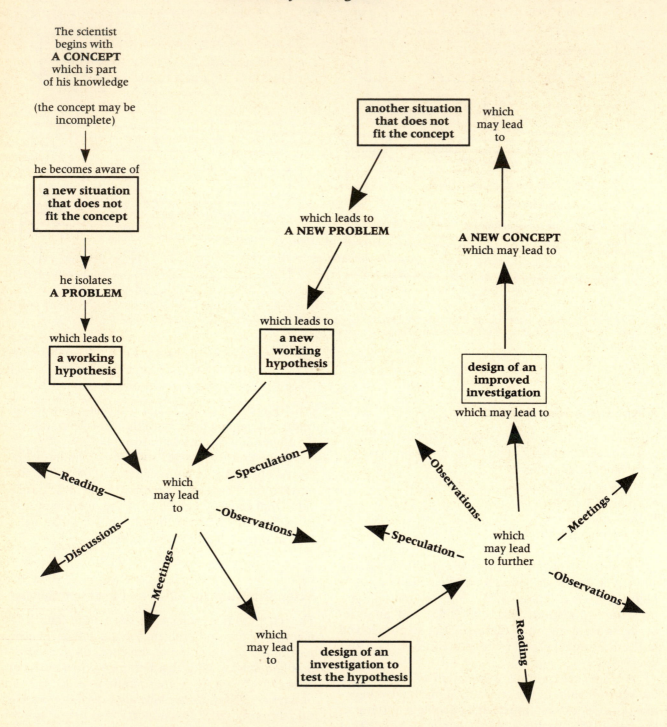

FIGURE 8.22   Steps in Problem Solving: A Scientist's View. Paul F. Brandwein, *Substance, Structure, and Style in the Teaching of Science*, new ed. (Orlando, Fl.: Harcourt, Brace, Jovanovich, 1971), pp. 19–20. Reprinted with the permission of Harcourt, Brace, Jovanovich.

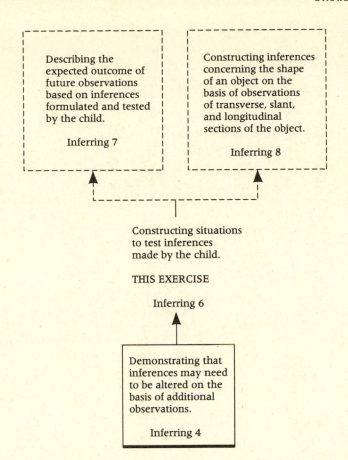

FIGURE 8.23   In the science-process approach, scientific thinking skills were broken down and presented in a series of hierarchically arranged lessons. American Association for the Advancement of Science, Science—A Process Approach (Washington, D.C.: American Association for the Advancement of Science, 1968), p. 6. Permission granted to reprint this material after. This work not endorsed by the American Association for the Advancement of Science.

terials, you will find some variation in the "lists" of skills, but in general they include the following:

- Basic thinking skills such as observing emphasize the foundations of science learning. The basic thinking skills are seen as prerequisites for the integrated thinking skills.
- Integrated thinking skills are related more directly to problem solving and are seen as the higher-order intellectual skills that problem solvers use.

Scientific Thinking Skills

Focusing on science thinking skills in the science curriculum is clearly one way of helping students become independent thinkers and problem solvers. How should secondary science teachers plan science courses to achieve this goal? Three approaches seem apparent. The first is a behavioral approach. In this approach the thinking skills of science are taught as separate skills or behaviors, for example, observing,

classifying, inferring, and hypothesizing. A second approach—a cognitive or constructivist approach—and one that is favored by most science educators, is to integrate the thinking skills of science with the content or concepts of science. In this plan, scientific thinking skills would be taught in the context of learning biology, chemistry, earth science, and physics concepts. A third approach is called the research or science project approach. In this plan, students would engage in a science research project using the thinking skills of science to inquire about natural phenomena.

The Behavioral Approach: Teaching the Thinking Skills of Science

Teaching the thinking skills of science as a set of skills emerged as a consequence of Science—A Process Approach (SAPA), a curriculum project developed during the 1960s for elementary and middle schools.[33] In this approach the complex set of skills a scientist uses was broken down into a number of skills that were to be mastered by the learner to develop a sound knowledge of science and its methods (see Figure 8.23). In this approach, the major goal of the curriculum is to teach the processes or thinking skills of science. Scientific concepts and facts were introduced in the overall framework of the curriculum but played a secondary role in the program. The SAPA program had a profound effect on other curriculum approaches, and a good deal of its impact remains today in science programs at the elementary as well as secondary level.

Most secondary science programs are not organized around a thinking skills approach, that is, texts and courses are organized around units or chapters of content. However, on closer inspection one discovers that middle school or junior and high school science programs include a series of "skill" activities spread throughout the textbook. In this approach, a skills approach becomes a strand in the science course. For example, in one science textbook the following skills are taught in separate lessons throughout the text:[34]

- Observations and inferences
- Determining variables and controls
- Constructing models
- Determining length, area, and volume
- Performing investigations
- Constructing graphs

[33]See Science—A Process Approach (Washington, D.C.: American Association for the Advancement of Science, 1968).
[34]Dale T. Hesser and Susan S. Leach, Focus on Earth Science (Columbus, Ohio: Merrill, 1989).

TABLE 8.5

SCIENTIFIC THINKING SKILLS

Basic Scientific Thinking Skills

Observing

Using the senses to gather information about an object or an event

Example: Describing a mineral as red

Inferring

Making an "educated guess" about an object or event based on previously gathered data or information

Example: Saying that a landform was once underwater because of the presence of brachiopod and trilobite fossils in the rocks

Measuring

Using both standard and nonstandard measures or estimates to describe the dimensions of an object or event

Example: Using an equal-arm balance to measure the mass of an object

Communicating

Using words or graphic symbols to describe an action, object, or event

Example: Describing the change in temperature over a month in writing or through a bar graph

Classifying

Grouping or ordering objects or events into categories based on properties or criteria

Example: Placing all minerals having a certain hardness into one group

Predicting

Stating the outcome of a future event based on a pattern of evidence

Example: Predicting the position of the moon in the sky based on a graph of its position during the previous two hours

Integrated Scientific Thinking Skills

Controlling Variables

Being able to identify variables that can affect an experimental outcome, keeping most constant while manipulating only the independent variable

Example: Controlling the type of soil or sand and the angle of incline when testing to find out what effect the amount of flow (water) has on the depositional rate of a model river in a stream table

Defining Operationally

Stating how to measure a variable in an experiment

Example: Stating that depositional rate will be measured in grams of sand deposited in the stream table's "ocean"

Formulating Hypotheses

Stating the expected outcome of an experiment

Example: The greater the amount of flow in a river the greater the depositional rate

Interpreting Data

Organizing data and drawing conclusions from it

Example: Recording information about weather changes in a data table and forming a conclusion that relates trends in the data to variables (such as temperature, pressure, cloud cover, precipitation)

Experimenting

Being able to conduct an experiment, including asking an appropriate question, stating a hypothesis, identifying and controlling variables, operationally defining those variables, designing a "fair" experiment, conducting the experiment, and interpreting the results of the experiment

Example: Describing and carrying out a process to find out the effect of stream flow on depositional rates in rivers

Formulating Models

Creating a mental or physical model of a process or event

Example: The model of how the processes of erosion, deposition, metamorphism, and igneous activity interrelate in the rock cycle

Based on Michael J. Padilla, "The Science Process Skills," *Research Matters...To the Science Teacher* (Cincinnati, Ohio; National Association for Research in Science Teaching, 1989). Used with permission of the National Association for Research in Science Teaching.

- Making a map
- Using laboratory equipment
- Using a globe
- Comparing and contrasting
- Making scale drawings
- Using star charts
- Reading a weather map
- Limiting the number of variables
- Forming hypotheses
- Interpreting a glacial map
- Classifying minerals
- Classifying rocks
- Using tables and charts
- Graphing data
- Interpreting data
- Sequencing events
- Designing an experiment
- Predicting outcomes

These 24 lessons constitute a behavioral approach to teaching the skills of science in the context of an earth science course. Students from time to time engage in a hands-on lesson specifically designed to teach a specific process of science.

Another approach that is used to teach the process of science is to include a separate chapter at the beginning of the course on science process skills. Usually the chapter is entitled "Problem Solving in Science," "The Nature of Science," or "How Scientists Think."

Lessons that focus on the processes or skills of science are evident by the stated objectives or the question that is asked. For example in the sample lesson shown in Figure 8.24, the students are asked to find out how observations differ from inferences. In the lesson students drop sheets of paper folded different amounts and record observations and inferences. There is no mention in the lesson of forces, gravity, air resistance—any one of which could have been the conceptual organization of the lesson. In the skills approach, the emphasis is on the process of science, not the concepts of science.

The Cognitive or Constructivist Approach: Integrating Process and Content

A second approach to dealing with scientific thinking skills is the cognitve or constructivist approach. In this method, scientific thinking skills are integrated with the learning of concepts. Joseph Novak, a leading proponent of this idea, points out that

> it should be evident…that there should no longer be a debate about the extent of emphasis on teaching the

content of science versus teaching the processes of science. If a constructivist perspective guides our work, and especially if we use compatible metacognitive tools, there is no reasonable way to teach the processes of science without simultaneously teaching its concepts and principles.[35]

In this approach scientific thinking skills are *learning strategies* that enable students to interact with the environment. If students are to learn about theories of rock formation, they should observe and make inferences about rocks. If they are to learn about the structure of the atom, they might construct models and test hypotheses with regard to different theories.

In this view the processes of science are an integral part of the learning cycle proposed by cognitive science educators. That is, when the teacher engages students in the "exploration" stage of the learning cycle to encounter a new scientific concept, the intent of the teacher is to have the students utilize the processes of science to explore objects, phenomena, and events. Thus, students would be involved in making observations and measurements, collecting and interpreting data, and making conclusions for the purpose of investigating a scientific object, event, or phenomena. In other cases, students might design experiments to test the efficacy of a particular scientific theory.

The processes of science become tools for the students to *understand* the concepts of science. In the cognitive perspective, scientific processes are the learning strategies for students to learn scientific concepts. Let's examine an example of a lesson that illustrates the convergence of scientific processes and scientific concepts.

In the initial stage of the constructivist's model of learning, the student gathers information about new scientific concepts. In the lesson shown in Figure 8.25, students gather data to test some ideas about heating rates of different materials—in this case, soil versus water. In this lesson students make observations, collect data, interpret data, and make conclusions in light of the main concept: convection currents.

Problem Solving in Practice: Science Projects and Fairs

Critical and creative thinking can be fostered by involving students in activities that involve the solving of problems in the context of science projects, re-

[35]Joseph D. Novak, "The Role of Content and Process in the Education of Science Teachers," in *Gifted Young in Science*, Paul F. Brandwein et al., eds. (Washington, D.C.: National Science Teachers Association, 1988), pp. 316–317.

SKILL 1–1

Observations and Inferences

Problem: How do observations differ from inferences?

Materials

identical sheets of paper (6)
paper
pencil

Procedure

1. With a partner, take two sheets of paper, place them together, and fold them in half one time.
2. Take the remaining four sheets and place them together, and fold them in half one time.
3. Now drop both sets of paper from eye level at the same time. Drop them several times and then record your observations and inferences in a data table. You and your partner should record your own observations and inferences separately.

4. Next, fold the two sheets in Step 1 in half again. Repeat Step 3. Record your observations and inferences.
5. Fold the two sheets in Step 1 in half again. They should now be folded three times. Repeat Step 3. Record your observations and inferences.
6. Fold the two sheets in Step 1 in half again. They should now be folded four times. Repeat Step 3. Record your observations and inferences.

Questions

1. Did your observations change as you tried new tests with the papers?
2. Did your inferences change as you tried new tests with the papers?
3. Was it easier to make observations or to make inferences?
4. When you kept folding the papers, what factors did you change?
5. Can you make inferences without first making observations?
6. What else do you need to have before you can make an inference?
7. Suppose you see little black lines in a rock. Is this an observation or an inference?
8. Suppose that you conclude that the lines in the rock were made when it broke away from a larger rock, or that someone scratched them into the rock with a pencil. Are these observations or inferences?
9. Why might your inferences be quite different from another person's?
10. Compare your data chart with your partner's. What did you both observe?
11. What did you both infer?
12. How are observations and inferences important in solving a problem?
13. List an observation that you made today.
14. List an inference that you made today.

Data and Observations

Trial	Observations	Inferences
1		
2		
3		
4		

FIGURE 8.24   A skills of science lesson. Dale T. Hesser and Susan S. Leach, *Focus on Earth Science* (Columbus, Ohio: Merrill, 1989), p. 11. Used with permission of the publisher.

INVESTIGATION 9–2

Heating Differences

Problem: How do soil and water compare in their abilities to absorb and release heat?

Materials

ring stand
thermometers (4)
soil
water
clear plastic boxes of identical size (2)
overhead light source with reflector

metric ruler
masking tape
graph paper
colored pencils (4)

Procedure

1. Fill one box two-thirds full of soil.
2. Place one thermometer in the soil with the bulb barely covered. Use masking tape to fasten the thermometer to the side of the box as shown in Figure 9–11.
3. Position a second thermometer in the box with its bulb about 1 cm above the soil. Tape the thermometer in place.
4. Fill the second box two-thirds full of water.
5. Position the remaining two thermometers the same way as you did in the soil box.
6. Attach the light source to the ring stand. Place the two boxes about 2 cm apart below the light. The light should be about 25 cm from the tops of the boxes.
7. Record the temperatures of all four thermometers with the light turned off.

FIGURE 9–11.

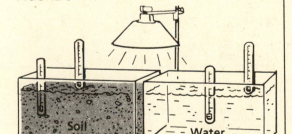

8. Turn on the light. Check to make sure that the bulbs of the thermometers are shielded.
9. Take temperature readings every two minutes for 14 minutes and record.
10. Turn off the light. Take temperature readings every two minutes for 14 minutes and record.
11. Using temperature units on the vertical axis and time in minutes on the horizontal axis, graph your data. Use a different colored pencil to plot the data from each different thermometer.

Data and Observations

Light on

Time in minutes	Thermometer reading			
	1	2	3	4
0				
2				

Light off

Time in minutes	Thermometer reading			
	1	2	3	4
0				
2				

Analysis

1. Did the air heat up faster over water or soil?
2. When the light was turned off, which lost heat faster, water or soil?
3. Compare the temperatures of the air above the water and above the soil after the light was turned off.

Conclusions and Applications

4. Explain how the information you gathered relates to land and sea breezes.

FIGURE 8.25 Science: A Process Approach—In this behavioral approach the complex set of skills a scientist uses were broken down and presented in a series of hierarchically arranged lessons. A lesson illustrating the cognitive approach in which scientific thinking skills are used as learning tools to help students explore and understand concepts. Dale T. Hesser and Susan S. Leach, *Focus on Earth Science* (Columbus, Ohio: Merrill, 1989), p. 11. Used with permission of the publisher.

search investigations, and science fairs. Although activities of this sort have in the past been carried out by individual students, many more students will benefit from these problem-solving activities if teachers would organize them as group projects. Bringing into the mainstream of the science curriculum an emphasis on science projects and research investigations expands students' concept of science and enables them to be immersed in problem solving. Let's take a closer look at how to implement science projects and science fairs in the school setting.

Science Projects

A ninth-grade physical science teacher during the first semester of a two-semester course organizes the students into four-member teams to investigate a problem, question, or topic. The teams have several weeks to complete their research and prepare for a class and a school presentation. Typical questions or topics that students investigate in this introductory physical science course include

- Which metals conduct heat best?
- Do magnetic fields affect the growth of beans?
- Do plants grow better with tap water or distilled water?
- How are earthquakes predicted?
- Which toothpaste is most abrasive?
- What is the acidity of rain and how has it affected the environment in selected sites in North America? How does this compare with other continents?
- How can the global warming trend be changed?

An effective strategy for carrying out science projects is described by Sharan and coworkers.[36] In their strategy, called group investigation (GI), students are organized into groups of four or five students, typically on the basis of heterogeneity.

Each group is responsible for investigating a topic or question that is (1) assigned by the teacher, (2) identified by the group, or (3) selected from a list generated by the teacher or the whole class.

The GI strategy lends itself to the implementation of science projects at all levels and in all subjects of the science curriculum. It is a form of cooperative learning, and its philosophy cultivates democratic participation and encourages the development of a classroom atmosphere conducive to inquiry and student exploration.

The GI strategy is organized into six phases (Figure 8.26), with the teacher assuming the role of

[36]Shlomo Sharan, et al., *Cooperative Learning in the Classroom*, pp. 4–6.

facilitator of learning. The teacher's role as facilitator is crucial in GI. Students need to be free to explore various questions, seek alternative methods and solutions, and collaborate with their peers, as well as with "experts" in the school and the community. As you will discover in the next section, students will be able to access data bases and students in other schools and cultures by means of computer telecommunications.[37]

Implementing science projects using the GI strategy can be facilitated by following the six phases:

1. Topic Selection. Students need to assume ownership for their science project, so it is best if some element of *choice* is built into the topic selection process. If the teacher has specific "projects" or "questions" that are integral to the goals and objectives of the course or unit of study, choice can still be provided by letting the groups choose from a preassigned list of topics.

For example, a physics teacher included in the course syllabus a list of topics that student teams would choose from during the fifth week of the course and then work on for three weeks. The teacher listed the following topics for team topic selection:

- How do the laws of reflection apply to driving an automobile?
- What was the "Hubble Trouble," and how can the problem be corrected?
- What are optical illusions, and are they real?
- How can mirrors be used as communication devices?
- How are the laws of reflection applied in everyday life?

[37]Jack Hassard et al., *Global Thinking: A Cross-Cultural and Interdisciplinary Problem Solving Curriculum and Telecommunications System* (Atlanta, Ga.: US-Soviet Global Thinking Project, 1990).

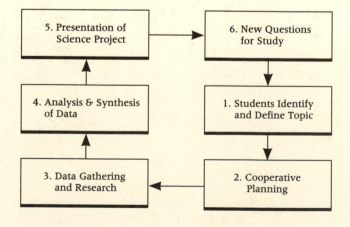

FIGURE 8.26   Steps in the group investigation process.

After presenting these questions as the foci for science projects, the teacher conducted a brainstorming session in which additional topics and questions were generated. From the combined lists, students in each group were asked to identify their first and second choice. Through a process of elimination, each group was involved in making a decision about the topic to investigate, although not necessarily its first choice.

Science projects, which might be a two-day to two-week affair to a full-blown science fair project that might involve some students for the major part of a semester, are tools that science teachers have used to encourage student thinking.

2. Cooperative Planning The students need time to analyze the topic they have chosen so that it can be broken into subtopics or questions. Some teachers conduct formal sessions in which each group meets and constructs a concept map of the topic or question to identify subquestions and subtopics. Each member of the group will eventually be assigned (by the group membership) to investigate one part (a subquestion) of the project, which will be shared with the small group.

3. Data Gathering and Research. Students carry out the plans formulated in phase 2. They can investigate their subtopic by gathering library information, doing a computer search, interviewing people, collecting materials, and even doing experiments. During this phase of the science project, students are engaged in problem solving and utilize the processes of science: observing, classifying, measuring, formulating hypotheses, and interpreting data.

During this phase it is important for the teacher to keep in touch with each group. Teachers meet with each group, and listen to reports on the group's progress on its science project research.

4. Analysis and Synthesis. The teacher arranges during class time for special sessions in which each group meets to analyze and synthesize the information and research data gathered during phase 3, and plan how it can be summarized in an interesting manner for a class presentation.

5. Presentation of the Science Project. Groups give an interesting presentation of the topics studied. There is an attempt to get students involved in each other's work and to expand the perspective on the topic. Experience has shown that student creativity can be facilitated if you encourage a variety of presentations and suggest that the least effective presentation is the lecture. Encourage groups to prepare presentations that involve the audience. Debates, demonstrations, hands-on learning activities, plays, videotape presentations, and computer simulations are effective ideas. Science projects can also be designed so that the results are displayed for others not only in the class but in the school at large.

6. Evaluation. You and the students should evaluate each group's presentation or display. An effective evaluation technique is to have the class evaluate each group presentation. A simple form similar to the one shown in Figure 8.27 can be used to improve future science projects.

Science Fairs

Teachers who involve students in science fair projects commit themselves to many months of planning for an event that typically takes place during two or three days during the spring of each year. Yet the rewards of science fairs for teachers and students far exceed the effort put into the event.

Science fairs can be the stimulus that is needed to motivate some students who otherwise might be turned off to science. Science fairs not only encourage critical and creative thinking, but they encourage students with a wide range of learning styles to become involved in a science fair project. Science fair projects also involve the community and the parents in science education. Parental involvement, which sometimes is seen as intrusive, can actually be a positive aspect of the science fair. School science must extend beyond the walls of the school; the science fair is the perfect event to bring science to the community.

Some school districts arrange with a shopping mall to use its space to display and conduct the judging of the science fair. Other school districts, such as the Atlanta Public Schools, conduct their science fair

---

Directions: Evaluate the group's presentation/display by checking a number for each of the questions below.

1. How effective was the presentation?
   1 2 3 4 5 6 7 8 9 10
2. How interesting was the presentation?
   1 2 3 4 5 6 7 8 9 10
3. How much did you learn from the presentation?
   1 2 3 4 5 6 7 8 9 10
4. What was the quality of the materials and demonstrations used in the presentation?
   1 2 3 4 5 6 7 8 9 10
5. What did you like about the presentation?

6. What would you suggest the group change in the presentation?

FIGURE 8.27 Science Project Feedback Form

(The Atlanta Science Congress) in one of the schools but involve hundreds of community agencies, universities, and businesses by soliciting prizes for various categories and asking professionals to participate as judges.

Teachers can integrate the science fair concept into the ongoing science curriculum by encouraging science projects, and by helping the students learn how to carry out research studies. Too often, students are not given enough guidance and lack experience in conducting a research study, or in preparing for one of a variety of science fair projects.

Three useful sources that you can use to guide you through the science fair process are

Robert C. Barkman, *Coaching Science Stars: Pep Talk and Play Book for Real-World Problem Solving* (Tucson, Ariz.: Zephyr, 1991).

Connie Wolfe, *SEARCH: A Research Guide for Science Fairs and Independent Study* (Tucson, Ariz.: Zephyr, 1987).

Barry A. Van Deman and Ed McDonald, *Nuts and Bolts: A Matter of Fact Guide to Science Fair Projects* (Chicago: Science Man, 1980).

To help students prepare for the science fair, teachers facilitate their work by establishing checkpoints along a time line, and explain the criteria used to judge their work. For example, the time line in Figure 8.28 outlines the procedures students should follow in completing a science fair project. An important aspect of the procedure is that the teacher must maintain a facilitative role by keeping in touch with the students, organizing planning and discussion sessions throughout the process, and being available for consultation.

Criteria to judge science fairs should be shared with the students at the beginning of the process. Typically the criteria include

1. Scientific thought, approach, thoroughness (30)
2. Originality or ingenuity (30)
3. Dramatic value or display (20)
4. Interview or oral presentation (20)

When evaluating scientific thought, judges typically look at the clarity of the problem statement, the sufficiency of background work, the appropriateness and thoroughness of the procedures used, the validity and reliability of the data, and the justification for the conclusions.

Originality refers to the uniqueness of the project, given the age and experience of the student(s). Originality can refer to the questions being asked, the procedures used, and how the data were analyzed and conclusions made.

Judging also takes into consideration the quality and aesthetic appeal of the student's display and presentation. Was the project presented clearly, and was the method or procedure clearly shown?

Week	Check When Completed	Procedure
1	_____	1. Identify project/question/research focus.
1	_____	2. Complete science fair entry form and turn it in for teacher approval.
2	_____	3. Initiate work on the science fair project.
2–4	_____	A. Organize and write out a procedure or plan for your work.
	_____	B. Identify hypotheses (if you are doing a research study).
	_____	C. Conduct your study.
	_____	D. Analyze your data.
	_____	E. Write your report.
4	_____	4. Begin work on your display. Present the information you collected in easy-to-read graphs or tables. If you did an experiment reserve special areas of your display for your problem, hypotheses, procedure, results, and conclusions.
5	_____	5. Prepare a 2–3 minute oral report.
5	_____	6. Prepare all written materials to be included with your display.
6	_____	7. Bring your project to school.

FIGURE 8.28 Science Fair Time Line

The interaction that students have with judges is important. Can the students explain the project and demonstrate their knowledge of the topic and the related concepts?

Science fairs have the potential of encouraging the habit of mind that the authors of Project 2061 so aptly put forward in their report, *Science for All*

Americans. As they point out, students can end up with richer insights and deeper understandings (by participating in an in-depth study) than they could hope to gain from a superficial exposure to more topics than they can assimilate.[38]

[38]*Project 2061: Science for All Americans*, p. 21.

THE COMPUTER AND
STRATEGIES FOR SCIENCE TEACHING

The science classroom can be a special place. It can be one in which students work together in learning teams to answer questions, to inquire, to pose questions, and to learn new ideas about the world. It can be a place in which critical and creative thinking are fostered. In this book we have emphasized the importance—based especially on the work of Piaget and the resulting theory of conceptual change teaching—that should be attached to the notion that students' knowledge about the world develops as a result of their interaction with physical objects, events and phenomena, and people. We have emphasized the importance of involving students in the process of learning as described by the learning cycle.

Since the mid 1970s, the computer has made its way into the educational scene. Can the microcomputer support and enhance the goals of involving students in the learning process, of encouraging inquiry and problem solving, and of fostering critical and creative thinking? Or is the computer simply another educational innovation that will have little effect on classroom learning?

On the one hand, some reports indicate that computers have not met the expectations that were raised when computers began appearing in classrooms. A report from the National Center for Technology in Education, a federally funded research center at Bank Street College, indicates that computers are not an integral part of *subject matter instruction*. Although students have more access to computers (125 students for each computer in 1984 versus 22 students for each computer in 1990), they are not used in ways that are productive (programming, word processing, telecommunications, problem solving). There is evidence that more middle class students have access to computers than poor students, and quite often when these students do have access, the instruction tends to be drill-and-practice software programs.[39]

On the other hand, there is evidence that some schools are using computers in ways that are productive and enhance high-level thinking in students. In some schools students use computers

- To write their own programs to solve problems
- To conduct microcomputer-based science labs (MBL)
- To search data bases for information on a wide range of science-related topics
- To interact via telecommunications networks with students in other schools within and outside their own country to collaborate on science investigations and topics

These and other applications of the computer represent new ways of thinking about how computers can be used in schools. Instead of using computers to "teach" what is currently in the science curriculum, these applications suggest rethinking what can be taught in the science curriculum.

Seymour Papert, in his book *Mindstorms: Children, Computers, and Powerful Ideas*, presents a vision in which students use computers to develop powerful ideas not only about the world but about their own thinking processes. To Papert, the kind of computer activities that are presented to students will affect profoundly the computer's impact on student learning and thinking. He makes the point that a computer can make a student's experience more like that of people in the real world. For example, if we consider the computer as a writing instrument—which all professionals today who write do—using the computer as a writing instrument in the science classroom helps students act and behave like real writers.

The computer can also be used as a tool to help "concretize the formal." One of the problems in science education is that students continue to have difficulty understanding scientific concepts. One difficulty that has perplexed teachers is how to provide students with experiences—activities—that will help students

[39]John O'Neil, "Computer 'Revolution' On Hold," *Update* 32 (9): 1–4 (Alexandria, Va.: Association for Supervision and Curriculum Development).

make conceptual changes. Researchers at Harvard's Educational Technology Center have been working for a number of years to develop teaching modules that combine effective hands-on activities with computer-based activities. One approach they have taken is to design activities in which students invent models of phenomena and then use the computer to examine computer-based models. They have found this approach to be useful in that it allows students to "see" their ideas and the conceptual relationships.

Papert puts these ideas this way:

> Stated most simply, my conjecture is that the computer can concretize (and personalize) the formal. Seen in this light, it is not just another powerful educational tool. It is unique in providing us with the means for addressing what Piaget and many others see as the obstacle which is overcome in the passage from child to adult thinking. I believe that it can allow us to shift the boundary separating concrete and formal. Knowledge that was accessible only through formal processes can now be approached concretely. And the real magic comes from the fact that this knowledge includes those elements one needs to become a formal thinker.[40]

Another aspect of science education that the computer should be tailored to is inquiry learning. We have shown that inquiry teaching provides the environment in which students can explore new ideas as well as testing their own ideas. The computer can be used as a vehicle—especially through the use of simulation software—for students to ask questions, to manipulate variables, and to examine materials, objects, and events in multiple situations. The computer used in this way enables students to utilize the processes of science in the context of computer problem solving.

The possibility of having at least one computer in each science classroom is a reality at the present time; the important notion is how that computer will be used, and what activities will be provided to augment the role of the student as an active inquirer and learner.

In this section we will explore how the computer can be used in the science classroom to enhance critical and creative thinking, and support the current consensus on how students learn—through an active involvement in the learning environment.

Establishing a Science Microcomputer Center

The microcomputer should take center stage in the science classroom; thus it is important to establish a science computer center in the classroom. The most advantageous approach is to set up the science computer center with the following items:

- Microcomputer
- Video monitor
- Printer
- Disc drive
- Microcomputer based lab (MBL, hardware and software)
- Modem
- Telephone line
- Software library
- Computer literature: books, articles, magazines

Students should have access to the microcomputer center; access can be controlled by organizing your class into learning teams and then assigning each learning team specific times in the center.

Computer Literacy: Goals for Computer-Based Science Teaching

A lot of confusion exists around the issue of how computers should be used in science education. We should start with the notion of computer literacy and then use this concept as we think about goals for computer-based science teaching.

Computer use in the science classroom should be based on how computers are used in the real world of science and society. For example I am writing these words on a Macintosh computer using a word processing program. Using the computer as a word processor should be as common as using a telephone, yet in our schools students rarely use the computer as a writing tool in the context of science education. Computers are used by scientists to study natural environments and phenomena. How much opportunity do students have to use the computer to explore models of atoms, clouds, weather patterns, and concepts as specific as density?

Computer literacy must be linked to the use of the computer in the real world. Thus, we might define computer literacy "as the skills and knowledge that will allow a person to function successfully in an information-based society."[41] As Upchurch and Lockhead point out, the computer's use has evolved from simple computing functions to the computer as a "knowledge manipulator." They point out that the computer as a knowledge manipulator allows students to use the computer as a tool to solve day-to-day problems.

[40]Seymour Papert, *Mindstorms: Children, Computers, and Powerful Ideas* (New York: Basic Books, 1980), p. 21.

[41]Richard L. Upchurch and Jack Lockhead, "Computers and Higher-Order Thinking Skills," in *Educators' Handbook: A Research Prospective*, Virginia Richardson-Koehler, ed. (New York: Longman, 1987), p. 152.

Joseph Abruscato, in his book *Children, Computers, and Science Teaching: Butterflies and Bytes*, suggests that computers can be used to foster an atmosphere of curiosity and learning and sees the computer as a knowledge manipulator as well. In reviewing the use of computers in science classrooms he has suggested four goal areas:[42]

1. *An object that the learners study as they would a butterfly or a light bulb.*
2. *A medium of instruction that the science teacher uses with other teaching strategies.* For example, the teacher might provide learners with "teaching programs" that help students learn.
3. *A homework and research tool that enables learners to more effectively carry out science activities, record observations, and produce charts or graphs of information gathered.*
4. *A science teacher's aide.* The computer can help teachers manage student records.

The computer can be an effective medium of learning in the science classroom. It is important to keep in mind that the computer can be used as a medium for learning. What follows are four different approaches or uses for the microcomputer in the context of science education. In each of these uses, computer software is an important aspect of the microcomputer.

We'll examine the use of the computer in the classroom in light of the following mediums: data base, tutor, communication link, science laboratory, and science writing tool.

The Computer as a Data Base

Suppose that students are working on one of the following topics:

> Migrating birds in North America
> Characteristics of mammals in the southwestern United States
> Properties of the ten most common elements
> Habitats of animals in North America
> Environmental facts and problems in third world countries

One use of the computer as a data base is to utilize the burgeoning number of data base software packages such as Scholastic's *Curriculum Data Bases in*

Life Science and Physical Science or *EARTHQUEST* (an environmental education data base; see Figure 8.29). A data base is a collection of related information stored in an organized, systematic manner. Data bases such as these provide the raw information students need to answer questions about topics such as those cited above. Students can find data to help them answer their questions and can easily access the information.

A more powerful use is the dial-up system, in which the modem enables the computer to become part of a telecommunications network. Data bases can be accessed from a number of commercial companies.

But a more powerful application of the concept of data bases is for students to create their own data bases. Software such as Appleworks, or the even more powerful (because it has graphics capabilities) Hypercard, enable students to design their own data bases and learn a methodology of organizing information. Some potential data bases that students could create in the science classroom would include

- Biographical information about members of the class
- The elements
- Commonly abused drugs
- Diseases
- Natural disasters
- Weather and climate data
- Mammals in your state

In each of these cases students will have to research the topic, determine how to organize the information and then enter it into the computer using a data base program. Data bases help students appreciate the complexity of information but enable them to learn to use the computer to manage large amounts of information. Data bases are nothing more than organizations of information. File cabinets, phone books, or a file of index cards of recipes are common examples of data bases. Introducing the students to computer data bases simply extends what they have already experienced.[43]

The Computer as a Communications Link

Imagine a science classroom in which a computer is connected to the telephone line by means of a modem and the students in the classroom are communicating on a day-to-day basis with students in

[42]Joseph Abruscato, *Children, Computers and Science Teaching: Butterflies and Bytes* (Englewood Cliffs: N.J.: Prentice-Hall, 1986), p. 6.

[43]For more information on the use of data bases, see *Databases in the Classroom: Teacher In-service Training Guide* (St. Paul: Minnesota Educational Computing Corporation, 1986).

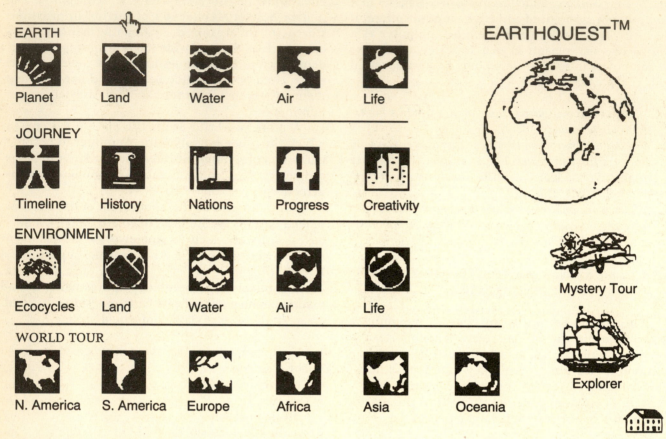

EARTH
Planet · Land · Water · Air · Life

JOURNEY
Timeline · History · Nations · Progress · Creativity

ENVIRONMENT
Ecocycles · Land · Water · Air · Life

WORLD TOUR
N. America · S. America · Europe · Africa · Asia · Oceania

EARTHQUEST™

Mystery Tour

Explorer

FIGURE 8.29   EARTHQUEST® Data Base, used with permission.

another city or country. Computer-based telecommunications networks enable individuals and groups to "go on-line" to send and retrieve messages, send files of information, and exchange graphics and video technologies.

One of the first programs of this nature in science education is the National Geographic Kids Network.[44] Although designed for students in grades 4 to 6, it illustrates the potential of the computer in the area of telecommunications. In the Kids Network students conduct original research, share data, and exchange letters and maps in a series of units including Hello (introduction to telecommunications), Acid Rain, and Weather in Action.

Another program is the U.S.–Soviet Global Thinking Project (GTP), a cross-cultural and interdisciplinary problem-solving curriculum and telecommunications system.[45] The GTP uses a Macintosh-based telecommunications system called AppleLink to en-

able students around the world to explore global problems and concerns and learn strategies of inquiry to solve local problems, as well as providing the knowledge and technological means to deal with them globally.

A third program is Global Laboratory, a science telecommunications program developed by Technical Education Research Centers (TERC) in Cambridge, Massachusetts.[46] The focus of Global Laboratory is to build a community of student researchers studying global ecological issues and sharing their experiences and data using telecommunications. Students around the world are involved in studies that include biological indicators as a means of monitoring the environment, ecosystem descriptions involving Solstice/Equinox Snapshots, and plots, physical/chemical monitoring, environmental plant growth, and modeling and climate change. School sites for the Global Laboratory project are located on several continents. Using Econet, a

[44]Write to NGS Kids Network, c/o National Geographic Society, Washington, D.C. 20036.

[45]U.S.-Soviet Global Thinking Project, *Global Thinking* (Atlanta, Ga.: Georgia State University, 1990).

[46]Technical Education Research Centers, *Global Laboratory Notebook* (Cambridge, Mass.: Technical Education Research Centers, 1991).

telecommunications system which is part of the Institute for Global Communications, participating schools send E-mail to each other, and participate in ongoing conferences on the Global Laboratory bulletin board.

The Computer as a Science Laboratory

The educational software that is now available will enable you to turn the computer into a science laboratory and activity center. The computer can be used to simulate nature quickly without the usual mess, and with the addition of a simple interface box and probes, it can be used to do sophisticated science experiments with ease.

It is helpful to organize the workspace in the computer area of the classroom so that teams of students can perform investigations. Students should be encouraged to think in terms of "What if..." and be provided with the materials and knowledge that will enable them to set up experiments that test their questions.

Simulations allow one or more variables in an experiment to be controlled. High-quality software is available in the areas of weather forecasting, space flight, flight simulations, ecology, volcanology, chemistry, physics, genetics, and many other topics. For example, a earth science simulation, called *Volcanoes*, from Earthware Computer Service, allows students to predict eruptions of mythical volcanoes. *The Galactic Prospector* from Walt Disney Personal Computer Software gives students access to data gathered from satellites and core drillings as they search for oil, gas, and various minerals.

Although simulations are no substitute for hands-on science activities, they do help students develop concepts and the ability to solve problems using logical methods. As you examine science teaching software, you should begin to think in terms of integrating it with the ongoing science curriculum. How can a particular software simulation be used in coordination with hands-on science instruction?

The computer can be a tool to promote conceptual change among students. At the Harvard Educational Technology Center a series of science units have been developed in which a hands-on approach and computer-based modeling and experimentation have been integrated to help students develop an understanding of science concepts. The unit entitled "Weight and Density" uses a conceptual change curriculum design (learning cycle) in which students explore concepts of weight and density qualitatively using hands-on experiences as well as computer-based simulations. The second phase of the unit introduces the students to a quantitative under-standing of weight and density, again using hands-on experiences and the computer. In the third phase of the unit students investigate thermal expansion. Researchers reported that a teaching approach integrating hands-on science with computer-based modeling was effective in bringing about conceptual change: students were able to differentiate between weight and density. However, it was also reported that the method was not effective with all students. One of the recommendations made was to design ways of making students more aware of their misconceptions and of exploiting the work of the computer and small group activities to promote more active dialogue among the students.[47]

One of the most powerful uses of the microcomputer in science education is using it as a data-capturing tool. Known as microcomputer-based labs (MBL), students can collect environmental data by connecting sensors (photometer, microphones, temperature probes, pH probe) to the computer. This approach was researched and developed at Technical Education Research Centers (TERC), and one of its first applications resulted in the development of the I.B.M. Personal Science Laboratories. TERC has continued its development of MBL, and has developed low cost monitoring technologies as part of its Global Laboratory Project enabling students to study environmental factors such as temperture, humidity, air pressure, haze, and ozone concentrations in the lower atmosphere.

Another program enabling students to conduct simulated experiments is the *Science Tool Kit* by Broderbund. It comes with a light and temperature probe and an interface box that connects to the computer's game port, and a well-written user's manual and experiment guide. The Science Tool Kit is supplemented with a series of modules including Force and Motion, Earthquakes, and the Human Body. In each of these modules, probes are used to explore phenomena, and in the case of Earthquakes, students can actually build an apparatus to simulate a seismometer and conduct experiments on "shock or earthquake waves."

The Computer as a Science Writing Tool

Science students are responsible for communicating what they are learning, for writing reports, and for keeping track of their progress. Recent research on

[47]Carol Smith, Joseph Snir, and Lorraine Grosslight, *Teaching for Conceptual Change Using a Computer-Based Modeling Approach: The Case of Weight/Density Differentiation* (Cambridge, Mass.: Harvard Educational Technology Center, 1987).

the use of the computer as a writing tool has shown that it can be used to stimulate written communication. Students ages 9 to 13 can improve their writing if they are given good reasons to write and if they are given appropriate tools. One valuable publication, *Writing and Computers*, by Collette Daiute (Addison-Wesley, 1985), has excellent discussions on how to use the computer as a tool to help very young writers, early adolescent writers, adolescents, and college students.

Word processing and desktop publishing programs are easy for students to use, and there are many to choose from. *Bank Street Writer Plus, Appleworks, Microsoft Word*, and *Word Perfect* are each powerful word processing systems that are applicable for the science teacher and student. Each of these programs is relatively easy to use and has options including a spelling checker and a thesaurus, two tools that are invaluable for the science writer-editor.

Writing in the science classroom, as we have discussed, can take on many forms and should be linked to the development of concepts, science projects, and research investigations. However, students' writing can be enhanced if they are given specific writing formats—newsletters, articles, stories, and book reports, for example. Once they learn a format they can apply it to different topics. Students should have an opportunity to engage in a variety of science writing on different topics using different formats. Writing a newspaper article, a science report, an advertisement, a fictional story, or a book report can be used to explore such topics as "Dinosaurs Found in the Galapagos Islands," "A Visit to Mars," "The Eruption of a Volcano," or "Problems with Hazardous Waste."

Case Study 1.
Computers: Boon or Bust[48]

The Case. Will computers produce the changes that many educators claimed they would, or will they simply be seen as another educational fad? Two educators present evidence to support their side of the issue. Lyn Chan is site coordinator of the State Model Technology School, Project TOPS, South San Francisco Unified School District. Stanley Pogrow, as associate professor of education at the University of Arizona and developer of the HOTS thinking skills program, specializes in instructional and administrative uses of technology.

Lyn Chan. Definitely a boon. In scenarios played over and over again, we see how computers have improved instruction and student achievement. For example, we see a student who has been self-conscious about his speech explaining the Logo Writer to his peers and teachers. Or consider a group of students I saw use a multimedia approach to create presentations on black Americans. They used the computer to search and gather images from a laser disc, accessed distant data banks through telecommunications, and used the word processor to write the script.

With the discovery of the capabilities and applications of computers, more and more teachers are empowered to be innovative and creative in their teaching. Teachers continue to search for resources and ideas outside the four walls of the classroom to increase their produc-

tivity and to better manage instruction, thereby widening the range of students' experiences. As a result, students are becoming actively involved in their own learning.

Like corporate America, which values the support of computers in everyday management and performance, teachers are determined to use computers in their profession. Of course, barriers like limited time to explore the world of computers and limited money for hardware and in-service training need to be addressed. And some administrators need to be convinced to support use of computers in the classroom. However, computers are here to stay.

Stanley Pogrow. A bust. Computers have not worked. Research summaries have never shown much effect beyond third grade. Yes, kids do like computers and are motivated by them. However, given education's tremendous problems, that is not enough.

Why the poor results? The recommendations of the technology movement have been wrong, contradictory, and without basis in demonstrated learning effects. The technology "experts" haven't the vaguest idea how to increase the learning of educationally disadvantaged students. Technology can *help* produce powerful forms of learning in these students if (a) intensive use of technology is provided by good teachers working

(Continued on p. 290)

Think Pieces

- Make a list of strategies you think will enhance critical thinking in science classrooms. Then make a separate list of strategies you think will enhance creative thinking. What criteria did you use to generate each list? How do the criteria compare?
- Construct an essay (no more than two pages) on the efficacy of using structured controversies in the science classroom to enhance critical and creative thinking.
- Find an article in the literature on the K-W-L reading strategy and write a brief report for the class.
- Prepare a think piece that defends the integration of reading and writing skills in the science curriculum, *or* argue against the integration of these skills in science teaching.
- Do you think science fiction has a place in the science curriculum? Why?
- What habits of mind do you think are enhanced by encouraging students to participate in either science projects or science fairs?
- What is the difference between the behaviorist and constructivist approaches to implementing the processes of science in science teaching? What is the rationale for each approach?
- How are the basic processes of science different from the integrated processes of science? Is there an implied hierarchy? Explain your response.

(Continued on p. 290)

(Continued from p. 289)
with students at key developmental points, and (b) use of technology is accompanied by sophisticated forms of conversation between teacher and students.

However, the technology movement believes the best way to get everyone using computers is to argue that computers can produce learning gains by themselves. As a result, the movement is not interested in models that combine technology with other techniques.

Instead of providing more training and programs, the many responsible educators interested in technology must start ignoring the recommendations of the "experts." We need to spend the next five years learning rather than disseminating. In particular, we need to learn how to design better ways of interacting with students. If we do, we will have something important to offer education in the future.

The Problem. Can electronic learning enhance creativity and critical thinking in the science classroom? In what ways do these two educators support or reject your point of view? What other evidence can you cite to support your position? ☐

[48]Based on *Update* 32 (9): 7. (Alexandria, Va.: Association for Supervision and Cur-riculum Development, 1990). Used with permission.

(Continued from p. 289)
* Write a brief essay entitled "New Ideas Discovered to Enhance Critical Thinking in Science Classes," basing your reflections on the essay by Roger von Oech (page 293).
* What role should the microcomputer play in enhancing critical and creative thinking in science teaching? ☐

Case Study 2. Questioning: Inquiry or Inquisition?

The Case. Joe Ellis, a high school biology teacher, was conducting a review session one day before the midterm examination. During the first four weeks of the course, he had covered the first three units in the text on biological principles, cells, and genetics. Joe was in his fourth year of teaching and had a reputation among the students for being "fair but tough." During the review period, Mr. Ellis asked questions based on the material in the text, and what was covered during the labs as well. One of the questions he asked was "What is cell theory?" He waited about a second and then called on Jack McKenna, a student who was struggling in the course and was not doing well in his math and history courses as well. Jack started to say something, but Mr. Ellis interrupted and said,"If you can't answer this question, then there isn't much hope for you on the test." He moved quickly to another student who answered the question easily. On two other occasions Jack McKenna tried to answer questions but was ignored by Mr. Ellis.

The Problem. Should students who don't know the answer be called on in class? If a student doesn't know the answer, how should the teacher respond? What feedback would you give Mr. Ellis? ☐

Science Teachers Talk

"What strategy of instruction do you find to be the most effective with your students?"

Jerry Pelletier. The technique I employ is a combination of "hands-on" experiences with inductive reasoning. Students are put in situations in which they must question what they observe, analyze data, and draw what they feel are logical conclusions. They are constantly questioned and encouraged to question.

John Ricciardi. Be honorable. Be equitable. Be open. When I can adhere to it, this basic strategy works well for me. Honoring your students is respecting their diversity and wholeness...their individuality and integrity. Honor their being always—whatever particular mental phase they may be in. Be equitable with your students. Rules must be fair and equal for all. No favoritism, belittlement, or force. Free choice should be the bedrock upon which all activities are constructed. Be open—and real to your limitations and weaknesses. If you do, your students will be open too, and grow with you. Be open to trust by believing in them. Be open... and aware of a learning that may be taking place in them that you don't fully perceive.

Mary Wilde. I have always had positive results with small group learning; however, I have really been able to enhance this teaching strategy by incorporating the cooperative learning format. There are many different cooperative learning models; however, the one

(Continued on p. 291)

(Continued from p. 290)

I find most successful is where each student within a group learns different material. Then each student is required to teach the others in the group what has been learned. The group is responsible for each other, for I often give individual tests and average them together to receive a group grade. I also like to organize small groups by assigning each member of a team a different task in order to achieve a single goal. For example, when we studied shoreline erosion, each group was responsible for building a papier-mâché model, painting and labeling depositional and erosional shoreline features, reading an article entitled "America is Washing Away," and writing an abstract or review on the article. Tasks were divided among the students and each had a responsibility to the group. One group grade was given for the entire project.

I really feel that small group work helps develop responsibility and commitment. Also, more can be accomplished and learned in small groups where a variety of skills and abilities are pulled together. The learner becomes active, not passive, and greater achievement results can be obtained.

Ginny Almeder. My teaching strategy is fairly traditional and linear. I begin with an assigned reading from the text or related literature. I believe that comprehension of the material is essential, provided it does not inhibit the curiosity necessary for problem solving. The reading is followed up with small group and class discussions and a hands-on activity which usually involves a lab or a postlab discussion. Occasionally, an audiovisual or student presentation is included. Where appropriate, I will reverse the plan and begin the unit with a lab activity. This

sequence allows us to proceed from the process level to the information level and shows students more clearly that it is the act of doing science that produces the body of scientific knowledge. Classwork also includes presentation and discussion of projects that students are working on independently. □

Research Matters: Using Questions in Science Classrooms by Patricia E. Blosser [49]

One objective of science teaching is the development of higher-order thinking processes in students. To achieve this objective, teachers need to facilitate communication with and among students. One of the methods for encouraging students to communicate is to ask them questions. Teacher questions can serve a variety of purposes:

- To manage the classroom ("Have you finished the titration?" "How many have completed problem 17?")
- To reinforce a fact or concept ("The food making process in green plants is called photosynthesis, right?")
- To stimulate thinking ("What would happen if…?")
- To arouse interest
- To help students develop a mind-set.

Any teacher can create his or her own list of additional functions questions can serve.

Science teachers are concerned about helping students to become critical thinkers, problem solvers, and scientifically literate citizens. If we want students to function as independent thinkers, we need to provide opportunities in our science classes that allow for greater student involvement

and initiative and less teacher domination of the learning process. This means a shift in teacher role from that of information-giver to that of a facilitator and guide of the learning process.

Central to this shift in teacher role are the types of questions that teachers ask. Questions that require students to recall data or facts have a different impact on pupils than questions which encourage pupils to process and interpret data in a variety of ways.

The differential effects of various types of teacher questions seem obvious, but what goes on in classrooms? In one review of observational studies of teacher questioning, spanning 1893 to 1963, it was reported that the central focus of all teacher questioning activity appeared to be the textbook. Teachers appeared to consider their job to see that students have studied the text. Similar findings have been reported from observational studies of teachers' questioning styles in science classrooms. Science teachers appear to function primarily at the "recall" level in the questions they ask, whether the science lessons are being taught to elementary students or secondary school pupils.

Why doesn't questioning behavior match education objectives? One hypothesis is that teachers are not aware of their customary questioning patterns. One way to test this hypothesis is to use a question analysis system. One commonly used system is that of Bloom's taxonomy of educational objectives, ranging from knowledge to evaluation. Other systems categorize questions as higher order or lower order. Lower-order questions are those of cognitive-memory thinking and higher-order questions involve convergent thinking, divergent thinking, or evaluative thinking.

(Continued on p. 292)

(Continued from p. 291)

Blosser developed a category system for questions used in science lessons. In this system, questions are initially classified as

Closed: limited number of
acceptable responses

Open: greater number of
acceptable reponses

Managerial: facilitate
classroom operations

Rhetorical: reemphasize,
reinforce a point

Questions which are classified as being either open or closed can be further classified relative to the type of thinking stimulated: cognitive-memory or convergent for closed questions and divergent or evaluative thinking for open questions. This system has been used successfully with both preservice and in-service science teachers to help them analyze their questioning behavior.

Investigations have been conducted to see if preservice teachers could improve their questioning behavior through question analysis. From these studies, it has been concluded that the use of models (audio, video) is helpful, that skill in the use of science processes appears to be related to the complexity of questions asked, that the use of a question category system can be learned, and that the number of divergent and evaluative questions asked in lessons can be increased.   □

[49]Patricia E. Blosser, "Using Questions in Science Classroom," *Research Matters...To the Science Teacher Monograph Series* (National Association for Research in Science Teaching). Used with permission of NARST.

On Problem Solving by Donald Robert Woods[50]

Many equate the teaching of problem solving with blackboard demonstrations and the assignment of homework problems. But research has shown this approach to be one of the least effective methods of developing problem-solving skills, which are among our most advanced mental skills, corresponding to the higher levels in B. S. Bloom's cognitive taxonomy. We teachers can do much more for our students than simply say, "Try solving these problems," or "Watch me while I write out this well-polished script of the highlights of the problem-solving process." What research has shown is that we can define problem solving and identify its cognitive and attitudinal components.

Students come to us with many misconceptions about problem solving, however, and with bad habits that students persist in using to approach problems. Students have been doing *exercises* for years and are generally quite good at it; along with their teachers, they have been calling this *problem solving*. Similarly, we teachers have been "working examples" on the board for years, and we think we're good at explaining and modeling the problem-solving process. In reality, it is only when we get stuck and feel panic at not knowing how to solve a problem that we really start to employ the skills this book addresses. When we use these important skills intuitively, we tend to dismiss them as "just experience."

The most challenging task in problem solving is to create a representation of the problem situation. Some call this "exploring the situation"; others describe it as

making connections between the problem situation and the subject's background experience. Whatever it is called, each person approaches it uniquely. Knowledge and problem solving are intimately connected: how we learn affects how we retrieve ideas and how we create representations.

Knowledge about the principles and laws of biology, chemistry, physics, and the earth sciences is not sufficient. Each discipline also has specific tacit information from experience. This kind of information is difficult to extract from our experience, let alone communicate to our students. Yet students need it for effective problem solving.

To develop problem-solving skills, a teacher needs to assume the role of facilitator and coach, rather than lecturer and provider of information.   □

[50]Donald Robert Woods, "Problem Solving in Practice," in *What Research Says to the Science Teacher: Problem Solving*, Dorothy Gabel, ed. (Washington, D.C.: National Science Teachers Association, 1989), pp. 97–98. Used with permission of the National Science Teachers Association.

On Computers and Science Teaching by Joseph Abruscato[51]

The bright, smiling students who bring "gifts" to you as you prepare the classroom for traditional science curriculum experiences are part of the first human generation to enter a world in which *work* has nothing to do with *perspiration*. They are to be twenty-first century adults, and their world will be as different from yours as sunrise is from sunset.

Your students will be the first citizens of postindustrial America. What you select as appropriate science concepts, processes, and attitudes for their attention must

(Continued on p. 293)

(Continued from p. 292)
respond to what *their* world is becoming—not yours. All teachers, including science teachers, must develop curricula that are tempered by our best guesses about what challenges and opportunities lie ahead.

You can acquire some clues about appropriate content for the science curriculum by simply observing modern society. Ask yourself, "What changes will affect today's students, and how can the science classroom help them prepare for adulthood?" The most important societal phenomenon of our time (aside from the potential for total nuclear destruction fifteen minutes from now) is that heavily industrialized societies such as ours are divesting themselves of the responsibility for manufacturing products that require an abundance of human labor. We are quickly learning that it is cheaper to purchase manufactured items from countries that have low labor costs—typically, the Far East and South America—than to continue to be involved in labor-intensive industries. The careers that your students will eventually have will probably be related to the production, management, and transfer of *information* and the provision of *services*.

What is bringing about this change from a "goods-oriented society" to an "information-and-service-based society"? There is one simple answer: the computer.

One of my favorite sources of absurd humor is a group of performers collectively known as Monty Python. A phrase that recurs in their television skits is "Now for something completely different." The best example of something "completely different" that I can think of is the computer. Its appearance in the midst of industrial society marks the end of the Industrial Revolution and the end of our now old-fashioned

views of what the students in our classrooms can and will become.

Perhaps Herbert A. Simon said it best:

Nobody really needs convincing these days that the computer is an innovation of more than ordinary magnitude, a one-in-several-century innovation and not a one-in-a-century innovation, or one of these instant revolutions that are announced everyday in the papers or on television. It really is an event of major magnitude.

The computer is changing more than industry. It is changing the very character of everyday life and our most fundamental ideas about what should be included in a quality education.   □

[52]Joseph Abruscato, *Children, Computers, and Science Teaching: Butterflies and Bytes* (Englewood Cliffs, N.J.: Prentice-Hall, 1986), pp. 3–4. Used with permission of the publisher

On Creative Thinking by Roger von Oech[52]

I once asked Carl Ally (founder of Ally and Gargano, one of the more innovative advertising agencies on Madison Avenue) "What makes the creative person tick?" Ally responded, "The creative person wants to be a know-it-all. He wants to know about all kinds of things: ancient history, nineteenth-century mathematics, current manufacturing techniques, flower arranging, and hog futures. Because he never knows when these ideas might come together to form a new idea. It may happen six minutes later or six months or six years down the road. But he has faith that it will happen."

I agree wholeheartedly. Knowledge is the stuff from which new ideas are made. Nonetheless, knowledge alone won't make a person creative. I think that we've all known people who knew lots of stuff and nothing creative happened. Their knowledge just sat in

their crania because they didn't think about what they knew in any new ways. Thus, the real key to being creative lies in what you do with your knowledge. Creative thinking requires an attitude or outlook which allows you to search for ideas and manipulate your knowledge and experience. With this outlook, you try various approaches, first one, then another, often not getting anywhere. You use crazy, foolish, and impractical ideas as stepping stones to practical new ideas. You break the rules occasionally, and hunt for ideas in unusual outside places. In short, by adopting a creative outlook you open yourself up to both new possibilities and to change.

A good example of a person who did this is Johannes Gutenberg. What Gutenberg did was to combine two previously unconnected ideas, the wine press and the coin punch, to create a new idea. The purpose of the coin punch was to leave an image on a small area such as a gold coin. The function of a wine press was, and still is, to apply a force over a large area in order to squeeze the juice out of the grapes. One day Gutenburg, perhaps after he'd drunk a glass of wine or two, playfully asked himself, "What if I took a bunch of these coin punches and put them under the force of the wine press so that they left their images on paper?" The resulting combination was the printing press and movable type.

Another example is Nolan Bushnell. In 1971, Bushnell looked at his television and thought, "I'm not satisfied with just *watching* my TV set. I want to play with it and have it respond to me." Soon after, he created "Pong," the interactive table tennis game which started the video game revolution.

Still another example of a person who did this is Picasso. One

(Continued on p. 294)

(Continued from p. 293)
day, Picasso went outside his house and found an old bicycle. He looked at it for a little bit, and then took off the seat and the handle bars. He then welded them together to create the head of a bull.

Each of these examples illustrates the power the creative mind has to transform one thing into another. By changing perspective and playing with our knowledge and experience, we can make the ordinary extraordinary and the unusual commonplace. In this way, wine presses squeeze out information, TV sets turn into game machines, and bicycle seats become bull's heads. The Nobel Prize winning physician Albert Szent-Györgyi put it well when he said:

> *Discovery consists of looking at the same thing as everyone else and thinking something different.*

Thus, if you'd like to be more creative, just look at the same thing as everyone else and "think something different."

[52]Roger von Oech, *A Whack on the Side of the Head* (New York: Warner, 1983), pp. 6–7. Reprinted with permission of Roger von Oech and Warner Books.

Problems and Extensions

- Present a demonstration, teaching tool, simple lab activity, creative homework assignment, or some strategy or technique related to one of the following areas of science teaching: earth science, life science, physical science, or science-technology-society.
- Create an EEEP (exciting example, everyday phenomena) for at least one concept from the fields of science identified in the previous problem. Remember that an EEEP should help make the unfamiliar familiar. Prepare your

EEEP and present it to the class, or videotape your EEEP, describing how you would use the EEEP in a science lesson. Present the video to the class.

- Prepare a microteaching lesson that focuses on one of the following teaching strategies and present it to your peers or a small group of secondary students. Videotape the lesson, review it, and write a brief report outlining the success you had in presenting the teaching strategy, and how you might change the lesson for a future presentation.

  - Using advance organizers
  - Elements of a stimulating classroom environment
  - Questioning
  - Using examples
  - Closure
  - Creating a positive learning environment

- Select a chapter from a secondary science textbook and analyze the questions posed in the chapter. You can use the system that we presented (low inquiry versus high inquiry) or some other system that you prefer. What is the ratio of low- to high-inquiry questions? To what extent is high-level thinking encouraged in the chapter?
- View videotapes of science teaching and analyze the lessons in terms of the teachers' use of questions. Use the coding system shown in Figure 8.30.
- Using the process outlined in this chapter on the use of structured controversies, prepare the necessary teaching materials to conduct a structured controversy either in your methods course or in a secondary science classroom.
- Select a chapter from a secondary science textbook and develop a science reading guide for the chapter using the example on page 261 as a guide. The reading guide should be interesting and can include experiential

"activities" such as a hands-on activity.

- Using the K-W-L reading strategy, design a series of lesson plans to help students comprehend the ideas in a chapter from a secondary science textbook or one of the chapters in this book. Have your plans checked by a peer in your class. Test your secondary science plans by field testing the lessons with your peers or a group of secondary students; if you choose this book, try them out on your peers and then send the results to me!
- Choose a partner in your class. Select a chapter from a secondary science textbook and draw a semantic map of the chapter. You and your partner should do this separately at first. Share maps with each other, and then create a cooperative semantic map. How do the individual and cooperative maps compare?
- Select one of the writing strategies presented in the chapter as the basis of a science lesson. Plan the lesson, then present it to a group of peers or students. How effective was the lesson in enhancing the students' writing skills?
- Prepare a lesson on the processes of science using either the behaviorist or constructivist approach. Write a short rationale, objectives, procedures, and evaluation for your lesson. Meet with peers in your class and compare your lesson plans, paying particular attention to the rationale and objectives. How are they different?
- Draw a map of a classroom showing how you would set it up to include a computer center. How would you enable students to gain access to the computer, when you have 28 students in each of your five classes?

Question Category	Lesson: Subject Grade	Lesson: Subject Grade	Lesson: Subject Grade
Closed Questions (low inquiry)			
Open Questions (high inquiry)			
Summary			

FIGURE **8.30**

RESOURCES

Abruscato, Joseph. *Children, Computers, and Science Teaching: Butterflies and Bytes*. Englewood Cliffs, N.J.: Prentice Hall, 1986.

This practical book shows science teachers how they can integrate computers into the regular science classroom.

Brandt, Ronald S. *Teaching Thinking*. Alexandria, Va.: Association for Supervision and Curriculum Development, 1989.

This is a collection of articles that appeared in the journal *Educational Leadership*. The articles convey ideas about critical thinking, how to teach thinking skills in the curriculum, and the evaluation of the thinking skills approach.

Bulman, Lesley. *Teaching Language and Study Skills in Secondary Science*. London: Heinemann, 1985.

This book gives practical advice to science teachers, based on the latest research into students' difficulties with scientific language. The book shows teachers how to help students develop the language and study skills necessary to be successful in science.

Gabel, Dorothy, ed. *What Research Says to the Science Teacher, vol. 5, Problem Solving*. Washington, D.C.: National Science Teachers Association, 1989.

This volume is intended to translate research on problem solving into practice. Several authors show how problem solving can be applied to the teaching of elementary science, middle level science, earth science, physics, biology, and chemistry.

Heimlich, Joan E., and Susan D. Pittelman. *Semantic Mapping: Classroom Applications*. Newark, Del.: International Reading Association, 1986.

Semantic (or concept) mapping is a strategy to help students tap their prior knowledge. Through teaching—from planning lessons to helping students learn concepts—semantic mapping can be a useful tool. This book provides step-by-step instructions and ways of applying it to the classroom.

Marzano, Robert J., and Daisy E. Arredondo. *Tactics for Thinking, Teacher's Manual*. Aurora, Colo.: Mid-continent Regional Educational Laboratory, 1986.

This is the training manual which accompanies Tactics for Thinking, a program designed for K–12 integration. *Tactics* is based on dividing thinking skills into three general areas: learning-to-learn skills, content thinking skills, and reasoning skills.

Marzano, Robert J., and Diane E. Paynter. *Tactics for Thinking*. Aurora, Colo.: Mid-continent Regional Educational Laboratory, 1989.

This large three-ring binder contains activities and blackline masters for the three categories of thinking skills for middle and secondary students.

Marzano, Robert, et al. *Dimensions of Thinking: A Framework for Curriculum and Instruction*. Alexandria, Va.: Association for Supervision and Curriculum Development, 1988.

This volume deals with thinking as a foundation for schooling and includes discussions on topics such as metacognition, critical and creative thinking, and the general thinking processes.

Oech, Roger von. *A Whack on the Side of the Head: How to Unlock Your Mind for Innovation*. New York: Warner, 1983.

This book will not only give you the permission to be creative but will show you how.

Rowe, Mary Budd, ed. *What Research Says to the Science Teacher, vol. 5, The Process of Knowing*. Washington, D.C.: National Science Teachers Association, 1990.

This volume uses cognitive science research to help science teachers understand the process of knowing.

Thelen, Judith N. *Improving Reading in Science*. Newark, Del.: International Reading Association, 1984.

This useful book shows science teachers how to diagnose reading problems in science, what prereading strategies science should use to help learners read science texts, and how to help students with text and vocabulary.

Tiedt, Iris McClellan, et al. *Reading/Thinking/Writing: A Holistic Language and Literacy Progam for the K-8 Classroom*. Needham Heights, Mass.: Allyn & Bacon, 1989.

This book is a useful reference for the science teacher who needs relevant information about how to integrate language skills into the science classroom.

Wilson, Carol, and Gary Krasnow. *hm Science Study Skills Program: People, Energy and Appropriate Technology.* Washington, D.C.: National Science Teachers Association, 1983.

This is a hands-on approach designed to help secondary students develop specific study skills appropriate for science courses. Skills include listening, building vocabulary, reading for meaning, taking effective notes, graphing, solving problems, and test taking.

9

DESIGNING AND ASSESSING SCIENCE UNITS AND COURSES OF STUDY

It's the night before the first day of school, and the teacher can't sleep! First days are full of anxieties. What will the students be like? Will they like me? Am I ready for them? Do I have all my textbooks? Will I have the supplies I need? Why did the department head assign me four sections of survey biology, and one section of survey earth science? Are my lesson plans written? One way to rid oneself of these anxieties is to invest quality time in the preparation and development of teaching plans.

Question: What do these concepts have in common?

- Concept planning maps
- Intended learning outcomes
- Working in cooperative planning teams
- Rationale
- Cognitions
- Cognitive skills
- Affects
- Psychomotor skills
- Brainstorming
- Performance assessments
- Portfolios

Answer: They are some of the elements that make up a dynamic and effective strategy that science teachers can use to design and assess learning materials for the science classroom.

Designing and assessing science units and courses of study is the focus of this chapter. The chapter is organized in such a way that you will be designing science teaching plans as you work through the chapter, and you will create assessment measures for the teaching plans that you design as well. Planning and assessment are integral aspects of the process of teaching. Too often assessment is thought of as something separate from instruction, and because of this, assessment and instructional planning have been integrated and placed in the same chapter. When you finish the chapter, you will have created the following products:

1. A rationale for a science unit, including general science education goals
2. A list of objectives (we'll call them intended outcomes) for a science unit grouped according to type of student learning
3. A concept map showing the relationships between the central ideas in your unit
4. An instructional plan (a set of lesson plans) describing the unit, including what learning objectives are intended, and the strategies you will employ to help students achieve the unit's objectives
5. An assessment plan describing measures to assess the major objectives of the unit to provide feedback to the students, and feedback for you on the effectiveness of your science unit

PREVIEW QUESTIONS

- What processes can be used to design an instructional plan?
- How should a teacher proceed to develop a mini-unit of instruction?
- What are intended learning outcomes? How do cognitions, affects, cognitive skills, and psychomotor skills differ?
- How can cognitive maps be utilized in the planning and development of teaching materials?

- What are the elements of the following types of lessons: direct-interactive, cooperative learning, inquiry or laboratory, and conceptual change?
- What are the elements of a course of study?

- What are some trends in the development of assessment strategies?
- What are the components of a comprehensive assessment strategy for student learning and instruction?

AN APPROACH TO UNIT DESIGN

Designing teaching plans is an active, creative, and time-consuming process, yet it is one that is underrated by many teachers, and indeed taken for granted as well. Yet as long as you are a teacher you will be involved in the design of teaching plans, even if you use the most current textbook or science curriculum project. Textbooks and curriculum projects need to be tailored to each group of students, and the person charged with this responsibility is you, the science teacher. Thus, it is advisable that you develop an approach to unit and course design that will enable you to modify existing teaching plans in textbooks and curriculum projects, as well as create original teaching units making use of the rich array of teaching materials available to science teachers.

Designing units and courses of study is a process that results in a product. The process of unit and course design is called instructional planning, and it will result in an instructional plan. The instructional plan that you will develop in this chapter will be a miniunit. What is important is for you to experience the process of planning, and by scaling the product down it will be more manageable, and you will also have a better opportunity to field-test your instructional plan with secondary students if it is relatively short.

Another concept that emerges from this discussion is curriculum. Curriculum is to be distinguished from instruction in the following way. Curriculum is not a process. It is a set of intended learning outcomes or goals. It is what science teachers hope students will learn. It focuses on the nature and organization of what science teachers want students to learn. Curriculum development therefore is a process in which the science teacher engages in the process of selecting and organizing learning objectives for a unit or course of study.

In this chapter you will be involved in both curriculum development and instructional planning. We might summarize this by saying that the curriculum you develop consists of what is to be learned, the goals indicate why it is to be learned, and the instructional plan indicates how to facilitate this.[1] Learning will only occur when you implement the instructional plan.

The general plan for designing units and courses of study is shown in Figure 9.1. In this curriculum and instruction design model, note that the boxes represent products, and that the lines show various processes in which you will be involved. Four questions will guide you through the process as you design a science miniunit, as well as complete science units and courses of study in the future.

[1]George J. Posner and Alan N. Rudnitsky, *Course Design: A Guide to Curriculum Development for Teachers, 3rd Ed.* (New York: Longman, 1986).

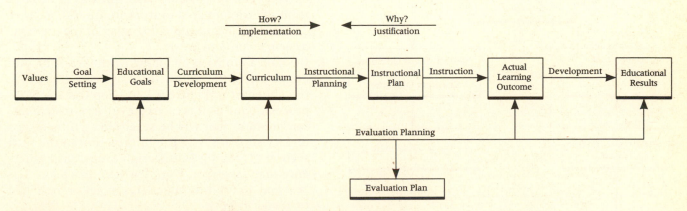

FIGURE 9.1   Curriculum and Instruction Design Model. From *Course Design: A Guide to Curriculum Development for Teachers 3rd Ed.* by George J. Posner and Alan N. Rudnitsky. Copyright ©1986 by Longman Publishing Group. Reprinted with permission from Longman Publishing Group.

1. *Why?* You need to consider why you are teaching the science unit, and this leads you to consider values and general science education goals and purposes.
2. *What?* Here you will consider the objectives of your unit and design a conceptional map to show the relationships between the major ideas of your miniunit.
3. *How?* The instructional plan for your miniunit—consisting of at least three lessons—will describe how you will engage the students to achieve the stated learning objectives.
4. *What did the students learn?* Your evaluation plan will help you provide information about what the students learned, and how successful the miniunit was.

INQUIRY ACTIVITY   9.1

....

Designing a Science Miniunit

....

In this inquiry activity you will proceed through the process of designing a science unit step by step.[2] The steps in the process are shown in Figure 9.2. You should take a moment to examine the figure before you go on. Note that the process is shown as a cycle, rather than simply a linear process. The feedback you obtain from the evaluation component of the model will enable you to revise your teaching units.

One of the major emphases in this book and in the science education community is the use of cooperative learning. We have found that cooperative planning is an effective tool in helping beginning science teachers design teaching materials. Student interns can be organized in heterogeneous (different majors, sex, age, experience) dyads (teams of two) or

triads (teams of three) for planning purposes. This will work especially well for designing teaching materials. Forming teams composed of different majors (e.g., an earth science major working with a biology major) will enable you to raise interdisciplinary questions for your teaching plans and allow you to play off each other's experience and knowledge. Although everyone will develop his or her own science miniunit, each will benefit from a cooperative team planning process. Now let's get on with the process of designing a science miniunit.

Materials

Teaching unit design steps, large sheets of newsprint, marking pens, secondary science textbooks in your field, hands-on materials as needed for your miniunit.

Procedure

1. Your task is to design a miniunit in a content area of your choice. Discuss with members of your team the content area to work in. Be sure to agree on the course and the grade level of the students to help focus your work.
2. Develop a miniunit by following the teaching unit design steps that follow this inquiry activity. Use the checklist in Figure 9.3 to keep track of your process. Your goal is to create a three-to-five lesson miniunit that will be taught to a group of secondary students. Use the information in the next section "Teaching Unit Design Steps," as a reference manual as you work together as a team. The best way to use this material is to get in there and let the ideas flow.
3. It will be obvious that you will be unable to do all your work in a team. There will be times when you will have to work individually to write objec-

[2]The material in this section is based on the work of Dr. Charles Rathbone, University of Vermont, and that of George J. Posner and Alan N. Rudnitsky, *Course Design*; and feedback from interns in the Alternative Teacher Preparation Institutes for Secondary Foreign Language, Mathematics and Science Teachers, Georgia State University.

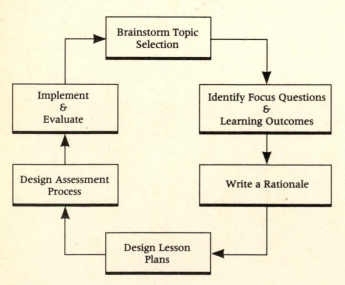

FIGURE 9.2   Design Cycle for Science Units

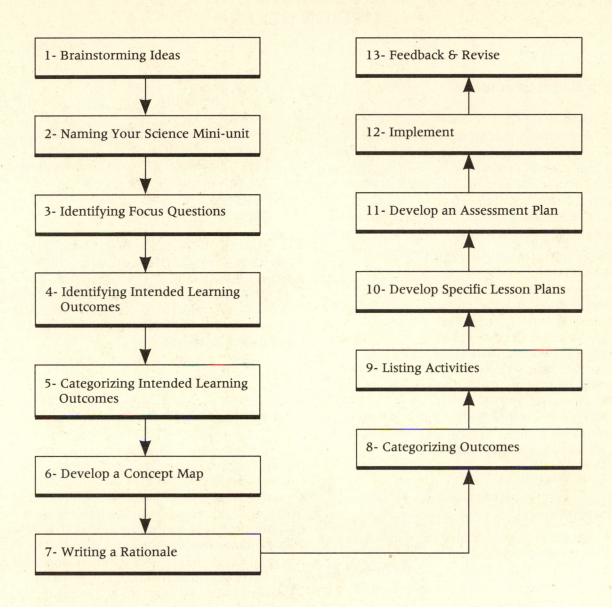

FIGURE 9.3   Miniunit Progress Checklist

tives, lesson plan details, and assessment items. However, you should take the time to explain to your design team what you have done in order to receive their feedback.

4. Teach your miniunit to a group of secondary students or peers. If you cannot teach the entire miniunit, try to implement at least one of your lessons.

5. You should report to the entire class by explaining the nature of the process, and the details of the miniunit. The report to the class should include feedback data gathered during the implementation of the unit either with peers or secondary students.

Minds on Strategies

1. How effective was your miniunit in helping "your students" understand and develop positive attitudes toward science?

2. What were some problems you encountered in developing the miniunit?

3. How would you modify your unit planning process the next time you engage in the process? Do you think the process you used was effective? To what degree?

4. What do you think is the most important aspect of instructional planning?

THE TEACHING UNIT
DESIGN STEPS

Step 1: Brainstorm Ideas

This should be a fast, free-flowing listing of terms, words, and phrases that you have for your miniunit of instruction. It is and should be considered an initial list of ideas. It will definitely be revised and changed as you continue the process. Therefore, make no restrictions on your self at this stage.

Here is an example of an initial topic brainstorming session for an earth science unit:

Faults and cracking of the earth's crust
Earthquakes
P-waves, S-waves, L-waves
How earthquake waves move
Epicenter
How to locate an epicenter
Damage caused by earthquakes and other natural disasters
Measuring the motion of the earth's crust
Seismographs
Field trip to a seismic station
Using the computer to measure earthquake waves or to simulate earthquakes
Plate tectonics
The rock cycle
Precautions people can take to lessen damage and save lives

Working in a dyad or triad, select a topic area and brainstorm a list of initial ideas for a potential teaching unit on the topic. Identify one of the members of the group as a recorder who will record the list on a large sheet of chart paper.

Step 2: Name Your Science Miniunit

Science units of instruction need focus, and naming them is a way to achieve this. Some teachers believe that science unit names are important because they can capture the importance or value of the unit, as well as the content of the unit.

The name you give to your unit is also a way of making sure that your initial ideas are logically connected to each other. As you can see from the list of science miniunits produced by science interns, there is no shortage of ideas:

Energy
Atoms, Molecules, and You
Chemical Reactions and Food
Chemical Bonding
Genetics
Living Things and How They Adapt
Properties of Matter
Drugs, Alcohol, and Tobacco and the Teenager
The Earth's Past
Energetics
Conserving Our Environment

Miniunit titles can communicate the essential meaning of a unit. Here are some titles of minicourses from the high school science project ISIS: Packaging Passengers, Kitchen Chemistry, Using the Skies, Salt of the Earth, Seeing Colors, Cells and Cancer, Food and Microorganisms, Birth and Growth, and Sounds of Music.

Look over your initial list of ideas and name your science miniunit.

Step 3: Identify Focus Questions

Too often we don't consider what the main points or questions for a science unit are, especially when the ideas have been formulated for us, as in a chapter of a textbook. Textbook chapters normally contain more information than students need to know. However, without focus or central questions it is difficult to help the students see the main point of the chapter, or in this case your science miniunit. Focus questions can also help students make a link with their prior knowledge, as well as establish a rationale for studying the science unit.

Here are some questions. Which of these would function as useful focus questions?

1. What is an atom?
2. How can knowledge about atoms affect my choice of foods?
3. What can families do to prepare themselves for natural disasters such as tornadoes, floods, hurricanes, and earthquakes?
4. How should people interact with the environment?
5. When was the first atomic bomb exploded? Where? By whom?

Posner and Rudnitsky point out that focus questions should help you define the heart of your unit or course.[3] As you think about your unit, you might ask yourself these questions: Which of the questions you have listed really get at the heart of your unit? What kinds of questions are you asking? Are they mostly *where* or *when?* Have you considered *how* or *why* questions? To what extent do your questions relate to the students, or to the relationship between the content of the unit and social issues?

Step 4: Identifying Intended Learning Outcomes

Your initial list of ideas and your thinking about focus questions should help to create a list or set of intended learning outcomes, or learning objectives. Intended learning outcomes are statements of what you want the students to learn. These statements can include skills, concepts, propositions, attitudes, feelings, and values.

The process of identifying learning objectives should begin with a consideration of your initial list of ideas from the brainstorming session. Learning objectives are skills, concepts, and values that you intend the students to learn. They are not activities, or things the students will do. If you have a favorite activity or field trip, consider what the students will learn from the experience to derive an intended learning outcome. Dominant and recessive genes, chromosomes, blood vessels, hallucinogens, psychological dependence, the precambrian age, fossils, and the cell wall are examples of science content that you might want students to learn. Computer keyboarding, focusing a microscope, measuring the temperature of the air, calculating averages, and graphing data are examples of skills you might want students to learn. Your list of intended learning outcomes can also include attitudes and values, such as caring for animals and plants, respect for the environment, and concern about the effects of research studies on the community.

Examine your initial list of ideas and decide which items are intended learning outcomes. Here is our original list. Those with a check mark best represent our classification of intended learning outcomes.

- ✔ Faults and cracking of the earth's crust
- ✔ Earthquakes
- ✔ P-waves, S-waves, L-waves
- ✔ How earthquake waves move
- ✔ Epicenter
  Drawing pictures of faults
- ✔ How to locate an epicenter

- ✔ Damage caused by earthquakes and other natural disasters
- ✔ Measuring the motion of the earth's crust
- ✔ Seismographs
  Field trip to a seismic station
  Using the computer to measure earthquake waves or to simulate earthquakes
- ✔ Plate tectonics
- ✔ The rock cycle
- ✔ Precautions people can take to lessen damage and save lives

Step 5: Categorize Intended Learning Outcomes

The next step is grouping intended learning outcomes into skill and nonskill categories. A science skill is something that students are expected to be able to do at the end of the unit of teaching. Skills include physical abilities such as measuring the volume of water using a graduated cylinder, as well as mental abilities, including describing the physical properties of rocks, inferring past events from fossil evidence, solving mechanics problems, and writing chemical equations. Skills represent the behavioral outcomes of learning and can often be measured by having students do what was intended, for example, writing chemical equations or measuring volumes.

There are nonskill outcomes that cognitive scientists and social psychologists claim are important in science courses. Nonskills are ideas (earthquakes, valence, momentum, force, uniformitarianism, dominance) and values (truth, openness, uncertainty) that are not necessarily directly observable outcomes yet to many represent the heart of science courses. Posner and Rudnitsky put it this way: "Generally speaking, nonskills comprise the knowledge and attitudes with which we think and feel. Skills comprise what we can do with this knowledge and how we can act on these feelings."[4]

Categorize your intended learning outcomes into nonskill and skill categories. You might have to add skills and nonskills to your original list. You should end up with two lists of outcomes as shown in Table 9.1.

Step 6: Develop a Concept Map

At this point you should use the technique of concept mapping to analyze the important ideas and words that you hope students will understand in your mini-unit. Novak and Gowin have advocated the use of

---

[3]Posner and Rudnitsky, *Course Design*, p. 15.

[4]Posner and Rudnitsky, *Course Design*, p. 20.

TABLE 9.1

ENVIRONMENTAL PROBLEMS IN OUR COMMUNITY

Nonskill Outcomes	Skill Outcomes
Respects the environment	Able to analyze a sample of water
Understands energy webs and food chains	Can measure the pH of liquids
Understands pollution	Can write equations for chemical processes
Knows how acids affect river water	
Understands "biodegradable"	

concept mapping as a tool for planning, and it is introduced here in the instructional planning process.

To develop a concept map, make a list of all the important concepts and terms that are related to the major concepts in your miniunit. The following steps should help you devise your concept map.[5]

1. Select the main concept from your list
2. Add concepts to your list if needed

[5]For additional information on developing concept maps, see Joseph D. Novak and D. Bob Gowin, *Learning How to Learn* (Cambridge: Cambridge University Press, 1984).

3. Write the concepts on index cards
4. Rank the concepts from the most inclusive to the most specific
5. Group the concepts into clusters. Add more specific concepts (on cards) if necessary
6. Arrange the concepts (cards) in a two-dimensional array
7. Write the concepts on a sheet of paper as they appear in the two-dimensional array
8. Link the concepts and label each link

For example, here is a list of concepts and terms for a miniunit on the "food chain," listed in order from most inclusive to least inclusive.

Consumers
Producers
Decomposers
Herbivores
Carnivores
Photosynthesizers
Algae
Bryophytes
Tracheophytes
Organic debris

Figure 9.4 shows an arrangement of these concepts and terms for the miniunit on food chains. Note the hierarchical nature of the map. The more inclusive concepts are located at the top of the map (e.g.,

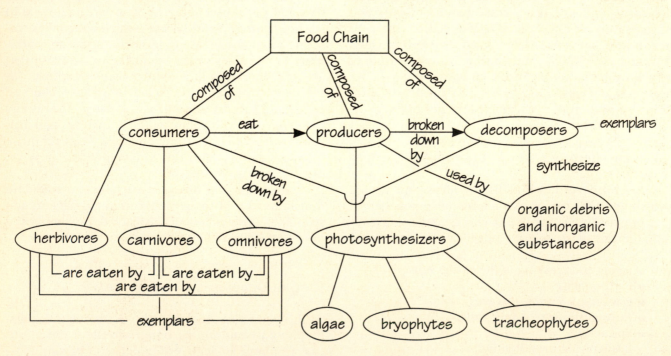

FIGURE 9.4   Concept Map for Miniunit on Food Chains. Joseph D. Novak and Bob Gowin, *Learning How to Learn* (Cambridge: Cambridge University Press, 1984), p. 122. Reprinted with permission of Cambridge University Press.

food chains, producers), followed by less inclusive concepts (bryophytes, tracheophytes) as you read down the map. Using this technique will help you analyze the nature of your miniunit at this stage. Does your concept map show inclusive concepts? Are there too many abstract concepts? What concrete concepts could you add to help the students understand the abstract concepts?

There is a good chance that you are planning your miniunit using a secondary science textbook. If you are working with a single chapter (miniunits typically represent only a part of a chapter, and on rare occasions a full chapter), make a list of the important concepts (no more than ten), and from this list create your concept map. You may discover that you will have to modify the map by adding concrete concepts that may have been "missing" from the textbook's approach to the topic, or you may have to de-emphasize and eliminate some concepts.

Step 7: Write a Rationale

You're probably wondering why writing a rationale wasn't the first step in the curriculum development process. It should be clear, however, that you have invested some time thinking about the initial ideas of the miniunit, categorizing them as skill and nonskill and analyzing the ideas in your unit by making a concept map. You are now in a much more powerful position to think about why you are teaching this miniunit, and how you will communicate this to your students.

In writing a statement describing the rationale for your miniunit you should consider these questions:

- How does the miniunit affect the future of the *students*, as well as their individual needs and interests?
- How does the miniunit contribute to *societal* issues and help students deal responsibly with science-related issues?
- How does the miniunit reflect the spirit and character of scientific inquiry, and the nature of the *scientific enterprise?*

Thus, a rationale statement will be determined by the values that influence your perception and conception of students (learners), society (science's relationship to society), and science (the subject matter).

Course, unit, and miniunit rationales can also be influenced by current trends and directions in science education. Since the 1970s there has been a shift toward a more student-centered (hands-on) approach to the learner. A greater emphasis has been, and will continue to be, placed on S-T-S (see Chapter 6).

Reports such as *Project 2061* have given greater impetus for a cognitive approach to science teaching, and NSTA's *Scope and Sequence Project* has influenced the organization of the curriculum.

The miniunit rationale that you write should also include a statement of science education goals (broad statements of intent) that reflect the integration of ideas concerning students, society, and the nature of science. For example, a rationale for a miniunit on "the earth's past" might contain a goal statement such as, "This miniunit is designed to give middle school students insight into, and an appreciation of, life as it existed in the past, and ideas concerning how life evolved on the earth."

A complete rationale should contain a goal statement, as well as of how the miniunit attends to conceptions of the student, society, and the nature of science. Here is a complete rationale for a unit on the environment. Notice that the rationale includes aspects of the student, society, and science, and how each is dealt with in the unit.

Rationale: Environmental Education Unit

To be truly educated, people need knowledge of, appreciation for, and skills relating to, the world around them. Schools, largely responsible for educating individuals in our society, must equip people with the knowledge and ability necessary for them to preserve the environment for the physical, psychological, and aesthetic needs of future generations. A person educated in this way knows enough to exhibit intelligent and reasonable concern for his or her environment and can presumably direct this knowledge and concern toward the preservation and improvement of the environment. A person's actions affect the environment as they occur in the context of the environment.

It is important, from the standpoint of enhanced self-image, personal enjoyment, and a meaningful life, for people to know basic facts about the environment and their relationship to it. They should, as well, have the capability and inclination to act in a positive manner toward the environment.

In a technological society, which tends to insulate its members from the environment, this subject area is second to none in its importance. Actual experience with the environment is necessary for the proper treatment of this subject. Technology in our society appears, at times, out of control, or at least controlled by factors not related to environmental quality. For the preservation of society and the enhancement of individual lives, people must interact with their world on an educated basis.[6]

[6]Posner and Rudnitsky, *Course Design*, p.47. Copyright © 1986 by Longman Publishing Group. Reprinted with permission from Longman Publishing Group.

Examine your list of intended learning outcomes, and write a rationale for your miniunit. As you think about the rationale, think about the students you will teach, and how the content will relate to them. Next formulate the relationship between the content in your unit and potential social issues that it brings up. And lastly, how will you approach the nature of science in this miniunit?

Step 8: Categorize Outcomes: Cognitions, Affects, and Skills

In this step of the instructional planning process, you will revise your initial learning outcomes, write these statements as objectives using a standard form, and categorize them into practical categories for teaching.

There are many ways to write learning objectives. You no doubt have learned how to do this in other educational experiences. Here are several samples of science objectives. Objectives should focus on what you want the student to learn, communicate clearly your intentions, and indicate how success can be achieved.

Sample Objectives

- The student will solve seven out of ten stoichiometry problems
- Given a box of ten rocks, the student will be able to sort the rocks into categories and subcategories based on physical appearance
- Given a problem, the student can design an experiment to test a hypothesis

We will return to stating objectives again as we examine the types of objectives that you will develop for teaching.

As you can tell from the title of this section, the categorization of objectives into nonskills and skills has emerged as four categories as shown in Figure 9.5. Cognitions and affects are nonskill learning objectives, while skill objectives are divided into the categories of cognitive skills and psychomotor skills.

Cognitions include the scientific concepts, principles, and generalizations (rock, mineral, force, matter, atom, erosion, photosynthesis) that form the substantive character of your miniunit and future units and courses of study. Affects are feelings about and attitudes toward the subject, or inner feelings concerning the student's self-concept, particularly with regard to his or her ability to think scientifically or be successful in science courses. Cognitive skill objectives refer to the student's intellectual abilities. Observing objects, inferring what happened from evidence, solving equations, and designing an experiment are examples of cognitive skill objectives. Psychomotor skill objectives include the obvious observable skills that students need in science courses. Using a meter rule, massing objects with a balance, focusing a microscope, and preparing a slide are examples of psychomotor skill objectives.

Let's look at each of these categories, and while you read through these sections, revise and rewrite your initial list of learning outcomes to produce a final list that includes examples from each of these four categories.

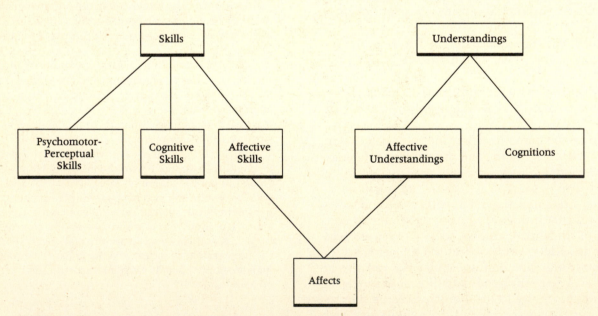

FIGURE 9.5 Concept Map of Intended Learning Outcomes. Based on Posner and Rudnitsky, *Course Design*, p. 62. Copyright © 1986 by Longman Publishing Group. Reprinted with permission from Longman Publishing Group.

TABLE   9.2

....

SUMMARY OF
LEARNING OUTCOMES
....

Cognitions	Concepts and propositions
Affects	Attitudes and feelings
Cognitive Skills	Cognitive abilities
Psychomotor Skills	Motor and laboratory abilities

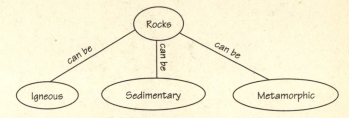

FIGURE 9.6   Concept Map of Rocks

Cognitions

Science textbooks are full of cognitions. Take a moment and look at the glossary (almost all science textbooks have one). These entries represent the major cognitions that the textbook authors felt were the most important science ideas that students should learn. Cognitions can be concepts. They can also be propositions or generalizations. Here are some examples of cognitions that interns wrote for science miniunits:

- The students will know the cell theory
- The students will understand the importance of a fossil
- The students will know the characteristics of an earthworm
- The students will know why chemical equations are used

It is helpful to differentiate between two types of cognitions, namely concepts and propositions. A concept is a synthesis of objects, events, or actions into a single idea or representation. Concepts are classifications enabling students to comprehend the vast array of experiences in science. Grouping rocks into categories of igneous, metamorphic, and sedimentary is a powerful tool in learning about earth materials. Students can learn these categories by first hand experiences with rocks of each class. A learning objective for understanding rocks might look like this:

The student should grasp the meaning of igneous (formed from the crystallization of molten materials), sedimentary (formed from the deposition of erosional material), and metamorphic (changed by heat or pressure) rocks.

To further define this learning objective, you should also show a concept map indicating the relationship between the concepts as shown in Figure 9.6. This approach to stating cognitions may be foreign to you if you have had previous experience writing "performance objectives." We will write performance objectives, but we will label them either cognitive skill objectives or psychomotor skill objectives.

Stating cognitions as we have done here recognizes the recent work in cognitive psychology. As Champagne and Klopfer point out, "the new cognitive perspective directs greater attention to the structure of the student's knowledge."[7] They recommend using networks or concept maps as an alternative way to state objectives for science teaching. For example, instead of writing objectives in the following manner, they propose that it is productive to write learning outcomes in terms of the cognitive processes and knowledge structures the students ought to learn, as shown in Figure 9.7.

- The student will be able to describe how waves are created and how they transfer energy
- The student will be able to distinguish between longitudinal and transverse waves
- The student will be able to measure the approximate speed of sound in air

Propositions are learning objectives in which more than one concept is linked together. Linking the concepts of mass and acceleration can give us the proposition of momentum. Or the concepts of force and acceleration can be combined to give the following proposition: The acceleration of an object is directly related to the force exerted on it.

Propositions, like concepts, can be learned through actual experience with the concepts, but in a more indirect manner. Propositions by their very nature are abstractions, and require the student to make associations and relationships between concepts. It is important to identify the concepts students will need to know to understand propositions. As you examine the concept map in Figure 9.7 note the array of concepts that must be understood to learn the major propositions in this miniunit on sound.

[7]Audrey B. Champagne and Leopold E. Klopfer, "Research in Science Education: The Cognitive Psychology Perspective," in *Research Within Reach: Science Education*, David Holdzkom and Pamela B.Lutz, eds. (Washington, D.C.: National Science Teachers Association, 1985), p. 175.

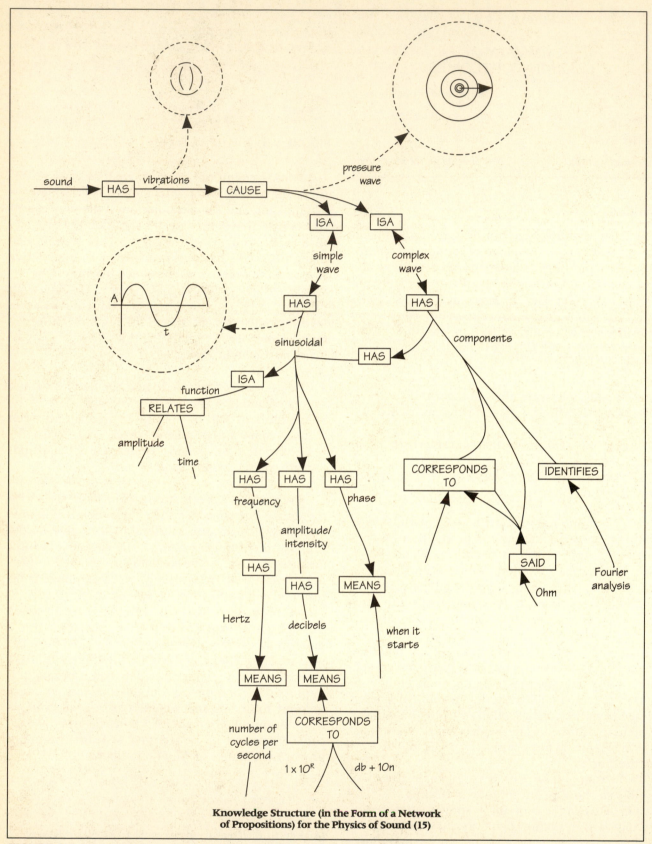

**Knowledge Structure (in the Form of a Network
of Propositions) for the Physics of Sound (15)**

FIGURE 9.7 Stating Objectives in the form of a Concept Map. Greeno, J.G. "Cognitive
Objectives of Instruction: Theory of Knowledge for Solving Problems and Answering
Questions." In *Cognition and Instruction*, D. Klahr, ed. Killsdale, NJ: Erlbaum Associates, Inc.,
1976. Reprinted with permission of the publisher.

At this point you should examine your list of outcomes and your concept map and rewrite your cognitions as either concrete concept objectives or propositional objectives. Here are some examples to help you in this task.

Sample Cognitions[8]

- The student grasps the concept of an ecosystem and its four subdivisions. (Grasping the concept here involves knowledge of what constitutes an ecosystem and what does not, as well as a knowledge of what phenomena are explained by "ecosystem.")
- The student understands the interrelationships of the component parts of ecosystem subdivisions. ("Understands" here includes the belief that these subdivisions are related, knowing what counts as evidence for this belief, and being aware of the subdivisions' action as one system.)

Affects

Science courses of study should also include objectives that deal with feelings, values, and attitudes. The following are examples of objectives that focus on feelings, values, and attitudes:

- The student should learn to expect and anticipate change
- The student should have confidence in his or her ability to plan and carry out research investigations
- The student should have respect for other living organisms
- The student should learn that knowledge is tentative
- The student is willing to listen to the opinions of others
- The student is willing to reject scientific facts on the basis of new and valid information

Affects are often examined in terms of David Krathwohl's system of categorizing values.[9] Affects need to be acted on to be truly congruent with Krathwohl's highest category, characterization (the student acts on a set of internalized values, is generally predictable, and is willing to act individually and shows self-reliance). If you include affects in your miniunits (and you should), realize that students can learn affects at different levels.

Useful verbs to help you write affects include *respond, choose, value, prize, trust, organize, tend, like,* and *respect.* Using the examples and suggestions, examine your initial list of outcomes, and write at least one affect for your miniunit.

Cognitive Skills

A cognitive skill is an objective that describes intellectual competencies that students will acquire in the science course. No doubt your list of original learning objectives contained cognitive skill objectives. Cognitive skills can be recognized quite easily because they are normally stated in performance or behavioral terms. For example, look at this list of cognitive skill objectives:

- The student will be able to describe the characteristics of sound
- The student will be able to distinguish between an atom and a molecule
- The student will be able to predict the location of the moon in the daytime sky

Cognitive skills are the most common form of objectives that teachers use to describe student learning. If you examine science textbooks, the objectives that are included in each chapter are generally cognitive skills.

When writing cognitive skills, you should consider the *action words* that you are using to describe the performance of the student. One convenient source of action words are the science processes that were discussed earlier. These include *observe, classify, infer, measure, collect* and *analyze* (data), *predict, formulate,* and *test* (hypotheses), and *design* (experiments). Other action words include *write, state, select, point to, name, match, list, interpret, identify, describe, distinguish, contrast,* and *compare.*

Cognitive skill objectives (as well as objectives for affects and psychomotor skills) should be written to include

1. What the learner is expected to do
2. What conditions are imposed when the student is asked to do it
3. How you will recognize the student's success in doing it

Finally, when designing science units, you should consider the intellectual level of the cognitive skills by using Bloom's taxonomy as a planning tool. hierarchy of six levels (see Table 9.3).[10] They are

[8]Posner and Rudnitsky, *Course Design,* pp. 97–98. Copyright © 1978 by Longman Publishing Group. Reprinted with permission from Longman Publishing Group.

[9]David Krathwohl et al., *Taxonomy of Educational Objectives: Handbook 2, Affective Domain* (New York: David McKay, 1965).

[10]Benjamin Bloom et al., *A Taxonomy of Educational Objectives: Handbook 1, The Cognitive Domain* (New York: David McKay, 1950).

TABLE 9.3

BLOOM'S TAXONOMY
OF COGNITIVE SKILLS

Category	Explanation	Illustrative Performance Behaviors and Learning Outcome
1.0 Knowledge	Knowledge is defined as the remembering of previously learned specific facts to complete theories, but all that is required is the bringing to mind of the appropriate information. Knowledge represents the lower level of learning outcomes in the cognitive domain.	*Performance behaviors:* Defines, describes, identifies, labels, lists, matches, names, reproduces, states. *Sample learning outcome:* The student will be able to *define* the following terms: *solute, solution.*
2.0 Comprehension	Comprehension is defined as the ability to grasp the meaning of material. This may be shown by translating material from one form to another (words to numbers), and by interpreting material (explaining or summarizing). These learning outcomes go one step beyond the simple remembering of material, and represent the lowest level of understanding.	*Performance behaviors:* Converts, explains, extends, generalizes, gives examples, infers, paraphrases, rewrites, summarizes *Sample learning outcome:* The student will be able to *summarize* the process of photosynthesis.
3.0 Application	Application refers to the ability to use learned material in new and concrete situations. This may include the application of such things as rules, methods, concepts, principles, laws, and theories. Learning outcomes in this area require a higher level of understanding than those under comprehension.	*Performance behaviors:* Changes, computes, demonstrates, discovers, manipulates, operates, prepares, produces, relates, shows, solves, uses. *Sample learning outcome:* The student will be able to *compute* the area of a forest given the formula for area, and a map of the targeted forest.
4.0 Analysis	Analysis refers to the ability to break down material into its component parts so that its organizational structure may be understood. This may include the identification of the parts, analysis of the relationships between parts, and recognition of the organizational principles involved.	*Performance behaviors:* Breaks down, diagrams, differentiates, discriminates, distinguishes, outlines, points out, relates, selects, separates, subdivides. *Sample learning outcome:* The student will be able to *diagram* in the form of a concept map the concepts related to photosynthesis.
5.0 Synthesis	Synthesis refers to the ability to put parts together to form a new whole. This may involve the production of a unique communication (theme of a report), a plan of operation (experimental design), or a set of abstract relations (scheme for classifying information). Learning outcomes in this area stress creative behaviors, with major emphasis on the formulation of new patterns or structures through combination.	*Performance behaviors:* Combines, compiles, composes, creates, devises, designs, generates, modifies, organizes, plans, rearranges, reconstructs, reorganizes, revises, rewrites, writes. *Sample learning outcome:* The student will be able to *plan* an experiment to test a hypothesis.

(Table continued on p. 311)

(Table continued from p. 310)

6.0 Evaluation

Evaluation is concerned with the ability to judge the value of material (science article or newspaper report, research study), for a given purpose. The judgments are to be based on definite criteria. These may be internal criteria (organization) or external criteria (relevance to purpose) and the student may determine the criteria or be given them. Learning outcomes in this category are the highest in the cognitive skill hierarchy because they contain elements of all the other categories

Performance behaviors:
Compares, concludes, contrasts, criticizes, describes, discriminates, explains, justifies, interprets, rates, relates, summarizes.

Sample learning outcome:
After viewing the videotape speeches by environmental activists, the student will be able to *rate* the speeches in terms of content accuracy, persuasion and tact.

Based on Benjamin Bloom et al., *A Taxonomy of Educational Objectives: Handbook 1, The Cognitive Domain* (New York: David B. McKay, 1950).

shown here with representative performance or intellectual behaviors characteristic of each level.

Examine your list of intended learning outcomes, and rewrite those that fall into the category of cognitive skills.

Psychomotor Skills

Science teaching should emphasize the learning of psychomotor skills, which certainly should include the work students do in the science laboratory, as well as while engaged in the variety of hands-on activities that you will prepare for them. There are a number of categories of psychomotor skills, including moving, manipulating, communicating, and creating.[11]

Psychomotor skill objectives are most obvious in the science laboratory. Although there is not full agreement on what outcomes are derived in laboratory activities in science, the following represent the major categories:[12]

1. Methodical procedure
2. Experimental technique
3. Manual dexterity
4. Orderliness

From this list it is obvious that laboratory science work involves cognitions and affects as well as psychomotor skills. Therefore, it is important to note that

most psychomotor skill objectives that you write will involve cognition as well. For example plotting a velocity-time graph with constant slope combines the psychomotor skill of graphing and cognitive skills and an understanding of velocity, time, and slope.

At this stage of your work, you should have a list of intended learning outcomes that includes cognitions, affects, cognitive skills, and psychomotor skills. You are now ready to move to the next step: listing potential activities.

Step 9: List Potential Activities

Potential activities should emerge from the list of learning outcomes that you have developed. One way to look at the activities you select for your mini-unit is to use the concept of "instructional foci" as used by Posner and Rudnitsky. Instructional foci are the means by which you will help students to attain the objectives you have created. For example, "understanding the nature of wave movement," is not an example of an instructional foci. It's an objective. However, using a toy slinky to design experiments through which wave movement might be observed is an example of an instructional foci. Instructional foci can be reading textbooks, field trips and experiences, case studies, debates, role playing or simulations, cooperative group activities, hands-on experiences, and guest speakers. Each serves as a focal point for a type of classroom experience designed to facilitate the learning of science objectives. Let's look at an example instructional foci for a miniunit on human genetics.

[11]Leslie W. Trowbridge and Rodger W. Bybee, *Becoming a Secondary School Science Teacher* (Columbus, Ohio: Merrill, 1990), p. 150.
[12]J. R. Eglen and R. F. Kempa. "Assessing Manipulative Skills in Practical Chemistry," *The School Science Review* 56 (1978): 261–273.

Hands-on and Minds-on Instructional Foci

Miniunit: Human genetics
Objectives

1. The students will understand the patterns of inheritance for several genetic diseases. (cognition)
2. The students will be able to construct pedigrees and use them to determine inheritance patterns. (cognitive skill)
3. The students will be inclined to express ethical and moral positions with regard to genetic diseases. (affect)

Potential Instructional Foci

- Reading text, *Modern Biology*
- *Brave New World* by Aldous Huxley
- Debate pro and con Huxley's book
- Computer simulations of genetics
- Lecture with visuals
- Pedigree analysis worksheets
- Field trip to a genetics research laboratory

Each of these instructional foci has an integrity of its own. Each can be used to achieve one or more different outcomes.

List potential instructional foci (activities) for the intended learning outcomes that you developed. Realize that some of your instructional foci may score low on a feasibility scale (e.g., field trip); however, it is important to investigate all possibilities at this stage before you actually write your lesson plans. And that's the next step.

Step 10: Develop Specific Lesson Plans

Your miniunit should contain between three and five lesson plans. Draft a set of plans based on the curriculum that you have designed, (e.g., the minicourse rationale,) and the set of learning objectives.

There are a number of formats that you can choose in designing lesson plans. Following are two models that you can use to write lessons for your miniunit.

Lesson Plan Template 1

This lesson plan template uses an outline format and is the most common form of lesson plan design. The major headings and subtopics are as follows:

A. Title of Lesson
B. Grade Level
C. Objectives
  1. Cognitive
    a. Cognitions
    b. Cognitive Skills
  2. Affects
  3. Psychomotor Skills
D. Materials/Media
E. Procedure
  1. Motivational Activity
  2. Development of Concepts
  3. Closure
F. Evaluation/Assessment

Sample Lesson 1

The following is a lesson developed by a teacher-intern during a summer institute. The lesson was the first of a four-lesson miniunit entitled "Drugs, Alcohol, and Tobacco".[13] Note how the author includes a variety of objectives and involves the students in whole class, small team, and dyad activities.

A. Title: Drug Abuse
B. Grade 8
C. Objectives
    1. The students will understand drug abuse.
    2. The students will be able to describe the effects of drugs on the body.
    3. The students will be able to compare psychological and physical dependence.
    4. The students will develop confidence in expressing opinions about drug usage based on an understanding of the effects on the body.
D. Materials/Media
    1. Overhead projector
    2. Transparencies showing the lesson objectives
    3. Pretest sheet/ one per student
    4. Variety of over-the-counter drugs: beer can, aspirin bottle, cold tablet package, cigarette pack, antacid sample, coffee can, tea bag
    5. Sample of generic drug
    6. Chart of commonly abused drugs, one per student
    7. Three-by-five note cards (for teacher use)

[13]Written by Mr. Mal Wallace, Intern at Georgia State University, currently teaching at the Cottage School, Roswell, Georgia.

(Continued on p. 314)

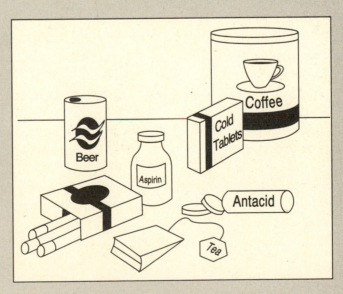

FIGURE 9.8   Over-the-counter drugs

(Continued from p. 313)

E. Procedure
1. Motivation
a. Show students a variety of over-the-counter drugs; aspirin, cold tablets, beer can, tea bag, coffee can, cigarette package. Have them describe what they can see; ask how these groups can be categorized; and ask how each of the items can be used.
b. Inform the students of the objectives by using the transparency.
2. Development
a. Students will be administered a pretest in order to determine their current knowledge of drugs and drug abuse. (Test contains five questions asking them to identify a drug, the effects of the following on the body: alcohol, cigarettes, marijuana, crack.
b. Students in small teams are asked to discuss and write a definition for each of these terms: counter drugs, generic drugs; to discuss what drugs can't be bought over the counter; and why some drugs require a doctor's slip.
c. Teacher asks students to compare the heroin or crack addict and the ten-cup-per-day coffee drinker.
d. Teacher discusses the dangers of drug abuse focusing on the followingconcepts: dependence, withdrawal, tolerance.
e. Teacher presents drug abuse chart, and asks students to work with a partner to answer a series of questions such as; What type of drug is crack? Whateffect does it have on the central nervous system? What drugs have severe withdrawal symptoms? Which type do not result in physical dependence when abused? Which type results in strong psychological dependence when used?
3. Closure
a. Teacher asks each partner-pair to write a short statement which defines drug abuse; compare psychological and physical dependence, and how this knowledge can be of benefit to people.
b. Teacher calls on partner-pairs at random and asks them to present their summary to the class.
c. Teacher assigns homework: find at least three advertisements in magazines in which drugs are being sold. Bring to class tomorrow.
F. Assessment
1. Teacher observes groups, and notes contributions of each member.
2. Teacher collects and evaluates summaries written by partner-pairs.

Lesson Plan Template 2

An alternative method for writing lesson plans is to align the lesson objectives, activities, materials, and assessment activities side-by-side:

Title: _____ Grade Level _____ Course _____

Objectives	Procedures	Materials/Media	Evaluation

Sample Lesson Plan 2

Title: *Make it light**   Grade Level: <u>Eight</u>   Course: <u>Physical Science</u>

Objectives	Procedures	Materials/ Media	Evaluation
1. The students will grasp the meaning of electric current, resistance, and potential difference.

2. The students will be able to predict how the elements of a circuit should be connected.

3. The students will be able to assemble an electric circuit. | *Motivation*
1. The students will simulate an electric circuit. Each student will push a box around the room along a path defined by masking tape. Desks will be placed in the path of the students.
2. Teacher will introduce the purpose and objectives of the lesson.

Development
1. The teacher will organize the students into teams of four. Each will select a name for its team from a list of electrical terms shown on the board: ohm, resistance, wire, battery, current. Each team will be asked to discuss and then to write "working definitions" of electric circuit, load, and source.
2. Teacher will ask students to suggest examples of electric circuits they have encountered and identify the load and source in each. | Masking tape
Shoe boxes
Desks

Chalkboard
Chalk
Per team: small light bulb, battery, one piece of wire |

Teacher monitors groups; records group behavior; asks questions. |

(Continued on p. 316)

*Written by Pamela Callahan Sandlin, teacher-intern, Georgia State University.

(Continued from p. 315)

Objectives	Procedures	Materials/ Media	Evaluation
	3. Teacher gives each group a bag containing a light bulb, battery, and two pieces of wire. They will be asked to draw a picture of how they think the items should be connected to make the bulb light. Each group will connect the items to test their electric circuit. They will experiment to find the correct connection. Each group will be asked to identify the source and the load.		
	Closure 1. Each group draws a picture of their circuit on a large sheet of chart paper. Pictures are assembled by the teacher, who asks each group to explain their results. 2. Teacher assigns homework. Each team member is given part of the homework assignment (each does two problems from the text), and explains that the whole team will turn in all problems tomorrow.	Chart paper Crayons	Teacher collects each group's chart and evaluates the work. Teacher explains that homework will be graded such that a group grade is given for the homework assignment.

Step 11: Develop an Assessment Plan

There are two elements of assessment that are important at this stage in planning a miniunit. Assessment should serve two broad purposes:

1. To answer questions and provide feedback with regard to student learning
2. To provide data with respect to the effectiveness of instructional plans

Your minicourse should contain elements of both in the assessment plan that you develop.

The most common or traditional approach to this step is to prepare a post evaluation instrument (a quiz or test) that would be administered when the students complete the miniunit. The instrument can be used to provide feedback on student learning and provide additional feedback from the students about the effectiveness of the unit.

The assessment instrument should be designed to evaluate each type of learning outcome that you included in the unit. You should then develop measures to evaluate

1. Cognitions
2. Cognitive skills
3. Affects
4. Psychomotor skills

You should also ask the students for feedback on how they reacted to the miniunit. One approach is to use the format you used in reflective teaching (see Chapter 7). Thus, you could include a separate form (Figure 9.9) allowing students to complete these questions anonymously:

---

1. During the miniunit how satisfied were you as a learner?
   _____very satisfied
   _____satisfied
   _____unsatisfied
   _____very unsatisfied
2. What could your teacher have done to increase your satisfaction?
3. What were your favorite activities? Why?
4. What were your least favorite activities? Why

FIGURE 9.9 Student Feedback Form

Assessment can also include other items and approaches. Informal and semiformal methods, as well as having students develop portfolios of their work, can be incorporated into an assessment plan.

Look ahead to the section on assessment, which provides specific suggestions for items, problems, questions, checklists, and questionnaires that you could use to evaluate student learning, and prepare an assessment plan for your miniunit.

Step 12: Implement the Miniunit

If it is possible, make arrangements to teach your miniunit to a group of secondary students. This can be done with either a whole class or with a small group of students. If this is not possible, try to arrange to teach one lesson from the miniunit to a group of peers. In either case, videotape one of your lessons to be used for reflection and feedback.

Step 13: Feedback and Reflection

Curriculum development and instructional planning are part of a large cycle. One of the most important parts of the cycle is a period of time devoted to gathering feedback on the unit, and reflecting on the effectiveness of the unit. Here are some questions to consider:

1. To what extent did students attain the learning objectives of the unit?
2. In what way did the students respond to the activities, assignments, and content of the miniunit?
3. Were the goals and content of the miniunit appropriate for the students in this class?
4. What activities did students seem to enjoy? Did these contribute to their understanding of the main cognitions of the unit?
5. What activities didn't students seem to enjoy? Do you think these activities are important? Do they contribute to the students' understanding of the main cognitions? How would you change these activities, if necessary?
6. Would you use this unit again in its present form? If not, how would you change it? What modifications would you make?

MODELS AND SAMPLE LESSON PLANS

This section is designed to provide examples of lessons based on the models that were presented in Chapter 7. Now that you have developed a miniunit, you might want to examine these lessons in more detail. Following you will find four types of lessons based on the models that were presented earlier:

1. Direct-interactive teaching lesson plan
2. Cooperative learning lesson plan
3. Inquiry or laboratory lesson plan
4. Conceptual change lesson plan

Direct-Interactive Teaching Lesson Plan Guide

The elements of a lesson based on direct-interactive teaching are shown in the left facing box. A sample lesson based on the model is presented in the right facing box.

Lesson Plan Guide: Direct-Interactive Teaching

1. Anticipatory set
   - Focus students.
   - State objectives.
   - Establish purpose.
   - Establish transfer.
2. Instruction
   a. Provide information.
      - Explain concept.
      - State definitions.
      - Provide examples.
      - Model.
   b. Check for understanding.
      - Pose key questions.
      - Ask students to explain concepts, definitions, attributes in their own words.
      - Have students discriminate between examples and nonexamples.
      - Encourage students to generate their own examples.
      - Use participation.
3. Guided practice
   - Initiate practice activities under direct teacher supervision.
   - Elicit overt responses from students that demonstrate behavior in objectives.
   - Continue to check for understanding.
   - Provide specific knowledge of results.
   - Provide close monitoring.
4. Closure
   - Make final assessment to determine if students have met objectives.
   - Have each student perform behavior.
5. Independent practice
   - Have students continue to practice on their own.
   - Provide knowledge of results.

Sample Direct-Interactive Teaching Lesson: Natural and Processed Foods

1. Anticipatory set
   a. Focus
      Teacher holds up both a carrot and frosting-filled cake. The teacher says, "You are what you eat."
   b. Objectives
      Today you will learn to understand food. You will learn the difference between natural and processed foods.
   c. Purpose
      This lesson will help you understand the foods you eat, and perhaps influence the foods that you eat in the future.
2. Instruction
   a. Provide information
      (1) Explanation: Teacher gives a minilecture on relevant facts concerning vitamins, calories, minerals, preservatives, package labels.
      (2) Model: Teacher uses the school lunch menu and lists the foods into two categories on the chalkboard: natural foods and processed foods.
      Teacher explains why each is placed in the appropriate category.
   b. Check for understanding
      (1) Key questions: What is the difference between a natural food and a processed food? What is a preservative? What is an example of a natural food? What is an example of a processed food? What can you learn from the label on an item you buy in the store?
      (2) Active participation: Teacher holds up real food items. Students indicate whether item is natural or processed by means of finger signals or colored cards.
3. Guided practice
   a. Activity
      Students work in small teams. Each group is provided with a stack of pictures of food items. Students divide pictures into two categories, natural foods and processed foods. Teacher monitors, assists, and provides feedback.
   b. Feedback
      Teacher moves from group to group and asks key questions to check for understanding. Also provides specific information as needed.
4. Closure
   Teacher designates one side of the room as "Natural Food" side, the other as the "Processed Food" side. Both kinds of foods are written individually on three-by-five cards. Students draw a card and line up on the appropriate side of the room.
5. Independent practice
   Students assess food in refrigerator or cupboards at home. Students make list with two categories (natural and processed) based on food in refrigerator or cupboard. Lists are analyzed in small groups the next day.

Cooperative Learning Lesson Plan Guide

The following lesson plan guide and sample lesson should help you develop cooperative learning lessons.

Lesson Plan Guide: Cooperative Learning

1. Preliminary information
   a. Course
   b. Grade level
   c. Lesson summary
   d. Objectives
   e. Materials
   f. Time required
2. Decisions
   a. Group size
   b. Assignment to groups
   c. Roles
3. Procedures
   a. Instructional task
   b. Positive interdependence
   c. Individual accountability
   d. Criteria for success
   e. Expected behaviors
4. Monitoring and processing
   a. Monitoring
   b. Intervening
   c. Processing

Sample Cooperative Learning Lesson: Identifying Skeletons[14]

1. Preliminary information:
   a. Subject area: life science
   b. Grade level: junior high
   c. Lesson summary
      Students are given pictures of skeletons of vertebrates and must figure out what they are.
   d. Objectives
      (1) Students will observe various vertebrate skeletons.
      (2) Students will identify animals by skeleton characteristics.
      (3) Students will recognize adaptations of skeletons that enable the animal to survive in its environment.
   e. Materials
      (1) Pictures of 14 vertebrates (per group): anteater, penguin, lion, hippopotamus, walrus, deer, kangaroo, frog, camel, bat, crocodile, elephant, giraffe, ape
      (2) Teacher transparency of the skeletons
      (3) Answer key transparency
   f. Time required: 45 minutes
2. Decisions
   a. Group size: three
   b. Assignment to groups
      Either random—counting off—or teacher assigned to groups with high, medium, and low students in each group.
   c. Roles: Recorder—records answers of the group; tracker—keeps the group on the task by watching the time and not letting the group dwell on one skeleton too long; encourager—makes certain each member is participating and invites any reluctant members in by asking them for their ideas
3. Procedures
   a. Instructional task
      Teacher says, "Your group will be given a handout containing pictures of 14 modern-day vertebrates' skeletons. You will have 15 minutes to identify the 14 animals by their skeletons."
   b. Positive interdependence
      Teacher explains; "I want one set of answers from the group. When you sign your name it means that you agree with the answers and can defend and explain them. Also, you will have a job to do to help the group achieve the task." At this stage the teacher assigns the students to groups, or has them formed randomly.
   c. Individual accountability
      "Each group member must be able to identify each skeleton and tell why it fits that classification. I may ask any one of you to come to the overhead and point out the features of the skeleton that led to your group's identification of it."

(Continued on p. 320)

[14]Susan Ward, "Identifying Skeletons," in *Structuring Cooperative Learning: Lessons Plans for Teachers,* Roger T. Johnson, David W. Johnson, and Edythe Johnson Holubec, eds. Edina, Minn.: Interaction, 1987), pp. 207–210. Used with permission.

(Continued from p. 319)
>   d. Criteria for success
>   Groups that correctly identify ten skeletons will have mastered the task. Groups that correctly identify more than ten skeletons will re- will receive bonus points.
>   e. Expected behaviors
>   "I expect to see all group members participating, each one performing his or her role, and everyone justifying answers by pointing out features on the skeletons."
>   4. Monitoring and processing
>   a. Monitoring
>   Teacher circulates to see how the groups are doing and give hints when groups are at a standstill. Hints should be general, such as which class the animal belongs to (amphibian, reptile, etc.). Also, watches and encourages participation from all group members.
>   b. Closing
>   Teacher identifies the skeletons at the overhead projector and has the groups assess how well they did. Has members of groups that successfully identified the skeletons explain the answers to the groups that missed them.
>   c. Processing
>   Teacher comments on the behaviors he or she observed, then has each student tell his or her group members what they did best.

Inquiry or Laboratory Lesson Plan Guide

Inquiry activities and laboratory activities can be approached by employing a simple learning cycle consisting of

> Prelaboratory (prelab)
> Laboratory
> Postlaboratory (postlab)

Lesson Plan Guide: Inquiry or laboratory

1. Prelab
   a. Objectives
   b. Overview of laboratory or inquiry
   c. Alternate demonstration, discrepant event, or inquiry session
   d. Safety precautions
   e. Group size
   f. Assignment to groups
   g. Roles
2. Lab
   a. Monitor
   b. Intervene
3. Postlab
   a. Whole class processing
   b. Application
   c. Clean up

Sample Inquiry Lesson: Investigating Mass, Volume, and Density

1. Prelab
   a. Objectives
      The student will devise methods to determine the density of an ice cube.
   b. Overview of laboratory or inquiry
      Review with students the laboratory that was done previously in which they learned to determine the density of various objects and substances. Students determined the density of solid objects such as pebbles, marbles, and paper clips. Explain that today they are going to apply their knowledge of density to solve an inquiry problem.
   c. Discrepant event.

Display two beakers containing equal amounts of clear liquids so all the students can see them. (One contains water and the other contains alcohol.) Do not tell the students what the liquids are, or that they are different. Ask the students what would happen if you were to put ice cubes in each liquid. Then place an ice cube in each beaker and ask the students to explain the results. Students might remark that the ice must be strange or funny. Interchange the ice cubes. The students might comment that the water must be funny or strange. Lead the stu-

(Continued on p. 321)

FIGURE 9.10   Inquiry Demonstration

(Continued from p. 321)

dents to realize that one liquid supports the ice the way water does and the other does not. Students might guess that the first liquid is water and the second liquid which does not support the ice must be different from water.

Ask the students to establish a relative density scale for the three materials, ice, water, and unknown liquid. (Answer: alcohol, ice, water.)

Now pose the problem: Find the density of the ice cube. Show the students where the materials are located that they can use to investigate the problem.

   d. Safety precautions

     Warn the students not to taste the liquids or the ice cubes.

   e. Group size

     Three

   f. Assignment to groups

     Random or high, low, average student per group.

   g. Roles

     Recorder: Keeps a record of the methods that the group attempts. Records any results made by group.

     Tracker: Keeps the group on task by watching the time, and not letting the group dwell on any one method.

     Materials handler: Is responsible for obtaining materials, and making sure they are returned at the end of the lab.

2. Lab

   a. Monitor

     Circulate around the room from group to group. Remember that the most important part of this laboratory is for the students to devise their own method of determining the density of an ice cube.

(Continued on p. 322)

(Continued from p. 321)

Some groups will need encouragement. As you visit groups ask questions, but frame them so that they will think about the information they really need to determine density.

    b. Intervene

      If you need to intervene in a group ask them to describe what they have done to determine the density of the ice cube, rather than tell them how.

3. PostLab

    a. Whole class processing

      Ask each group to give a one-minute or less description of their method. You can expect the following methods:

        (1) Used a metric rule to calculate volume, and balance to determine mass.

        (2) Pushed the ice cube into water in a graduated cylinder and then used balance to determine mass.

        (3) Ice is placed in alcohol and it will sink. As water is added and the mixture stirred, the ice will rise. When the density of the liquid is equal to the ice, the cube will be suspended. Density of liquid is determined by measuring volume and mass of a sample of the liquid.

        (4) Since ice sinks in alcohol, mass is determined on a balance, then volume is determined by using the displacement procedure in alcohol.

      Have the groups analyze the results they got for the density of the ice cube. How do their results vary? What is the average or central tendency of the groups' results? How does this compare with the known value of the density of ice?

    b. Application

      Have each team select one of the following and devise a method to determine its density : an elephant, a teenager, an automobile, a flea.

    c. Clean up

      Materials handler should return materials to storage area.

Conceptual Change Lesson Plan Guide

Conceptual change teaching employs another form of learning cycle. Three phases are included: the exploration phase, the concept development phase, and the concept application phase. The exploration phase includes not only the exploration of the ideas and concepts of the lesson but is used to ascertain student misconceptions.

Lesson Plan Guide: Conceptual Change

1. Exploration phase
   a. Ascertain student misconceptions.
   b. Engage student interest through motivating activity or experience.
   c. Facilitate discussion among students in small teams.

(Continued)

Conceptual Change Lesson: What's in Soil?[15]

1. Exploration phase
   a. Ask the class questions similar to
      (1) What is soil?
      (2) What's in soil?
      (3) Why are soils important?

These questions can be used to determine what students know about soil and what misconceptions they might have.

   b. Each student team is given a Baggie containing dry soil, one sieve set, one magnifier, white paper, and one balance.

(Continued on p. 323)

[15]Based on *It's Just Dirt, Isn't It?: A Model Science Unit* (Columbia, S.C.: Office of Research, South Carolina Department of Education, 1990), pp. 16–18. Used with permission.

(Continued from p. 322)

d. Pose open-ended questions.

e. Monitor and observe student groups.

2. Concept development phase

a. Help students invent or discover concepts.

b. Give information.

c. Guide whole class discussion.

d. Provide explanations.

e. Define appropriate concepts and terms.

3. Concept application phase

a. Students apply newly learned concepts.

b. Students interact and apply concepts in small teams (or individually).

c. Facilitate discussion by means of open-ended questions.

d. Observe and monitor students in learning teams.

(Continued from p. 322)

c. Students complete an activity sheet as shown in Figure 9.11

2. Concept development phase

a. Discuss answers to the lab activity with the whole class.

(1) Are soils made up of one particle size?

(2) What is the largest particle size?

(3) What is the smallest particle size?

b. Introduce the following concepts: soil texture classification scheme, inorganic fraction, organic fraction.

3. Concept application phase

a. Working in teams, have students classify their soil samples using the classification scheme introduced in phase 2.

b. Ask students in their teams to answer these questions:

(1) Why do gardeners sometimes need to put additional sand into the soil?

(2) Why is it important to till and mix up the soil before planting?

(3) How has technology changed the way a farmer tills the soil?

c. For homework, have students work on these "think about" questions:

(1) If you have a soil sample that contains 20 grams of organic material and 30 grams of inorganic material, what percentage of the soil is organic?

(2) What is the difference between the organic and inorganic parts of soil?

(3) Why does water form a puddle on some soils and sink into the ground on others?

LESSON 3	**ACTIVITY #2**

"What's in Soil?"

Date: _____   Student Name: _____

A. Getting Started:

–After you are in your groups, select one person to come to the materials area and collect the following:

- –"Baggie" containing dry soil   –1 magnifier   –1 balance
- –1 sieve set   –white paper

B. Complete the following steps:

1. Without opening the baggie, describe the soil in your own words. _____
_____

2. Record the mass of your soil sample: _____ **grams.**

3. Make sure the sieve sets are arranged so that the sieve with the largest holes is on top, next largest holes, next to top, smallest holes next to bottom and the solid pan on the bottom.

4. Carefully empty the dry soil from the baggie into top sieve.

5. Replace sieve cover and gently shake sieve set BACK AND FORTH, NOT UP AND DOWN for 1 minute.

6. Empty the material that remained on the sieve with the **largest holes** onto a white piece of paper.

Describe this material. _____
_____

Mass _____ gms.   Percentage _____ %

8. Repeat steps f and g for material retained on the sieve with the **next largest holes**.

Describe this material. _____
_____

Mass _____ gms.   Percentage _____ %

9. Repeat steps f and g for material retained on the sieve with the **next-to-smallest holes**.

Describe this material. _____
_____

Mass _____ gms.   Percentage _____ %

10. Repeat steps f and g for material retained on the sieve with the **smallest holes**.

Describe this material. _____
_____

Mass _____ gms.   Percentage _____ %

11. Repeat steps f and g for material that collected in the solid pan at the bottom.

Describe this material. _____
_____

Mass _____ gms.   Percentage _____ %

12. Answer the following questions:

a. Are soils made up of one particle size? _____

b. What is the largest particle size? _____

c. What is the smallest particle size? _____

FIGURE 9.11 Conceptual Change Activity Worksheet. *It's Just Dirt Isn't It?: A Model Science Unit* (Columbia, SC: Office of Research, South Carolina Department of Education, 1990), p.10. Used with permission.

DESIGNING A
COURSE OF STUDY

Designing a course of study involves the same principles that you employed to develop a miniunit. The difference is that for course design you will be clustering learning outcomes, and sequencing the resulting units of instruction. Planning interesting courses of study starts with the consideration of goals and rationale for science teaching, just as it did when you planned a miniunit. The following discussion explores some of the elements that should be considered in designing a science course.

Elements of a Course of Study

Some students were overheard making these remarks about one of their teachers: "I'm really looking forward to taking Mrs. D'Olivo's science course next year." "Yes, she plans interesting projects for the students, and her tests are not hard but they are very interesting." Science courses can be interesting to students if they are carefully planned, and if the needs of the students are taken into consideration.

Course planning involves the same processes that were used to develop a unit of teaching. In the section that follows, we will examine a high school science program entitled "Global Science," by John

Christensen,[16] to illustrate how to describe the elements of a course of study. There are several reasons for choosing this program. First, this is a program based on a textbook, and the odds are that most of your courses will be based on the use of a textbook. Second, Global Science is an example of an S-T-S course and reflects an interdisciplinary approach. There are five elements that should be considered in the course planning process:

- Rationale or philosophy
- Learning outcomes
- Units of study
- Instructional strategies (foci)
- Evaluation

These elements should be shared with the students as part of a course syllabus and distributed to them at the beginning of the course. An abbreviated form of them should be prepared for the students' parents.

To make the elements of a course plan concrete, here is a course plan for Global Science based on the curriculum written by John Christensen.

[16]John W. Christensen, *Global Science: Energy, Resources, Environment.* (Dubuque, Iowa: Kendall Hunt), 1991.

Sample Course Plan
Global Science: Energy, Resources, Environment

Rationale/Philosophy   The rationale for Global Science is described in a letter to the students, and in a list of assumptions about secondary school science education. The letter follows and you can use it as a model for writing your own course rationales.[17]

Dear Student:
You are living at an exciting time. In the next several years extremely important decisions are going to be made, and you will play a role in making them. These decisions will affect the position of your country in the world of nations, your feeling of who you are and how you relate to other people around you, the standard of living you will have, and the amount of personal freedom you will enjoy. Many of these decisions are related to energy, resources, and environment.

How well we make these decisions in large part depends upon how well we un-

---

[17]John W. Christensen, *Global Science: Energy, Resources, Environment* (Dubuque, Iowa: Kendall Hunt), 1991. Reprinted with permission of Kendall Hunt.

(Continued from p. 325)

derstand the issues. It is the purpose of this course to build basic background for understanding energy-resource-environmental problems. This is not just another science course. The problems we will be dealing with are in the here and now. You will find that the road you travel as you work through these pages can be an exciting journey—if you have the proper attitude.

Science is a tool at our disposal. It is a powerful tool, and it will play an important role at this turning point in our history. What is exponential growth? How bad is the energy-resource-environmental problem? Does the earth have a carrying capacity? Can we live better with less? What are our alternatives? How do we get there from here?

The study of this course won't provide all the answers, but you'll be much better prepared to face many issues because of your experiences with these materials.

In addition to this statement, Global Science is based on the following four assumptions about secondary school science education:

1. The study of science should be a meaningful endeavor for all students in a modern society.
2. Science is best learned by experimenting and analyzing data—not just by reading and doing problems. This is how basic science skills are gained.
3. Society is best served, and student interest held, when relevant material is emphasized.
4. If only one science course is required at the high school level, that course should emphasize the ecosystem concept and resource use.

Learning Outcomes   The following objectives constitute the learning outcomes for Global Science.

- To build a firm understanding of the ecosystem concept: the basic components of an ecosystem, how the living and nonliving interact, and the cyclic activities that take place (cognition)
- To build an understanding of the concepts and laws that govern our use of energy and mineral resources (cognition)
- To examine our present energy sources and consumption patterns and to apply this knowledge as a component of future planning (cognitive skill)
- To build an understanding of the fundamentals of exponential growth: the phenomenon, growth rates, doubling times, graphical representation, applications in resource depletion (cognition)
- To relate the fundamentals of exponential growth to population and resource situations at national and global levels (cognitive skill)
- To explore the many energy sources not extensively used today (cognition)
- To appreciate the complexity of developing useful new resources by examining problems related to their wide-scale use (affect)
- To examine how humans relate to their economic environment and understand how economics affects social and scientific decision making (cognition)
- To develop an awareness of the "spaceship earth" ethic and the interrelatedness of resources, economics, environment, food production, and population growth (affect)

- To develop basic scientific skills and attitudes as useful tools for problem solving (cognitive and psychomotor skills)
- To develop the skills and attitudes needed to continuously clarify and modify personal values and goals as we react to a changing world (affect)
- To enable students to observe, analyze, and draw conclusions from situations related to resources and the environment and to use this information to take effective action as responsible citizens (cognitive skill, affect)
- To acquaint students with careers and challenges in our resource industries and industries which will depend crucially on our resource decisions (affect, cognitive skill)

Units of Study   The approach to designing and sequencing the units of study in Global Science are based on the following themes:

1. The earth and its resources are finite.
2. Humans are partners with nature, *not* masters of nature.

The course is organized around ten units of study as follows:
1. The Grand Oasis in Space
2. Basic Energy/Resource Concepts
3. Energy and Society
4. Growth, Population, and Food
5. Energy Supply and Demand
6. Energy for the Future
7. Mineral Resources
8. Making Peace with Our Environment
9. The Economics of Resources and Environment
10. Options for the Future

Instructional Strategies (Foci)   Global Science involves the students in a number of instructional strategies as follows:

- *Global Science* (course text)
- Films (a detailed list is correlated by unit)
- Field trips
- Laboratory activities (every unit's development is focused around several laboratory, hands-on science activities)
- Discussion
- Problem-solving activities
- Debates
- Case studies

The instructional plan is organized on a week-to-week basis. Figure 9.12 shows the instructional plan and focal instructional activities for the first nine weeks.

Evaluation   A comprehensive evaluation plan is employed in the Global Science course. Since students are involved in a wide variety of learning activities, the evaluation plan reflects this diversity. Evaluation is based on the following criteria:

36 Week Schedule: Global Science

Wk	Topic	M	T	W	R	F
1	Orientation to Course and Themes	Orientation to Course and Requirement	*Lab:* A Voyage to Mars	Postlab— Components to an Ecosystem	*Film:* Cry of the March—or— Field trip to local ecosystem	*Summary:* Fundamental Laws of Human Ecology
2	The Ecosystem Concept	Labor Day	Microorganism Lab. S4—The Concept of Ecosystems.*	Lab observa. *Film:* Web of Life	*Activity:* All Tied Up	Lab observa. Test #1
3	Analyzing Ecosystems	Lab observa. *Lab* Analy. of Ecosystems	Observing ecosystems: What to look for	Observing ecosystems. *Activity:* The Social Req. of Survival	Observing ecosystems. Complete Microorg. lab	Complete observing ecosystems. Ch. 1 Review
4	Basic Energy/Resource Concepts	Microorganism postlab. Food preserv.	Ch. 1 Exam	*Lab:* Basic Ideas about Matter/ NRG	Basic Ideas (Day 2)	Conversion Factors
5	The Concept of Efficiency	*Lab:* Let's Have Tea	The Second Law of Thermodyn.	*Film:* Toast. Efficiency	*Activity:* Determ. Syst. Eff.	S9—En. Flow in Ecosystems
6	Energy Flow	Productivity of Ecosystem. *Film:* The Salt Marsh	En. Flow in a Modern Society	En. Quality Strategies for Energy Users	Ch. 2 Review	Ch. 2 Exam
7	Modeling Exponential Growth	*Lab:* Modeling Exponential Growth	Day 2 of lab	Day 3 of lab. Graphing instructions	Collect dice lab *Lab:* Analyzing Exponential Growth	Postlab dice lab. *Film:* World Population
8	Analyzing Population Growth	Analyzing Population Growth	Birth Control Methods	Info. on STDs. Fill out demographic survey	AIDS Info.	The Abortion Issue
9	Family Planning	A & D of Children	Demographic Transition	*Film:* Teenage Father	Ch. 3 Review	Ch. 3 Exam

FIGURE 9.12 First Nine Weeks' Teaching Schedule: Global Science. From *Global Science (Teacher's Guide), Third Edition* by John Christensen. Copyright 1991 by Kendall/Hunt Publishing Company. Used with permission.

1. Course portfolio: Each student maintains a portfolio of his or her work, including laboratory reports, a log in which the student makes weekly entries, copies of all homework, and reports.
2. Tests and quizzes: Weekly quizzes and chapter tests (provided with the curriculum) are administered.
3. Cooperative team investigation: Each student works with a team to investigate a global problem and present results to the whole class.
4. Readings: Students make concept maps of each chapter and write brief summaries of the main concepts of each map.
5. Laboratory activities: Labs are done in small teams (although membership changes during the course). Students are evaluated on the basis of their cooperativeness and contribution to the group's process, as well as on all activities. The principles of cooperative learning are employed in all laboratory activities.

INQUIRY ACTIVITY   **9.2**

Designing a Course of Study: The Course Syllabus

In this activity you identify and describe the major elements of a course of study for some area of middle or high school science. Your product will be a course syllabus that you would give to your students on the first day of class.

Materials

Secondary science textbooks, newsprint, marking pens

Procedure

1. Select an area of science from the following list to develop a course of study for a one semester or quarter course:

   Introductory biology
   Survey physical science
   Advanced chemistry
   Introductory chemistry
   Geology
   Introductory earth science
   Unified science (seventh grade)

2. Design a course of study by preparing a course syllabus that should contain the following elements:

   Rationale or philosophy
   Learning outcomes
   List of the units of study
   Instructional strategies
   Evaluation

3. Work with a partner to brainstorm the content of the elements of the course of study. Use large sheets of newsprint to record your ideas from the brainstorming session. On your own use the results of the brainstorming session to create the course syllabus

Minds on Strategies

1. Compare your product with those created by other students in your class. What problem areas were identified? How were they resolved?
2. How were the courses characterized? What philosophies guided the development of the courses?
2. Why are course syllabi important? Are they necessary?
3. What modifications in the plan presented here did you make?

ASSESSING LEARNING,
UNITS, AND COURSES OF STUDY

There are several aspects of assessment that we will explore in this section, each of which is pertinent to the notion of evaluation. Assessment is a process that will involve you in making judgments about the progress of your students and the effectiveness of your teaching plans. Assessment involves measurement and testing to gather data useful in making judgments. However, in the past several years new forms of assessment and evaluation have emerged from the emphasis on the theory and research in cognitive and motivational psychology, and cooperative learning.[18] This recent trend, which is incorporated in the discussion of assessment that follows, is characterized by the following elements:

1. Some of the new assessment strategies involve performance assessment. In this method students are required to actually perform the skills and strategies in the form of hands-on assessment questions

2. Assessment strategies provide teachers with better knowledge of their students strengths and weaknesses by giving teachers insights into students' process skill abilities

3. Assessment strategies rely on cooperative learning. In this approach students actually work together on assessment problems

4. Many assessment tasks are conceptual and therefore involve students in problem solving, higher-level reasoning, critical thinking, and creativity

5. Evaluation should be authentic. Assessment is authentic if it is "congruent with the results needed from science education; that is, if it asks students to demonstrate knowledge and skills characteristic of a practicing scientist or of the scientifically literate citizen."[19] Authentic evaluations involve the students in real experiences, for example, doing science activities, solving problems, and thinking critically and creatively

We will explore three general methods of evaluation—informal, semiformal, and formal—and relate each of these to the central issue of providing feedback to students about their learning, and to the teacher about the effectiveness of instruction.[20] Finally, we will conclude this section by showing how these methods can be integrated into a holistic form of evaluation called the portfolio.

Informal methods

It's unfortunate, but many teachers do not make use of informal evaluation methods to give students feedback, as well as using them as a measure of their own effectiveness. Informal methods involve the direct interaction of the teacher with the students, sometimes during classtime, but also at prearranged times for an informal session, such as a conference with a student. Let's examine a few informal techniques you can use in the science classroom to assess student learning.

Observing Students

David Berliner points out that it is not always obvious from student nonverbal behavior (frowns, puzzled looks, shaking head) whether the student does or does not understand. However, observation of social behavior is an effective tool to determine the level of involvement of students in groups. Since it is important to involve students in small group activities, paying attention to their behavior, verbal as well as nonverbal, is a helpful way to gain insight into their learning. One useful device is to create an observation form that efficiently enables the teacher to watch student behavior during cooperative learning activities, and to record instances of social or interpersonal skills that are being encouraged. For example, the interpersonal skills such as "active listening," "staying on task," "asking questions," and "everyone contributing" can be observed by using an observation form as shown in Figure 9.13. The teacher records the names of the students in each group, spends a few minutes watching each group individually, and then records instances of the interpersonal skill. Later the teacher returns to the group, and provides specific feedback to the group of their interpersonal skill development.

[18]For an in-depth examination of this see Audrey B. Champagne, Barbara E. Lovitts, and Betty J. Calinger, *Assessment in the Service of Instruction* (Washington, D.C.: American Association for the Advancement of Science, 1990).

[19]Barbara E. Lovitts and Audrey B. Champagne, "Assessment and Instruction: Two Sides of the Same Coin," in *Assessment in the Service of Instruction*, Audrey B. Champagne, Barbara E. Lovitts, and Betty J. Calinger, eds. (Washington, D.C.: American Association for the Advancement of Science, 1990), p. 6.

[20]This scheme is based on a paper by David C. Berliner, "But do they understand?" in *Educators' Handbook: A Research Perspective*, Virginia Richardson-Koehler, ed. New York: Longman, 1987, pp. 259-293.

Group Members

Interpersonal Skill	Group 1	Group 2	Group 3	Group 4	Group 5
Active Listening					
Staying on Task					
Asking Questions					
Contributing Ideas					

FIGURE 9.13   Interpersonal Skills Observation Form

Observing student behavior and interaction can be enhanced by using questioning strategies and listening carefully to student questions. Let's look at these two informal methods of assessing learning.

Asking Questions

In Chapter 8 we introduced and examined questioning as a teaching strategy. Asking classroom questions can also be employed to informally assess student learning. Let's examine how.

One of the most powerful uses of classroom questions as an assessment tool is when students are engaged in cooperative learning activities or laboratory activities. The role of the teacher during small group and laboratory activities takes on a monitoring function. During this time, the teacher can visit individual groups to explore the content and methods that students are using in their investigations and small group work.

The following techniques are useful in helping you assess student understanding:

- *Use a variety of questions.* Try to strike a balance between low-order (recall) and high-order (application, synthesis, evaluation) questions. The use of higher-order questions has been shown to be motivational, where as low-order questioning is useful for a probing strategy
- *Use wait time.* Recall from Chapter 8 that Mary Budd Rowe found that most teachers wait less than a second after asking a question. Science teachers who practice waiting at least three seconds can create a classroom atmosphere beneficial to student cognitive, as well as affective, learning. Rowe found that the length of student response increases, failures to respond decrease, confidence increases, speculative responses increase, student questions increases, and the variety of student responses increases

- *Ask probing questions.* If a student, after the teacher asks a question and waits at least three seconds, gives an incorrect answer, the teacher should probe the student answer with other questions. Probing provides a second opportunity for the student, and gives the student a chance to express his or her understanding. Berliner explains that probing to help students clarify and improve answers is more effective than probing to get an answer in increasing student achievement.[21]
- *Redirect questions.* If you are working with a cooperative group, redirecting the same question is a useful assessment tool for the group. Suppose that you ask one student in the group a question, and this student is unable to answer or gives an incorrect answer. You then could redirect by asking another student in the group the same question.

Student Questions

You will have many experiences saying to your students, "Any questions?" This is usually followed by silence. For student questions to be useful tools for assessing student learning, an environment must exist in which students will be willing to ask questions.

The kinds of questions that students ask can inform teachers directly about student understanding of what has been taught. In general students do not ask very many questions, unless the teacher encourages this to happen in the classroom. One of the skills that should be taught in small group learning is asking questions. Asking questions is not only a way to learn but a tool for the teacher to gauge student understanding. Some researchers have found that the level (using Bloom's taxonomy) of student questions can

[21]Berliner, "But Do They Understand?" p. 270.

FIGURE 9.14   Observing students, asking them questions, listening to their questions, and conferencing with them are examples of informal assessment strategies. © Daemmrich.

be increased by creating a more favorable or positive climate through the use of positive reinforcement.[22]

One technique science teachers can use to increase the chance that students will ask questions is to use a lot of silence. The use of wait time to induce students to ask questions can work if you are willing to "wait out the silence." Eventually, after some period of uncomfortableness, the students might ask a question.

Listening to the kinds and levels of questioning is another informal assessment tool.

Conferencing

Meeting with students individually or in small teams is a powerful informal method of assessment. Many teachers find that the student who appears shy and reticent in class, is open and talkative in a private meeting, and is willing to answer questions and share information. One of the things students appreciate about conferences is how special they feel that they have your undivided time, even for a brief period of time.

Conferences should be used to build rapport between the teacher and the student. It is a wonderful time to strengthen a bond between you and the student, and at the same time to "clear the air" over a problem that might persist if not attended to.

Conferences can be used as a time to review a student's portfolio (see the section The Portfolio). The portfolio can become the "agenda" for the conference and an opportunity for the teacher to ask questions, and a chance for the student to do the same.

Semiformal Methods

Semiformal methods of assessment, although similar to informal methods, require more thoughtfulness than the informal methods just discussed. We will look at two semiformal methods that you could use to gather assessment information and evidence about student learning: classroom practice and homework.

Monitoring Classroom Practice

The type of science classroom advocated in this book suggests that students should be actively involved in learning activities. In some activities students work in small teams tutoring each other, discussing concepts and the results of laboratory activities, or reviewing for a quiz, as well as completing activity and work

sheets. In all of these cases the teacher should be actively involved in monitoring the students as they interact with each other or work individually.

Structuring classroom activities for diagnostic purposes, and as a way to check student understanding, is an important aspect of student learning. Involving students as active learners is a way of perceiving and observing student abilities. Recent research has criticized the traditional approach to assessment as being overly static (e.g., relying heavily on end of the week quizzes or unit tests). A more viable approach is to observe students "at work" as learners, and use this experience as a valid "measure" of student learning.

For instance, the Russian psychologist Lev Vygotsky advocated the creation of an atmosphere (zone of influence) in which students could act on what they learned. For instance, instead of asking children to describe a picture, they were instructed to act it out. In these cases, students were able to show that they had a grasp of concepts and relationships, even at a very young age.[23] Using static quizzes and tests may not provide a legitimate opportunity to assess what students know. This is one of the main reasons for the movement toward dynamic evaluation and performance testing.

One technique that can be employed here is to have a section in a student portfolio devoted to classroom practice. That is, students can be instructed to collect and file samples of their classroom work as viable evidence of their learning, not simply a chore to be carried out. If value is placed on this aspect of student performance, it will be valued by the student.

Anecdotal comments and observations of individual students and teams of students is another technique that you can use to assess student performance in class. Teachers can use matrix charts in which specific behaviors and evidences of competence are noted. For instance, suppose that you want to assess students' process abilities. A chart such as the one shown in Figure 9.15 can be used by the teacher as a monitoring device to note evidence of student process skills.

Homework

In general the reason for homework is to give the student an opportunity to practice what has been presented or started in class, or perhaps prepare for class. Homework can also be used as an assessment device. Berliner points out that with homework that is not evaluated (or perceived as being evaluated) or that is seen as busywork by the student, the teacher

[22]Berliner, "But Do They Understand?" p. 271

[23]Mikhail Yaroshevsky, *Lev Vygotsky* (Moscow: Progress, 1989), p. 267.

Students/ Groups	Observing	Classifying	Inferring	Data Analysis	Hypothesizing

FIGURE 9.15 Anecdotal Assessment

misses an opportunity to assess student competence and enhance student learning.

Formal Methods

- On a weekly quiz, students are asked to answer ten questions, each similar in form to the following question:
  What bearing does the fossil record have on Darwin's theory of evolution?
  a. It shows that the theory has serious defects.
  b. It provides conclusive proof that the theory is correct.
  c. It provides supporting evidence for the theory.
  d. It neither supports nor conflicts with the theory.
- A team of students is asked to design and conduct an experiment to calculate the distance that Hot Wheels cars can jump between ramps when released from a given height on an inclined plane.[24]
- Students are given the results of five students who perform in three athletic events (frisbee toss, weight lift, and 50-yard dash). They are asked to evaluate and decide which of the five students

would be the all-around winner. Students need to devise their own approaches to reviewing and interpreting the data, applying it, and explaining why they selected a particular "winner."[25]

Formal assessment methods generally involve the use of a test. What we want to emphasize in this section is that tests can vary in their form and effectiveness in assessing student learning. You are probably quite familiar with forms of tests such as multiple choice, true-false, fill-in-the-blank, analogies, matching, and short answer or essay. The first of the previously listed items is an example of one of these forms: the multiple choice test. I'm sure you are quite familiar with it! But the other two items are assessments, as well. Notice that they look more like activities than test questions. Also note that it appears as if one assessment is administered to a group. These two are examples of performance assessments.

Formal evaluation has been undergoing a transformation. The trend is toward increasing use of performance assessments, using questions that look more like activities, and using student portfolios. Let's look at these ideas, first by examining performance assessment, then by examining a plan for organizing assessment of cognitions, cognitive skills,

[24]Joan Boykoff Baron, "Performance Assessment: Blurring the Edges among Assessment, Curriculum, and Instruction," in *Assessment in the Service of Instruction*, Audrey B. Champagne, Barbara E. Lovitts, and Betty J. Calinger, eds. (Washington, D.C.: American Association for the Advancement of Science, 1990), p. 127.

[25]Educational Testing Service, *Learning by Doing: A Manual for Teaching and Assessing High-Order Thinking in Science and Mathematics* (Princeton, N.J.: Educational Testing Service, 1987), pp. 20–21.

affects and psychomotor skills, and finally by looking at the concept of student portfolios.

Performance Assessments

Performance assessments typically involve students, either individually or in small teams, in the act of solving a problem or thinking critically about a problem, data, or observation. Performance assessments also involve assessing students on their ability to use science skills such as sorting and classifying, observing and formulating hypotheses, interpreting data, and designing and conducting an experiment. What follows are some examples of performance assessment items used in a recent test.[26] Notice that each item is a sort of task, requiring the student to "do science," reflecting the hands-on, minds-on approach advocated by science educators.

1. *Observing:* Students watch a demonstration of centrifugal force and then respond to written questions about what occurred during the demonstration. Students need to make careful observations about what happens as the teacher puts the steel balls in different holes on the Whirlybird arms and then infer the relationship between the position of the steel balls and the speed at which the arm rotates.

2. *Formulating hypotheses:* Students describe what occurs when a drop of water is placed on each of seven different types of building materials (equal-sized pieces of plastic, painted wood, brick, metal, roof shingle, glass). Then the students are asked to predict what will happen to a drop of water as it is placed on the surface of an unknown material (piece of porous cinder block), which is sealed in a plastic bag so that they can examine it but not test it.

3. *Classifying:* Students are asked to sort a collection of small-animal vertebrae into three groups and to explain how the bones in the groupings are alike. To complete the task, students need to make careful observations about the similarities and differences between the bones and to choose their categories according to sets of common characteristics. (Figure 9.16)

Performance assessments are creative approaches that you can employ in an assessment plan. They are creative because the emphasis is on the methods as well as on the ideas that students generate. They place the student in situations that are in accordance with what science instruction should look like. There is a high correlation between performance assessment

and a hands-on, conceptual approach to science teaching.

Following are some of the characteristics of performance assessments:[27]

- Typically involve students in real-world contexts
- Involve students in sustained work, sometimes over several days
- Focus on the "big ideas" and major concepts, rather than on isolated facts and definitions
- Are broad in scope, usually involving several principles of science
- Involve the students in using science processes, the use of scientific methods, and manipulation of science tools
- Present students with open-ended problems
- Encourage students to collaborate and brainstorm
- Stimulate students to make connections among important concepts and ideas
- Are based on scoring criteria related to content, process, group skills, and communication skills

Formal Assessment Scheme

Performance assessments should be complemented with a systematic strategy of student assessment. In this section we will examine a scheme that integrates learning theory, instruction, and assessment.

An effective assessment strategy should be based on the philosophy or rationale of the course or unit that students are studying, as well as on the intended learning outcomes of the experience. Advocating a rationale based on scientific thinking skills and hands-on learning but assessing outcomes using true-false and fill-in-the-blank tests is incongruent. We will advocate an assessment strategy that is correlated with rationale and intended outcomes.

A valid assessment system must incorporate each of these outcome areas: cognitions, cognitive skills, psychomotor skills, and affects. One plan that meets this requirement is the Florida Assessment Plan, which was developed at Florida State University and Georgia State University.[28,29] Based on separate research and development projects at these universities, the system incorporates learning outcomes and as-

---

[26]*Learning by Doing.* pp. 5–6.

[27]Boykoff Baron, "Performance Assessment," pp. 133–134.
[28]Educational Research Institute, Florida Science Assessment Project, Final Report (Tallahassee, Fla.: Educational Research Institute, College of Education, Florida State University, 1971).
[29]Jack Hassard, Stanley Rachelson, and Dinah Basket, *Florida Elementary Science Assessment Project Final Report* (Atlanta, Ga.: Science Education Unit, School of Education, Georgia State University, 1974).

Classifying

The question with successful responses

WHAT IS THE SAME ABOUT THE BONES IN EACH GROUP?

Here's what you do:

1) Look at the collection of labelled bones. These bones are from the backbones of different animals.

Activities to Conduct

2) Put the bones into three groups. Make sure that there is something the same about all the bones in each group. You must use all the bones.

What did you find:

Record Findings

3) Write the letters of the bones in your three groups.

   Group A: _C, D, K_____

   Group B: _A, E, J, L_____

   Group C: _B, F, G, H_____

4) What is the same about the bones in each of your three groups?

Account for Findings

   Group A: _all have one long piece projecting ; all_
   have a hole in middle of central part

   Group B: _all have a central large area with_
   hole and two long pieces projecting out

   Group C: _all are essentially a central structure_
   with a hole in the middle and no long thin
   pieces projecting off them

(Grade 11) ▶

FIGURE 9.16   Sample performance assessment item showing a successful response by an eleventh grade student. Educational Testing Service, *Learning by Doing*. Used with permission.

sessment items based on cognitive thinking, affective learning, and psychomotor learning.

In this system, eight categories of intellectual skills or type of learning were identified and used to create a matrix of outcomes and assessments in earth science, life science, and physical science. Science teachers have access to the system via a telecommunications network housed at Florida State University.[30] Let's take a look at the system and then use it as part of an assessment strategy for science miniunits, units and courses of study.

The eight categories of intellectual skills or type of learning incorporate cognitions, cognitive skills, affects, and psychomotor skills. The categories include

1. Motor chains
2. Verbal chains
3. Discrimination
4. Conceptualization
5. Rules
6. Problem solving
7. Cognitive strategy
8. Attitudes

As you can see, the first category describes psychomotor skills; categories 2 through 7 include cognitions and cognitive skills, and affects are identified in category 8. By using this system you will be able to take a systematic approach to developing assessment items that will constitute quizzes and tests. Take a moment and look at Table 9.4, which shows how the intellectual skills are correlated to learning outcomes, human performance, and assessment items.

[30]George Dawson, professor of science education, Florida State University, has created the computer network system. In this system, teachers can access the pool of outcomes and assessment items via a computer dial-up system.

TABLE 9.4

THE FLORIDA ASSESSMENT PROJECT SYSTEM

Intellectual Skills	Learning Outcome (Action Word)	Human Performance	Assessment Example
1. Motor Chains	Manipulates	Executes a skilled motor performance	Weighs substance on a balance
2. Verbal Chains	Recalls	States fact, generalization, or descriptions	Lists minerals in Mohs's scale of hardness
3. Discrimination	Discriminates	Distinguishes objects or object features as same or different	Tells whether photographs of galaxies are same or different
4. Conceptualization	Identifies or classifies	Classifies an object or situation in accordance with a definition	Classifies granite as an igneous rock
5. Rules	Demonstrates	Applies a rule, law, or concept to specific example	Determines a density of a mineral
6. Problem Solving	Generates	Generates a solution to a novel problem	Determines effect of velocity on erosion in stream
7. Cognitive Strategy	Originates	Originates a novel problem and solution	Gets an answer to "I wonder what would happen if . . ."
8. Attitudes	Chooses	Chooses a course of action, expresses a feeling toward a person, object, or event	Writes a letter to congressional representative supporting air quality

Florida Science Assessment Project, Final Report (Tallahassee, Fla.: Educational Research Institute, College of Education, Florida State University, 1971).

Assessment Strategies

Let's look at some examples of assessment items given several learning outcomes. You can refer to these examples as model assessment items and make use of them as you design your own.[31]

Psychomotor Skills (Motor Chains)

In an inquiry-oriented, hands-on science course, students will be involved in the manipulation of labora-tory apparatus, scientific instruments, and the tools necessary to carry out investigations and activities. Although the assessment of motor chains may seem to include only psychomotor skills, it is important to note that cognitive skills are usually involved in any activity involving this form of learning. However, providing students feedback through psychomotor as-sessments is a powerful way to reinforce the impor-tance of motor skills and learning in the science classroom.

Example: Motor Chains

Learning Outcome   Given a graduated cylinder, colored water, and an empty con-tainer, and asked to put 40 mL of liquid into the empty container, the student manip-ulates the beaker and the graduated cylinder by measuring the 40 mL of the liquid and pouring this measured amount into the empty container.

Assessment Items   Here is a graduated cylinder, some colored water in a beaker and an empty container. Pour 40 mL of the colored liquid into the empty container.

Cognitions and Cognitive Skills

There are six categories of assessment items that when combined provide measures for cognitions and cogni-tive skills. There is a heirarchy implied in the cate-gories, as there was a heirarchy in Bloom's taxonomy, which was discussed in chapter 9. Many test develop-ers use Bloom's taxonomy to develop assessment items. Developing items using the taxonomy is a valid approach to assessment. However, in this section we will use the Florida approach because it is more closely aligned with the recent emphasis on cognitive thinking and cognitive science. The six categories of cognitive thinking will be presented in the following sequence: verbal chains, discrimination, conceptual-ization, rules, problem solving, and cognitive strategy.

Verbal Chains

In this form of thinking, students are involved in recalling information, either verbally or in writing. This indicates that the student knows the proper se-quence of words in response to a request for verbal information. This category represents the lowest level of cognitive thinking: the recall of information.

Discrimination

When students discriminate, they decide whether or not things are identical. Discrimination does not imply that they can identify the specific properties of the things that are the same or different.

Example: Verbal Chains

Learning Outcome   Given a volume of a liquid (melted paraffin) that is changing from a liquid to a solid, the student identifies the change by stating that the liquid is "freezing."

Assessment   When the liquid paraffin forms solid paraffin, what process has taken place? (Answer: freezing.)

---

[31]The examples of assessment items are based on *The Florida Catalog of Science Objectives,. Series 13, Earth-Space and Physical Science Series* (Tallahassee, Fla.: Florida Department of Education, 1973).

Example: Discrimination

Learning Outcome   Given granite and a set of igneous rocks, the student discriminates between granite and other plutonic rocks.

Assessment   Indicate with a "d" or "s" whether the other rocks are the same or different from granite for each characteristic listed.

Characteristic	Color	Crystal Size	Hardness	Texture
Rock 1				
Rock 2				
Rock 3				

Conceptualization (Concept Learning)

In conceptualization, or concept learning, students recognize a class of objects, object characteristics, or events. In this form of learning, students have learned the concept of class to the extent that they can classify examples of it by instant recognition. Recognition can involve any one or all of the senses: sight, hearing, smell, taste, or touch. Concept learning also involves classification. In this case, students use a definition to put something into a class (or some things into classes). Students' behavior in this case indicates that they know the parameters of the class or classes and that they can either verbalize them or use them when asked to do so (see Example 2).

Example 1: Conceptualization

Learning Outcome   Given the names of parts of a typical plant cell, the student draws a cell and names the given parts.

Assessment   Draw and label a cell with these parts present.
   a. cell wall
   b. nucleus
   c. chloroplasts

Example 2: Conceptualization

Learning Outcome   Given a map of the moon with features labeled, the student classifies the labeled moon features according to age.

Assessment   The map (Figure 9.17) shows a region of the moon with three areas marked. In boxes to the right of the map place the letters of the features in order by age, listing the youngest first.

Rules

Rule learning involves students in applying specific concepts, rules, or procedures to a specific task. The key notion is the process of applying rules. Application is an important part of the learning cycle that was developed earlier in the text. Students need opportunities to apply what they are learning, and they need assessment feedback on their progress.

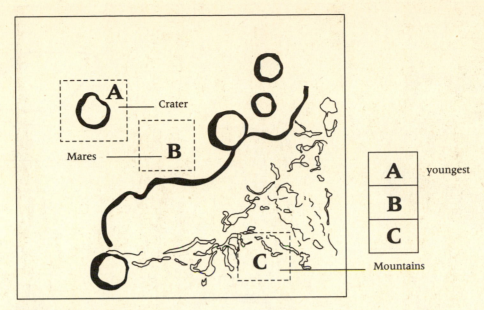

FIGURE 9.17   Map of the Moon

Example: Rules

Learning Outcome   Given a hypothetical situation in which the student crashes on the moon and certain items (see Figure 9.18) survive the crash, the student chooses survival items in terms of importance for reaching a moon base.

Assessment   Imagine you are a member of a landing crew and your space ship has crashed on the lighted side of the moon. Your survival depends on reaching a moon base 200 miles away. Below are listed 8 items left intact after landing (these items are also shown as cards). Rank them in order of importance in terms of reaching the moon base. Place a number 1 by the most important, a number 2 by the second most important, and so on.

_____50 feet nylon rope

_____.45 caliber pistol

_____star map

_____5 gallons of water

_____box of matches

_____two 100-pound tanks of oxygen

_____food concentrate

_____life raft

_____first-aid kit

_____solar-powered heater

_____magnetic compass

_____signal flares

_____case of powdered milk

_____solar-powered FM Walkie-Talkie

_____parachute silk

Problem Solving

When students put together two or more rules, definitions, or concepts to solve a problem, they are generating ideas, solutions, and procedures. Problem solving is another application process of scientific concepts. Students need to be given many opportunities to solve problems, and assessment strategies need to include problems to solve.

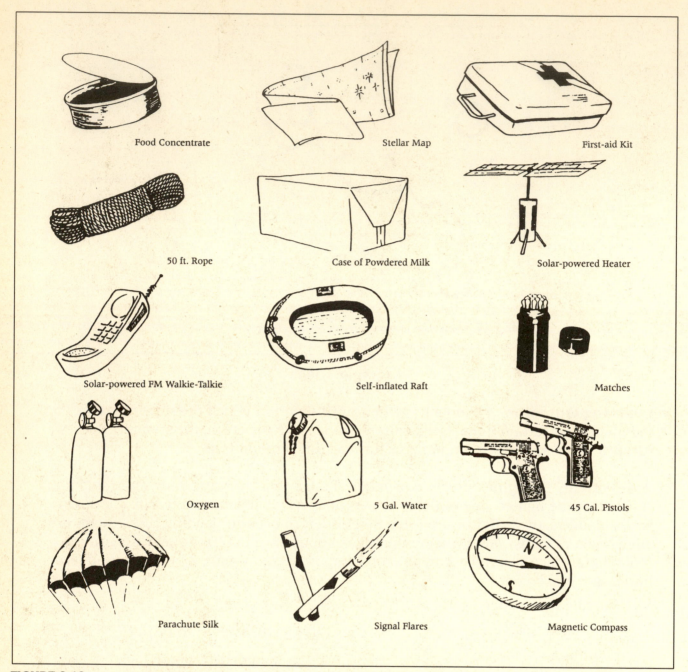

FIGURE 9.18 Survival Item Cards

Example: Problem Solving

Learning Outcome   Given a hypothetical situation describing an individual's energy requirements and given materials that model these needs, the student generates a system to fulfill the energy requirements.

Assessment   Imagine this situation. Your neighbors are planning to add a garage to their house. Their energy needs include lighting, heating, and running a motor for the garage door. The energy they will use is electricity. You have been given a dry-cell battery, a light bulb, a piece of nichrome wire, and a small motor. Using any or all of these materials, make a system that shows how your neighbors can meet their energy needs.

Cognitive Strategy

Cognitive strategy refers to student creativity. The key notion here is originating a plan or an idea. Cognitive strategy involves the combination of ideas to propose and solve problems. In this case, students know the material well enough to identify a problem area and organize the proper concepts and procedures to solve it.

Example: Cognitive Strategy

Learning Outcome   Given an organic and inorganic fertilizer, soil, containers and seeds, the students designs and conducts an experiment to find out which fertilizer is more effective.

Assessment   Before you are the materials you may use in order to carry out an experiment. You are to design it yourself. Its purpose is to determine which fertilizer is better for making the plants healthier and produce more peppers.

Rewritten for groups: Before your team are the materials you may use in order to carry out an experiment. You are to work together with your team members to design the experiment. Its purpose is to determine which fertilizer is better for making the plants healthier and produce more peppers.

Affects

Affective outcomes can be assessed, just as we have shown how to assess psychomotor and cognitive outcomes. In the model being presented here, affects are classified as attitudes.

Attitudes

Attitudes involve, among other behaviors, choosing. If students make a choice, they are deciding to behave in a certain way in repeated situations. Choosing, however, does not guarantee that a student will act in a certain way. (See the discussion on values and attitudes in Chapter 6.) A student might choose to say that smoking is harmful but after school be seen smoking a cigarette. Or a student might say that recycling is an important part of a family's responsibility for the environment and then not use trash stream separators at restaurants or in the school cafeteria.

Example: Attitudes

Learning Outcome   Given a list of endangered animal species and a statement that they are being killed never to be seen again by humans, the student chooses to speak in favor of protecting the animals.

Assessment   The bald eagle, blue whale, California condor, Everglades kite, red wolf, Key deer, cougar, alligator, and whooping crane are all species of animals in the United States that are endangered species. These animals may all be killed and may never be part of the earth's ecosystem. What should be done about this problem?

There are other techniques that you can use to assess attitudes. One technique that you might find useful to include is the semantic differential.[32] To use the technique the teacher selects a concept or idea and then develops a set of relevant bipolar adjectives or adjective phrases. For example, suppose that you wanted to assess your students' attitude toward science. You could use the semantic differential scale shown in Figure 9.19.

The students would simply check along the continuum what their attitude is with respect to each bipolar pair. Doran recommends summarizing the entire class's responses and then calculating a mean for each bipolar adjective pair.[33]

Other "concepts" can be assessed using this technique. For example, you could use this technique to assess student attitudes about the following:

Chemicals
Rocks
Science courses
This unit
Alcohol

For each of these you would identify relevant bipolar adjectives and then make a form similar to the one shown in Figure 9.19. There are other techniques that can be utilized to assess attitudes, including writing term papers, debates, discussions of issues, interviews, and reports.

The Portfolio

An assessment strategy employed by some teachers for many years is to have students collect and file their course work in a loose-leaf notebook or a folder. This collection of student work is called a portfolio. In recent years educators have been advocating that school science assessments include portfolios of students' work. What are the characteristics of a portfo-

lio? To find out, read the following description of portfolios used in a high school science class.

This year in science, you will put your work in a three-ring binder. This notebook, which we will call a portfolio, will belong to you. It is to be kept in Room 25 at all times. In the first section of the portfolio you will keep your personal record sheet and a calendar of what we do each day in class. Here are blank record sheets and a blank calendar for September, to begin your portfolio development process. You will also put all the handouts from this class in this first section of your portfolio. The second section will be a journal. Each day at the end of class, you will have a few minutes to write about what you did and learned and felt in your science class. I will look at your journal and make sure you have written something every day, but I will not grade what you write in your journal. In the third section of your portfolio, you will keep copies of all the activities we do in science class—worksheets, laboratory notes, and group work, for example. The fourth section will be for homework. The fifth and last section will be for review sheets and quizzes. You will have to buy dividers for your portfolio.[34]

Portfolios have at least three characteristics:[35]

1. *They contain real documents.* Portfolios contain authentic representations of students' work, such as homework, laboratory data sheets, and journal entries.
2. *They contain a range of material.* The science portfolio contains a full range of student work in order to demonstrate a range of performance and ability.
3. *They demonstrate growth over time.* The portfolio should include samples of work over time. For example, the daily journal entries will provide insight into the students' experience over time.

Using portfolios as part of an assessment strategy enables students to be evaluated on a broad set of competencies, skills, performances, and products, and may, in the opinion of some educators, be closer to an authentic assessment of student performance. Portfolios will contain student work that can be evaluated and graded with checklists and test scores, as well as work that lends itself to oral and conference evaluation. Portfolios encourage a dynamic approach to assessment in which teacher notes, and samples of students' work—(lab reports, logs, worksheets, video tapes), as well as quiz and test papers, are included.

---

[32]For many examples of affective assessment (as well as assessment of cognitive and motor skills) see Rodney L. Doran, *Measurement and Evaluation of Science Instruction* (Washington, D.C.: National Science Teachers Association, 1980), pp. 59–70.

[33]Doran, *Measurement and Evaluation*, p. 61.

[34]Angelo Collins, "Portfolios for Assessing Student Learning in Science: A New Name for a Familiar Idea?" in *Assessment in the Service of Instruction*, Audrey B. Champagne, Barbara E. Lovitts, and Betty J. Calinger, eds. (Washington, D.C.: American Association for the Advancement of Science, 1990), pp. 158.

[35]Collins, "Portfolios for Assessing Student Learning in Science," pp. 157–166.

Science is

Meaningful	___:___:___:___:___:___:___:	Meaningless
Bad	___:___:___:___:___:___:___:	Good
Useful	___:___:___:___:___:___:___:	Useless
Confusing	___:___:___:___:___:___:___:	Clear
Unimportant	___:___:___:___:___:___:___:	Important
Simple	___:___:___:___:___:___:___:	Complex

FIGURE 9.19   Semantic Differential Scale

Science Teaching Gazette

Volume Nine   Designing and Assessing Science Units and Courses of Study

Science Teachers Talk

"What tips would you give beginning teachers in planning and preparing lessons?"

Jerry Pelletier. The first tip I would give any new teacher is that they must have good classroom management. Students must understand why they are in the class and what the teacher expects of them. When planning a lesson teachers must always keep in mind who their audience is. The lessons must be geared to the level and understanding of their students. They should never assume that students have mastered a skill For example, if students will be measuring distances with rulers a teacher should never assume that the students fully understand how a ruler should be used These skills should always be integrated as part of the lesson. Reinforcementof key skills and concepts should always be made part of any lesson. It is also important to have closure in order to summarize the main ideas developed during the lesson.

John Ricciardi. Be able. Be directed. Be diverse. Be clear and be yourself! Be able—by allowing yourself ample time to prepare daily lessons. Choose the time of day or night when you are most alert and rested. For me, the most ideal prep period occurs between 3:30 and 5:30 in the morning. Be directed—by identifying and being aware of your daily goals and objectives. Know exactly what you want your students to learn— before you begin constructing a lesson. Be attuned to the science

content material that is least likely and most likely to be remembered and understood. Be diverse—by providing choices. Incorporate a variety of instructional mediums and multisensory activities. Make your lessons interdisciplinary by integrating art, drama, prose, and poetry into your science. Be clear—and your students will understand you and know what you expect from them. Assignment responsibilities should be in writing and easily accessible. Be yourself— and more real to them by weaving your personality and character into your lessons. Present and instruct in ways that feel right— rather than in ways that you think are right.

Mary Wilde. Advice to beginning teachers or even experienced teachers is that good planning and preparation should be at the top of a teacher's priority list of duties. Good planning and preparation saves a teacher lots of time and creates an environment that you and your students can enjoy. Years of teaching have taught me that this planning time increases with experience, not decreases. It is very important to plan and prepare new activities and to introduce innovating ideas that excited the teacher in the classroom. If a teacher begins to loose that excitement and enthusiasm, one should look at the lesson plans. To keep lesson plans full of creative, innovative ideas, continuing education, formal or informal, is very important. The quickest way to loose that "spark"

in the classroom is to repeat old lessons over and over, with little variation. I feel lack of preparation and planning, and lack of professional growth to help one become innovative in the classroom, is the major contributing factor to "teacher burnout."

Something that helps me with instructional planning is allowing myself some "brainstorming" time. My running time alone allows me to think of creative ideas that I can implement in the classroom. This can be a walk or a bicycle ride, but if time is designated to "brainstorming," one can be quite surprised with the results. It is important for me to be outside for this, for sitting at a desk does not seem to produce the same results. I do take

Continued on p. 344

Think Pieces

- What are the advantages and disadvantages of stating objectives in behavioral or performance terms?
- How would the following theorists respond to the preceding question? Piaget, Bruner, Skinner, McCarthy, and Dann.
- How is a cognition different from a cognitive skill? How do the two notions relate to each other?
- Explain why affects should be part of science instruction.
- How do the following lesson plan models differ? Which one do you prefer? Why? Lesson plan models: direct-interactive teaching, cooperative learning, inquiry or laboratory, and conceptual change teaching. ☐

Continued from p. 343
advantage of whatever "alone" time I have. When I was returning from St. Louis after the National Science Teachers Association conference, I occupied the 12-hour drive by developing a unit centered around topographical maps, and planned how I could share this unit at the next NSTA conference. I jotted my ideas down on a fast-food restaurant napkin, and when I returned to Atlanta, I began working with those ideas. Lesson planning and creativity do not come to one without effort. If this is made a

priority, all other teaching responsibilities will flow smoother.

Ginny Almeder. In terms of planning and preparation for lessons, I would suggest the following:

1. Be well informed about your subject. Read the text and relevant literature. Making reference to a recent magazine or newspaper article or television science program can make the material more relevant and directly involve students who may have been exposed to the same material.

2. Write out your objectives and cross-reference them to your activities. This will enhance the clarity of your lessons and facilitate learning.

3. Set up your labs ahead. Allow extra time to hunt for equipment or to make last minute modifications.

4. Have alternative plans in case your original one is not workable.

5. Plan to incorporate some "real-life" applications into your presentations. This will help students to appreciate the personal implications of science. ☐

Planning Activities: Earth Science

Shake, Rattle, and Quake: Earthquake Waves

In this activity, students investigate the differences between primary and secondary earthquake waves by simulating waves with a toy called a Slinky. The activity can be done as a demonstration, or as a small group activity in which students work in pairs or groups of four.

Objectives
1. To observe and differentiate earthquake waves
2. To understand the effects of earthquake waves

Concepts
- Earthquake
- Waves (P and S)
- Energy

Materials
Slinky (double length)

Procedure
1. To explore primary earthquake waves, with a partner, stretch a Slinky to about 5 meters in

length. While holding one end, gather in about 15 extra coils and let them go. Repeat several times while watching the coils. Draw a picture showing your observation.

2. To explore secondary earthquake waves, with a partner, stretch the Slinky to about 5 meters in length again. Quickly move your hand to one side and back again in a snapping motion. Repeat several times while watching the coils carefully. Draw a picture showing your observation.

3. To find out what happens when earthquake waves meet each other or bounce off objects, stretch the Slinky to about 5 meters again. Both you and your partner should quickly move your hands to one side and back again.

Application to Science Teaching
1. Draw a concept map for "earthquakes" to describe the conceptual nature of a lesson or miniunit on earthquakes.

2. Rewrite the objectives of this activity to include a cognitive skill, a psychomotor skill, and an affect.

3. Use this lesson as a central instructional foci for a miniunit on earthquakes. What other instructional foci would you include?

Don't Take It for Granite: Rock Classification
In this activity students use a simple dichotomous key to classify rocks as either igneous, metamorphic, or sedimentary.

Objectives
1. To identify the properties of rocks
2. To classify rocks according to their physical properties

Concepts
- Rock
- Igneous
- Metamorphic
- Sedimentary
- Mineral
- Interlocking crystals (minerals)
- Noninterlocking crystals (minerals)

Materials
Box of rocks: granite, obsidian, pumace, basalt, limestone, shale,

Continued on p. 345

Continued from p.344
sandstone, conglomerate, slate, schist, gneis: small bottle; dilute hydrochloric acid; hand lens

Procedure

1. Observing one rock at a time, use the classification key to determine whether the rock is igneous, metamorphic, or sedimentary.
2. Make a chart of your results including the rock sample, specific properties, and the class it belongs to.
3. Use reference books to determine the name of each of the rock specimens.

Rock Classification Key[36]

1a. If the rock is made up of minerals that you can see, go to 2a.
1b. If the rock is not made up of visible minerals, go to 5a.

2a. If the rock is made up of minerals that are interlocking ("melted together"), go to 3a.
2b. If the rock is made up of minerals that are noninterlocking ("glued together"), go to 6a.
3a. If the minerals in the sample are of the same kind, the rock is metamorphic.
3b. If the minerals in the sample are of two or more different types, go to 4a.
4a. If the minerals in the sample are distributed in a random pattern (not lined up), the rock is igneous.
4b. If the minerals in the sample are not distributed randomly but show a preferred arrangement or banding (lined up), the rock is metamorphic.
5a. If the rock is either glassy or frothy (has small holes), it is igneous.

5b. If the rock is made up of strong, flat sheets that look as though they will split off into slatelike pieces, it is metamorphic.
6a. If the rock is made of silt, sand, or pebbles cemented together (it may have fossils), it is sedimentary.
6b. If the rock is not made of silt, sand, or pebbles but contains a substance that fizzles when dilute hydrochloric acid is poured on it, it is sedimentary.

Applications to Science Teaching

1. Design a concept map that includes all the concepts in the concept list given previously. Make sure you show how the concepts are linked together. Use your map to answer this
Continued on p.346

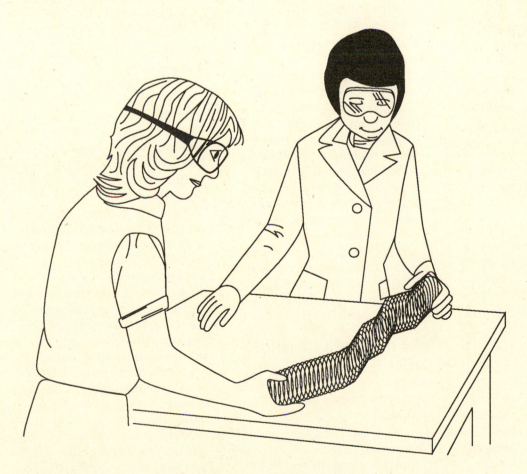

FIGURE 9.20   Pairs of students can explore earthquake waves by means of a toy slinky.

Continued from p.344
question: What prerequisite concepts do students need to know before they can do the rock classification activity? Add these concepts to the map.

2. What central or focus cognition will this activity help students understand?

3. What is the basis for this statement: In this activity students will be involved in con-cept learning, not proposi-tional learning? ☐

[36]Based on an activity in *Geocycles*, Intermediate Science Curriculum Study, Tallahassee, Fla. Copyright 1973 Florida State University.

Planning Activities: Life Science

Light On: Responses of Earthworms

In this activity students will explore the way earthworms respond to environmental changes.

Objectives
1. To generate hypotheses regard-ing the interaction of earthworms and changing envi-ronmental conditions
2. To design safe experiments to find out how earthworms respond to environmental changes
3. To gather data to test hypothe-ses about earthworm behavior

Concepts
• Environmental change
• Environmental factors (touch, smell, sound, gravity, temperature, light)

Materials
Earthworms, pieces of paper, vine-gar or household ammonia, damp sheet of newspaper, small box, flashlight, and other materials as required by experiments

Procedure
1. Tap the head of an earthworm gently with a finger or a pencil eraser. How did the earthworm respond?
2. How do you think earthworms will respond if any of the follow-ing environmental factors are changed: light, temperature, smell, gravity, sound?

3. Write one or more hypotheses that incorporate how you think earthworms will respond to various environmental condi-tions.
4. Design an experiment to test each hypothesis.
5. Conduct the experiment, and use the data to "test" the hypotheses.
6. What do you conclude about earthworm behavior?

Application to Science Teaching
1. This is an open inquiry science activity. Students will have to design experiments to test hypotheses. What cognitive skills are required to enable stu-dents to complete this activity?
2. Handling living things in the science classroom can be the medium to teach important at-titudes and values about science. How would you ensure that these attitudes and values were indeed part of the activity, and that the students were assessed on them as well? Describe your plan.

Who Done It? A Genetics Caper

In this activity students will apply their knowledge of genetics to solve a "crime."

Objectives
1. To understand the concept of dihybrid

2. To solve a problem involving two genes
3. To appreciate the value of genetic knowledge in every-day life

Concepts
• Homozygous dominant
• Heterozygous
• Homozygous recessive

Materials
None

Procedure
1. Have students read the "Case of the Hooded Murderer." (Option: Read the case aloud to the stu-dents as they follow along with their own text. Make sure you read the case, not one of the students.)
2. Conduct a brief discussion of the case with the students.
3. Have the students work in teams to complete the chart (Figure 9.24) and solve the mystery of the hooded murderer.

The Case of the Hooded Murderer[36]
Lord Robert Lancaster's body— with a long dagger protruding from his chest—lay sprawled in his library. A draft of Lord Robert's new will, which would have disin-herited his family and left his vast fortune to charity, was still on his desk. The will was not signed and so his nieces and nephews would inherit his money and property.

    The Lancasters were a large, wealthy British family. Lord Robert's brothers and sisters had all died before him, and he never married. But he was scarcely alone. His twelve nieces and nephews had moved into the houses on the family estate.

    One of the police officers who came to investigate the mur-der was Inspector Watson, a shrewd sleuth who had once planned to be a biologist. His spe-cial interest was genetics, and he was particularly interested in the

Continued on p. 347

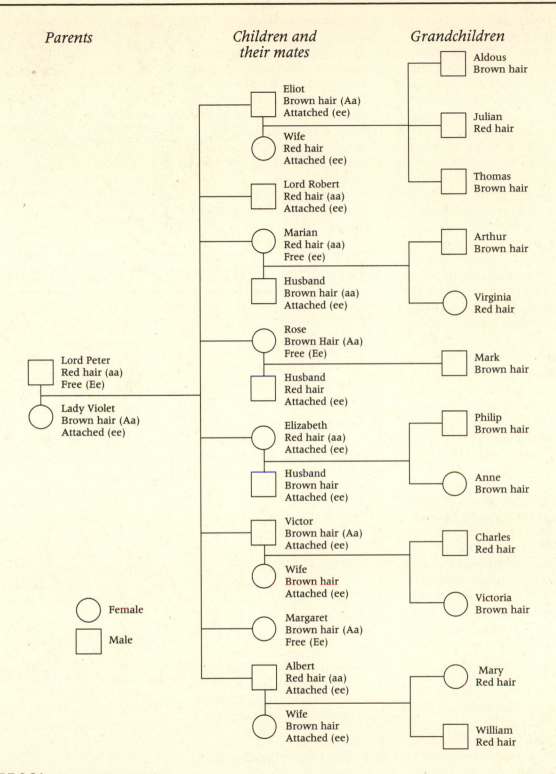

FIGURE 9.21   Inspector Watson's Chart

Continued from p. 346
Lancaster murder because of certain patterns of inherited traits in the family.

As Watson explained to Holmes, "Old Lord Peter (Lord Robert's father) is shown in that portrait over the fireplace. As a young man he had bright red hair. His wife, Violet, was a brunette. Half their children, including the late Lord Robert, had red hair; the others were brunette. As only a recessive pair of genes (aa) will produce red hair, each of Lord Peter's children had received an a gene from him."

Continued on p. 348

Continued from p. 347

Watson went on, "We know Lady Violet had A genes because she was a brunette, and even one A gene will produce brown hair. But Lady Violet must have been heterozygous (Aa) because half her children had red hair."

In questioning the family and servants, Inspector Watson found a witness to the murder, a maid who had heard a groan from the library. Afraid to go in, she had peeped through the keyhole and seen someone in a long, hooded cape. "I couldn't even tell whether it was a man or a woman, sir. But I did see a bit of red hair sticking out from under the hood. The person had a nervous habit of pulling on one earlobe, which I noticed was not an attached lobe."

"Aha!" said Watson. "Earlobes, also, owe their attachment to one pair of genes. A person who is homozygous dominant (EE) or heterozygous (Ee) has free ear lobes, and someone who is homozygous recessive (ee) has attached ear lobes.

The inspector began drawing up a chart of the Lancaster family, using portraits and family albums. Some information was not available, but he learned three important pieces of information. First, old Lord Peter Lancaster had free earlobes. Second, Lady Violet had attached earlobes. Third, some of their children had attached earlobes.

By strange coincidence, Lord Robert's brothers and sisters had all married persons having attached earlobes.

Unfortunately, no pictures of the suspects were available, and Inspector Watson had not yet met them in person. The servants could not remember whether the suspects had free or attached earlobes, but of course they knew which had red hair and which were brunettes. Inspector Watson added that information to the chart.

Application to Science Teaching

1. The learning cycle includes three phases: exploration, concept development, and concept application. In which phase would you recommend using this activity? What is your rationale?[37]
2. What prerequisite concepts are needed for students to solve this problem?
3. What are the cognitive skills that students need to engage in this activity? ☐

[36]The Case of the Hooded Murderer, © 1984 Addison-Wesley Publishing Company. Reprinted with permission of Addison-Wesley.

[37]Most science educators would recommend using this in the application phase of a sequence of lessons. They claim that the students would need to be familiar with several concepts before being able to solve the problem. On the other hand, some science teachers would recommend using this as an exploration activity, especially if it was carried out in small groups. The activity would also expose student misconceptions, which would be useful in planning future lessons.

Planning Activities: Physical Science

Chemistry in the Bag

In this activity students will investigate a chemical reaction occurring in a plastic bag.[38]

Objectives

1. To observe a chemical reaction
2. To describe the evidence indicating that a chemical reaction has taken place

Concepts

- Chemical
- Chemical change
- Heat
- Gas
- Indicator

Materials

Calcium chloride, sodium bicarbonate (baking soda), red cabbage juice (boil red cabbage for 5 minutes, pour off liquid) or phenol red indicator, Ziplock plastic bags (quart size), plastic spoon

Procedure

1. Ask students what they think is a chemical. Write their answers on the board.
2. Ask students what they think might happen if two chemicals are mixed together. Write their responses on the board.
3. Explain to the students that they are going to explore chemicals and chemical reactions using relatively safe chemicals, but that they should keep chemicals off their clothes and skin, rinsing with water if chemicals do make contact, wiping up spills as they happen, and washing hands at the end of the activity.
4. Have teams of students obtain small cups containing baking soda and calcium chloride, small bottle of cabbage juice, wood splints, a vial, goggles, and 1 plastic Ziplock bag.
5. Give students a handout that contains the instructions and questions for investigation.

Inquiry Procedure

1. Measure one spoonful of calcium chloride and place it into the Ziplock bag.
2. Add one spoonful of sodium bicarbonate (baking soda) to the bag. Zip the bag closed and shake it to observe for any evidence of a chemical change.
3. Pour 10 mL of cabbage juice into a small vial. Carefully put the vial into the bag without spilling the indicator.
4. Zip the bag closed.
5. Tip the vial of indicator.

Continued on p. 349

Continued from p.348
Inquiry Questions
1. What happened when the indicator mixed with the baking soda and calcium   chloride?
2. What are at least five observations?
3. Do you think a chemical reaction occurred?
4. How would you define a chemical reaction?

Application to Science Teaching
1. What safety precautions will you take doing this activity? Can safety precautions be included as intended learning outcomes? How would you phrase a learning outcome that addressed safety in the science classroom?
2. Draw a concept map using the concepts previously listed. Be sure to include the linking phrases.
3. What are some additional activities that you could do with these chemicals that would build on this initial activity? List ideas for at least three to five lessons.

An Eggzact Experiment
In this activity, students observe a teacher demonstration (a discrepant event) and then theorize possible explanations for the event.

Objectives
1. To describe the effects of air pressure on an object
2. To explain how an egg is able to be forced into a bottle

Concepts
- Air pressure
- Molecules
- Heat
- Cooling

Materials
Glass quart milk bottle, hard-boiled egg with shell removed, match, crumpled-up piece of paper

Procedure
1. Tell the students that you are going to do a demonstration. You are going to try to put an egg (show it to them) into a bottle (show them) without touching or forcing the egg in the bottle. Ask them if they have any ideas about how this could be accomplished. Record their ideas.
2. Carefully drop a small wad of burning paper into the milk bottle. Just before the flame goes out, place the egg, the smaller end down, in the opening of the bottle. Have the students watch the egg squeeze into the bottle.
3. Arrange the students in pairs to make a diagram and write explanations for what they think caused the egg to go into the bottle.
4. After ten minutes, have students explain their ideas to the whole class.

Applications to Science Teaching
1. Could this "activity" be used as a performance assessment? How would you set it up, and what criteria would you use to "evaluate" the students' performance?

2. What are some additional activities that you could use to help students understand air pressure, or other properties of air?
3. What cognition does this activity help students understand?   □

[38]This activity is an adaptation of one in Jacquieline Barber, *Chemical Reactions* (Berkeley, Calif.: Lawrence Hall of Science, 1986); and Beth Baker et. al., CHEM-PACS (Denver, Colo.: Practical Activities with Common Substances, 1989).

PROBLEMS AND EXTENSIONS

- What is a performance assessment? Give one example for earth science, life science, physical science, and S-T-S.
- Write an objective for each of the levels in Bloom's taxonomy in one of the following content areas: rocks, electricity, ecosystem.
- Write an assessment item for each of the objectives you designed.
- Select a chapter from a science book of your choice and create eight assessment items using the Florida Assessment Project plan.
- Prepare concept maps for one of the following:
  a. A chapter from a high school biology, chemistry, earth science, or physics text.
  b. A potential miniunit on electric circuits, physical and chemical changes, reproduction, erosion, or groundwater pollution.   □

RESOURCES

Abruscato, Joe, and Jack Hassard. *The Whole Cosmos Catalog of Science Activities for Kids of All Ages.* Glenview, Ill.: Scott, Foresman, 1991.

This activity book contains a wealth of ideas for science lesson plans in life, earth, and physical sciences, and science and technology, as well as games, biographies, and activities relating science to social studies, music, and art.

Champagne, Audrey, Barbara E. Lovitts, and Betty J. Calinger. *Assessment in the Service of Instruction.* Washington, D.C.: American Associa-tion for the Advancement of Science, 1990.

This volume contains papers on assessment from the 1990 AAAS Forum for School Science. Papers cover topics including hands-on assessment, performance assessment, portfolios for assessment, and examples of programs in practice.

Doran, Rodney L. *Basic Measurement and Evaluation of Science Instruction.* Washington, D.C.: National Science Teachers Association, 1980.

This book provides practical examples of assessment strategies in cognitive thinking, affective learning, and laboratory activity. The book also presents ideas on item and test analysis, as well as on how to grade students.

Learning by Doing: A Manual for Teaching and Assessing Higher-Order Thinking in Science and Mathematics. Princeton, N.J.: Educational Testing Service, 1987.

This manual is based on the NAEP's pilot study of higher-order thinking skills assessment techniques. The manual contains a rationale for "hands-on assessment," and examples of assessment items for the following thinking skills: classifying, observing and making inferences, formulating hypotheses, interpreting data, designing an experiment, conducting a complete experiment.pollution. □

10

FACILITATING LEARNING IN THE SCIENCE CLASSROOM

- "Jim, put your history book away and help Alice with the assignment."
- "Class, you have ten minutes to complete the activity."
- "Material handlers, be sure everyone has a pair of goggles before going into the lab."
- Speaking quietly to a cooperative team, the teacher says, "Your team should be proud of the way you reported the results of yesterday's lab—nice work!"
- "Today we'll be organized into six groups; count off by 4 in each group; the ones will take the role of facilitator; twos, material handler; threes, recorder; and fours, reporter."
- "Stop the action!...There's one more piece of equipment each group needs. Here it is. Send your material handler to get it."
- "When you finish working with the mice today, be sure to put them carefully in their cages."

These are samples of teacher statements each of which focuses on the role of the teacher in the science classroom. If you reread the statements you will note that in each case the teacher is directing the students in some aspect of classroom behavior or performance. These are management behaviors that all teachers engage in during every science lesson. Knowing how to manage science classrooms is the focus of this chapter.

Effective classroom management incorporates teacher behaviors that result in high levels of student involvement in classroom activities and minimizes student disruptions and other behaviors that interfere with student involvement in science learning. The classroom manager can also be viewed as a facilitator of learning. Facilitators help other persons learn by creating a classroom environment that is conducive to learning. Facilitators take into consideration the individual needs of students and at the same time realize that they are working with groups of students and therefore must be focused on group dynamics and social behavior. Group management behaviors that result in effective learning environments will be explored and related to science teaching.

Effective classroom management requires that teachers get off to a good start each year, or whenever a new course begins. We will explore effective science teaching behaviors for the beginning of the year and examine sample lessons that effective science teachers use to begin the year.

We will also examine the physical dimensions of classroom management by focusing on room design and materials in the science classroom, and consider some of the issues related to safety in the science classroom.

THE FACILITATIVE SCIENCE TEACHER

The facilitative science teacher is one who understands and knows how to manage groups of students to produce high levels of involvement. There is a wide body of research and a number of "management" programs that indicate the importance of issues related to classroom management. When we ask our beginning science interns what their number one concern is about teaching, invariably it is how to deal with discipline problems. No one would deny that discipline problems can create havoc for the beginning teacher. Instead of dealing with discipline problems as the problem, our approach here will be to examine management behaviors that contribute to increasing levels of learning and involvement in the classroom. Putting out the fires of discipline might seem at first appearance as an effective method of "controlling" student behavior. However, the research on classroom behavior shows this not to be true.

A researcher by the name of Jacob Kounin identified a cluster of proactive teacher behaviors that distinguished effective classroom managers from ineffective ones.[1] But the way he discovered this is worth recounting. Initially Kounin was interested in how teachers handled misbehavior. That is, he wanted to know what effect the teacher's action to stop the misbehavior (called a desist) had on other students in the class. Some of the questions he considered included, Does a desist "serve as an example and restrain behavior in other students? Does a desist cause other students to behave better, or pay more attention? Which students are affected by desist actions, off-task students, or on-task students?"

Kounin found that the manner in which teachers handled misbehavior made no difference in how the students reacted. The only exception he found was that punitive desists tended to create emotional discomfort among other students in the class. Kounin's research did not imply that teacher's desists were not effective. For example, he reported that one teacher used the technique of flicking the lights on and off as a signal for the students to stop talking. For this teacher the technique worked. However, he also reported another teacher in the same school who used the same technique; it didn't work. Kounin realized that there was some aspect of teacher behavior other than the teacher's desist techniques that was influencing student behavior.

Kounin discovered by studying videotapes of over eighty teachers that there were group management strategies that teachers used that created a class environment characterized by high levels of student involvement. These strategies were proactive behaviors on the part of the teacher and taken as a whole created a class climate that prevented or discouraged behavior problems before they started. Let's take a look at Kounin's strategies and relate them to managing science for whole-class instruction. Then we'll turn our attention to management strategies appropriate for small group work, and finally we'll look at some very recent suggestions for managing science classrooms that attempt to engage students in high-level tasks.

Effective Management Behaviors[2]

To be an effective facilitator of learning requires that you understand something about effective group management behaviors. We will explore six group management behaviors that effective teachers appear to incorporate into their style of teaching. These six behaviors, which are summarized in Table 10.1 are variables that have been correlated with student involvement in learning activities. They are behaviors that are evident in classrooms of teachers who have few student misbehaviors (or if they do have student misbehaviors, they are taken care of swiftly and fairly). Since they are correlations, teachers should realize that these behaviors do not necessarily cause student behavior.

With-it-ness

With-it-ness is the teacher's ability to communicate to students that he or she knows what they are doing in the classroom at all times. With-it-ness is a monitoring behavior, not only during small group work but when you are making a presentation, or if students are doing individual seat work.

Teachers who are successful in monitoring students appear to have "eyes in the back of their head." They are able to spot misbehavior via a sixth sense—almost as if they are able to see every student all of

---

[1] Jacob Kounin, *Discipline and Group Management in Classrooms* (Huntington, N.Y.: Robert E. Kreiger, 1970).

[2] This section is based on American Federation of Teachers, "Research on Effective Group Management Practices," in *Educational Research and Dissemination Program*, Lovely Billups, ed. (Washington, D.C.: American Federation of Teachers).

TABLE   10.1

EFFECTIVE GROUP MANAGEMENT PRACTICES

Practice	Explanation	
With-it-ness	Teacher's ability to communicate to students what they are doing in the classroom at all times. It involves nipping problems in the bud before they escalate.	Do you have eyes in the back of your head?
Overlapping	The teacher's ability to effectively handle two classroom events at the same time as opposed to becoming so totally glued to one event that another is neglected.	Can you deal with working with a small group, while at the same time a student returns from the counselor, another student drops a cup containing a mixture of water and sand, and the teacher across the hall sends a student in with a message for you?
Smoothness	Teacher's ability to manage smooth transitions between learning activities.	Do your activities have clear endings before moving on to new ones?
Momentum	Teacher's ability to maintain a steady sense of movement or progress throughout a lesson.	Are your activities conducted a. At a brisk pace? b. In logical steps? c. Without lengthy directions?
Group Focus and Accountability	Teacher's ability to: • Keep the whole class involved in learning so that all students are actively participating • Hold students accountable for their work • Create suspense or high interest	Would students in your class say they were kept on their toes?

Based on "Research on Effective Group Management Practices," in *Educational Research and Dissemination Program*, American Federation of Teachers.

the time. For teachers to communicate with-it-ness to the students, they must indicate this awareness through some action indicating an awareness of student behavior. The easiest way to exhibit this is to stop misbehavior in a timely and appropriate manner. This means nipping behavior problems in the bud—before they manifest themselves and spread to other students like a virus.

Eye contact, asking questions, physically moving toward students who are about to misbehave, and redirecting students to prevent misbehavior are some individual teacher behaviors that will convey to students the teacher's sense of with-it-ness.

Overlapping

In the real world of the classroom, multiple events occur simultaneously, and the effective manager is able to deal with them. In this management practice, the science teacher does not get totally immersed in one event (helping a team with their titration apparatus) at the expense of other pending situations. Throughout the day, there is a very high chance of interruptions from students entering your class from the outside, or from announcements over the PA system. The skilled manager is able to maintain the flow of instruction by holding the entire class accountable for continuing, while at the same time dealing with the intrusion.

Smoothness

This management practice refers to the teacher's ability to manage smooth transitions between learning activities. Kounin identified a number of classroom behaviors that tended to impede smoothness. Here in summary are some behaviors to be on the look out to reduce:

1. Bursting in on a group or the whole class with new information or instructions when the students are really not ready for it. For example, suppose that the teacher told the class that groups had ten minutes to complete an activity in which they were classifying vertebrate bones. With four minutes of the students' time left, the teacher bursts in with these instructions: "In addition to what you are doing, now I want each group to name the bones and to report their finding to the whole class." It might be better to have said this as part of the original instructions or to wait until the ten minute period is up and then announce the new instructional procedures, allowing necessary time for completion.

2. Some teachers have a tendency to start an activity and leave it dangling by starting another activity. For example, a physical science teacher begins the lesson by going over the homework and asks three students to go to the board to write the answers to the first three problems. While they are on the way to the board, the teacher asks the class if they are ready to review yesterday's laboratory activity. Many students raise their hands and start talking about the lab. Meanwhile one student at the board is having difficulty with one of the homework problems. The teacher's attention is now drawn to the class talking about the lab.

3. Sometimes activities are never completed. The activity is truncated. As in the second case, there is a chance that the teacher might not finish going over the homework assignment.

4. Sometimes teachers call attention to a problem in the middle of an activity that could have been dealt with later. The interruption stops the flow of instruction. Examples of this are minor misbehaviors, such as a student in chemistry class looking over an English term paper. An incident like this can be handled by simply walking to the student and pointing or touching the paper. The paper is put away with no incident. A problem occurs when the teacher goes up to the student and asks why he or she is reading an English paper in

chemistry class and gets into a discussion. By this time the whole class is interrupted.

Momentum

Effective managers move their lessons at a brisk pace and appear to have very few slowdowns in the flow of activities. Maintaining momentum, or a steady sense of movement throughout the lesson, helps engage the learners in activities. Slowdowns or time not well utilized between activities tends to cause students to lose interest.

Slowdowns are generally caused by teachers overdwelling on a task, and fragmenting activities into trivial steps when it might be better to formalize them as a single activity. Overdwelling can kill a good activity. Teachers who spend too much time giving detailed instructions on a laboratory activity can reduce the students' initial interest. Writing out the instructions or reconceptualizing the activity can eliminate overdwelling. Another practice to be aware of is lecturing for too long a period of time. Unless there is brisk movement during the lecture, it can be a turnoff to student interest. Lecturing students' about misbehavior can also impede the flow of instruction.

According to Kounin's work, momentum appeared to be the most important management behavior for promoting active involvement among students and reducing misbehaviors.

Group Focus and Accountability

We might have called this practice group focus and "individual" accountability. As a teacher you will always be involved with a large group of students, and at the same time you must hold each student accountable for learning. Maintaining group focus—keeping students on their toes—as well as holding each student in the class responsible for learning are key management practices.

One of the key behaviors in this practice is the format that you choose for student involvement. Which of these formats do you think will result in greater student involvement?

- Teacher-led large class format
- Individual seat work
- Small group format

Reports show that teacher-led large class and small group work are more effective in promoting focus and involvement.[3] Individual seat work appears to be less motivating than the pace set by the teacher with the whole class, or work established for small cooperative teams.

Group focus also can be achieved by "keeping students on their toes." Here are some strategies to achieve this goal:[4]

1. Attracting students' attention by asking a question before calling on a student to respond
2. Holding attention by pausing to look around the group to bring the students in before calling on someone to respond or recite, by asking for a show of hands before selecting someone, or by using other high interest cues such as "Be ready, this might fool you"
3. Keeping students in suspense as to who will be called upon next by avoiding a predictable pattern for selecting students
4. Calling on different students with sufficient frequency so that students don't tune out because the same group is always called upon
5. Interspersing individual responses with mass unison responses
6. Alerting nonperforming students in a group that they may be called upon in connection with the performer's response or to recall something the performer recited
7. Using a random number technique (having students in each group number off) to call on students in the class

Group focus is also dependent on conveying to students that they are each accountable for their academic and social behavior. If you convey to the class that you expect each person to be ready to respond or to complete assignments, the chances are that they will remain academically involved. The following are some ways to maintain accountability in a group format:[5]

1. Teacher checks students' answers or other performances by asking them to hold up answers or some prop indicating an answer
2. Teacher requires group to recite in unison while actively listening for individual responses
3. Teacher checks for understanding of a larger number of students by asking some students to com-

ment on whether another student's answer or performance was right or wrong
4. Teacher circulates around the group and checks the answers or performance of students at their seats while another student is asked to perform aloud or at the board
5. Teacher asks for the raised hands of students who are prepared to demonstrate a skill or problem and then requires some of them to demonstrate it

The management practices that have been presented here are reflected in the classrooms of teachers who have high rates of student engagement and low numbers of misbehaviors. These practices appear to be essential in the variety of tasks that teachers plan to involve students in science activities. However, we will highlight two special aspects of science teaching and examine the management or facilitative skills and behaviors that appear to be important. First we will examine management of laboratory and cooperative learning activities and then investigate facilitative skills important to teaching high-level tasks.

Facilitating Laboratory and Small Group Work

Most laboratory work is planned for small teams of students. It is therefore important to keep in mind the management strategies that researchers on cooperative learning have found to be essential and effective in maintaining student involvement. Johnson and Johnson, as was discussed in Chapter 7, have identified five management practices that teachers should employ in cooperative learning experiences.[6]

[6]David W. Johnson and Roger Johnson, *Circles of Learning: Cooperation in the Classroom*, (Alexandria, Va.: Association for Supervision and Curriculum Development, 1986).

FIGURE 10.1 Schema for managing cooperative learning tasks

[3]Edmumd T. Emmer and Carolyn M. Evertson, "Synthesis of Research on Classroom Management," *Educational Leadership*, January, 1981, pp. 342–347.
[4]Lovely Billups, "Research on Effective Group Management Practices," p. 14.
[5]"Research on Effective Group Management Practices," pp. 14–15.

These include positive interdependence, individual accountability, face-to-face communication, interpersonal skills, and processing (Figure 10.1). Briefly, here is a review of these behaviors.

Positive Interdependence

Students need to value the performance of each member of the group, as well as their own. A sense of mutual dependence is established by agreeing on a goal, dividing up the work load or materials, resources or information, differentiating roles, and providing joint rewards. Each of these contributes to creating an environment of positive interdependence. Specific management strategies that will help create positive interdependence during small group or laboratory activities include the following:[7]

1. Establishing Common Goals

- Having each group complete and turn in a single product (worksheet, report, project, data sheet)
- Structuring a group discussion so that it results in agreement or consensus

2. Structuring Joint Rewards

- Giving rewards to individual groups such as a grade or score, or a privilege or treat
- Giving groups extra rewards for individual efforts of their members (bonus points, improvement scores)
- Structuring intergroup competitions (e.g., games, tournaments)

3. Structuring the Task

- Providing roles for each group member (facilitator, materials handler, encourager, recorder-reporter)
- Making turn taking a part of the activity
- Using experts—division of labor (e.g., dividing content so that each team member becomes an expert: igneous, metamorphic, sedimentary)
- Dividing resources or information (limiting materials or access to information)

Individual Accountability

There is always the fear that group work will result in one or two students doing all the work while the rest get a free ride. The structure of cooperative learning is also dependent on each student's mastery of the material being learned and taking responsibility for sharing in the attainment of the group's goal. Individual testing, grading, and feedback are part of the cooperative learning approach. Here are some additional management strategies that can be used to structure individual accountability into cooperative learning and laboratory activities:[8]

1. Focusing on the Contribution of Individuals

- Using roles (communicator, materials handler, praiser, recorder-reporter)
- Dividing the task (minitopics, subdivision of text, part of assignment)
- Using experts or resource persons
- Coding individual contributions (markers, individual papers)
- Getting feedback on individual student behavior
- Providing opportunity for group or individual reflection

2. Focusing on the Outcomes of Each Individual

- Individual tests, quizzes
- Individual homework assignments
- Individual reports, data, essays
- Group rewards for individual behavior (each person receives bonus points for every person who turns in homework)

Face-to-Face Communication

Students need to be put in situations in which they interact with each other face to face. Learning in small groups is dependent on students talking with each other. Paying attention to room arrangement is crucial here. You may have to take the time to rearrange the furniture of your room so that small groups of students can sit facing each other.

Interpersonal Skills

Just as students need to learn the skills of doing science, you will discover that students, if placed in cooperative learning groups, will need to learn some communication skills. One effective strategy is to teach students the interpersonal skills that they will be using in the small group activities. A technique that is effective is the use of an interpersonal T-chart (Figure 10.2). T-charts are created by a teacher-led discussion of the interpersonal skill that will be uti-

---

[7]Nancy Graves and Ted Graves, *What Is Cooperative Learning? Tips for Teachers 'n' Trainers* (Santa Cruz, Calif.: Cooperative College of California, 1989), p. 2.2.

[8]Graves and Graves, *What is Cooperative Learning?* p. 2.3.

ACTIVE LISTENING	
Sounds Like	**Looks Like**
• Say "uh-huh" as speaker talks. • Use open-ended questions to keep the speaker talking. • Paraphrase what the speaker says. • Use encouragement to keep the speaker talking. • Accept what the speaker says rather than give your opinion. • Summarize the speakers comments.	• Nod • Eye contact • Lean forward • Smile • Relaxed postures • Hands unclenched • Arms not crossed

FIGURE 10.2 Interpersonal Skills T-Chart. Linda Lundgren, *Cooperative Learning with Biology: The Dynamics of Life* (Columbus, Ohio: Merrill, 1991), p. 13. Used with permission.

lized in the small group activity or laboratory activity. Students are asked to brainstorm what the skill "sounds like" by coming up with phrases they would use while working in the group to encourage "active listening" (or other identified skills; everyone contributes, questioning, staying on task). Then students are asked to brainstorm what the skill "looks like." The results are written as a large poster that is mounted in a place where all students can see it. During the cooperative activity or lab, students "practice" using the skill.

Processing

Providing time for groups to process their work is important management strategy. Too often students work together but never have the opportunity to reflect on their work, make suggestions for improvement, or give each other feedback. The effective facilitator of small groups provides each team with a few minutes to process. The teacher organizes this by presenting to the groups one or more problems or questions to discuss after the activity has been completed. For example, a science teacher might give the groups any one of these questions to discuss:

1. Write positive statements about how each member contributed ideas in the activity. Share these statements with the other group members.

2. What skills did you effectively use? Which need to be worked on?
3. What behaviors contributed to learning in your group?
4. In what ways did your group encourage the contributions of all group members?
5. How did you feel about participating in this task?
6. What did you learn from this activity?
7. How would you do this activity differently?
8. Did you participate fully in the activity?

In addition to knowing these five management strategies, it is important to put them together in a smooth-flowing plan or series of phases. The next section describes a series of phases that constitute a management plan for small group and laboratory activity.

A Management Plan for Small Group or Laboratory Work

The following phases should be used when implementing small group or laboratory activities.[9] The phases enable the science teacher to integrate positive management strategies with the elements of lesson planning developed in Chapter 9.

Phase I: Preparation

There are a number of management decisions that have to be made in advance if cooperative group work and laboratory work is to be effective. You should decide upon the goals for the activity. Once goals have been established and you have chosen the activity, task, or laboratory exercise or problem in which students will be engaged, you will need to make these decisions:

1. Decide upon the size of each group, and how students will be assigned to them. Generally, keep groups from two to four students. Student teams should be heterogeneous by ability, sex, and ethnicity.

[9]Based on Johanna Strange and Stephen A. Henderson, "Classroom Management," *Science and Children*, November/December, 1981, pp. 46-47.

Phase I Preparation	**Phase II** Preactivity Discussion (Prelab)	**Phase III** Equipment and Safety Needs	**Phase IV** Activity Monitoring	**Phase V** Postactivity Processing

FIGURE 10.3 Schema for management process for cooperative learning and laboratory activities

2. Identify roles for each member of the team. Examples include facilitator, recorder, reporter, and materials handler.
3. Obtain all the materials needed for the activity. Organize them either at stations around the room or in the laboratory, or put them in containers that can easily be distributed to individual teams.

Phase II: Preactivity Discussion

Effective facilitators of small group work prepare students to work in small teams and in the laboratory. Students need to have a clear understanding of the task, whether it is a problem to investigate, a phenomenon to observe and analyze, or a small group activity that is accomplished sitting together around tables pushed together. The teacher's ability in establishing group focus is important at this stage. Here are some of the management procedures that should be accomplished during this phase:

1. Explain the task, pointing out the goal and purpose of the activity
2. Explain how students will work together to achieve positive interdependence, and how each group member is individually accountable
3. Identify the interpersonal skill(s) that will be emphasized in the activity
4. Explain time constraints and when the students should be completed with their work

Phase III: Equipment and Safety Needs

After students are in their assigned teams and arranged as small groups at their workstations (around desks, in the lab), take time to explain the equipment needs of the activity. Equipment should be picked up by a materials handler from each group (who is also responsible for facilitating cleanup and return of the equipment at the end of the period). If equipment is already set up at workstations in the classroom, or at laboratory tables, someone from each group should be designated as being responsible for the equipment.

You need to anticipate any safety problems associated with the equipment and materials that students will be handling. At all times, students should wear safety goggles when handling materials in a science activity. One principle to keep in mind is to weigh an activity's science educational value against its hazard. School liability experts recommend that if a potentially hazardous activity cannot be changed or altered to reduce the risk, it should be eliminated.[10]

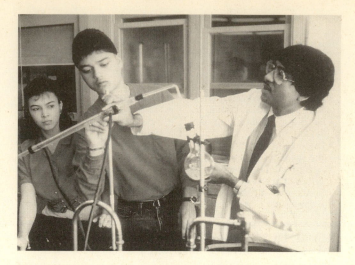

FIGURE 10.4 The management of small group and lab activities requires the with-it-ness of active monitoring. ©Paul Conklin.

Take the time to demonstrate the use of materials or equipment before the students begin the activity. Discovering how to use equipment and materials is not one of the goals of the science laboratory.

Phase IV: Activity Monitoring

Effective science teachers actively monitor the work of students in cooperative activities and laboratory work. Active monitoring involves observing the groups as they work on the problem or task. Some teachers make notes on each group's performance, especially noting progress on interpersonal skills. The teacher should intervene by asking students, from time to time, to explain what they are doing or to answer questions. An important role in monitoring is to exhibit with-it-ness by preventing any misbehaviors before they begin. Laboratory work, especially if students are standing in a laboratory environment, can "facilitate" misbehavior among students. Your careful monitoring will head off most of these problems.

Phase V: Postactivity Processing

The facilitator role of the teacher is maximized during postactivity processing. There are two aspects to postactivity facilitating that are crucial. On the one hand, the teacher is responsible for helping the students understand the concepts or the results of the

[10]Gary E. Downs, Timothy F. Gerard, and Jack A. Gerlovich, "School Science and Liability," in *Third Sourcebook for Science Supervisors*, LaMoine L. Motz and Gerry M. Madrazo, Jr., eds. (Washington, D.C.: National Science Teachers Association, 1988), pp. 121–125.

experiment carried out in the laboratory. The teacher should be center stage to facilitate verbal communication among the entire class. Reports from students selected randomly from each group can be given. The teacher can ask high-level questions to extend the reasoning ability of students. The teacher might arrange for the results from each group to be posted for all to observe. This dimension of the postactivity processing should focus on the cognitive and affective outcomes associated with the activity.

On the other hand, postactivity processing should also be an opportunity for the cooperative groups to process their work as a team. This is an opportunity for the team to sit together to reflect on how well they worked as a team, and on what steps they need to take to improve their team's ability to function. The teacher can suggest a question or two for the team to consider and discuss, for example, What skills did the group use effectively? What are some areas that need to be worked on?

Facilitating High-Level Thinking Tasks

In their book *Windows into Science Classrooms*, Tobin, Kahle, and Frasher address a problem that is associated with science teachers' attempts at creating higher-level cognitive learning environments. Many of the issues discussed in their work focus on the classroom management behaviors exhibited by science teachers. These authors, who take a strong stand in favor of constructivism (students build or construct their own knowledge structures), point out that in the two classrooms that they studied, the level of cognitive thinking tended to hover near the bottom of Bloom's taxonomy in the cognitive domain.

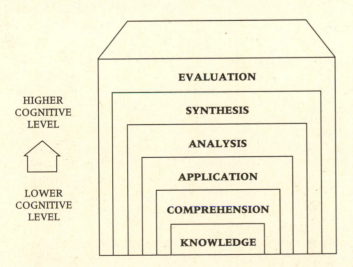

FIGURE 10.5 Effective management behaviors can raise the cognitive level of the science classroom.

What can teacher's do, and specifically, what management strategies might science teachers employ to raise the cognitive level of thinking of students? Let's explore this a bit. Floyd H. Nordland, one of the contributors to *Windows into Science Classrooms*, describes how teacher behavior can influence the cognitive level of learning activities. He describes this scene in one teacher's classroom:

Peter (a high school biology teacher) is positioned in front of the demonstration desk expounding on the human repiratory system. The students are seated behind long horizontal benches. They are quiet and attentive to Peter's lecture presentation and many of them are taking notes.

Peter: The nervous system has the most critical oxygen requirements of any tissue of the body. In fact, brain tissue deprived of an oxygen supply for as short a time as one or two minutes will produce irreparable brain damage.

At the front bench to Peter's right, Jeffrey's hand shoots up. Peter continues lecturing either unaware of Jeffrey's insistent hand-waving or studiously avoiding it.

Peter: Are there any questions?

Peter carefully scrutinizes the entire class before somewhat reluctantly calling on Jeffrey.

Jeffrey: I was watching *60 Minutes* on the television recently and they talked about an American kid who was under water for a long time. I think that it was about fifteen or twenty minutes. When they pulled him out, they were able to revive him and apparently there was very little brain damage. How can you explain this?

Peter: Well, I don't know anything about that as I don't watch *60 Minutes*.

Peter continues with some disparaging comments about the negative aspects of watching too much television and Jeffrey's excellent question is never acknowledged intellectually. Thus an opportunity to teach and learn at the application, synthesis or analysis level is lost and the instruction continues at the lowest possible cognitive level.[11]

A simple management decision could have resulted in a high-cognitive-level discussion and exploration

[11]Floyd H. Nordland, "The Cognitive Level of Curriculum and Instruction: Teaching for the Four Rs," in *Windows into Science Classrooms*, Kenneth Tobin, Jane Butler Kahle, and Barry J. Fraser, eds. (London: Falmer, 1990), p. 135.

of the case that Jeffrey described. Unfortunately this opportunity was lost.

In a study to investigate the management of comprehension and other higher-level cognitive tasks, Julie P. Sanford found that there are a number of management strategies that appear to foster higher-level thinking in the science classroom.[12] Sanford found that for teachers to successfully engage students in comprehension-level tasks, the teachers should

1. Create an aura of accountability around the task to force students to attempt it
2. Provide a variety of safety-net devices to keep students from failing at the task

Sanford reported that to make students accountable, "teachers raised the price of noncompliance," by making the following types of management decisions: making the task count more toward a final grade and sending messages home to parents indicating little or no progress on research reports.

Sanford's safety nets appear to be interesting management devices to encourage students to take risks while attempting higher-level cognitive tasks. In general she explained that these safety nets were designed to reduce the individual's risk of failure, thereby increasing the chance that students would at-tempt the tasks. Here are some management strategies that were found to be effective:

1. *Small group work:* Team learning in which the individual students' work was pooled was more effective than individual students turning in their own "lab reports."
2. *Peer assistance:* Allowing students to help each other.
3. *Test procedures:* One technique was to balance difficult material with very familiar or easy material on the test. Another technique was to count the higher-level questions on tests less than the easy questions so that lack of success on high-level questions did not result in failure.
4. *Revision option:* Giving students the option of revising papers and products before being turned in for a final grade.
5. *Teacher assistance:* Teacher made time to answer questions and help clear up difficulties.
6. *Extra credit assignments.*
7. *Models:* Providing models of successful products for students to examine.
8. *No-risk pop quizzes:* Students receive extra credit for perfect papers or for each correct answer; no penalty for incorrect ones.
9. *Review sessions.*
10. *Flexible grading system:* Allows the teacher to reduce the value of assignments on which the whole class did poorly.[13]

[12]Julie P. Sanford, "Management of Science Classroom Tasks and Effects on Students' Learning Opportunities," *Journal of Research in Science Teaching* 24, no. 3, (1987) 249–265.

[13]Sanford, "Management of Science Classroom Tasks," 249–265.

INQUIRY ACTIVITY   10.1

Windows into Science Classrooms

In this inquiry activity you will either observe teachers in action in their classrooms or observe science teaching videotapes. In either case, you will be looking into these classrooms to find out about how these teachers manage the classroom and apply the principles of group management behavior.

Materials

Videotape of science teacher (or a visit to a science classroom)

Procedure

1. Secure permission from a science teacher to visit his or her classroom. Explain that you are interested in observing the way he or she manages his or her classroom. Tell the teacher that you will be using a form to look for certain examples of classroom behavior.
2. Observe the science teacher (or a video of a science teacher) for at least one complete lesson. Keep a narrative record of the verbal behavior of the teacher. That is, keep a running record of what

Management Behavior	What the Teacher Did or Said	How the Student Responded
With-it-ness		
Overlapping		
Smoothness		
Momentum		
Group Focus and Accountability		

FIGURE 10.6   Windows Observation Form

the teacher does and says. This will help you discuss the management behavior of the teacher and enable you to ask better questions.

3. Use the Windows Observation form (Figure 10.6) to record examples of the management behaviors.

Minds on Strategies

1. To what extent did the teacher show evidence of each management behavior? What was the teacher's strength in management?

2. How did the teacher handle small group activities?
3. How did the teacher handle laboratory activities?
4. Did the students show evidence of being engaged in high-level thinking tasks? How did the teacher encourage and manage high-level thinking tasks?
5. How did the teacher manage misbehaviors?

EFFECTIVE TEACHING FOR THE BEGINNING OF THE YEAR

According to Donna Bogner, author of *Starting at Ground Zero: Methods for Teaching First Year Chemistry*, the first day of a course is "ground zero." This is the day that sets the tone for the year and "is your opportunity to quell the fears of the hesitant, to capture the interest of the curious nad to introduce young minds to the fascinating world of chemistry."[14]

In this section we will deal not only with management practices on the first day of school but for an

extended period of time, which we will call the "beginning of the year." We will examine several aspects of classroom management that will help get the year off to a good start. These include effective room arrangements, the establishment of rules and procedures, first-day lesson plans, and plans for the first two weeks of school.

Room Arrangements

Although teachers typically have no control over the size or shape of the classroom, and in some cases the

[14]Donna Bogner, *Starting at Ground Zero: Methods of Teaching First Year Chemistry*, Vol. 1, (Hutchinson, Kansas: Genie Publications, 1989), p. 49.

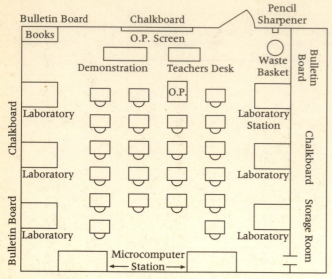

FIGURE 10.7 Effective room arrangement for whole class instruction. After Edmund T. Emmer et al., *Classroom Management for Secondary Teachers,* © 1989 Prentice-Hall, Inc. Used with permission of the publisher.

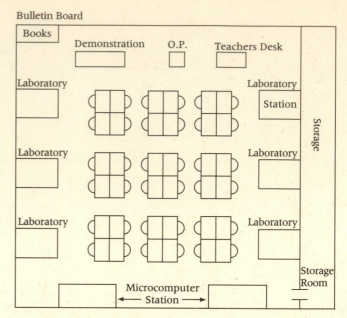

FIGURE 10.8 Effective room management for small group instruction and cooperative learning. After Edmund T. Emmer et al., *Classroom Management for Secondary Teachers,* © 1989 Prentice-Hall, Inc. Used with permission of the publisher.

things that can be done to arrange the classroom for effective science teaching. There are a number of management considerations to guide the way you arrange the classroom. Research by Emmer, Evertson, and Anderson[15] has indicated that effective managers of classrooms had good room arrangement that helped eliminate potential distractions for students and opportunities for inappropriate behavior, and permitted easy monitoring of students at all times. Figure 10.7 and Figure 10.8 show effective room arrangements for whole class and cooperative learning instructional formats, respectively.

In these cases furniture is arranged to facilitate easy-flowing traffic patterns, avoiding congestion in areas such as the pencil sharpener, trash can, laboratory stations, demonstration table, and work areas. Desks should be arranged based on the instructional goals and needs of the students. In the arrangements shown, it is a relatively easy management function to stop the action and have the students move their desks together for cooperative learning activities. Incidentally, this can work even if the students have slanted desks. Effective teachers, according to these researchers, took stock or control of their room and created an environment that was conducive to learning.

Establishing Rules and Procedures

Good management begins with a set of classroom rules and procedures that are clearly articulated and taught to the students. The effective set of rules and procedures tend to be general and require thinking and interpreting on the part of the students. Naturally, there might be exceptions to this, such as "You may not work in the laboratory without safety goggles" or "Only one person speaks at a time." Let's take a look at some of the dimensions of establishing rules and procedures.

Rules for the Science Classroom

Here are some of the findings from research with regard to the establishment of classroom rules and procedures.[16]

- Effective teachers created rules and procedures to guide students' behavior with respect to appropriateness of student talk; movement within and outside the classroom; getting the teachers attention; storing personal belongings; use of equipment (computer); work during laboratory
- Effective teachers also created strategies to be used to positively reinforce good student behavior or sanction misbehavior
- Effective managers taught the rules and procedures just as they would teach any concept or idea
- Effective managers presented their rules over a period of days or weeks, and reteach the rules on a regular basis

[15]E.T. Emmer, C.M. Evertson, and L.M. Anderson, "Effective Classroom Management at the Beginning of the School Year," *The Elementary School Journal,* 1980, Vol. 80, No. 5, pp. 219–231.

[16]American Federation of Teachers, "Research on Effective Group Management Practices."

FIGURE 10.9 Effective science teachers establish rules and procedures that will facilitate student learning. Roger T. Cross.

- Teachers also integrated rules and procedures into the ongoing classroom instruction by teaching rules appropriate to an instructional activity, for example, handling acids, using a microscope, using dissecting equipment
- Rules tended to be stated positively, rather than negatively

What rules and procedures and how many should science teachers establish in their classroom? First, a rule is a statement that describes a general expectation of behavior, whereas a procedure typically applies to a specific activity.[17] Typically teachers should try to keep the number of rules between four and seven. For example, one set of rules that many beginning teachers use in their classroom is this simple list of 4-Ps:

- Prepared
- Prompt
- Productive
- Polite

These rules lend themselves to discussion and interpretation, and they encourage students to think about classroom behavior. Other teachers prefer rules to be spelled out in a little more detail. Here is a list of rules that effective teachers put into practice:

1. Bring all needed materials to class
2. Be in your seat and ready to work when the bell rings
3. Respect and be polite to all people
4. Do not talk or leave your desk when someone else is talking
5. Respect other people's property
6. Obey all school rules

Many teachers see rules as an opportunity to teach students responsibility and try to involve students in the development of rules. To do this some teachers suggest stating rules in terms of students' rights (e.g., to learn respect for other people and property). This has the added advantage of giving students an opportunity in the ownership of rules. In the science classroom that is attempting to foster inquiry and high-level cognitive thinking, this aspect of democracy seems rather important.

On the first day of class time should be devoted to teaching the rules and involving the students in a discussion of the interpretation of the rules. Some would argue, "Why do you allow interpretation, why not be specific and eliminate any argument?" The problem with this approach is that too many rules would have to be stated, and student misbehavior would tend to increase given the lengthy list of rules.

What happens if students break a rule? In general, the best approach is to state consequences, in advance, if rules are broken, and relate the consequences to a hierarchical system. For example:[18]

1st time student breaks rule:	name on board	= warning
2nd time student breaks rule:	Name ✔	= One detention
3rd time student breaks rule:	Name ✔✔	= Two detentions
4th time student breaks rule:	Name ✔✔✔	= Two detentions
5th time student breaks rule:	Name ✔✔✔✔	= Two detentions (call parents, student sent to vice-principal)

Consequences ought to be used to teach responsibility. Consequences, if used properly, are not punishments. For example, a consequence of not cleaning up the lab table is to be told to clean it up. A punishment would be to apologize to the teacher in front of the whole class. This approach to positive discipline accepts the notion that dealing with student misbehavior is part of the teaching role; that students should be treated with dignity; and that discipline works best when it is integrated with effective teaching and management practices.[19]

[17]Edmund T. Emmer et. al., *Classroom Management for Secondary Teachers* (Englewood Cliffs, N.J.: Prentice-Hall, 1989), p. 19.

[18]Based on Lee Canter and Marlene Canter, *Assertive Discipline: Resource Materials Workbook, Secondary, 7–12* (Santa Monica, Calif.: Lee Canter & Associates, 1984), p. 5.

[19]Richard L. Curwin and Allen N. Mendler. *Discipline with Dignity* (Alexandria, Va.: Association for Supervision and Curriculum Development, 1988).

These consequences are only examples. Consequences should be chosen that are appropriate for you and have an impact on the students. According to researchers, it's not the severity of the consequence that has the impact on behavior but the inevitability of receiving the consequence. Other consequences include[20]

- Last to leave classroom
- Citation
- Clean up the lab
- Letter to parents
- Campus cleanup
- In-school suspension
- Sent to vice-principal
- Sent to another room
- Assigned seat
- Demerits

Good management systems also include provisions for rewards—feedback that conveys to students that the teacher appreciates the cooperation they are showing. It is especially important that rewards become part of the instructional system. For example, one common form of reward is *praise:* "Class, I appreciate how clean that lab was today when you finished the experiment." Praise can be focused on an individual, a small group, or the whole class. To be most effective, praise should be given after student performance, not as a carrot to induce behavior. The praise should be sincere and genuine, and should describe the particulars of the behavior ("I liked the way you handled the microscopes yesterday in the lab"). For praise to be seen as credible, students must be convinced that the teacher has considered the performance carefully and means what he or she says about it: "Yesterday, when we went outside to observe the soil and collect samples, I noticed how careful you were not to disturb plants, and how careful you were not to make distractions for other classes who were inside the building. That was great. Thanks."

Some teachers use a formalized system of rewards, thereby applying behaviorism by establishing a set of positive reinforcers that can be earned for appropriate behavior. The teacher would establish in advance a system of points that could be earned for "good behavior." As the class accumulates the points within a specified period of time, they can earn privileges such as[21]

- No homework for one night
- Free time in class
- Popcorn or pizza while watching a video
- Class trip
- Outdoor activity
- Free homework pass redeemable anytime throughout the year
- No homework over weekend

Procedures or routines apply to more specific activities, as was stated previously, and need to be established and maintained. In the science classroom there are a number of special activities that need attention, in addition to the general procedures all teachers need to attend. These include routines at the beginning and end of class, the use and handling of materials, equipment, and living things, and routines for whole class instruction, cooperative learning work, and laboratory work.

Rules, consequences, and rewards should be included as part of the course syllabus, and each student should get a copy. The system should also be posted in a prominent place in the classroom.

The Beginning and the End of Class

Communicating clearly your expectations regarding the beginning and ending of class is an important aspect of classroom management. Effective teachers communicate what is expected of students when class begins. Some teachers begin class by taking attendance by means of a seating chart (especially at the beginning of the year). This is done quickly, and routinely. At the beginning of the year, teachers prefer to call the names from the roll in order to learn the names of students and use it as an alternative device throughout the year. You will also have to establish a routine to deal with students absent the previous day: some teachers have a folder or bin containing handouts and assignments from previous lessons so that students who were absent simply obtain their work from this place. Whatever routine is established, it should be the student's responsibility to find out what was missed on the absent days.

The beginning of the period for many teachers is when homework is collected. If you are going to go over the homework, hold off collecting it until you have dealt with it in the lesson. Once the routines have been established, you can formally begin the lesson. Some teachers have a routine for this: walking into the center of the room or standing in front of the chalkboard or wall containing a planning schedule for the lesson. Other teachers begin the lesson with a stimulating inquiry demonstration. Letting students

[20]Based on Canter and Canter, *Assertive Discipline*, p. 6.
[21]Based on Canter and Canter, *Assertive Discipline*, p. 11.

know that it's "show time" establishes a businesslike yet friendly atmosphere of learning.

Routines are needed to end a class. You need to establish from day one a procedure for ending the class. If during the class period the students were involved in a lab or hands-on activity, you need to establish a cleanup routine. Normally, the assignment of materials handlers can facilitate this routine. However, you need to build in time at the end of the lesson—two or three minutes—for this to happen. Let students know how you want them to be dismissed. Most teachers prefer to dismiss the students directly, rather than letting them dash out the door on the bell signal.

Handling Materials and Equipment

In activities that require the use of materials, you should appoint materials handlers for each team. This person should be given the responsibility of picking up the materials for the group and distributing them within the group, and of coordinating the cleanup of the group and returning the materials to a designated area.

Special-equipment use, such as microscopes, electronic balances, and microcomputers, should be preceded by a procedural minilesson. If a microcomputer is housed in your room, you should post procedures and rules for its use.

If you have plants, terraria, aquariums, and animals in your classroom, you should plan to teach lessons on the use, care, and observation of living things. The use of living things in the classroom, especially animals, is a controversial issue in science education. The procedures that you develop concerning living things in the science classroom can be used to teach cognitive as well as affective outcomes of learning.

How should the science teacher begin the year? How should lesson plans during the beginning of the year be structured? We now turn our attention to "the first day" and "beyond," and examine the management strategies that will promote high levels of involvement among the students.

INQUIRY ACTIVITY   10.2

....

Developing a Classroom Management Plan

....

In this inquiry activity you will be thinking ahead and designing a classroom management plan suitable for a middle or high school science classroom. Your plan should include classroom rules, rewards, and consequences.

Materials

Poster board, marking pens, ruler

Procedure

1. List the rules that you would use in a science class. The rules should be based on the principles discussed in the chapter.
2. Identify the consequences for breaking the rules.
3. List a series of rewards and how students can earn points to achieve the rewards.
4. Design a large classroom management poster that could easily be read by all students in a class.
5. Present your classroom management plan to the class.

Minds on Strategies

1. What general principles guide the management of student behavior?
2. How effective, in your judgment, are your plans?
3. Show your plans to a practicing science teacher for feedback. What suggestions does the teacher make?

The First Day

The first day's lesson should be planned to establish interest in the course, but perhaps more important, to incorporate activities that will help you establish contact with the students and establish you as the leader of the class. One way to begin is to greet students at the door and hand them a one- or two-page syllabus of the course, and tell them that today they can sit anywhere they wish. If you were going to do a small group activity on day one, you could give each student a color-coded card that would be used later to form groups. This procedure helps to establish the teacher as being in charge as soon as the students pass through the door.

TABLE 10.2

FIRST DAY'S ACTIVITIES
IN THREE CLASSES

Teacher A		Teacher B		Teacher C	
Introduction of teacher and roll	5 minutes	Filling out information cards and roll call	9 minutes	Introduction of teacher and roll call	2 minutes
Presentation of rules and procedures	21 minutes	Presentation of rules and supply requirements	8 minutes	Presentation of rules and procedures	12 minutes
Election of class officers	2 minutes	Diagnostic test	21 minutes	Filling out information cards	7 minutes
Preview of week's activites	7 minutes	Oral review of rules and supply requirements	2 minutes	Seatwork	33 minutes
Seatwork	18 minutes	Free time: students talking or waiting	16 minutes		
Closing	1 minute				

From Julie Sanford and Carolyn Evertson, *Classroom Management in a Low SES Junior High (Research and Development Report No. 6104)* Austin: Tex: Research & Development Center for Teacher Education, The University of Texas at Austin, 1980) as quoted in "Effective Classroom Management for the Beginning of the School Year," *Educational Research and Development Program* Lovely Billups, ed. (Washington, D.C.: American Federation of Teachers), p. 18.

Contact with the students is an important aspect of day one. As soon as the students are seated and the bell has rung, effective teachers begin with roll and then introduce students to the room. Some teachers take a few minutes for students to introduce themselves to each other.

What should the lesson structure for day one consist of? Table 10.2 compares the lesson plans of three junior high teachers. Look the lesson plans over. Which lesson is that of an effective classroom manager? What are your reasons? Study the lesson plans, then read the section that follows the table to get more insight into these three patterns.

According to Julie Sanford and Carolyn Evertson,[22] teacher A was very effective in terms of student on-task behavior and student disruptive behavior, teacher B was less effective, and teacher C was effective at the beginning of the year, but then problems began to escalate during the year. On the first day teachers A and C had cooperative classes, while B had disruptive problems, especially at the end of the period.

Teacher A reinforced the rules on a fairly consistent basis over the next three weeks, whereas B and C did not. Although disruptive behavior was very little for C at the beginning, it increased as the year went on. Teacher A maintained a constant leadership role, and provided no dead time during lessons. Teacher B continued to allow free time, and also had the most misbehaviors.

Effective teachers prepare first-day lessons that

- Establish the teacher as the leader of the class
- Provide as much opportunity for teacher-student contact as possible
- Present the class rules, consequences, and reward system
- Involve the students in an interesting activity
- Establish appropriate opening and closing lesson routines

Let's look at a couple of first-day science lessons and then examine a two-week schedule put into place at the beginning of the year.

First Lessons

Getting off to a good start requires careful planning not only of the first lesson but of the first two or three weeks of school. In this section several first-day lessons are presented.[23] Notice the management practices that are included in each, and how the teachers establish themselves as classroom leaders, engage the students, and have the students leave the class knowing that this teacher is with it!

[22]Julie and Carolyn Evertson, *Classroom Management in a Low SES Junior High (Research and Development Report No. 6104)* Austin: Tex: Research & Development Center for Teacher Education, The University of Texas at Austin, 1980).

[23]These first-day lessons are modeled after ones in Emmer et al., *Classrom Management for Secondary Teachers,* p. 78.

Physical Science: Day One

This lesson could serve as an example of a lesson in either an eighth-grade physical science class or first-year chemistry or physics.

Greeting Students

As students enter the classroom, Mrs. Broadway greets the students at the door and tells them to take a seat near the front of the room, and answers students' questions.

Introduction (1 minute)

When the bell rings Mrs. Broadway moves to the front of the demonstration table and sprays a mist on a piece of newsprint. The words, Chemistry I, appear in a vivid orange color. She tells the class that this is first-year chemistry and to check their schedules to make sure they are in the right room. She extends a warm welcome to the class and tells them that she hopes they will like chemistry.

Roll Call (3 minutes)

She explains that when she calls a name, she wants the students to raise their hand, and to tell her the name they wish to be called. After roll call, she records the names of the two students not present.

Course Syllabus and Overview (6 minutes)

Mrs. Broadway distributes copies of the course syllabus, which contains the title of the course, the rationale, a few course objectives, and the topics for the quarter. She has on her demonstration table about eight household items (baking soda, bleach, mineral water, etc.) and says that chemistry is all around them, and in this course she hopes that they come to appreciate the world of chemistry. She uses an overhead transparency, which lists the first two topics that will be studied: "How Chemists Find Out about the World," and "Atoms, Building Blocks of the World." She goes over the syllabus, and answers a few questions.

Presentation of Class Rules and Procedures (12 minutes)

Mrs. Broadway distributes a mimeographed sheet that summarizes the rules and requirements for Chemistry I. The sheet lists five rules, a section on the method of grading and evaluation, and information on keeping all their work in a three-ring notebook, which Mrs. Broadway calls a portfolio. She tells them to put their name and the date on this sheet, and to place it as the first page (behind the cover sheet) of their portfolio. She explains that the classroom rules are very important. They are in Chemistry I and they will be doing experiments that require safety precau-

tions and she must have their cooperation at all times. She then describes the grading system and then shows the students an example of a completed portfolio.

She then takes a few minutes to go back over the rules and the consequences for not following the rules. She asks the students if they have any questions about the rules. One student asks about rule 1, which is "Bring all needed materials to class." He asks,"What materials are needed?" Mrs. Broadway smiles and says to the class, "Let's make a list of the things that are needed." The class makes this list: textbook, notebook, pencil. Mrs. Broadway explains that there will be additional procedures, especially when they start doing activities. She will teach these procedures when they are needed. She also tells the students that she will review the rules again and again.

Activity: Burning Candle (20 minutes)

Mrs. Broadway presents a very large candle to the class (it is about 15 inches tall). She walks around the room so that the students can observe it closely. She gives each student a sheet of paper and asks them to write their name and the date and period on the top of the paper, and asks each student to write at least ten things about the candle. So the students can observe it more easily, she mounts the candle on the demonstration table for all to see. After two minutes she says to stop writing. Now she lights the candle and asks the students to observe the candle, and to write five more observations of the burning candle. After two minutes, she tells the students to stop writing and she blows the candle out. She then goes to the board and asks for one student to give at least three observations. A student raises her hand; Mrs. Broadway calls on her. She continues this, until she has written about twenty-five observations of the candle on the board. Mrs. Broadway explains to the class that this chemistry activity is important because this is where chemistry begins—with observing things in the natural world. She collects the papers and tells the students that during the course, they will do a variety of activities.

End of Class (4 minutes)

After collecting all the papers, Mrs. Broadway tells the students that she would like them to find pictures of examples of chemicals in magazines and newspapers and bring at least one into class tomorrow. She also explains that they should write ten observations of the "chemical" that they find. Mrs. Broadway explains that it is her procedure to dismiss the students and they are not to leave even if the bell rings. She tells them that before they leave, she expects that the lab (if they used it) or their desks if they did a hands-on activity at their desks, must be clean before dismissal. She compliments the class on their behavior and says that she looks forward to seeing them tomorrow.

Life Science: Day One

This day-one activity could be used in a middle school life science class or in Biology I.

Greeting Students

Mr. Rose greets the students standing outside the door of his life science classroom. The students are coming from across the hall, where they have been in math. He smiles and says hello to the students as they enter the classroom.

Introduction (4 minutes)

Mr. Rose introduces himself at the front of the room. He says that he enjoys teaching life science and was greatly influenced by where he grew up—in the Colorado Rockies—and as a result has always loved the outdoors. He tells the students that this course is called Life Science and that they should be sure they are in the correct room.

Routines (6 minutes)

Mr. Rose tells the students to raise their hands when he calls out their name and to tell him if they should be called by a different name. He then passes out four by six cards by giving the person at the head of each row cards for the row and tells them to take one and pass the rest back. He asks the students to fill out the card as shown on the overhead projector, which shows a sample card with this information: name, address, telephone number, pets, how many brothers or sisters, favorite animal and plant.

Presentation of Rules and Course Requirements (20 minutes)

Mr. Rose has four rules: be prompt, polite, productive, and prepared, and they are listed on a sheet of paper which he gives to each student. Some examples of behaviors for each rule are listed and Mr. Rose uses these to help the students understand the rules. He encourages questions and a few students ask him about the rules. Mr. Rose points out the plants, aquariums, and terraria in the room. He explains that these are there for the class' enjoyment, but also to help them learn about life science. He tells the students that they will be using the computer center during this class, but he will explain the rules for its use when they begin using it in three days. Mr. Rose explains that the students should obtain a three-ring notebook like the one he shows them, and they

should bring it to class tomorrow, and they should purchase a set of dividers (which he shows them) for the notebook. They will be keeping their work in these notebooks. He gives them a handout describing the objectives and activities for the first unit (entitled "Ecology") of the course and goes over the handout with the class. Mr. Rose collects the cards.

Activity (12 minutes)

Mr. Rose gives each student a small brown paper bag containing one object (a pencil, an eraser, a marble, a rock, a leaf, or a pine cone). He also gives each student a sheet of paper and tells them to put their name, the date and the period on the paper. He tells the students to lift the bag up, and move it about, but not to look in the bag. He asks the students to write at least three things about the object in the bag. Now he tells the students to open the bag and look inside, but not to touch the object. Without naming the object, he tells them to write three more things about the object. Finally, he tells them to take the object out of the bag and write three more observations of the object. Mr. Rose tells the students to compare their observations with the person sitting near them with the *same* object. Finally, Mr. Rose tells the students to put the objects back in the bag and to place them on the desks. Mr. Rose asks for student volunteers to describe some of their observations. He makes the point that learning about biology begins with careful observations. He asks one student in each row to collect the bags and bring them to the demonstration table.

End of Class

Mr. Rose gives each student a handout containing pictures of animals and plants. He explains that he wants the students to look around their environment, on the way home, and at home, and check off each animal or plant that they can observe. They should put their name and the date on the paper, and bring them to class tomorrow. Finally, the bell rings, and Mr. Rose dismisses them.

Beyond Day One

If we assume that the "beginning of the year" includes the first two or three weeks or perhaps even a month of the course, it is important that you carefully plan these weeks to establish routines that will help your classes run smoothly. In the preceding first-day examples, both teachers established a routine, but more importantly established themselves as the leader in charge of the class. They also involved the

TABLE   10.3

THE FIRST TWO WEEKS
OF CHEMISTRY CLASS

Monday	Tuesday	Wednesday	Thursday	Friday

Introductions
  Rules
  Burning Candle
  Activity
	Go over homework
	
	Chemical Observation
	  Activity I
	
	Textbook: ChemCom
		Chemical Activity
		  Part II
			Introduce ch.
			
			1, "Quality of Water"
			
			Role play of water
			  emergency in
			  Riverwood
				Prelab procedures
				
				Lab: Foul Water
				
				Postlab discussion

Monday	Tuesday	Wednesday	Thursday	Friday

Cooperative Learning
  Activity (rules for
  group work):
  Students in teams
  study chapter 1 and
  answer worksheet
  problems.
	Introduce survey
	  activity: "Water use
	  in your home"
	  Presentation on
	  earth's water: the
	  water cycle
		No-risk pop quiz
		
		Prelab
		
		Lab: Classifying mix-
		  tures
			Postlab (mixtures)
			
			Introduction to using
			  symbols and formu-
			  las: Student practice
			  in teams.
				Introduction to use of
				  computer center in
				  class: Students will
				  use program on
				  symbols and formu-
				  las. Each team will
				  have 10 minutes
				  today in center.

students in an interesting learning activity and extended it by giving them an activity-oriented homework assignment to continue the approach. Here is what each of them did on day two.

On day two, Mrs. Broadway started class by taking roll. She then used the pictures the students brought from home to discuss chemistry in the environment. She distributed the textbooks and spent some time showing the students the sections of the book, the glossary, and some tips on using the book. She introduced an activity that would take two days (chemical observations), and assigned a homework reading and problems for each student to complete.

Mr. Rose, on day two, began by reviewing the rules, then followed this with an activity in which students in pairs observed seashells. They measured the shells and drew diagrams showing the shapes and the environment in which the shells live. He then distributed the textbooks, and had the students look over the first chapter, "Life in the Sea." Mr. Rose read (rather dramatically) the first section of the book and then asked the students to study the first chapter and come in with three questions about the first chapter, each written on a card.

For the next two weeks, both teachers set in place the character of their course that would continue throughout the year. Table 10.3 shows the first two weeks of Mrs. Broadway's chemistry class, highlighting the activities and procedures. Notice that she introduced the students to different aspects of her chemistry course (lab, small group work, use of the computer, the textbook) over the two-week period of time. She took the time to teach and reteach the rules and routines in a proactive approach to class management.

MANAGING SCIENCE CLASSROOM
MATERIALS AND FACILITIES

As a science teacher you will be responsible for the materials that you will need to teach science—from textbooks to string, microscopes, and rock samples. Where are materials for the science classroom obtained, and how should they be organized in the classroom facility?

Materials for Science Teaching

If you ever have gone to a national science conference, or even a state science conference, you come away knowing that there is an enormous supply of materials—textbooks, hands-on materials, and equipment—available for science teaching.

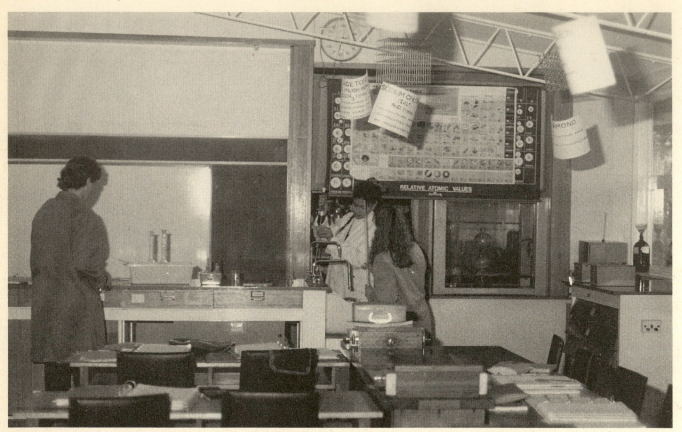

FIGURE 10.10   A science classroom: What does the beginning science teacher
need to know about science materials and facilities? (Credit: Roger T. Cross)

Inventorying Science Materials

It is important to have a broad view of the nature of
science materials and equipment. As a science teacher
you will use textbooks, supplemental books, computers, audio-tutorial materials, day-to-day teaching supplies, equipment, and various technological projectors.
Two aspects of science materials and equipment are
considered here. First, an evaluation of the nature of
equipment and materials should be made, and then
an inventory of existing equipment and materials so
that an accurate accounting can be made.

Evaluating the teaching or learning equipment and
materials needs of a science department requires collaboration between members of the science department.
As a student intern or student teacher, you will have
the opportunity to gain some insight into equipment
and materials needs by discussing the checklist shown
in Figure 10.11 with a member of a science department.

A checklist evaluation such as this provides an
evaluation of a science department's inventory of
equipment and materials. Keeping up with equipment
and materials needs to be a cooperative effort among
members of the department. In a middle school, in
which teachers are not organized into content departments, the science teachers across teams need to collaborate to evaluate the equipment and materials, and
then move to the next step, keeping an accurate inventory so that orders can be made for each year.

The budgets for purchasing science equipment
and materials are typically very limited. Textbooks for
science are often purchased on a rotating basis every
four to six years. These funds are provided by the
state government in half of the states in the United
States, and by the local schools in the remaining
states. The purchase of teaching materials for laboratory work is cited as a problem in many schools, although this varies considerably from one district to
another. The equipment and materials needed can be
evaluated more effectively if an inventory is made
and kept up-to-date. A paper-and-pencil inventory
can keep up with existing materials, and can be used
to make decisions when ordering time rolls around.
Some science departments use one of several computer programs to inventory science materials.

Obtaining Science Supplies

Science equipment and materials can be obtained locally, or from textbook companies, computers suppliers,
and science suppliers. A worthwhile activity for you

Directions: Rate each item on a scale of 1 (low) to 5 (high). For each item, make comments about the teaching/learning potential for the science department.

Science Equipment/Materials Item	Rating				
	Low				High
1. Up-to-date textbooks and laboratory manuals (or equivalent materials in courses designed not to have textbooks or lab manuals) are provided.	1	2	3	4	5
2. Supplementary books, reference books, and other printed mateirals representing a considerable range of sophistication and diversity of student interest are provided.	1	2	3	4	5
3. Various programmed learning materials are available to facilitate independent student learning of specific skills and content.	1	2	3	4	5
4. Students are provided access to a computer and appropriate software and are instructed in the use of this equipment.	1	2	3	4	5
5. The school or department library includes an adequate selection of books, periodicals, and pamphlets on the sciences, the social impact of science, and the history, philosophy, and sociology of science and technology.	1	2	3	4	5
6. Filmloops, transparencies, filmstrips, motion pictures, and videos are available and obtainable when needed by students or teachers.	1	2	3	4	5
7. Projectors for filmloops, filmstrips, motion pictures, videos, and transparencies are available and obtainable when needed.	1	2	3	4	5
8. The recorders, cassettes, and VCRs are available and obtainable when needed.	1	2	3	4	5
9. Equipment and materials for laboratory experimentation and individual student projects are adequate and available when needed.	1	2	3	4	5
10. The system of distribution of laboratory supplies and equipment is reasonably simple and efficient.	1	2	3	4	5
11. Provisions are made for prompt replacement of equipment that wears out or is lost, stolen, or damaged beyond repair.	1	2	3	4	5
12. An effective, continuous inventory of science equipment and supplies is maintained.	1	2	3	4	5
13. Catalogs of science equipment and supplies are readily available to science teachers.	1	2	3	4	5
14. Procedures for requesting and ordering supplies and equipment are reasonable, simple, and efficient.	1	2	3	4	5
15. Science teachers are centrally and effectively involved in the selection and purchase of all instructional equipment and materials for use in the science department.	1	2	3	4	5

FIGURE 10.11 Science Equipment and Materials Checklist. After *Guidelines for Self-Assessment of Secondary-School Science Programs, no. 4, Science Facilities and Teaching Conditions in Our School* (Washington, D.C.: National Science Teachers Association, 1984), pp. 8–10. Reproduced with permission from *Guidelines for Self-Assessment of Secondary-School Programs, no. 4 Science Facilities and Teaching Conditions in Our School.* Copyright ©1984 by the National Science Teachers Association, 1742 Connecticut Avenue, NW, Washington, D.C. 20009.

is to write to some of the companies listed Figure10.12 and request their current catalog.

It is important to find out which companies can be relied on for purchasing science teaching materials. Other teachers can, of course, be of help, but it is worthwhile to investigate this on your own. Writing and requesting a catalog is one way to begin. Visiting the vendors at science conferences is another. For instance, if you are a biology teacher and you want to order live organisms, you will want to be sure that the company will deliver when they promise. Establishing a relationship with the company, either

American Instrument Co.
8030 George Ave.
Silver Springs, MD
20910

Bausch & Lomb
Scientific Optical Products
    Division
1400 North Goodman St.
Rochester, NY 14602

Carolina Biological
    Supply Co.
2700 York Rd.
Burlington, NC 27215

Central Scientific
    Company
11222 Melrose Ave.
Franklin Park, IL 60131

Connecticut Valley
    Biological Supply
82 Valley Rd.
Southhampton, MA
    01073

Delta Biologicals
P.O. Box 26666
Tucson, AZ 85726

Denoyer-Geppert
    Science Co.
5215 N. Ravenswood Ave.
Chicago, IL 60640

Difco Laboratories
920 Henry St.
Detroit, MI 48232

Edmund Scientific Co.
1010 E. Glouster Pike
Barrington, NJ 08007

Fisher Scientific Co.
Education Materials
    Division
4901 W. LeMoyne Ave.
Chicago, IL 60651

Frey Scientific Company
905 Hickory Lane
Mansfield, OH 44905

Hubbard Scientific
    Company
1946 Raymond St.
Northbroke, IL 60062

Nasco
901 Janesville Ave.
Fort Atkinson, WI 55538

Sargent-Welch
    Scientific Co.
7300 N. Linder Ave.
Skokie, IL 60076

Science Kit and Boreal
    Laboratories, Inc.
777 East Park Dr.
Tonawanda, NY 14150

Ward's Natural Science
    Establishment
P.O. Box 1712
Rochester, NY 14603

FIGURE 10.12

via the mail or through a sales representative, is a way to ensure good service.

Science Kits

Science teaching materials cannot only be purchased as individual items but can also be purchased as kits of science materials. A number of science suppliers have designed science kits around single topic areas such as weather, mechanics, magnetism, electricity, sound, aeronautics, plant growth, energy alternatives, the human body, and ecology. The kits contain the hands-on materials that you can use to design inquiry activities, experiments, and demonstrations. Most kits are supplemented with a teacher's guide or resource book describing suggested activities and experiments.

Microchemistry kits are a recent innovation of kit makers and have provided chemistry teachers an opportunity to plan chemistry activities in a safety-enhanced environment. For instance, kits focus on concepts and principles such as acids, bases, and salts, rates of reaction, electrochemistry, and organic chemistry. There are several advantages to the micro-science kits. For one, the quantities of chemicals are minimal, yet the chemistry remains the same as if larger quantities were used. Lab setup time is reduced, since the teacher does not need to prepare as much material. The activities also lend themselves to desktop lab experiences, and in some cases movement into the lab is unnecessary.

A number of curriculum projects in recent years have designed science units or modules of study and have packaged the teaching materials into kits. CEPUP (Chemical Education for Public Understanding Project—see Chapter 6) is a good example. The CEPUP materials incorporate the same principle as the microchemistry kits; investigative hands-on science activities are performed by students but with smaller quantities of materials.

Science kits can be ordered from a number of science suppliers. Here are several that provide kits that can be used at the middle school and high school level.

- Carolina Biological Supply Co.
  2700 York Rd.
  Burlington, NC 27215
  1-800-334-5551
- Fisher Scientific-EMD
  4901 W. LeMoyne Ave.
  Chicago, IL 60651
  1-800-621-4769
- Frey Scientific
  P.O. Box 8101
  Mansfield, OH 44905
  1-800-225-FREY
- Science Kit and Boreal Laboratories
  777 East Park Drive
  Tonawanda, NY 14150
  1-800-828-7777

Science kits provide the science teacher with a unit approach to ordering teaching materials, and they support the hands-on, minds-on philosophy developed in this book. Since most kits contain a generic supply of teaching materials for a topic, the science teacher can be creative and innovative when planning lessons. For example, a kit on "kitchen chemistry" would contain materials enabling you to design lessons on mixtures, compounds, solutions, crystals, acids, bases, electrolysis, and chromatography. Instead of having to order items separately, the innovative teacher can order a kit and then tailor it to the kinds of experiences deemed appropriate for his or her students.

Science kits can also be an organizational structure for storing and maintaining science materials. Once you have purchased several kits, you can begin to design and develop your own kits, basing them on your own teaching units.

Science teachers have taken the concept of science kits and adapted it to create what commonly are called "shoe-box" science learning kits. These learning kits have the same characteristics as commercial science kits; they contain smaller quantities of science materials and they focus on science concepts and principles. Shoe-box kits can be easily stored and facilitate the management of the science teaching environment.

Facilities for Science Teaching

The facilities used in science teaching should reflect the purposes, goals and trends in science education. The principal recommendations made by the most prestigious groups influencing science education today, the National Science Teachers Association and the American Association for the Advancement of Science, suggest an approach to science teaching in which

- Students are given opportunities to inquire into science phenomena.
- Students are actively engaged in collecting, sorting, observing, interviewing, graphing, and experimenting
- Students solve problems of relevance to themselves and the world around them
- Students work together to solve problems and inquire using a team approach

These recommendations have implications for the nature and philosophy upon which decisions about science facilities should be made, and what the resulting science classrooms should look like.

Spaces for Learning: Guidelines for Science Facilities and Classrooms

We have already discussed room arrangements (see pages 361-362). In this section we will explore some guidelines that you should be familiar with as you think about, and then take over responsibility for, your own science teaching environment. We'll approach this from an inquiry view; that is, you will investigate the science facilities of a local school and use some guidelines to assess the facilities and then, based on your experiences, try your hand at designing a science classroom facility.

The National Science Teachers Association has recommended a number of principles that should guide decision makers with regard to science spaces and the furnishings within them. Science educators must take into account[24]

- The great diversity of students who will use them
- The nature of present day science and technology
- Current trends and approaches to science teaching
- The safety of the students
- Flexibility, for adaptation to future uses

Designs for Engagement

Science classrooms can be designed to engage students, as well as provide a teaching space conducive for a variety of teaching models, for example, direct instruction, inquiry, and cooperative learning. Figure 10.13 shows several classroom designs for middle school and high school science classrooms. Naturally, the physical and life sciences have different requirements in terms of specialized equipment. For instance, the life science classroom will include options for living things, greenhouses, terraria, and aquariums.

Note that in the designs shown in Figure 10.13 the science classrooms are shown as combination laboratory-classrooms.[25] Specific modifications can be made for biology, chemistry, earth science, and physics. The peripheral laboratory provides an excellent design because the teacher has good visibility in either a lab or whole class mode, the flow of traffic is good, and the room lends itself to rearrangement quite easily, especially for small group learning. Desks can be moved to attain this goal.

Other models use rectangular tables, island designs, and row designs. What advantages and disadvantages do you see for each of these classroom designs?

[24]Victor M. Showalter, ed. *Conditions for Good Science Teaching* (Washington, D.C.: National Science Teachers Association, 1984), p. 3.
[25]See "What Should Science Supervisors Know About Laboratory or Classroom Facilities and Safety?" in *Third Sourcebook for Science Supervisors*, LaMoine L. Motz and Gerry M. Madrazo, Jr., eds. (Washington, D.C.: National Science Teachers Association, 1989), pp. 100–136.

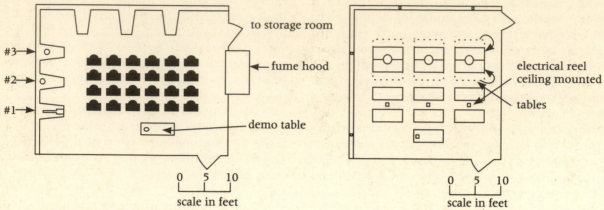

Figure #1 PERIPHERAL LABORATORY

Optional Sink Locations;
#1 Chemistry; tub sink with optional
   trough in countertop
#2 Physics or Biology
#3 Physical Science

Figure #3 TYPICAL BIOLOGY
LABORATORY/CLASSROOM

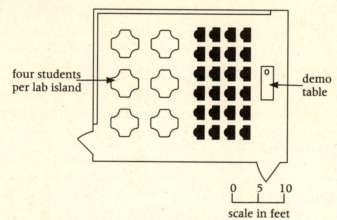

Figure #4 ISLAND DESIGN

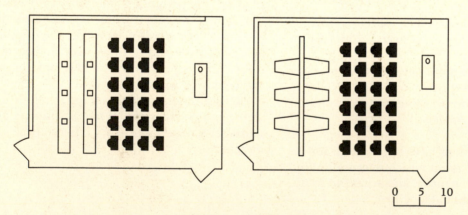

Figure #5 ROW DESIGNS

FIGURE 10.13   Some Science Classroom Designs. Gary E. Downs, Timothy F. Gerard, and Jack A. Gerlovick, "School Science and Liability," in *Third Source Book for Science Supervisors*, LaMoine L. Motz and Gerry M. Madrazo, Jr., eds. (Washington, D.C.: National Science Teachers Association, 1989), pp. 121–125. Reproduced with permission from *Third Source Book for Science Supervisors*, Copyright 1989 by the National Science Teachers Association, 1742 Connecticut Avenue, NW, Washington, DC 20009.

INQUIRY ACTIVITY   10.3

Designing and Evaluating Science Facilities

This inquiry activity is designed to involve you in thinking about science facilities in general, and the design and arrangement of a science classroom in particular. In the first part you should visit either a middle school or a high school science department to conduct an interview to find out about their science facilities. In the second part you should design a science classroom and then defend it during a "science spaces and places" seminar. This activity is best done in teams of two or three.

Materials

Science facilities questionnaire (Figure 10.14), newsprint, marking pens, rules

Procedure

Part I: Assessing Science Facilities

A. Make arrangements to visit either a middle school or a high school science classroom to interview one or more teachers. Tell them that the interview will focus on science classroom and lab facilities, and that you will be using questions from an NSTA questionnaire.

B. After visiting the school, rate each item on the questionnaire.

Part II: Blueprints for a Science Classroom

A. Design a scale drawing of a science classroom for either a middle or junior high school science

---

Directions: Evaluate each of the following items using a scale of 1 (low) to 5 (high) based on your observations and discussions with the faculty of a middle school or high school science department.

Item	Low				High
			Assessment		
1. The science facilities include classroom and laboratory activity areas that adapt well for each course that is taught.	1	2	3	4	5
2. The classrooms and laboratory areas are adequate in number and size for the number of students who take science courses.	1	2	3	4	5
3. The classrooms and laboratory areas are designed to accommodate or facilitate full-class instruction and discussion.	1	2	3	4	5
4. The classrooms and laboratory areas are designed to accommodate or facilitate full-class laboratory work.	1	2	3	4	5
5. The classrooms and laboratory areas are designed to accommodate or facilitate small-group activities (say, four to six students; committee activity, discussion, laboratory activities).	1	2	3	4	5
6. The classrooms and laboratory areas are designed to accommodate or facilitate individualized or independent study projects, for example, student lab projects, library or reference reading, making displays.	1	2	3	4	5
7. The classrooms and laboratory areas are designed to accommodate or facilitate student-teacher conferences and counseling.	1	2	3	4	5
8. The total laboratory area is adequate in size for real experimentation, that is, it provides at least 5 square meters per student in a combined classroom-laboratory or 4 square meters per pupil in a separate laboratory area.	1	2	3	4	5
9. The size of individual laboratory work surfaces are adequate, that is, approximately 1 square meter per student.	1	2	3	4	5

(Figure continued on p. 376)

FIGURE 10.14   Spaces and Places for Science Learning Questionnaire.

(Figure continued from p. 375)

Item	Assessment
	Low **High** 1 2 3 4 5

10. Laboratory tables and work surfaces are adequately heat- and chemical-resistant for the courses for which they are used.
    1   2   3   4   5

11. Student laboratory working spaces include ready access to utilities and other services.
    1   2   3   4   5

12. Laboratory areas are adequately provided with devices, means, and direction for proper, safe disposal of waste materials, that is, broken glass, chemicals, plant and animal specimens.
    1   2   3   4   5

13. Each classroom area includes a demonstration table provided with essential services and located so that demonstrations can be viewed by the entire class.
    1   2   3   4   5

14. Each classroom is well adapted for the use of audiovisual equipment and techniques, including adequate chalkboard space, bulletin or display boards, room-darkening blinds, display cases, projection screens, computer access in the classroom, and video technology.
    1   2   3   4   5

15. A professional departmental library is maintained by, and is accessible to, the science faculty with funds provided within the budget.
    1   2   3   4   5

16. The departmental professional library includes a good selection of professional books, periodicals, and newsletters or other current releases of information relating to science teaching; for example, recent as well as "classic" books on method of science teaching and newsletters or other reports of recent and current curriculum development projects.
    1   2   3   4   5

17. Adequate equipment and space have been provided for growing plants under controlled conditions of light, temperature, and humidity.
    1   2   3   4   5

18. Adequate equipment and space have been provided for the maintenance of animals in a humane manner.
    1   2   3   4   5

19. An outdoor nature study area is available or near the school grounds or campus.
    1   2   3   4   5

20. Each laboratory area or each course taught (for example, chemistry, biology, earth science) is provided with storage space for the supplies and equipment used (approximately 15 percent of the total area.)
    1   2   3   4   5

21. Storage areas are adjacent or readily accessible to the classrooms or laboratories where the materials are used.
    1   2   3   4   5

22. Special storage facilities and security measures are provided for items and materials that are particularly costly, delicate, or that are hazardous, poisonous, or flammable.
    1   2   3   4   5

23. Adequate preparation, maintenance, and repair facilities are provided for teachers, including space, utilities, and limitations on student access.
    1   2   3   4   5

24. The materials, equipment, and services in the science department are kept in good working order.
    1   2   3   4   5

25. Adequate office facilities are provided for each teacher, including desk and drawer space, file cabinet space, book shelves or case, and access to computer facilities.
    1   2   3   4   5

---

course or a high school science course. If you choose a high school, indicate whether it is a biology, chemistry, earth science, general science, or physics classroom.

B. In designing the room make sure you consider work areas, various types of instruction that will occur in the room, technology, ways of reducing congestion, storage areas, safety, reference materials, teacher's desk(s), students' desks, file cabinets, and trash disposal. Add others as you deem necessary.

C. Draw the blueprint on a large sheet of newsprint. Label items, and prepare your team for a presentation. Be able to defend the design of your room based on the principles of classroom management developed in this chapter, as well as on current trends and goals of science education.

Minds on Strategies

1. In what ways does philosophy influence the physical design of science learning environments?
2. What suggestions would you make to the school(s) you visited in terms of improving their science facilities and room designs?

SAFETY IN THE SCIENCE CLASSROOM

A hands-on, minds-on science program might run the risks associated with potential accidents while students are handling chemicals, working with glass, or heating things with open flame burners. However, Downs, Gerard, and Gerlovic[26] investigated school related accidents and found that "science environments are generally among the safest places in schools." For example, they reported that of 1042 Iowa school accidents reported during a two-year period, only 9 percent of them were science-related. Science teachers appear to provide relatively safe places for students to work.

One thing you'll want to become familiar with is the Laboratory Safety Workshop which was established at Curry College. The goal of the workshop is to help science teachers become informed about safety issues and make safety an integral aspect of science education. You can receive information about the program, or purchase the program's newsletter "Speaking of Safety Newsletter," by contacting

The Laboratory Safety Workshop
Curry College
Milton, MA 02186
(617/333-0500, ext. 2220)

In this section we will explore some of the principles and practices that you should consider as you plan science lessons for science classroom environments.

[26]Gary E. Downs, Timothy F. Gerard, and Jacks A. Gerlovick, "School Science and Liability," in *Third Source Book for Science Supervisors*, LaMoine L. Motz and Gerry M. Madrazo, Jr. eds. (Washington, D.C.: National Science Teachers Association, 1989), pp. 121–125.

A Safe Science Environment

A safe classroom environment begins at the beginning of the school year, and should be one of the aspects of planning and preparation for the opening of school. Here are some aspects of the environment that you should use as a checklist to determine the safety-worthiness of the science classroom. Not all science rooms contain labs, so you will find that when a teacher is about to do a lab activity, he or she will make arrangements to use the lab. Therefore, when considering safety, the teacher will need to include all rooms he or she uses for science instruction.

- The school has an automatic sprinkler system
- Each lab has two unobstructed exits
- Each lab is provided with adequate ventilation, fume hoods, and heat-removal systems
- Safety-glass screens or shields are available at each demonstration table
- Labs are equipped with readily accessible fire blanket and fire extinguishers
- Labs are equipped with a readily accessible shower and an eye-wash fountain
- Each lab has a readily accessible, properly equipped first-aid cabinet
- Safety goggles are available for each student and teacher for hands-on and lab activities
- Laboratory coats or aprons are provided for each student and teacher
- All possibly hazardous materials (that is, chemical reagent bottles) are labeled with safety precautions and used only under direct supervision

FIGURE 10.15   Safety in the science classroom means providing a safe environment for students as they explore science.

- Printed, easily accessible instructions for handling laboratory accidents are available and known by the teacher
- Printed, easily readable safety precautions are posted in each laboratory[27]

A safe science learning environment occurs when the teacher assumes leadership for the general safety regulations that must be put into place in the classroom. Signs should be posted listing safety precautions, and the rules of behavior while doing science lab activities. The teacher should describe any safety issues related to the activity, for example, the handling of glass, chemicals, or biological materials.

The first time that a hands-on activity or a laboratory experiment is conducted, the teacher should point out and show how the following safety equipment should be properly used.

- First-aid kit
- Fire blanket
- Eye-wash facilities
- Safety goggles
- Laboratory gloves
- Laboratory aprons

[27]After *Guidelines for Self-Assessment of Secondary-School Science Programs, no. 4, Science Facilitates and Teaching Conditions in Our School* (Washington, D.C.: National Science Teachers Association, 1978), pp. 5–8. Used with permission of the National Science Teachers Association.

General classroom safety can be accomplished by integrating safety rules into your lesson plans. If, for example, you are dissecting fruit, reinforce the safety rule that no items in the science classroom can be tasted. Here are some general safety rules that you might post on a large sign in place to be easily read by the students:

General Safety Rules

1. Wear safety goggles when doing hands-on laboratory activities
2. Wash hands thoroughly at the end of each laboratory period
3. Clean up work areas after each activity or laboratory period
4. Do not eat, drink, or engage in boisterous play in the classroom
5. Lab activities can only be done under the supervision of the teacher
6. Do not touch face, mouth, or eyes when working with plants, rocks, soil, or chemicals

Chemical Safety

The use of chemicals in the science classroom requires knowledge of the potential danger associated with any chemical. The Occupational Safety and Health Administration (OSHA) and the Environmental Protection Agency (EPA) have produced guidelines and recommended lists of chemicals that should not be used in middle and high school science classrooms. In this section we will consider two aspects of chemical safety: rules for chemical usage and assessment and storage of chemicals.

Rules for Chemical Usage

Regardless of the chemical, it should be treated as potentially hazardous, and the teacher should always alert students to this notion. Students should be instructed to handle all chemicals with care. The following rules, suggested in the use of ChemCom, a high school chemistry program, outline an approach to chemical safety.

1. Perform lab work only when your teacher is present
2. Always read and think about each laboratory procedure before starting
3. Know the locations of all safety equipment, for example, safety shower, eye wash, first-aid kit, fire extinguisher, and blanket.

4. Wear a lab coat or apron and protective glasses or goggles for all lab work. Wear shoes (rather than sandals) and tie back loose hair.

5. Keep the lab table clear of all materials not essential to the lab activity.

6. Check chemical labels twice to make sure you have the correct substance.

7. Do not return excess material to its original container unless authorized by the teacher.

8. Avoid unnecessary movement and talk in the lab.

9. Never taste laboratory materials. If you are instructed to smell something, do so by fanning some of the vapor toward your nose. Do not place your nose near the opening of the container.

10. Never look directly down into a test tube; view the contents from the side. Never point the open end of the test tube toward yourself or your neighbor.

11. Any lab accident should immediately be reported to the teacher.

12. In case of a chemical spill on your skin or clothing rinse the affected area with plenty of water. If the eyes are affected water-washing must begin immediately for 10 to 15 minutes or until professional assistance is obtained.

13. When discarding used chemicals, carefully follow the instructions provided.

14. Return equipment, chemicals, aprons, and protective glasses to their designated locations.

15. Before leaving the laboratory, ensure that gas lines and water faucets are shut off.

16. If in doubt, ask![28]

Audiovisual programs are available from some suppliers in the form of filmstrips and videos that demonstrate and discuss safety issues in the science class. Materials such as these can be used to teach specific lessons on safety (also refer to the Curry College Safety Program in this chapter).[29]

Assessment and Storage of Chemicals.

Chemical safety is dependent upon the teacher's knowledge of the chemicals used in lab activities. Suppliers of chemicals have devised various systems to inform purchasers of chemicals about the nature of the chemicals and how to store them.

For, example Fisher Scientific has devised a chemical safety and alert system that includes

- Safety symbols (which can be used with the students)
- Chemical storage codes
- National Fire Protection Association (NFPA) diamond hazard code.

For each chemical, information is provided that immediately informs the teacher about the hazards associated with the item, how and where it should be stored, and what precautions should be taken when using the chemical.

The safety codes (Figure 10.16) identify the equipment and gear that need to be worn when handling a specific chemical. The teacher is informed and warned whether goggles, gloves, an apron, or a hood are needed with each chemical.

The NFPA hazard code (Figure 10.17) rates hazards numerically inside a diamond on a scale of 0 to 4. In practice, the diamond shows a red segment (flammability), a blue segment (health, i.e., toxicity), and a yellow one (reactivity). The fourth segment is left blank. It's reserved for special warnings, such as radioactivity or hazardous when in contact with moisture.

The numerical ratings are:

4 = extreme hazard
3 = severe hazard
2 = moderate hazard
1 = slight hazard
0 = according to present data: none

A rating of 4 (most severe hazard) indicates that goggles, gloves, protective clothing, and a fume hood should be used.

A rating of 0 (no special hazard) indicates that a substance may be safely handled and used in the lab with no special protection required other than safety glasses.

The storage of chemicals should not be done alphabetically (as some schools do). They should be organized according to a classification system designated the U.S. Department of Transportation (DOT). Materials are rated as

- *Flammable:* Store in areas segregated for flammable reagents.
- *Reactive and oxidizing:* May react violently with air, water or other substances. Store away from flammable and combustible materials.

---

[28]*ChemCom: Chemistry in the Community, Teacher's Guide* (Dubuque, Iowa: Kendall/Hunt, 1988), pp. xi–xii.

[29]See Frey Scientific, 905 Hickory Lane, P.O. Box 8101, Mansfield, OH 44901 (1-800-589-1900). A-V material includes *Safety in the Science Lab* videotape, *Science Safety Program* (filmstrip or video), *Science Safety for Elementary* (appropriate for middle schools), as well as a lab safety sign set and handbooks on safety.

FIGURE 10.16 Safety Codes (Source: Fisher Scientific)

- *Health hazard:* Toxic if inhaled, ingested, or absorbed through the skin. Store in secure area.
- *No more than moderate hazard:* May be stored in general chemical category.
- *Exceptions:* Some reagents may be incompatible with other reagents of same category and must be stored separately.

These ratings can be used to organize a storage plan for the chemical stockroom. Figure 10.17 shows a

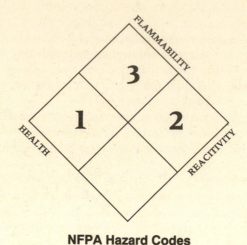

NFPA Hazard Codes

FIGURE 10.17 NFPA Hazard Codes (Source: Fisher Scientific)

graphic depiction of the relative amounts of chemicals in each of the chemical groups in the average high school chemical storage area. This plan can be helpful in organizing a schools' storage room. Oxidizers must be isolated from flammables, and all flammables should be stored in a dedicated flammable storage cabinet. If highly toxic chemicals or poisons are part of the inventory, they should receive special security. In general, chemicals should not be stored on the floor or above eye level. Finally, the chemical storage room should be locked.

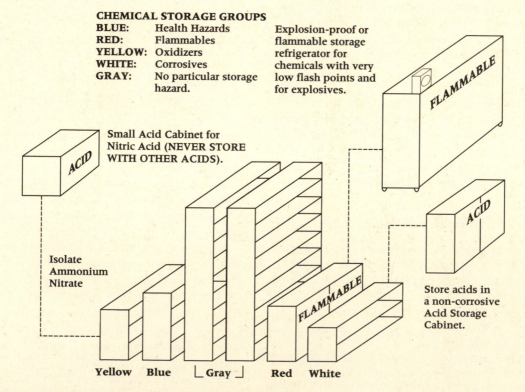

FIGURE 10.18 Chemical Storage Groups and Relative Space Required in a Typical High School Science Department (Source: EMD 91/92 *Biology, Chemistry, Physics Catalog,* (Chicago, Ill.: EMD, Division of Fisher Scientific).

Living Organisms in the Classroom

Living organisms should be part of life science programs at the middle and high school levels. The inclusion of living organisms also results in a number of safety and humane issues that the science teacher needs to deal with. According to a study done by Mayer and Hinton, the most common use of animals in the classroom is for dissection.[30] In fact, they reported that the higher the grade level, the greater the chance the teacher included dissection. For instance, they indicated that 74 percent of the middle school teachers and 94 percent of the high school teachers had their students dissect at least one animal.

They reported that the use of live animals in the classroom showed the opposite trend. Sixty-seven percent of the elementary teachers used live animals in the classroom, whereas 43 percent of the middle school teachers used them, and 54 percent of the high school teachers used live animals.

If live animals such as rabbits, guinea pigs, hamsters, gerbils, mice, and rats are used in the classroom they should be handled carefully with gloves. Teachers should not bring wild animals into the classroom, since the likelihood of zoonotic diseases is high.

Figure 10.19 outlines the requirements for various animals that teachers might keep in the classroom. It is of the utmost concern that all animals in the classroom be treated in the most humane manner possible. In general animals should be used for behavioral studies, not for invasive studies. In such cases, bacteria, protozoans, and insects can be used to achieve these objectives.

Animal dissection should be done with the utmost care, and after careful consideration of goals and objectives. Mayer and Hinton suggest that teachers consider the following questions before choosing to do a dissection activity with their students:[31]

1. What are the goals and objectives in the course you are teaching?
2. How does the use of dissection support these goals?
3. Should live dissections be used?
4. How many dissections are necessary?
5. Can this number be reduced by using videotapes or demonstrations?
6. Can items from the supermarket that are "past their prime" be substituted?
7. Are there alternatives such as computer programs and videodisc programs that would be as effective or more effective in reaching your goals?

Care must be taken when students do dissections. Only clean and sterilized instruments should be used, and the students should be warned about cutting themselves using the equipment. Students should thoroughly wash their hands after dissections.

Teachers should be aware of the regulations of the state in which they teach with regard to the use of animals in the classroom. Teachers need not only to protect the students from potential diseases but to protect themselves from violating state regulations.

[30]Victor J. Mayer and Nadine K. Hinton, "Animals in the Classroom," *The Science Teacher*, March, 1990, pp. 27–30.

[31]Mayer and Hinton, "Animals in the Classroom," pp. 27–30.

REQUIREMENTS FOR VARIOUS ANIMALS

Food and Water Plants (for fish)	Rabbits	Guinea pigs	Hamsters	Gerbils	Mice	Rats
Daily pellets	rabbit pellets: keep dish half full		large dog pellets one or two			
or grain	keep dish half full	corn, wheat, or oats:	1½ tablespoon	or sunflower or 2 teaspoons	canary seeds or oats: 2 teaspoons	3-4 teaspoons
green or leafy vegetables, lettuce, cabbage and celery tops	4-5 leaves	2 leaves	1 leaf	1/8 leaf or less	1/8 — 1/4 leaf	1/4 leaf
or grass, plantain, lambs' quarters, clover, alfalfa (or hay, if water is also given)	2 handfuls	1 handful	1/2 handful	—	—	—
carrots	2 medium	1 medium		—	—	—
Twice a week apple (medium)	1/2 apple	1/4 apple	1/8 apple	1/8 apple or 1/2 core and seeds	1/2 core and seeds	1 core
iodized salt (if not contained in pellets)	or salt block	sprinkle over lettuce or greens				
corn, canned or fresh, once or twice a week	1/2 ear	1/4 ear	1 tablespoon 1/3 ear	1/2 tablespoon or end of ear	1/4 tablespoon or end of ear	1/2 tablespoon or end of ear
water	should always be available	necessary only if lettuce or greens are not provided				

	Goldfish		Guppies	
Daily dry commercial food	1 small pinch		1 very small pinch; medium size food for adults; fine size food for babies	
Twice a week shrimp — dry — or another kind of dry fish food	4 shrimp pellets, or 1 small pinch		dry shrimp food or other dry food: 1 very small pinch	
Two or three times a week tubifex worms	enough to cover 1/2 area of a dime		enough to cover 1/8 area of a dime	
Add enough "conditioned" water to keep tank at required level	allow one gallon per inch of fish add water of same temperature as that in tank — at least 65° F		allow ¼ - ½ gallon per adult fish add water of same temperature as that in tank — 70-80° F	
Plants cabomba, anacharis, etc.	should always be available			

	Newts		Frogs	
Daily small earthworms or mealworms	1-2 worms		2-3 worms	
or tubifex worms or raw chopped beef	enough to cover 1/2 area of a dime enough to cover a dime		enough to cover 3/4 area of a dime enough to cover a dime	
water	should always be available at same temperature as that in tank or room temperature			

	Water turtles	Land turtles	Small turtles
Daily worms or night crawlers or tubifex or blood worms and/or	1 or 2	1 or 2	1/4 inch of tiny earthworm enough to cover 1/2 area of a dime
raw chopped beef or meat and fish-flavored dog or cat food	1/2 teaspoon	1/2 teaspoon	
fruit and vegetables fresh		1/4 leaf lettuce or 6-10 berries or 1-2 slices peach, apple, tomato, melon or 1 tablespoon corn, peas, beans	
dry ant eggs, insects, or other commercial turtle food			1 small pinch
water	always available at room temperature; should be ample for swimming and submersion		
	3/4 of container	large enough for shell	half to 3/4 of container

FIGURE 10.19   Requirements for Various Animals. Grace K. Pratt-Butler, *How to…Care for Living Things in the Classroom* (Washington, D.C.: National Science Teachers Association, 1978), p. 11. Reproduced with permission from *How to…Care for Living Things in the Classroom,* Copyright 1978 by the National Science Teachers Association, 1742 Connecticut Avenue, NW, Washington, D.C. 20009.

Science Teaching Gazette

Volume Ten

Facilitating Learning in the Science Classroom

Case Study 1.
Misbehavior in the Lab

The Case. Mrs. Jones, a first year teacher, is conducting one of the first laboratory exercises of the year. One of her students, John, throws rubbing alcohol into a burning Bunsen burner. Brad, the student working next to John, gets burned on his hand.

The Problem. How do you handle this situation? How could this incident have been prevented? What safety violations do you think might be part of this case?

☐

Case Study 2.
The Smiths Come to School

The Case. Doreen Smith is a student in Jeff Murdock's first period Biology I course. Doreen has visited Mr. Murdock quite often for detention after school. Doreen has been breaking the rules for the eight weeks since school began. Mr. Murdock has decided it's time for a parent-teacher conference. The meeting has been set for before school (since it is more convenient for the Smiths). On the day of the conference, the Smiths arrive, and as soon as they come into Mr. Murdock's class, they accuse Mr. Murdock of picking on their daughter, and treating her differently than the other students in the class.

The Problem. How do you proceed from here? What do you say to the Smiths? How can you resolve this dilemma?

Think Pieces

- Do some research on this issue: Should dissection by used in middle and high school science classrooms? Find at least two sources in science education journals that discuss the issue. What are arguments for the use of dissection? What are arguments against the use of dissection?
- The following are considered to be important classroom management behaviors: with-it-ness, overlapping, smoothness, momentum, group focus, and accountability. Which one do you think is the most important? Why? How do these behaviors relate to each other?
- What special safety precautions should be taken in each of the following situations?

1. a middle school life science course
2. a high school physics course
3. a high school chemistry course
4. a middle school earth science course ☐

Science Teachers Talk

"How do you manage your classroom, and what is the most important piece of advice you would give a prospective teacher concerning classroom management?"

Jerry Pelletier. I would say that I am always in charge. I try to create an atmosphere in which students can question, move about, and converse with each other within a structured environment. Students understand that if they work well within this environment that there will be rewards throughout the year. The reward for eighth graders is a trip to Great America for Physics Day. The most important piece of advice I would give to teachers regarding classroom management is that they must be consistent. Students must understand the goals of the classroom and the consequences of not attaining these goals.

Bob Miller. Keep tight control of the classroom until you have had

(Continued on p.384)

(Continued from p. 383)

a chance to assess personalities and group dynamics. Distribute responsibilities within the classroom on a reward system: daily work, extra credit, and so forth.

Ginny Almeder. I am most comfortable in a relaxed but structured classroom. I try to give the students as much freedom and responsibility as they can manage. Seating is open. I am fairly flexible and try to remain open to student input. Short class discussions are held to deal with classroom procedures. I typically follow a set of rules consistent with school policy.

If behavioral problems occur, I deal with these immediately and directly in class. If the behavior persists, I speak with the individual after the class and describe the situation as a mutual problem that both of us need to solve. I ask for suggestions and offer suggestions, and we arrange a strategy to deal with the situation. If the problem persists, I will use a seating change, parent conference, or rarely, an after detention for further dialogue.

Mary Wilde. Classroom management is a very individual aspect of teaching, for what works for one teacher may make another teacher miserable. However, if teaching science is done through small groups, using hands-on activities and the inquiry approach, one must be flexible but well organized. If you are a teacher that needs a quiet classroom to maintain sanity, science is probably not the area for you. Science is not a discipline where all students should be quietly sitting in their seats, reading or note-taking about concepts they should be experiencing. Therefore, to organize a student- and activity-centered classroom, there are some general guidelines that lead to greater success: These are just a few suggestions that

might help classroom management in the science classroom. The most important aspect to remember is that you have to look at yourself and develop a classroom system you are comfortable with. Allow for flexibility, but be organized so that the time you spend on planning and preparation is reflected through the achievement of your students.

1. Establish class rules in the beginning before you begin any group activity. They need to know your expectations and their limitations on classroom order.
2. Make sure that whatever rules are made, they are enforced consistently. Before discipline problems arise, know how you will handle them, and communicate this with your students.
3. Particularly at the middle school level, don't give assignments you aren't going to grade or discuss in the classroom. If you want them to fulfil the responsibility to complete work, then you need to make sure they see a purpose for their effort.
4. Teach laboratory safety procedures before beginning any lab activities so that the students will know what to be cautious about, and how to deal with the situation.
5. Provide ample time for students to complete laboratory or hands-on activities. It is of the most importance that all materials, equipment, and lab procedures are available to students.
6. Be sure all students understand the laboratory procedures before allowing them to continue.
7. Develop lab assistants who are trained to set up labs and gather materials. This can be organized through the "science club."

Dale Rosene. Understand your students' needs and development. This will allow you to better understand and deal with their behaviors. Make your rules appropriate—enough to maintain order, but not too many, which might create a hostile environment. Be open and friendly and admit your mistakes. Use humor, and finally be yourself so that students can see you as a person.

John Ricciardi. I sense a classroom of students as being a unified, but independent entity unto itself—an awesome ecosystem of thought and feeling—a kind of greater being of multi-body mind and spirit. If a classroom is perceived as such a creature, *then its management can be like maintaining the healthful life of an organism* [Italics mine].

Here are three helpful "care and feeding" hints for the classroom:

1. The classroom organism must be comfortable in its physical environment—changing and using a variety of lighting levels, furniture positions, wall decorations, background music is important to maintaining a stimulating "mind space" for the creature to grow in.
2. The classroom organism must not be harnessed and controlled. Learn instead to coax and nurture it with reflexive input and response. Distractions and disruptive "order imbalances" are normal and natural. Know that the creature by itself, will quickly find its equilibrium again.
3. The classroom organism must be treated humanly—with dignity and respect at all times. The integrity of all individuals must be equally honored within the wholeness of their own identity and unity.

Legal Liability
**by Gary E. Downs,
Timothy F. Gerard, and
Jack A. Gerlovich**[32]

Safety should be everyone's concern. People who behave in an unsafe manner not only expose themselves to a greater risk of accident or ill health, but increase their chance of being held legally liable for injury to the property and health of others. For example, a careless driver endangers more than his or her own life; by also endangering the lives of others, he or she increases the chances of being legally liable in an accident. There is a direct correlation between safety and the conduct the law expects of us all.

The civil law expressed throughout the 50 states presumes that people have an obligation to behave reasonably where their conduct affects other people. This is an extension of the criminal law, which is designed to protect the lives and property of citizens. The proscriptions embodied in criminal law are considered undisputed specific dictates of reason. But, in a free society, it is neither possible nor desirable to specify everything that citizens ought to do to preserve order within society.

How, then, should science educators conduct themselves in order to avoid being found negligent and held liable for damages arising out of an accident? Although the law does not say that accidents always result from

negligence (they often happen through the fault of no one), there is a tendency in a complex technological society to try to assign fault and apportion liability accordingly. This is especially true in education, where students are presumed to be under the care and protection of their teachers. A science educator, then, must conform to what other science edu-cators would consider reasonable in order to minimize their risk of legal liability. This recommendation is largely common sense. For example, a science teacher who asked a student to pour water into concentrated acid to demonstrate its explosive effect would surely be found negligent. More technical matters, however, require judgement and testimony by other science teachers who are familiar with the given situation. The operative question revolves around what consensus opinion would be as to the reasonableness of this conduct by others in the science education profession. For this reason it is of the utmost importance that educators keep abreast of the developments in their profession in terms of safety and the content they teach.

Thereafter, educators must accept their responsibility to implement safe conduct in three main areas: first is the duty to adequately supervise students, to "keep a watchful eye." Improper supervision is the single greatest cause of legal liability in science education. Proper supervision will prevent most accidents. The second duty is to inform students of all foreseeable and reasonable hazards. A teacher must tell students all that they need to know to participate safely in any given learning project; the appropriate and accurate dissemination of information relevant to student safety should be an integral part of lesson planning. The third duty is to maintain safe equipment and

a safe environment in which students learn. All three duties pertain to the teacher's responsibility when in charge of students in a learning situation. Again, the duties must be performed according to the acceptable standards of the profession to insure protection from legal liability. □

Problems and Extensions

- Make a list of classroom rules and identify the disciplinary consequences for students who break each rule.
- Consider the following classroom routines. Select one or two from the list or a routine of your choice and try to think of ways in which these rou-tines could be used to provide academic experience for your students: attendance, passing out materials, taking care of living organisms, cleanup.
- How should a class begin? What routines or transitions will facilitate students coming to your class and being ready to begin?
- What are some "activities" or "events" that could prevent a class from starting on time?
- What are some routines that you could use at the beginning of each class or while students are entering that could help to ready students to focus their attention while you take care of necessary administrative tasks or prepare to begin? (Hint: journal or log writing, picking up materials,…)
- Brainstorm some potential things that you can do at the end of a class if the activities you planned don't take you to the end of the period. What are some successful ways of utilizing this "end of class" time?
- What safety precautions should you take if you were to use cal-

(Continued on p. 386)

[32]Excerpted from Gary E. Downs, Timothy F. Gerard, and Jack A. Gerlovich, "School Science and Liability," in Third Source Book for Science Supervisors, LaMoine L. Motz and Gerry M. Madrazo, Jr., eds. (Washington, D.C.: National Science Teachers Association, 1989), pp. 121–125. Reproduced with permission from Third Source Book for Science Supervisors, Copyright 1989 by the National Science Teachers Association, 1742 Connecticut Avenue, NW, Washington, D.C. 20009.

(continued from p. 385)
cium in a lab activity? Include a discussion of the safety gear and equipment that students would need, and hazards associated with its use.[33]

- Prepare a lesson plan for the first day of one of the following classes: earth science, biology, chemistry, physics, or physical science. Your lesson plan should detail and account for each minute in the lesson.

- Design a set of "Safety Person Says" signs for a secondary science classroom. Some ideas include Danger—Flammable Materials, Danger—Acid, Danger—Poison, Eye Wash, Wear Your Goggles. ☐

Calcium

Ca

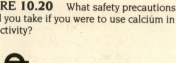

FIGURE 10.20   What safety precautions should you take if you were to use calcium in a lab activity?

FIGURE 10.21   Safety Person Says

[33]Calcium is highly flammable in finely divided form. Contact with water or moisture releases flammable hydrogen gas. Irritant to eyes and skin. Wear eye guard, proper gloves, and safety clothing. Use with adequate ventilation, and keep away from moisture and oxidizers. (Source: Fisher Scientific)

RESOURCES

Bogner, Donna. *Starting at Ground Zero: Methods of Teaching First Year Chemistry. Vol. 1–4.* Hutchinson, Kans.: Genie, 1989.

If you are going to teach chemistry, you should purchase this four volume set of teaching ideas written by an outstanding high school chemistry teacher. The book is based on the integration of learning styles research, and many of the activities in the program were developed Dreyfus/Woodrow Wilson Master Teachers.

Canter, Lee, and Marlene Canter. *Assertive Discipline: Resource Materials Workbook, Secondary 7-12.* Santa Monica, Calif.: Lee Canter & Associates, 1984.

This book presents Canter's discipline plan and provides the teacher with a set of classroom suggestions and reproducible pages for immediate use.

Curwin, Richard L., and Allen N. Mendler. *Discipline with Dignity.* Alexandria, Va.: Association for Supervision and Curriculum Development, 1988.

This book presents a postive approach to the issue of classroom discipline and management. The book's humanistic approach and comprehensive model offers an alternative to packaged discipline programs.

Emmer, Edmund J., et. al. *Classroom Mangement for Secondary Teachers.* Englewood Cliffs, N.J.: Prentice-Hall, 1989.

This excellent book is a comprehensive and well-organized presentation of the essential aspects of classroom management for secondary teachers.

CHAPTER

11

SCIENCE
FOR ALL

"Science for All" is the concluding chapter of this book and represents the final unifying theme upon which the book is based. "Science for All" as a theme "implies there's something in it for everyone. It also acknowledges that science education in the past has not been science for all, that large numbers of persons have indeed been failed or passed over by the science curriculum."[1]

The "Science for All" theme implies that science education should be based on several fundamental characteristics of learners and what school science should be:

1. Each learner is unique
2. There is enormous diversity among learners, and this diversity should be respected and accepted
3. School science should be designed to meet the needs of all students, not just the science prone or those who will pursue careers in science
4. School science should be inclusive, not exclusive. Those who have been traditionally turned away from science should be encouraged to come in

In this chapter we will explore the theme "Science for All" from four interrelated perspectives: global, multicultural, gender, and exceptionalities. Each of these perspectives requires teachers to deal with the notion of uniqueness and diversity at the same

time; in so doing the teacher approaches science teaching from a holistic point of view. For example, Mary M. Atwater in writing about multicultural education in science education, says:

> In order for science teachers to most effectively instruct culturally diverse students, they need exposure to certain philosophies, knowledge, and skills. Multicultural science teachers must have democratic attitudes and values, and pluralistic ideology, a belief that society can be composed of different cultural groups which can function and interact for the good of society. Diverse ethnic groups are not expected to merge into one generic group with all of its members embracing the same culture.[2]

This chapter is designed to heighten your awareness of, as well as provide some background information on, the global, multicultural, and gender perspectives of science teaching, as well as help you explore the range of exceptional students who will be part of all of your science courses.

A Global Perspective

Should science teachers incorporate a global perspective in the goals and objectives of science courses? If they should, what is the rationale for doing so? Up until recently, a global perspective was within the

[1]Peter J. Fensham, "Science for All," *Educational Leadership,* January, 1987, pp. 18–23.

[2]Mary M. Atwater, "Including Multicultural Education in Science Education: Definitions, Competencies, and Activities," *Journal of Science Teacher Education* 1, no.1 (Spring 1989): 17.

perview of social studies teachers. Why should science educators become involved? How can a global perspective be of service to the "Science for All" theme?

Global Events

* In the summer of 1991, about seven hundred science teachers met in Moscow to exchange ideas about science education. At the conference, teachers from these two nations presented workshops, engaged in panel discussions, and heard presentations by distinguished teachers and scientists. But beneath all of this was something more profound. The teachers in attendance were drawn together by the deep human bond of wanting to understand and find out about each other.
* Since 1983 to the present a group of science educators focused in Atlanta but involving teachers from various parts of the United States have been engaged in a collaborative project (the U.S.-Soviet Global Thinking Project) with teachers from Moscow and St. Petersburg to develop a computer network and environmental curriculum.
* TERC's Global Laboratory—a computer network and science program—has linked together schools in over ten nations to study global ecology.
* Schools in California and Alaska have been linked with schools in Mexico and Japan to explore the uses of computer technologies in various subject areas.
* Individual science teachers have designed programs in their own schools in which students have looked outward to engage in collaborative studies with students in other countries. For example Brian Slopey, a junior high science teacher in Vermont, has led a project in which his students and students in St. Petersburg are investigating water quality in the rivers that run through their respective cities.

These are examples of ways in which people from different parts of the world are engaged in collaborative projects and events that will result in students gaining a global perspective.

A Rationale for a Global Perspective

In recent years there has been an effort to globalize education in the United States. This is being done in a variety of ways, and it might be worthwhile to discuss the underlying reasons for this effort.

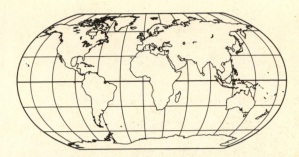

FIGURE 11.1   A global perspective, among other factors, enhances state-of-the-planet and cross-cultural awareness.

Robert Hanvey, in discussing a rationale for a global perspective, suggested that it

> involves learning about those problems and issues that cut across national boundaries, and about the interconnectedness of systems—ecological, cultural, economic, political, and technological. Global education involves perspective taking—seeing things through the eyes and minds of others—and it means the realization that while individuals and groups may view life differently, they also have common needs and wants.[3]

Perspective taking—"seeing things through the eyes and minds of others"—is an important point that Hanvey makes. To create learning environments for all students requires this philosophical position. Not only is this position supported by global education activities such as international business, the Peace Corps, foreign student exchange programs, and international activities, but by changes in the social structure of the world and the globalization of America. Let's examine how these affect the perspective that science teachers take as they think about the goals and objectives of science education in global terms.[4]

Global Education Activities

Teachers and students in today's classes are already involved in global activities. Many teachers have traveled to other countries and share their experience with their students. In some classrooms, a number of students were born in a foreign country and bring with them the perspective of their native country. Students may belong to clubs, churches, and civic groups that are involved in international activities. All of these activities are examples of global education activities. In the past few years a number of

[3]Robert G. Hanvey, *An Attainable Global Perspective* (Denver: Center for Teaching International Relations, 1976).

[4]This discussion is based on Lee F. Anderson, "A Rationale for Global Education," in *Global Education: From Thought to Action*, Kenneth A. Tye, ed. (Alexandria, Va.: Association for Supervision and Curriculum Development, 1990), pp. 13–34.

technology-related projects have brought distance learning into the classroom of teachers in many parts of the world. This is still another example of how global activities are affecting science education.

The Changing Structure of the World

Another factor supporting global education is the way in which the social structure of the world has changed. The first, and perhaps most obvious, is the accelerating growth of global interdependence. This is evident in fields such as economics, politics, geography, ecology, and culture, to name a few. For example, in the realm of economics, world markets dominate many companies, and indeed, the emergence of the multinational corporation (MNC) is an indicator of the social changes in the economic arena. A product produced by an MNC might be assembled in the United States, but it may consist of parts from as many as 15 different countries.

Changes in transportation and communication has lessened the isolation of people from each other. This people and electronic traffic flow has affected money, information, ideas, goods and services, technology, diseases, and weapons.

The science teacher can easily help students understand the ecological implications of changes in the social fabric of the world. Diseases spread rapidly from one continent to another—the AIDS epidemic is a good example of this, as well as the yearly fear of viruses invading the United States. Environmental problems, which in the 1970s were examined as if they were national in scope have now taken on a global perspective, given concerns over the depletion of the ozone, global warming, thermonuclear war, toxic and nuclear waste disposal, deforestation, and the extinction of animal and plant species.

The Globalization of America

Perhaps one of the most powerful rationales for a global perspective is the fact that the United States has become globalized. This can be seen in terms of America's economy, political life, and culture, as well as in a growth in global consciousness.[5]

For instance, according to Anderson, American workers make products of which more than one in five are exported, and more than 50 percent of the most important industrial raw materials are imported.[6]

Many U.S. firms have branches abroad, and an increasing number of U.S. firms are foreign owned. Anderson makes a powerful point when he reminds us to think about many of the products we use every day. What do they have in common? (Answer: many were made by companies that are foreign owned: Pillsbury flour, Hills Brothers coffee, Carnation creamer, Friskies cat food, Bic pens, Foster Grant sunglasses, Burger King Whoppers, Michelin tires, Shell gasoline, and a Seagram's gin martini.)

America's political life has become globalized as well. Anderson writes that "the accelerating globalization of the American economy is paralleled by an increasing internationalizing of American government and politics."[7] America's politicians look beyond their own constituents, and many travel abroad. The action is at the state level. In recent years nearly all the governors traveled abroad seeking markets for their state's products, as well as recruiting foreign firms to their cities. There are other signs showing how American politics has become globalized. Lobbying groups representing multinational firms now exist in Washington to influence politicians. These and other things attest to the globalization of U.S. governments and politics.

Global Consciousness

Because of the business opportunites at a global level, the awareness or consciousness, especially of the business community, has become increasingly global.[8] The scientific community has also influenced the global consciousness of Americans. International conferences, reports by science panels and institutions, and even science curriculum efforts have raised the awareness level of people so that now terms like global warming, ozone depletion, and acid rain are not unfamiliar.

The Goals of Global Education: Implications for Science Teaching

There is a strong support base for including a global perspective in science teaching. The global perspective is an underlying basis for an interdisciplinary and multicultural approach to science teaching. This can imply that individual courses—say, chemistry—can involve students in interdisciplinary and multicultural thinking, as well as school districts developing indi-

[5]Lee F. Anderson, "A Rationale for Global Education," in *Global Education: From Thought to Action*, Kenneth A. Tye, ed. (Alexandria, Va.: Association for Supervision and Curriculum Development, 1990), p. 21.
[6]Anderson, "A Rationale for Global Education," p. 21.
[7]Anderson, "A Rationale for Global Education," p. 27.
[8]Anderson, "A Rationale for Global Education," p. 31.

vidual science courses. Some potential goals of the global perspective include[9]

1. *Perspective consciousness:* An awareness of and appreciation for other images of the world. Students can learn to understand how people in Eastern as well as developing nations conceptualize the world. Global thinking will enrich the activities of students by involving them with a world that is multicultural.
2. *State-of-the-planet awareness:* An in-depth understanding of global issues and events, including the identification of the important global problems facing the planet, and what it will take to get them solved.
3. *Cross-cultural awareness:* A general understanding of the defining characteristics of world cultures, with an emphasis on understanding similarities and differences.
4. *Systemic awareness:* A familiarity with the nature of systems and an introduction to the complex international system in which state and nonstate actors are linked in patterns of interdependence and dependence in a variety of issue areas.
5. *Options for participation:* A review of strategies for participating in issue areas in local, national, and international settings. Technology-related systems of communication enable all students to communicate with people in remote places.

There are a number of ways that the global perspective can be applied to science teaching: infusing global education into the science curriculum, integrating it into planning, using technology, and developing curriculum models. The first three implications can be implemented by individual science teachers, whereas the fourth will require the collaboration of state science curriculum planners, or district-level science coordinators.

Infusing Global Education into the Science Curriculum

Aspects of global education can easily be infused into individual science courses, from grades K to 12, thereby helping students make connections between their study of science (biology, chemistry, earth science, physics) and global issues and problems. To some science educators, global education is important enough that it should be a theme or strand running through the curriculum. Mayer, describing a rationale for teaching science from a global view, says that "global education should be a thread running through the science curriculum. Our future leaders and voters (today's students) must understand our interrelationships with peoples around the world and how our daily activities affect our planet and its resources."[10]

An example of a thread of global education running through the science curriculum is the Pennsylvania Department of Education's strategy of providing supplementary curriculum or study units in global education (Table 11.1).

Integrating Global Science Education Activities into Planning

Another strategy is to integrate global science activities, such as the Global Food Web into chapters of textbooks and units of study. The Global Food Web[11] is a science program focusing on the environment, food supply, and human nutrition. It consists of activities designed to expose Georgia students to issues of both national and international relevance, especially in the interrelated areas of the environment. The challenge is to relate science concepts and units of study to global themes such as environment, pollution, natural resources, energy, food, population, war technology, and human health and disease. There is a wealth of sourcebooks and curriculum projects that teachers use to develop ideas and lesson plans (see Chapter 6 on S-T-S for specific curriculum projects).

Using Technology to Globalize the Classroom

By means of on-line communication via computer networks, a classroom in Bexley, Georgia, can be connected to a classroom in Zaria, Nigeria, and classrooms anywhere can tap into large data bases on mainframe computers.

The globalization of the science classroom enables students in one culture to communicate with students in another. Students can use the computer network to send E-mail (electronic mail) to students in distant lands. Students can investigate problems together, such as comparing the levels of pollution in streams and rivers, share information on acid levels in rivers, or conduct a joint study to analyze trash removal systems in different cities.

Many schools around the globe have turned some of their rooms into global classrooms by joining networks enabling them to "talk" to teachers and students

[9]Steven L. Lamy, "Global Education: A Conflict of Images," in *Global Education: From Thought to Action*, Kenneth A. Tye, ed. (Alexandria, Va.: Association for Supervision and Curriculum Development, 1990), p. 53.

[10]Victor J. Mayer, "Teaching from a Global Point of View," *The Science Teacher*, January, 1990, pp. 47–51.
[11]Cooperative Extension Service, *Global Food Web* (Athens: The University of Georgia, Cooperative Extension Service, 1990).

TABLE   11.1

CURRICULUM STRATEGY FOR INFUSING
GLOBAL EDUCATION INTO SCIENCE TEACHING

Grade Level	Global Topic/Unit	Purpose/Goals
Sixth grade	Windows on the World: Student Perceptions	• Focus activities on value systems—individual, group, societal, cultural, or planetary. • Involve all student in simulation studies centering on the global way of life from the viewpoint of various topic areas.
Seventh grade	Stewardship of the Spaceship Earth	• Focus activities on environmental issues at a local and/or state level and examine how they contrast with national and/or global issues. • Develop activities centering on the types of alternatives that are avail-able with regard to environmental issues.
Eighth grade	Citizen Responsibilities Concerning the Environment	• Focus investigations and activities on local, state, and national channels of government and the techniques they use to respond to environmental needs and/or issues. • Investigate and compare the U.S. system of government with that of a foreign government. In addition, develop activities exploring the United Nations and other international efforts to respond to global environmental concerns.
Ninth grade	Understanding Human Choices	• Focus activities on the problems confronting individuals, nations, continents, and the human species as global concerns expand. • Focus activities on students' abilities to understand difference between pre-global and global perspectives.
Tenth grade	Opinion and Perspective	• Focus activities on awareness of varying perspectives with regard to the individual and the world, followed by investigative research about the different perspectives. • Focus activities on discovering and recognizing global perspectives that differ profoundly from those of this country.
Eleventh grade	The World in Dynamic Change	• Focus on research and investigative activities revealing present key traits, mechanisms, or technologies that assist in the operation of global dynamics. • Conduct activities on awareness of theories and related concepts regarding current global change.

(Table continued on p. 393)

(Table continued from p. 392)

Grade Level	Global Topic/Unit	Purpose/Goals
Twelfth grade	State of Planet Earth	• Focus activities on the most recent worldwide environmental conditions—migration, political change, war and peace, economic conditions, and so forth. • Develop activities on awareness of students' roles and their responsibility to become involved in one or more of these world conditions and to work towards a resolution.

Frank H. Rosengren, Marylee Crofts Wiley, and David S. Wiley, *Internationalizing Your School* (New York: National Council on Foreign Language and International Studies, 1983), pp. 17–18.

in other countries. An example is AGE: Apple Global Education network, which can be accessed via AppleLink, a telecommunications network. Others have purchased curriculum projects such as KidsNet, published by National Geographic, to accomplish similar aims. In this system, information is shared and students explore an environmental problem (acid rain) by conducting research experiments. Classes in one region compare their measurements and data with students in different parts of the world. KidsNet also has the provision of providing an expert: a scientist to whom the students can communicate to ask questions.

Another program that uses technology to help geographically diverse students interact is the Global Lab project, a collaborative effort between Technical Education Research Centers (TERC), a Cambridge Massachusetts-based think tank, and the Institute for New Technologies (INT), a Moscow-based research center. The Global Lab project enables students to share information about phenomena unique to their local environment with students in the other country. Russian students, for instance, who live close to Chernobyl measure radioactivity and share the data across the network.

A team of teachers and science educators at Georgia State University have collaborated with teachers and researchers in Moscow and St. Petersburg to develop a computer network and curriculum materials called Global Thinking Project. The project links classrooms to study common environmental problems and is designed to foster global thinking.

These and other efforts make the notion of a global science classroom a reality and therefore a powerful tool in helping students connect with peers in other countries, thereby developing an awareness of other people.

As Margaret Riel points out, networking activity, not network connections, should be the driving force in the creation of global classrooms.[12] The technical connections need to be used by creative teachers to provide another vehicle for human communication. In this case, human communication is between students from diverse cultures, continents, and ethnic groups.

The World (Global) Core Curriculum

In response to the question, Is global education needed? Robert Muller, assistant secretary-general of the United Nations, responded with a proposal that became known as the world core curriculum. Muller's curriculum is based on his 33 years experience in the United Nations, and the knowledge of the

FIGURE 11.2 Technology enables students in remote places to join computer networks to produce a global classroom. Projects such as Global Laboratory and the Global Thinking Project enable students in geographically diverse locations to share scientific information as well as work together to solve global environmental problems. ©Daemmrich.

[12]Margaret M. Riel, "The Educational Potential of Computer Networking," paper presented at the Annual Meeting of the American Educational Research Association, San Francisco, April 16–20, 1986. ERIC: ED 303 155.

state of the planet's educational system. He says this:

Let me tell you how I would educate the children of this planet in the light of my 33 years of experience at the United Nations and offer you a world core curriculum which should underlie all grades, levels and forms of education including adult education.

Alas, many newly-born will never reach school age. One out of ten will die before the age of 1 and another four percent will die before the age of 5. This we must try to prevent by all means. We must also try to prevent that children reach school age with handicaps. It is estimated that ten percent of all the world's children reach school age with a handicap of a physical, sensory or mental nature. In the developing countries, an unforgivable major cause is still malnutrition.

Thirdly, an ideal world curriculum presupposes that there are schools in all parts of the world. Alas, this is not the case. There are still 814 million illiterates on this planet. Humanity has done wonders in educating its people: we have reduced the percentage of illiterates of the world's adult population from 32.4 percent to 28.9 percent between 1970 and 1980, a period of phenomenal population growth. But between now and the year 2000, 1.6 billion more people will be added to this planet and we are likely to reach a total of 6.1 billion people in that year. Ninety percent of the increase will be in the developing countries where the problem of education is more severe. As a result, the total number of illiterates could climb to 950 million by the Bimillennium.

Education for all [italics mine] remains, therefore, a first priority on this planet. This is why UNESCO has rightly adopted a World Literacy Plan 2000. With all these miseries and limitations still with us, it remains

important, nevertheless, to lift our sights and to begin thinking of a world core curriculum.

As I do it in the United Nations, where all human knowledge, concerns, efforts, and aspirations converge, I would organize such a curriculum, i.e, the fundamental life-long objectives of education, around the following categories:

I. Our Planetary Home and Place in the Universe
II. Our Human Family
III. Our Place in Time
IV. The Miracle of Individual Human Life[13]

A world core curriculum, as Muller envisions it, would enable all students to understand the planet upon which they were born, and to know their relationship to human groups all over the planet. Ethnic and cultural understanding is an important aspect of Muller's core curriculum. He also suggests that students need to understand their place in time—where humanity has come from, and where it is going. Finally, the Muller proposal suggests that "individual human life is the highest form of universal consciousness on the planet," and that the curriculum should be based on a holistic integration of physical well-being, mental development, moral development and spiritual development.

The Muller world core curriculum is distinctly interdisciplinary and multicultural. The global perspective for education as envisioned by Muller is organized around four "harmonies," as shown in Figure 11.4. Although traditional topics are part of the plan (students study our relationship to the sun, the atmosphere, biosphere, mountains, minerals, nutrition, etc.), they are viewed in relationship to a larger consciousness—the global perspective.[14]

---

[13]Robert Muller, "A World Core Curriculum," *New Era,* January, 1982.
[14]The Robert Muller School, *World Core Curriculum Manual* (Arlington, Tex.: Robert Muller School, 1986), p. 10.

FIGURE 11.3 The World Core Curriculum envisions a program in which students understand their connections to the planet as well as to people and other living things. NASA.

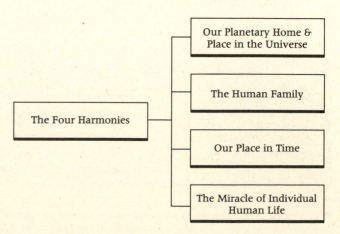

FIGURE 11.4 Muller's Curriculum based on Four Harmonies. Used with permission of the Robert Muller School.

INQUIRY ACTIVITY   11.1

Exploring the Global Perspective

This activity is designed to engage you in a global science activity, and then in considering some applications of the global perspective to science courses of study.

Materials

One set of "global problem" cards for each team, sheet of newsprint, marking pens

Procedure

1. You should join with several class members to do this activity to create several "global" teams, each of which should represent a different country. Form your groups, and then as a large group decide which country each team will represent.
2. Take a few minutes and make a list of information about the country you represent. You may need reference books to provide the teams with background information about the countries for each team.
3. Each team should have a set of global problem cards (Figure 11.5). These problems represent some of the most important environmental problems facing the earth today.[15] Your team should investigate these problems from the point of view of citizens from the country you represent.
4. To prepare for a "cross-cultural discussion," your team should engage in the following activities:

a. Rank-order the cards from most important or significant problem to least important or significant. Make a record of your team's decision.
b. Select the most pressing problem, and then arrange the other problems in a "web" or "map" and indicate how the global problems interrelate with each other. Make a record of your team's work on a sheet of newsprint.

5. Bring all the teams together to simulate an international conference on global environmental problems. Choose a moderator from the group, and then have each team prepare a short (no more than 5 minutes) report.
6. The moderator should use the questions in the conclusions section to facilitate a group discussion.

Minds on Strategies

1. In what school science context would you use this activity? What would the goals and purpose be of using this activity? Suggest appropriate follow-up activities that this activity would stimulate for students.[16]
2. To what extent did you feel competent understanding global problems from the point of view of a citizen from another country? Does this influence decision making, and deciding upon the significance of a problem? In what way?
3. Try this activity with a group of secondary students. Evaluate their performance in light of your own.
4. How does an activity like this help develop a global perspective for students? For teachers?

[15]From Rodger W. Bybee, "Global Problems and Science Education Policy," in *Redesigning Science and Technology Education*, Rodger W. Bybee, Janet Carlson, and Alan J. McCormack, eds. (Washington, D.C.: National Science Teachers Association, 1984), pp. 60–75.

[16]For a source of global science activities, see Betty A. Reardon, *Educating for Global Responsibility* (New York: Teachers College Press, 1988).

A MULTICULTURAL PERSPECTIVE

In April 1989 the Association for Multicultural Science Education (AMSE) was founded and held its first meeting at the National Science Teachers Association conference in Atlanta. In the bylaws of the AMSE, the following statement defined its purpose and objectives:

The purposes for which this Association is organized are to stimulate and promote science teaching to students of culturally diverse backgrounds and to motivate such students to consider science-related careers;

to explore and promote the improvement of science curriculums, educational systems and teaching methods in schools to assist such stimulation; to recruit and involve teachers of all minorities in science education and to initiate and engage in activities and programs in furtherance of improving the science education of culturally diverse students.[17]

[17]Bylaws, Association for Multicultural Science Education, Washington, D.C., 1989, p. 1.

Land Use

Energy

Nuclear Reactors

Air Quality

Mineral Resources

World Hunger

Water Resources

Human Health

Population Growth

War Technology

Hazardous Substances

Extinction

FIGURE 11.5 Global Problem Cards. (Note: Copy this page and cut into 12 global problem cards to be used for Inquiry Activity 11.1)

In this section we will explore multicultural education and indicate the central place that multicultural science education should play in planning science lessons, choosing content, selecting teaching materials, and teacher behavior in the classroom.

Multicultural education is a movement that began in the 1970s, but as James Banks points out, has expectedly suffered from some confusion.

> Since multicultural education...deals with highly controversial and politicized issues such as racism and inequality [it] is likely to be harshly criticized during its fomative stages because it deals with serious problems in society and appears to many individuals and groups to challenge established institutions, norms and values. It is also likely to evoke strong emotions, feelings and highly polarized opinions. As it searches for its raison d'être, there is bound to be suspicions and criticisms.[18]

What Is Multicultural Education?

According to Fred Rodrigues, the acknowledgment of the importance of multicultural education was reflected in the American Association for Colleges of Teacher Education's (AACTE's) 1973 statement, No One Model American. The statement has become the basis for most other definitions of multitcultural education. Multicultural education, according to the AACTE statement, involves the following:

- Multicultural education is education which values cultural pluralism
- Multicultural education rejects the view that schools should seek to melt away cultural differences or the view that schools should merely "tolerate" cultural pluralism
- Multicultural education programs for teachers are more than "special courses" or "special learning experiences," grafted onto the standard program
- The commitment to cultural pluralism must permeate all areas of educational experience[19]

Another organization, the National Council for the Accreditation of Teacher Education (NCATE), advocated that multicultural education should be part of all teacher training programs. The NCATE definition of multicultural education is as follows:

Multicultural education is preparation for the social, political and economic realities that individuals experience in culturally diverse and complex human encounters.... This preparation provides a process by which an individual develops competencies for perceiving, believing, evaluating and behaving in different cultural settings.[20]

Multicultural education is seen not as a course of study or a subject, nor as a unit of study, nor for just minority students, it is rather a way of teaching and learning, thereby offering a perspective on education.[21] The teacher who embodies a multicultural approach creates a classroom environment that respects cultural diversity, and presents lessons that draw upon the cultural diversity implicit in the content being presented.

Status and Goals

In a report entitled *Education That Works: An Action Plan for the Education of Minorities*, by the Quality Education for Minorities Project (QEMP) at the Massachusetts Institute of Technology, the authors propose a national effort to create a new kind of learning system that recognizes the value and potential of all students. They set forth the following goals for the year 2000:[22]

Goal 1: Ensure that minority students start school prepared to learn

Goal 2: Ensure that the academic achievement of minority youth is at a level that will enable them, upon graduation from high school, to enter the workforce or college fully prepared to be successful and not in need of remediation

The QEMP authors pointed out that this goal will be achieved when it is ensured that tracking does not occur at any point along the educational pipeline; when any performance gap is bridged between nonminority and minority students by the fourth grade; when minority students leave elementary school with the language, mathematics skills, and self-esteem that will enable them to succeed; and when minority youth are excelling in core academic courses (science, math, history, language) by the eighth grade that keep their college and career options open, and achieve, at a minimum, the same high school graduation rates for minority and nonminority students.

---

[18]James A. Banks, "Multicultural Education and its Critics: Britain and the United States," in *Multicultural Education: The Interminable Debate*, Sohan Modgil, Gajendra Verma, Kanka Mallick, and Celia Modgil, eds. (London: Falmer, 1986), p. 222.

[19]Fred Rodrigues, *Mainstreaming a Multicultural Concept into Teacher Education* (Saratoga, Calif.: R & E, 1983), p. 12.

[20]Cited in Rodrigues, *Mainstreaming a Multicultural Concept*, pp. 13–14.

[21]Rodrigues, *Mainstreaming a Multicultural Concept*, p. 15.

[22]Quality Education for Minorities Project, *Education That Works: An Action Plan for the Education of Minorities* (Cambridge: Massachusetts Institute of Technology, 1990), pp. 7–9.

Goal 3 : Significantly increase the participation of minority students in higher education, with a special emphasis on the study of mathematics, science, and engineering

Goal 4: Strengthen and increase the number of teachers of minority students

Goal 5: Strengthen the school-to-work transition so that minority students who do not choose college leave high school prepared with the skills necessary to participate productively in the world of work and with the foundation required to upgrade their skills and advance their careers

Goal 6: Provide quality out-of-school educational experiences and opportunities to supplement the schooling of minority youth and adults

As is pointed out in the QEMP report, the school population in the United States is changing such that the term "minority" will, if it hasn't already, lose its meaning. About 20 percent of the U.S. population is Alaska native, American Indian, black, or Hispanic, and by 2020, over a third of the nation will be minority. Some states (New Mexico and Mississippi) have "minority-majority" enrollments in schools, and California and Texas will soon be in the same category (Figure 11.6).

One of the points that the authors of the report make is that many minority and low income students not only begin school without the skills required to succeed in the present school system, but teacher expecta-

tions of students' abilities are often low, thereby resulting in poor performance and low levels of success.[23]

The high dropout rate among minority students can be traced to factors such as being behind in school, low teacher expectations, having to work, becoming teenage parents, being involved in gangs, and boredom.[24] Although dropout rates for black youth have declined over the past decade (from 27 percent in 1978 to 15 percent in 1988), the rate for Hispanic youth was 36 percent, or over three times that of white youth. This was the same rate for American Indian and Alaska native (Table 11.2).

To ensure a quality education for minority youth, the QEMP focused on a number of strategies and suggestions for middle and high school curricula and instruction. It is important to take note of these recommendations because, although the QEMP authors are moving toward a national focus and agenda, individual schools and teachers can implement their strategies as part of a multicultural approach. Here are some of their recommendations, especially those that relate to the teaching of science.

1. Focus on team learning by establishing "schools-within-schools" and families of learning, and implement cooperative learning strategies in the science classroom.

[23]Quality Education for Minorities Project, *Education That Works*, p. 17.
[24]Quality Education for Minorities Project, *Education That Works*, p. 18.

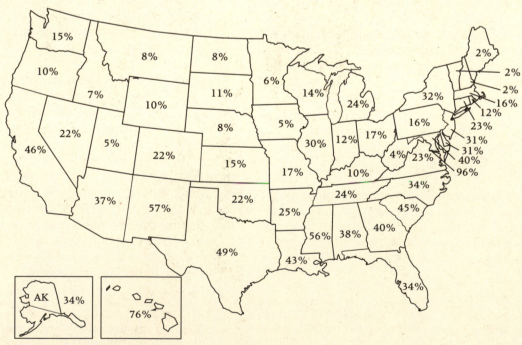

FIGURE 11.6   Minority Enrollment in K-12 Schools, Quality Education for Minorities Project, *Education Test Works*, p. 12.

TABLE 11.2

1980 HIGH SCHOOL SOPHOMORES WHO DID NOT GRADUATE AND WERE NOT IN SCHOOL

Group	By 1982 (%)	By 1986 (%)
White	14.8	7.6
Black	22.2	11.4
Hispanic	27.9	18.0
American Indian/ Alaska Native	35.5	27.1
Asian American	8.2	2.0
Total	**17.3**	**9.2**

Quality Education for Minorities Project, *Education That Works: An Action Plan for the Education of Minorties* (Cambridge: Massachusetts Institute of Technology, 1990), p. 18.

2. Implement science curriculum reforms, including Project 2061, the NSTA Scope and Sequence, and curriculum efforts such as the BSCS Human Biology Project. Of special note, the report emphasized a science for all theme in the sense that it favored a science education program that was hands-on, and that made an explicit connection between science and familiar events. The committee also supported National Council of Teachers in Mathematics (NCTM) standards, which recommend an emphasis on problem solving, the understanding of under-lying concepts, the use of tools such as calculators and computers, and the application of mathematics disciplines to real life.

3. Replace the tracking curriculum in high school with a core academic curriculum to prepare students for college and the workplace. The committee puts strong emphasis on making sure that all students arrive at high school with a course in algebra, and that all students take mathematics and science in high school as part of the core.

4. Provide on-site health services and strengthen health education. Schools should work with students to deal with drug and alcohol dependency and help reduce the number of teenage pregnancies. Drug and sex education programs should be part of the school curriculum.

5. To increase the number of minority students in higher education, especially in the study of mathematics, science, and engineering, high schools and colleges should work together to provide programs to not only encourage but provide academic programs to support this goal. Examples include offering six-week summer science residential programs for high school juniors and seniors on college campuses to study mathematics, science, and engineering. The project also recommended expanding the funds available for recruitment.

These were not all the recommendations of the QEM Project. However, these provide insight into the directions that science education should move in. Let's explore multicultural education as it specifically pertains to science education.

Multicultural Science Teaching

According to Mary M. Atwater, multicultural science education should be considered an integral part of the discipline of science education. She points out that a body of knowledge about multicultural science education exists, and that this knowledge includes understandings of group identification, culture, and science. She goes on to say that multicultural science education relates to science learning and achievement, science instruction, and the involvement of different human beings in the science.[25]

Steven J. Rakow and Andrea B. Bermudez suggest that meeting the needs of Hispanic students in science teaching requires that teachers go beyond the simplistic view of the group differing only in language.[26] They outline a holistic approach in which communication is only one factor that may influence science learning. Their broader approach considers the influence of learning style, family, and historical culture as an influence on learning. Using Banks's six elements, which can be used to understand the needs of various ethnic groups, they suggest examining multicultural science education in terms of values and behavioral styles, language and dialects, cultural cognitiveness, identification, nonverbal communication, and perspectives, world views, and frames of reference.

What should science teachers do to implement a multicultural approach in their own school and classroom? Three levels of decision making and action taking are presented here. As a beginning science teacher, the science department and classroom recommendations should be pertinent.

[25]Mary M. Atwater, "Multicultural Science Education: Definitions and Research Agenda for the 1990's and Beyond," paper presented at the National Association for Research in Science Teaching, April 1990, p. 4.
[26]Steven J. Rakow, and Andrea B.Bermudez, "Science is 'Ciencia': Meeting the Needs of Hispanic American Students," paper presented at the Annual Meeting of the National Association for Research in Science Teaching, April 1990.

At the School Level

Effective change should involve the entire school. Too often teachers who act as individual trailblazers get discouraged because of the lack of support and assistance. To increase the chances of a successful multicultural approach in the classroom, the following processes should be implemented:[27]

1. Establishing a whole-school multicultural policy and then ensuring that the science department, or members of interdisciplinary teams in middle schools, reflect this policy
2. Using all appropriate opportunities to challenge prejudice as and when it arises in the school community
3. Ensuring a consistent approach to dealing with racist incidents in the school
4. Providing, whenever possible, opportunities for teachers to gain some experience of different cultures

At the Science Department or Team Level

Members of a department or a team should engage in discussion and agree on a policy of multicultural science education. Some steps, strategies, and ways to accomplish this are as follows:[28]

1. Develop a policy for the teaching of science from a multicultural perspective in a way that will seek to counter the causes of inequality and prejudice.
2. Ensure that all members of the department or team are provided with in-service training to aid them in implementing this policy.
3. Review the approaches to language in science, taking into account the role of science education in language development. In this regard, it is important to use the research on the ways students learn. You have learned that teachers should start with the ideas students themselves have and then use a variety of ways to help students express their ideas: talking, writing, reading, and listening. In Chapter 8 a number of strategies were developed to aid in student learning. The strategies we use in the science classroom need to reflect our attempt to create a multicultural environment. By respecting students' ideas in both monolingual and bilingual settings, students will be encouraged to express themselves, thereby helping them understand science concepts.

4. Collaborate with ESL (English as a second language) teachers to share expertise.
5. Discuss the values underlying the assessment schemes being used. In Chapter 9, a broad-based model of assessment was presented that placed value on a wide variety of student work and encouraged the use of performance test procedures, as well as small group problem-solving assessments.
6. Ensure that resources do not contribute to stereotypical views or carry racist implications (see Inquiry Activity 11.2).
7. Use and develop links with parents and other members of the local community to gain a better understanding of different customs, diets, and religious beliefs.

In the Science Classroom

Science teachers can enrich their approach to teaching by taking a multicultural perspective in the preparation and carrying out of lessons, units, and courses of study. The suggestions that follow will influence the way you view science, and, combined with the next section on gender issues in science education, give you a broader view of the field.[29]

1. Adopt the view that sees cultural diversity as a positive advantage. Students from different cultural backgrounds will bring varied experiences to the classroom. Valuing this diversity can enrich the examples that will come forth during class discussions of concepts and phenomena.
2. Make use of science topics and concepts for teaching in a multicultural context. Choosing content topics such as genetics, evolution, foods, and communication, for example, can provide a rich starting point in the multicultural context.
3. Science teachers can challenge notions that ethnic origin places limitations on individual potential. Activities can be developed that show students that the stereotyping of individuals is an inaccurate view of the world, and they can explore this via case studies, field trips, and guest speakers. As is the case in the research on women in science, people of color have been forgotten in the sciences as well. The history of science is full of examples of ethnic minorities' contributions to the field, and this historical perspective can be important in this regard (see Figure 11.7).
4. Demonstrate the validity of technology in particular cultural contexts. Viewing technology as a response to human need means investigating the notion of appropriate technology. Teachers can provide oppor-

---

[27]Based on Cristine Ditchfield, *Better Science: Working for a Multicultural Society* (London: Heinemann Educational Books/Association for Science Education, 1987), p. 38.
[28]Based on Ditchfield, *Better Science*, p. 38.

[29]Items 1 to 6 based on Ditchfield, *Better Science*, pp. 12–38.

BENJAMIN BANNEKER, EARLY AMERICAN SCIENTIST

Benjamin Banneker was born and raised on a farm in Patapsco Valley, Maryland. He attended school until he was fifteen and went on to teach himself algebra and geometry. While still at school, he developed a particular interest in mathematics and astronomy.

His interest in the natural sciences was brought to the attention of a man named George Ellicott, a businessman and amateur scientist. Banneker's friendship with Ellicott stimulated his interest in astronomy, and at the age of fifty-six he began the study of astronomy. In the years that followed Banneker mastered the fundamental principles of astronomy, much to the surprise and pleasure of Ellicott. He could, for example, project forthcoming eclipses with extreme accuracy.

Banneker began preparation of an almanac to be used in the states of Maryland, Virginia, Delaware, and Pennsylvania. He published his first almanac in 1792, and for the next ten years he continued to publish this indispensable pamphlet. Almanacs were first introduced in colonial America in 1639 and played a very significant role in a society in which nearly everyone lived on a farm. In addition to calendars and information on heavenly bodies, these early almanacs contained information on astrology, history, literature, and poetry. Everybody in colonial America relied on almanacs for accurate information.

Because it was so important in colonial life, Banneker's almanac was a most welcome contribution. It soon became as popular as Benjamin Franklin's *Poor Richard's Almanac*, published years earlier.

The achievements of Benjamin Banneker in science, mathematics, and astronomy have often been considered as symbolic of what black Americans could have accomplished if given the opportunity.

Benjamin Banneker became so absorbed with his scientific research that he found farming too demanding and eventually leased his land to several farm families to help supplement his meager income. Wealth did not accompany the fame he achieved.

Banneker was often compared to Benjamin Franklin, who also engaged in scientific research and published almanacs. Banneker died in 1806, an American scientist whose color made him a "forgotten man."

1. Obtain a copy of a current almanac. What type of information does it contain?

2. Research contributions made by other black American scientists.

3. Racial prejudice has been, and still is, one force preventing people from developing fully and making contributions to all of humanity. What are some of the other forces?

Edward S. Jenkins et al., *American Black Scientists and Inventors*, copyright © 1975 National Science Teachers Association, Washington, D.C., pages 23–26. Used with permission.

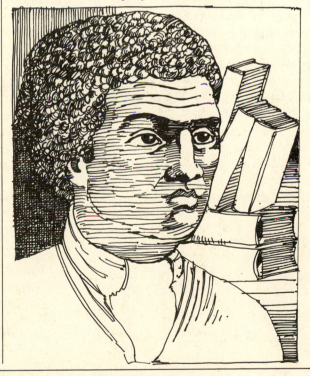

FIGURE 11.7 The history of science is full of examples of ethnic minorities' contributions to the field of science. Joe Abruscato and Jack Hassard, *The Whole Cosmos of Science Activities for Kids of All Ages*. Glenview, IL: Scott Foresman and Company.
© 1991 Scott, Foresman. Reproduced by permission of the publisher.

tunities for students to study the positive and negative implications of particular technologies. This content area can result in powerful discussions, and topics for research. For instance, in the British report,[30] science teachers suggest that at one level there is the issue of technological change and the proposition that such changes, introduced to enrich the lives of some, may also impoverish the lives of others. An example might be the construction of a highway that might bring some the benefits of travel but for others create havoc in their environment.

5. Illustrate that science has different implications for individuals in different parts of the world.

6. Set Western science in a historical context that illustrates that it is relatively young in terms of the history of science. Western science emerged in the sixteenth century and can be compared with older science practiced in China and the Arab world.

[30]Ditchfield, *Better Science.*

7. Develop an awareness and knowledge of student learning styles, and plan lessons that provide a variety of approaches. Within any group of students there will be a range of learning styles, for example, field dependent versus field independent; visual, auditory, kinesthetic, tactile; left brain, right brain; active, reflective; concrete, abstract. Using a variety of learning activities will meet the learning-style needs of students and provide the experiences students need to develop concepts and an interest in science.

8. Arrange the classroom physical environment so that it clearly communicates to students that you value and respect cultural diversity. Involve students in the development of bulletin boards that recognize the contributions of ethnic and women scientists in the development of science. Provide books and magazine articles so that students can read about the diversity of scientists and careers in science.

INQUIRY ACTIVITY   11.2

....

Investigating Images Portrayed in Science Teaching Materials

....

This inquiry activity is designed to help you detect the extent to which materials are biased in terms of ethnicity. It has been estimated that 90 percent of all science teachers use textbooks, and that the textbook is what school districts use, by and large, to determine the science curriculum. For students, the textbook is their contact with science. The image it portrays of science will have a powerful impact on them.

Materials

Science textbooks, the multicultural textbook evaluation checklist (Figure 11.8)

Procedure

1. Work with a team (or individually) to evaluate the images of science and people as portrayed in more than one secondary science textbook. You might want to compare a contemporary book in a field with one published 20 to 30 years ago.

2. Use the checklist in Figure 11.8 to determine the multicultural character of the books.

Minds on Strategies

1. How valid are the images that are portrayed in the books?

2. What effect do you think the images portrayed will have on secondary students?

3. If you were assigned one of these books to use as the text for a course, how could you compensate for any deficiencies in the materials?

4. Investigate other checklists and evaluation forms for the content of textbooks. Examples include

   a. *Ethnic Authenticity Checklist:* Fred Rodrigues, *Mainstreaming a Multicultural Concept into Teacher Education* (Saratoga, Calif.: R & E, 1983).

   b. *Sexism in Textbooks:* Sue V. Rosser, *Female-Friendly Science* (New York: Pergamon, 1990), pp. 73–91.

	Yes	No

Section A. The Image of Science

1. Is the application of science illustrated with examples from all over the world?
   Comments:

2. Is up-to-date technology only illustrated by Western examples?
   Comments:

3. Is science set in a context that relates to people?
   Comments:

4. Does the way in which the material is presented suggest that only Western ways are valid?
   Comments:

Section B. Images of People

5. Are the people featured in the material from more than one cultural background?
   Comments:

6. Does the material present a positive view of people from different cultural backgrounds? For example, are they illustrated performing tasks of equal esteem?
   Comments:

7. Are the negative images confined only to people from a particular cultural background?
   Comments:

8. Are people who are considered eminent scientists from different parts of the world represented?
   Comments:

9. Does the material present equal images of women and men?
   Comments:

Section C. General

10. Is the material presented in such a way as to eliminate damaging feelings of superiority, based on race, in the European person?
    Comments:

11. Are the customs, life styles, and scientific traditions of people presented in a way that helps to explain the value, meaning, and role of customs in their lives?
    Comments:

FIGURE 11.8   Multicultural Textbook Evaluation Checklist. After Christine Ditchfield, *Better Science: Working for a Multicultural Society* (London: Heinemann Educational Books/Association for Science Education, 1987), p. 40.

GENDER ISSUES

- In the United States women comprise approximately 50 percent of the work force, yet only 9 percent are employed as scientists and engineers
- Only 12 percent of students enrolled in high school technology programs are female[31]
- In the recent National Assessment of Educational Progress science study, girls continue to score below the national mean on all science achievement items and express negative attitudes toward science[32]
- Many more females are less interested than males in fields with a significant mathematical component such as physics, chemistry, and engineering[33]
- Research studies suggest that teaching styles and other school-related factors are important in encouraging girls (as well as boys) to continue courses and careers in science[34]

The issue of gender is a real one in the mathematics and science classroom, and in this section we will investigate the research that has been done to discover ways to improve the participation of girls in middle school and high school science, and to increase the career participation of women in science. A number of researchers and teachers, not only in the United States but in many countries around the world, have

- Conducted research to document and explore the problem of gender differences in the classroom
- Compiled teacher training courses and in-service programs to improve teachers' ability to create environments conducive to "female-friendly" science.[35]
- Developed curricular and instructional materials designed to increase equity and encourage girls and minorities in science courses and careers

The Participation of Women in Science

"I wonder whether the tiny atoms and nuclei or the mathematical symbols or the DNA molecules have

any preference for either masculine or feminine treatment."[36] As Anne Haley-Oliphant points out, it is doubtful that inanimate objects have a preference for being studied by males or females, but the statistics on the participation of women in science is much less than the percentage of women in the population.[37]

The history of scientific ideas reads like the history of men scientists. Yet women have participated in science since antiquity; but their status and role was always in the shadows of men, and much of their work was unrecognized by the scientific community, stolen, or lost because it was never published.[38]

In her book, *Hypatia's Heritage*, Margaret Alic provides a historical glimpse of the accomplishments of women in science from antiquity through the nineteeth century. She describes the research, theories, and observations made by a long list of female contributors to the field of science. Alic sounds an important challenge to science teachers when she documents the contributions of women in the sciences, yet in school science, the discoveries, theories, and writings of women are left out of the pages of most school textbooks.

> Science is that body of knowledge that describes, defines and, where possible, explains the universe—the matter that constitutes it, the organisms that inhabit it, the physical laws that govern it. This knowledge accumulates by a slow, arduous process of speculation, experimentation and discovery that has been an integral part of human activity since the dawn of the race. Women have always played an essential role in this process.
>
> Yet we think of the history of science as a history of men. More than that, we think of the history of science as the story of very few men—Aristotle, Copernicus, Newton, Einstein—men who drastically altered our view of the universe. But the history of science is much more than that. It is the story of the thousands of people who contributed to the knowledge and theories that constituted the science of their eras and made the "great leaps" possible. Many of these people were women. Yet their story remains virtually unknown.[39]

The impact of keeping this aspect of the history of science from science courses is to convey a sexist character of science. Although blatant racism and sexism

---

[31]Mary Jo Strauss, *Recommended Resources for Use in Developing Programs to Achieve Sex Equity in Mathematics, Science and Technology* (Baltimore: Maryland State Department of Education, 1990), p. 3.

[32]Ina V. S. Mullis and Lynn B. Jenkins, *The Science Report Card: Elements of Risk and Recovery* (Princeton, N.J.: Educational Testing Service, 1988), pp. 23–30.

[33]Daryl E. Chubin, *Elementary and Secondary Education for Science and Engineering* (New York: Hemisphere, 1990), p. 17.

[34]Jane Butler Kahle, "Retention of Girls in Science: Case Studies of Secondary Teachers," in *Women in Science: A Report from the Field*, Jane Butler Kahle, ed. (Philadelphia: Falmer, 1985), pp. 49–76.

[35]See Sue V. Rosser, *Female-Friendly Science* (New York: Pergamon, 1990)

[36]Chien-Shiung Wu, cited in Jane Butler Kahle, *Women in Science*, (Philadelphia: Falmer, 1985), p. 169.

[37]Anne E. Haley-Oliphant, "International Perspectives on the Status and Role of Women in Science," in *Women in Science*, Jane Butler, ed. (Philadelphia: Falmer, 1985), p. 169.

[38]Margaret Alic, *Hypatia's Heritage: A History of Women in Science from Antiquity through the Nineteenth Century*, (Boston: Beacon, 1986).

[39]Alic, *Hypatia's Heritage*, p. 1.

have been eliminated from textbooks, there is evidence to suggest that women and people of color are absent from the written content of the textbooks.[40] Stories and anecdotes about the accomplishments of women and minorities are missing, in general, from textbooks. Providing vignettes and involving students in an exploration of the contributions of a wide range of scientists—not just Western white males—can have a positive impact on students' perceptions of science.

The current status of women in medicine and engineering in Western, Eastern and some developing nations reveals a pattern that provides the basis for an analysis of the status of women in science and engineering (Table 11.3). The comparison between Western countries such as the United States and Britain reveal differences in participation compared with some Eastern and developing countries.

Because of this disparity in the participation of women in science, a number of Western nations have developed intervention programs to increase participation. The programs are quite varied. For example, in Britain several programs have been developed. One program, Girls into Science and Technology (GIST) focused on

1. Raising teachers' awareness of girls' underachievement and helping them realize that they can do something about it
2. Testing children's entering attitudes and knowledge in science and technology for comparison with later performance and choice
3. Arranging a series of visits to schools by women working in scientific and technological jobs who could act as role models for girls[41]

The GIST program followed a cohort of approximately 2000 students from the time they entered secondary school until they made their option choices at the end of the third year. The purpose of GIST was to alter girls' attitudes toward physical science and technical subjects by the end of year 3 so that in years 4 and 5 they would choose these options. Although the results were not clear-cut, the project had an impact on students' and teachers' attitudes toward science, and helped create a climate of discussion, debate, and investigation of issues surrounding gender in the science classroom in Britain.

TABLE 11.3

PERCENTAGE OF WOMEN PHYSICIANS AND ENGINEERS IN NINE COUNTRIES

	Physicians	Engineers
Western Countries		
Britain	17	0.5
Canada	33	8.2
France	15	2.0
USA	10	1.0
West Germany	30.6	7.8
Eastern Countries		
Hungary	42	no data
Poland	44	11
Soviet Union	75	44
Developing Countries		
Burma	50	15
Mexico	34.7	6.7

Haley-Oliphant, "International Perspectives," p. 170.

The programs that were designed to improve the status of women and minorities in science careers, as well as participation in science courses of study in secondary schools, investigated the reasons for this phenomenon. For instance, Joan Skolnick, Carol Langbort, and Lucille Day indicate that gender socialization negatively affects girls' attitudes toward mathematics and science. They point out that female socialization promotes fears about competence, reliance on the judgment of others, an interpersonal and verbal orientation, and an unfamiliarity with toys, games, and activities that stimulate the learning of spatial relationships and basic mathematics concepts.[42] They think that the strategies used to teach mathematics and science don't favor confidence building, rely on abstract knowledge, and are generally devoid of activities and group work. They suggest a problem-solving strategy in which students' confidence and competence in problem solving is built through active, hands-on experiences.

Alison Kelly, in investigating why girls don't study physical science, suggests also that girls' lack of confidence may be a reason but also suggests that the mascu-

[40]Rosser, *Female-Friendly Science*, p. 74. Especially refer to chap. 6, "Sexism in Textbooks."

[41]See Alison Kelly, Judith Whyte, and Barbara Smail, "Girls into Science and Technology: Final Report," in *Science for Girls?* Alison Kelly, ed. (Milton Keynes, England: Open University Press, 1987), pp. 100–112.

[42]Joan Skolnick, Carol Langbort, and Lucille Day, *How to Encourage Girls in Math and Science*, (Englewood Cliffs, N.J.: Prentice-Hall, 1982), p. 47.

line image of science and the impersonal approach of science contribute to the problem. In studies of students' attitudes toward science Kelly found that a large number of boys but very few girls agree with statements like "A woman could never be a great scientist" or "Girls don't need to learn about electricity or light." She points out that if boys believe that science is a man's world, then this attitude will affect the science classroom; for example, boys will dominate the lab and discussions.[43]

Teacher classroom behavior may also contribute to girls' not continuing with science in secondary school. One particular example of this is reported by Kahle in a study of gender issues in the science classroom. In two in-depth investigations of science classrooms, Kahle found that the grades teachers gave influenced student science options. This is not surprising, since grades contribute to one's self-confidence. However, in the cases that Kahle studied, one of the teachers gave twice as many high grades to boys as girls; the other teacher gave equal numbers of girls and boys the highest grades. In the cases of the teacher who gave more boys the highest grade, the percentage of females taking courses in physics, chemistry and physical science ranged from 14 to 33 percent, whereas 45 percent and 50 percent of the other teacher's female students took physics and chemistry, respectively.[44]

The way in which teachers interact with students in the classroom can have an impact on students attitudes toward science and future career choices. Kahle suspects that the role or metaphor the teacher uses to describe his or her teaching style may be a factor. For example, teachers who see themselves as a facilitator of learning provide a classroom climate more conducive to nonsexist teaching. Taking on a facilitative role as teacher may be due to prior educational experiences. Kahle says this about a teacher who was perceived as a facilitator of learning: "Sandra, on the other hand, clearly saw her role as change agent. She felt that she had experienced discrimination and she consistently endeavoured to be a transformer, not reproducer, of sex-role stereotypes which promulgated gender differences."[45]

It is interesting to note that another teacher's metaphor for describing the role of the teacher was a dispenser of knowledge and information. In this role, science is perhaps perceived as impersonal, and the teacher perhaps sees the role as one of passing on information rather than of helping students find out and use their own resources to learn.

What can science teachers do to reduce bias in teaching? What strategies can be employed in the classroom that will reduce sex stereotyping and encourage girls in science courses and careers?

Strategies to Encourage Females in Science Courses and Careers

One of the characteristics of successful strategies to increase the participation of females in science courses and careers has been to focus on the skills identified in research that might cause the problem. Thus, there is a strong emphasis on developing spatial ability, problem-solving ability, and mechanical skills. Another trend in successful strategies is the emphasis on mathematics. Researchers realized that students needed to take mathematics in secondary school to pursue science courses of study in college. This focus on math became known as the critical-filter idea. Science teachers can play an important role by counseling students into math, as well as integrating math in an appealing way into science classes.

Let's take a closer look at the strategies and characteristics of successful programs that are applicable to the science classroom.

Successful Program Characteristics

In summarizing a report published by the American Association for the Advancement of Science on programs that increased access and achievement of females and minorities, Cole and Griffin point out that the results are hopeful in producing positive outcomes. They listed the following as characteristics of programs that raised the achievement level and increased science career choices of females and minorities:

- Strong academic component in mathematics, science, and communications, focused on enrichment rather than remediation
- Academic subjects taught by teachers who are highly competent in the subject matter and believe that students can learn the material
- Heavy emphasis on the applications of science and mathematics and careers in these fields
- Integrative approach to teaching that incorporates all subject areas, hands-on opportunities, and computers
- Multiyear involvement with students
- Strong director; committed and stable staff who share program goals
- Stable long-term funding based with multiple funding sources

[43]Kelly, *Science for Girls?* p. 15.
[44]Jane Butler Kahle, "Real Students Take Chemistry and Physics: Gender Issues," in *Windows into Science Classrooms: Problems Associated with Higher-Level Cognitive Learning,* Kenneth Tobin, Jane Butler Kahle, and Barry J. Fraser, eds. (London: Falmer, 1990), pp. 92–134.
[45]Kahle, "Real Students Take Chemistry and Physics," p. 127.

- Recruitments of participants from all relevant target populations
- University, industry, school cooperative program
- Opportunities for in-school and out-of-school learning experiences
- Parental involvement and development of base of community support
- Specific attention to removing educational inequities related to gender and race
- Involvement of professionials and staff who look like the target population
- Development of peer support systems
- Evaluation, long-term follow-up data collection
- "Mainstreaming"—integration of program elements supportive of women and minorities into the institutional program[46]

Successful Teaching Strategies

A number of strategies have proven to be successful in influencing female attitudes and achievement in science and mathematics. In general these strategies support a hands-on approach, in which students work in small learning teams and have opportunities for "academic discussions." But there is a great deal more than this general approach that seems to enhance female participation in science. Let's look at some specific stategies.

Content and Activities

Barbara Smail has suggested that science classes need to provide more experiences that she calls "nurturative" to provide a balance to the "analytical/instrumental" activities that she suggests characterize most school science (Table 11.4).

The comparison of analytical-instrumental with nurturative characteristics is similar to the left-brain, right-brain dichotomy that was discussed in Chapter 2. Proponents of left brain and right brain thinking such as McCarthy suggest an integration resolution to this dichotomy, as Smail recommends. Students should be exposed to both analytical as well as nurturative as suggested in Table 11.4. A balanced selection of content and activities would result in an array of creative activities.

For example, one school she worked with suggested changing writing assignments and examinations to reflect the following:

[46]Michael Cole and Peg Griffin, *Contextual Factors in Education: Improving Science and Mathematics Education for Minorities and Women* (Madison: Wisconsin Center for Educational Research, 1987), pp. 12–13.

TABLE 11.4

SOME CHARACTERISTICS OF STUDENTS AND SCIENCE TEACHING

Analytical/Instrumental	Nurturative
Interest in rules	Interest in relationships
Interest in machines	Interest in people
Interest in fairness and justice	Pragmatism
Views world as hierarchy of relationships (competitive)	Views world as network of relationships (cooperative)
Emphasis on analytical thought	Emphasis on aesthetic appreciation
Interest in controlling inanimate things	Interest in nurturing living things

Barbara Smail, "Organizing the Curriculum to Fit Girls' Interests," in *Science for Girls?* Alison Kelly, ed. (Milton Keynes, England: Open University Press, 1987), p. 83.

- Instructions on how to make something, such as an electric motor
- Descriptions of solutions to problems in unusual situations (building a distillation apparatus from bits and pieces available on a desert island)
- Letters to friends describing experiments or concepts
- Letters to newspapers expressing ideas on the social implications of science (e.g., the dumping of nuclear wastes, seal culling, vivisection)
- Newspaper reports (e.g., of an acid tanker accident, hazards, and cleanup)
- Advertisements (e.g., for Bunsen burners and other laboratory equipment)
- Imaginative accounts (e.g., the carbon or nitrogen cycle from the point of view of an atom involved, expeditions into ears or the blood stream)
- Diaries
- Script for TV interviews of inventors
- Poems on scientific phenomena

Social Arrangements

Another strategy that appears to enhance females' interest in science and contribute positively to academic performance is altering the work arrangements in the classroom. There is considerable evidence that cooperative, mixed-ability group processes enhance learning and cognitive development in some circumstances.[47] However, there is a potential problem in the use of student-led small groups, as has been pointed

[47]Cole and Griffin, *Contextual Factors in Education*, p. 33.

out by Cohen. Although learning of content in a small peer group is positively related to frequency of interaction, frequency of interaction is correlated with social status in the classroom.[48] There is the danger that the high-status students will dominate peer groups, leaving out the low-status students, who often are students with low socioeconomic status and low ability.[49]

Proponents of cooperative learning have devised management practices in which leadership roles in peer groups are rotated, thereby giving students of varying status the opportunity to play out leadership roles.

Confidence Building

Skolnick, Langbort, and Day have described strategies that teachers can implement to build confidence in the ability to solve science and (mathematics) problems. As they point out, expert problem solvers *believe they can do it*. It is crucial that middle and junior high school science teachers employ these methods, since this is the time that students will be influenced into or out of mathematics and science courses at the high school level. They suggest five strategies that they think will build the self-esteem that both boys and girls need to have in being science and mathematics problem solvers:

- Success for each student
- Tasks with many approaches
- Tasks with many right answers
- Guessing and testing
- Estimating[50]

Sex-Role Awareness

One way of discovering that students perceive science roles as masculine is to have students draw pictures of scientists (see Inquiry Activity 1.3). Kahle reported that only 10 percent of rural American boys and 28 percent of the girls drew women scientists.[51] Males and females think that science is a male profession. To help students see new possibilities the teacher must take an active role in a number of awareness-type activities so that the options are multiplied for students. Skolnick, Langbort, and Day suggest two categories of strategies:

- Content relevance
- Modeling new options

Content relevance means choosing and using science content to help students see the relevance of science in their lives now, as well as a potential career option later. For instance, Smail reports that science programs should capitalize on the fact that girls who had a positive view of the effects of science on the environment and humans were more likely to choose courses in the physical sciences. She points out that in Britain, new curriculum goals emphasized

- A study of those aspects of science that are essential to an understanding of oneself, and of one's personal well-being. The renewed emphasis on "human biology" for the middle school science curriculum is an example of putting this goal into practice.
- A study of key concepts that are essential to an understanding of the part science and technology play in the postindustrial and technological society.
- Appreciation that technologies are expressions of the desire to understand and control the environment and that technologies change in response to changing social needs. This goal and the previous goal underscore the importance of emphasizing the science-technology-society theme (Chapter 6) in science education.[52]

Content relevance and modeling new options can be directly tied to career awareness. Connecting science courses to careers in science and engineering is an important content dimension of all science courses. Skolnick, Langbort, and Day suggest how the content in a science or mathematics course can do this:

> The content of math and science problems can provide a fundamental kind of career education. It can develop new perspectives on work and on the abilities of both sexes. Word problems can inform children about occupations unfamiliar to them and provide information about the labor market, such as jobs and salaries, women's position in the work force, and math and science courses they will need for particular careers. Illustrating many kinds of content relevance, creative math and science problems encourage girls' pursuit of nontraditional occupations.[53]

Although there are a number of excellent resources to help plan activities, the teacher in the final analysis is the most important. The day-to-day modeling that the teacher provides will overshadow any "special" program that is provided by the school or science department.

One program that teachers should investigate for ideas about planning activities in their own classroom is

---

[48]Elizabeth Cohen, *Designing Groupwork* (New York: Teachers College Press, 1988).

[49]See Cole and Griffin, *Contextual Factors in Education*, pp. 24–42.

[50]Skolnick, Langbort, and Day, *How to Encourage Girls*, pp. 49–50.

[51]Kahle, "Real Students Take Chemistry and Physics," p. 131.

[52]Smail, "Organizing the Curriculum to Fit Girls' Interests," in *Science for Girls?* Alison Kelly, ed., p. 84.

[53]Skolnick, Langbort, and Day, *How to Encourage Girls*, p. 56.

SPACES (Solving Problems of Access to Careers in Engineering and Science).[54] This program was developed at the Lawrence Hall of Science and includes math and science activities for elementary and secondary students designed to stimulate students' thinking about scientific careers, develop problem-solving skills, promote positive attitudes toward the study of mathematics, increase interest and knowledge about scientific work, strengthen spatial visualization skills, and introduce language and familiarity with mechanical tools. The topics developed in the SPACES program include

- Design and construction
- Visualization
- Tool activities
- Attitudes and personal goals
- Job requirements and descriptions
- Women in careers

SCIENCE FOR ALL
EXCEPTIONAL STUDENTS

The science teacher can have a powerful impact on the exceptional student, whether physically disabled, behaviorally disordered, at risk, or gifted and talented. Understanding, indeed empathizing with people with disabilities, is the first step in expanding the theme that science is for all. For instance, read what Barbara Mendius, who has poliomyelitis, has to say about her experiences in school science.

I acquired my entire formal education while in a wheelchair. Since sixth grade my major academic interest has been science; this culminated with a M.S. of Biology from the University of Illinois in October, 1977.

In considering my own education, I firmly believe that my laboratory experience was most important in shaping my scientific ability. The first major obstacle for the handicapped student of science is getting the hands-on laboratory experience so important to engender scientific expertise.

I feel lucky indeed as I recall the variety of lab work which I performed in school. I have thought carefully about the factors which contributed to my successful scientific education; it all comes down to people—parents, administrators, teachers—willing to cooperate in my behalf. Realizing the value of scholarship, my parents took an active interest in my education. Active, but not pushy. Beginning in fourth grade I attended the local public school. For class field trips, mom would drive one of the groups to the museum, or to the nature center—for laboratories are not only found in schools. Dad went to all the parent's nights, talked with my teachers, and came home glowing about my progress. Impressed with my parents' interest and my ability, school administrators were wonderfully cooperative. For some of the administrators, I was their first and only handicapped student.

That all changed in high school. There, all of my parents' interest, and all of the administrators' cooperation would have been wasted were it not for enthusiastic science teachers who gave me the freedom to do as much as I could of what everyone else was doing. Sometimes it only meant putting the microscope or analytic balance on a low table. Sometimes it meant rearranging the greenhouse so I could get down the overgrown aisles. In one case it meant encouraging this shy student to enter the state science fair and helping me choose an appropriate experiment which I could carry out myself. My teachers were the ones who ultimately placed science within my reach.

But we worked together, so my stubbornness and perseverance deserve some credit also. Science had piqued my brain; I was determined to learn as much as I could, actually doing as much as I could. I realized that if I wanted to do the acid/base experiments I would have to show that I could carry solutions around in my lap without spilling. If I wanted to fire-polish my own glassware I had to show I could use a Bunsen burner without setting myself aflame. If I needed to move a microscope to a lower table, I had to show that I could do that without smashing it to smithereens. I had to prove myself all along the way, but my teachers accepted my physical abilities and, although I often caught a watchful eye on me, they did not stifle my enthusiastic investigations.

In summary, my major recommendation for science education is to involve the orthopedically disabled student in laboratory experiments. Visual, auditory, tactile, olfactory, and gustatory clues can elucidate scientific principles; ingenuity and perhaps some extra work are all that are required.[55]

Exceptional Students in the Regular Science Classroom

Public Law 94-142, the Education for All Handicapped Children Act, was enacted in 1975. It is the law that most school districts use as the legal requirement for mainstreaming disabled students into the regular classroom. Although this law was preceded by Public Law

[54]Sherry Fraser, *SPACES: Solving Problems of Access to Careers in Engineering and Science* (Palo Alto, Calif.: Dale Seymour, 1982).

[55]Greg P. Stefanich et al.,"Addressing Orthopedic Handicaps in the Science Classroom," Educational Resources Information Center, ED 258 802, 1985, pp. 13–15.

FIGURE 11.9 Disabled students actively engage in science activities. © Daemmrich.

93-380 (Safeguards for the Rights of Handicapped Individuals) and Public Law 93-112 (the Rehabilitation Act of 1973, which maintained that handicapped individuals must not be excluded from the benefits of any organization receiving federal funds), it is Public Law 94-142 that is of most importance in this discussion of exceptional students in science education. The law was inacted to ensure that:

> all handicapped children have available to them...a free appropriate public education which emphasizes special education and related services designed to meet their unique needs, to assure that the rights of handicapped children and their parents or guardians are protected, to assist states and localities to provide for the education of all handicapped children, and to assess and assure the effectiveness of efforts to educate handicapped children.[56]

The federal regulations for disabled students require that all disabled students be provided the most appropriate education in the least restrictive environ-

[56]Public Law 94-142 (Education of the Handicapped Act of 1975). *Federal Register* 42, 163 (August 23, 1977).

ment. There are a number of principles that you should be aware of when dealing with exceptional students in the science classroom.

The Right to Due Process of Law

All students have a right to a least-restrictive learning environment, and the federal laws protect students and parents by insisting on a specific procedure that must be followed in all cases. This includes notification in writing before evaluation begins, parental consultation, the right to an interpreter or translator if needed, the right of the parent to inspect all educational records, parental consultation, and a procedure for appeals.

Protection in Evaluation Procedures

Each state and school district must ensure that a nonbiased, meaningful evaluation procedure is in place. For instance, tests and other evaluation materials must be administered in the student's native language, unless it is clearly not feasible to do so.

Education in the Least Restrictive Environment

According to Public Law 94-142, it is required "that special classes, separate schooling, or other removal of handicapped children from the regular educational environment occurs only when the nature or severity of the handicap is such that education in regular classes...cannot be achieved satisfactorily."[57] The concept of least restrictive environment must be determined for each student on an individual basis.

Individualized Educational Program (IEP)

The requirement for an IEP is a way of ensuring that each exceptional student is considered as an individual and is not simply placed in a class and not treated in accordance with his or her needs. The IEP is regulated by the following procedures:

1. A statement of the present levels of educational performance
2. A statement of annual goals, including short-term instructional objectives
3. A statement of the specific educational services to be provided to this handicapped student and the extent to which such student will be able to participate in regular educational programs
4. The projected date for initiation and anticipated duration of such services
5. Appropriate, objective criteria and evaluation procedures and schedules for determining, on at least an annual basis, whether instructional objectives are being achieved[58]

Parental Involvement and Consultation

According to the laws surrounding the exceptional student, parents should play an important and integral part in the education of their children. The law makes it necessary for schools to involve parents in planning educational programs for students with special needs, and to become aware and be sensitive to the value of parental involvement.[59]

Cooperation and Joint Planning with Special Educators

Since the law requires that students with special needs be placed in the "least restrictive" environment this means the regular classroom for most exceptional students. Joint planning between the regular classroom teacher and the special educator is an intent of the law and is evident in the requirement that both educators be involved in the development of the student's IEP.

The Exceptional Student in the Science Classroom

All students that enter the science classroom should be accepted for themselves, be included, and be allowed to participate as fully as everyone else. One of the most significant factors mitigating against full participation of disabled students in the mainstream is lowered expectations. Lowered expectations of teachers, combined with lowered expectations of parents, can contribute to the problem. Stefanich puts it this way:

> Parents of the severely disabled support this lowered expectations of teachers for two basic reasons: (1) a lack of confidence that their handicapped child may someday be able to find employment and become a useful part of society and (2) a hesitancy to place increased demands on the child because they perceive that the hard work and time necessary for adequate academic functioning would deprive the child of leisure time activities or physical therapy.[60]

In a report by the Task Force on Women, Minorities, and the Handicapped in Science and Technology, it was estimated that 22 million Americans of working age have some physical disability, yet only 7.2 million of these are employed. The National Science Foundation found only 94,000 disabled scientists and engineers working in 1986. The report stressed that people with disabilities live longer and are able to pursue careers because of improving medical technology. People with disabilities are a large and growing segment of the population. In fact, in 1987 over 1.3 million of the 12.5 million students (or 10.5 percent) enrolled in postsecondary education institutions reported having at least one disability, which makes them the largest "minority."[61]

The report also pointed out that low expectations and lack of encouragement are keeping students with disabilities from participating fully in mathematics and science, particularly in science laboratory

---

[57]Bill R. Gearheart, Mel W. Weishahn, and Carol J. Gearheart, *The Exceptional Student in the Regular Classroom* (Columbus, Ohio: Merrill, 1988), p. 27.

[58]Gearheart, Weishahn, and Gearheart, *The Exceptional Student*, p. 29.

[59]Gearheart, Weishahn, and Gearheart, *The Exceptional Student*, p. 33

[60]Stefanich et al.,"Addressing Orthopedic Handicaps", pp. 1–2.

[61]*Changing America: The New Face of Science and Engineering, Final Report* (The Task Force on Women, Minorities, and the Handicapped in Science and Technology). ERIC Document ED 317386, 1989, p. 26.

courses. Parents, science and mathematics teachers, and counselors must encourage students with disabilities to pursue the study of, and careers in, science and engineering.

What can the science teacher do to provide for the special needs of exceptional students? How can students with handicaps be encouraged to pursue science courses and careers? What can the science teacher do to build the self-esteem of students? There are a number of broad guidelines that the science teacher can follow:

1. Obtain and read all relevant information and health background information available on the student
2. Educate yourself on the physical or psychological nature of the handicap, and how it affects the student's potential for learning
3. Contact the special education expert in your school to determine the source of help available
4. Determine the special equipment needed by the student[62]
5. Talk with the student about limitations due to his or her handicap and about particular needs in the science class
6. Establish a team of fellow teachers (including resource teachers and aids) to share information and ideas about the special students
7. Be aware of barriers, both physical and psychological, to the fullest possible functioning of the disabled student
8. Consider how to modify or adapt curriculum materials and teaching strategies for the disabled student without sacrificing content
9. Encourage the disabled student. Teachers' perception of students' abilities can be self-fulfilling prophecies
10. Educate other students about handicaps in general, as well as about the specific handicaps of students in their class. It is wise to confer with the disabled student before making a decision to take this action[63]

Exceptional students should be encouraged to participate fully in science experiences, and the science teacher must make special provisions for this to happen. If one of the important goals of school science is to develop scientific thinking skills to inquire

about natural phenomena, the exceptional student must be included in inquiry and laboratory activities, just as the other students.

In the sections to follow we will explore the following exceptionalities: physically impaired students (motor or orthopedic, visual, and hearing disabilities), speech impaired students, learning-disabled students, gifted and talented, and at-risk or low-achieving students.

Physically Impaired Students

Physical impairments represent a broad category of conditions including nervous system disorders (cerebral palsy), mental disorders (mental retardation), muscular-skeletal disorders (rheumatoid arthritis), cardiovascular disorders (coronary heart disease), respiratory disorders (emphysema), digestive disorders (cancer of the colon), urogenital disorders (kidney disease), endocrine-metabolic disorders (diabetes mellitus), and sense organ disorders (auditory and visual impairment). Only three of the impairments will be discussed here: motor impaired or orthopedically disabled, visually impaired, and hearing impaired.

Motor or Orthopedically Impaired Students

The orthopedically disabled student is one who has an impairment that interferes with the normal functioning of the skeletal system, the joints, the connective tissue, and muscles.

An assessment technique that allows the science teacher to determine an appropriate mitigative strategy for a particular student is the "life function impairment assessment," which interrelates general disability areas (motor-orthopedic, behavior, chronic disease, auditory, and visual) with five categories of functional impairment (health, mobility, communication, social-attitudinal, and cognitive-intellectual).

Figure 11.10 shows a life function impairment assessment for a particular male student with spina bifida. The student is wheelchair-bound, although he can use crutches for short distances and can transfer to a chair or bench from the wheelchair. The chart shows a problem in the social-attitudinal life function. In this case it is observed he has trouble keeping friends; a urine bag is necessary and this has contributed to an odor problem; he tends to use the wheelchair aggressively. In terms of cognitive-intellectual life function, he has lowered grades, despite an average IQ.

The following are some strategies that teachers can use to help physically disabled students in the science class:

[62]E.C. Keller et al. *Teaching the Physically Disabled in the Mainstream Science Class at the Secondary and College Levels.* (Morgantown, W. Va.: Printech, 1983).

[63]Based on Rodger Bybee, "Helping the Special Student to Fit In," *Science Teacher,* October, 1979, pp. 23-24.

Life Function	General Disability Area				
	Motor/ Orthopedic	Behavior	Chronic Disease	Auditory	Visual
Health	Some Problem [a]				
Mobility	Some Problem [b]				
Social/ Attitudinal	No Problem				
Communication	Significant Problems [c]				
Cognitive/ Intellectual	Some Problems [d]				

FIGURE 11.10 Life Function Impairment Assessment for a Male Student with Spina Bifida.

a. Use of urine bag is necessary. Time must be scheduled for emptying.

b. Most of the time is wheelchair bound. Can use crutches for short distances and can transfer to chair or bench from wheelchair.

c. Has great difficulty making and keeping friends. Part of the problem is that urine odor causes avoidance by peers. Uses wheelchair aggressively, makes inappropriate use of vulgar language. Disrupts class with outbursts and unrelated information. General lack of self-discipline and respect for authority.

d. Problem of depressed grades, despite average I.Q. No indication that this problem is directly due to disability, but he was a home-bound student through the 4th grade.

Based on an assessment of a student in a marine science summer program, from E. C. Keller et al., *Teaching the Physically Disabled in the Mainstream Science Class at the Secondary and College Levels* (Morgantown, W.Va.: Printech, 1983), p. 41.

1. Provide for assistance (if necessary) from other people for such things as pushing a wheelchair or taking notes for the disabled student
2. Be aware of and use mechanical devices as needed such as a wheelchair lapboard, a tape recorder, a voice synthesizer, electric hot plates instead of Bunsen burners, or pencil holders
3. Employ teaching strategies such as allowing more time to complete assignments, nonmanual types of teaching techniques, breaks for stretching, and storing materials so as that they are accessible to all
4. Evaluate the architectural facilities of the classroom. Consider access to wheelchairs, lowering chalkboards, altering the height of desks and lab tables, and the use of nonskid floors for students with crutches
5. Use a variety of student presentations including oral and written. Use nonverbal signals such as blinking, head nodding, or a pointer[64]

Visually Impaired Students

In working with students with visual impairments, Kenneth Ricker points out that the emphasis should be on the student, not the visual impairment. He points out that we should place the emphasis on teaching science to students who happen to have some sort of visual impairment.[65]

Visually impaired students include both students with limited vision and the totally blind. Ricker suggests that science teachers can be aided if they know how the student functions effectively in learning situations that involve four different levels of tasks:

1. Close-up tasks such as reading a book, drawing a diagram, or using a microscope
2. Nearby tasks at an arm's length, such as manipulating objects or handling lab equipment
3. Distant tasks such as reading what is on the chalkboard or viewing what is on the screen
4. Mobility tasks such as moving to different places in a lab or going on field trips[66]

One of the important aspects of working with students who are visually impaired is to make sure that they get to perform hands-on science activities. A number of science teachers and researchers have developed a variety of strategies to encourage hands-on science teaching for the visually impaired. Here are some ideas that science teachers can use to help visually disabled learners become science learners.

[64]E. C. Keller et al., *Teaching the Physically Disabled in the Mainstream Science Class at the Secondary and College Levels* (Morgantown, W.Va.: Printech, 1983), pp. 57–59.

[65]Kenneth S. Ricker, "Science for Students with Visual Impairments," Educational Resources Information Center (ERIC), Document ED258800, 1985.

[66]Ricker, "Science for Students with Visual Impairments," p. 11.

1. Keep materials, supplies, and equipment in the same place, whether it is on a laboratory table or in a field kit
2. Describe and tactually-spatially familiarize the student with all equipment
3. Describe and tactually-spatially familiarize the student with the classroom and laboratory
4. Have the student do a trial run before the activity
5. Enforce safety rules rigidly, since the visually impaired student is more apt to work closer to the activity and laboratory table
6. Use simple adaptive procedures on equipment
7. Use tactile models to show visually impaired students what you visually show nonimpaired students
8. Convert color change indicators in chemistry laboratories to tactile indicators by filtering the precipitate and having the student touch the precipitate on filter paper
9. Use a thermoform machine to make multiple, reproducible tactile diagrams
10. Make arrangements for tactile examination if touch is not normally permitted
11. Present examinations in a form that will be unbiased to visually impaired students
12. Use various mechanical devices such as a tape recorder, an overhead projector, a slide previewer, and raised-line drawings
13. Nondominant laboratory partners can facilitate the work of the visually impaired during hands-on activities, and while working in the lab[67]

Hearing-Impaired Students

Hearing-impaired students may have either partial or total hearing loss. One of the implications of having a hearing loss is the constant problem of not being able to hear, either completely or partially, human voices and environmental sounds. Hearing loss can also affect the individual's language development, depending on the time of the onset of the hearing loss. Speech and voice quality may also be affected by hearing loss. Therefore, the teacher should be aware that cognitive, personal, and social development might also be affected by a hearing impairment.[68]

The teacher can help the hearing impaired student in science class by being aware of the following.

1. In considering the physical environment, place the student within 15 feet of the teacher. The student should also be placed with his or her better ear toward the class, and try to avoid placing the student near congested areas, or where there are excessive vibrations and noise.
2. To help the student with communication, ensure the presence of an interpreter where needed. Communicate directly with the student, and use note writing and the chalkboard for one-to-one discussion. Speak clearly, at a moderate pace, and without exaggeration or overenunciation, while making eye contact.
3. Teaching strategies such as providing an outline of the lesson or activity, as well as step-by-step directions prior to a lab activity, are helpful. Use multiple sensory activities, especially visual, to reinforce vocabulary and science concepts.
4. Encourage students to ask questions, make use of glossaries, and take notes on major points of the lesson or activity.
5. Arrange for a nondominant lab partner for helping with communication.[69]

Learning-Disabled Students

According to the U.S. Department of Education, the number of students receiving special education through federal programs rose from 3.7 million in 1976–1978 to 4.4 million in 1986–1987, a 19 percent increase. The number of students receiving services for learning disabilities, currently the largest category of handicap, increased by more than 140 percent.[70]

A specific learning disability means a disorder in one or more of the basic psychological processes involved in understanding or in using language, spoken or written, which may manifest itself in an imperfect ability to listen, think, speak, read, write, spell, or do mathematical calculations. It is estimated that 2 percent of adolescents have some type of learning disability, making this the most common type of handicap. Learning disabilities also include perceptual handicaps, brain injury, minimal brain dysfunction, and dyslexia. The term does not include students who have learning problems that are fundamentally caused by visual, hearing, or motor handicaps, or by mental retardation, emotional disturbance, or environmental, cultural, or economic disadvantage.[71]

However, a controversy exists around the identification of learning disabled. Some specialists claim that at least half the students labeled as learning disabled could be more accurately labeled as slow learners, as students with second-language backgrounds,

---

[67]Keller et. al., *Teaching the Physically Disabled*, pp. 19–20.
[68]Keller et. al., *Teaching the Physically Disabled*, p. 27.

[69]Keller et. al., *Teaching the Physically Disabled*, pp. 38–39.
[70]John O'Neil, "How 'Special' Should the Special Ed Curriculum Be?" *ASCD Curriculum Update*, September, 1988, p. 1.
[71]Gearheart, Weishahn, and Gearheart, *The Exceptional Student*, p. 271.

as students who misbehave in class, as those that are absent often or move from school, or as average learners in above-average school districts.[72]

Another controversy exists surrounding what type of intervention program should be used to help learning-disabled students. One concern that specialists at the secondary level have is the emphasis recent reforms have placed on academic goals. Secondary teachers working with learning-disabled students will have to work harder to help these students keep up with the demands of the curriculum. One remedy is to balance the curriculum for learning-disabled students with what is called a "functional" approach, in which students are taught life or living skills that will aid the student in employment opportunities beyond high school.

What can be done in the science classroom to help learning-disabled students? Charles Coble and colleagues suggest that focusing on structure in the science classroom will benefit the learning disabled-student. They suggest that structure is important because of the perceptual and cognitive difficulties that the learning-disabled student has in being unable to mask out extraneous stimuli to deal with the task at hand. They suggest a number of ways to promote structure:[73]

1. Establish a routine
2. Limit choices
3. Ensure that the student is attending
4. Give the student clues to facilitate remembering
5. Sequence instruction
6. Be specific in criticism and praise
7. Provide visible time clues
8. Confer with the LD teacher
9. Develop empathy

Science education has much to offer the learning-disabled student. Providing a hands-on curriculum within a structured environment can provide the twin needs of the learning disabled: structure and motivation. For example, Coble and colleagues outline some ways to modify science instruction for some learning-disabled characteristics (Table 11.5).

The Gifted and Talented in Science

Gifted students, like students with learning disabilities, or students with physical disabilities, have special learning needs that require a special education. Giftedness

has to do with people who have special gifts that are in some way superior to others of the same age. In science education, gifted education has been variously known as gifted in science, science talented, science prone, or students with high ability in science. The Westinghouse Talent Search, science fairs, and advanced placement

TABLE 11.5

MODIFYING SCIENCE INSTRUCTION FOR LEARNING-DISABLED STUDENTS

LD Student Characteristic	Modification in Science Classroom
The LD student may lack coordination: drop equipment or may cut self using strippers.	Use plastic containers; teach the use of wire strippers.
Equipment and materials may distract the impulsive student.	Clear desk of extraneous equipment. Use one item at a time and collect each after use. Package materials for activities that reduce distractions.
Impressed with the end product, the LD student misses the scientific concept.	Ask questions that emphasize cause-effect relationships. Ask a series of sequential questions.
LD student generalizations may be incorrect.	Anticipate possible misconceptions and discuss differences before errors are made.
The LD student may have difficulty comprehending abstract terms and concepts.	Begin at the concrete level and move step-by-step to the abstract level. Use demonstrations, models, and pictures.
The LD student may have a reading disability.	"Rewrite" the text using a highlighter to emphasize important information. Teach "mapping" strategies to give students an alternative way of "reading" text material. Have volunteers tape the textbook.
The LD student may lack exploratory drive.	Used forced/interesting choice strategies. Give students structured options.
The LD student may explore in a random, purposeless manner.	Use the direct-interactive teaching model.

After Charles R. Coble, Betty Levey, and Floyd Matheis, "Science for Learning Disabled Students," *Educational Resources Information Center (ERIC)*, ED258803, February, 1985, p. 20.

[72]O'Neil, "How 'Special' Should the Special Ed Curriculum Be?" pp. 1-3.
[73]Charles R. Coble, Betty Levey, and Floyd Matheis, "Science for Learning Disabled Students," *Educational Resources Information Center (ERIC)*, ED258803, February, 1985, pp. 7–10.

courses in science are examples of the emphasis on gift-edness in science education.

There are a lot of myths, and there is as well a low level of tolerance for giftedness in society. These myths and problems are similar to the kinds of myths and problems surrounding students with disabilities. First, let's examine what giftedness is, and the myths surrounding giftedness, then move on to characteristics of giftedness in science and some examples of ways of working with gifted students in science education.

A Definition of Giftedness

There are a number of definitions that focus on the notion of performance. One of the earliest definitions states that gifted and talented students are those identified by professionally qualified persons as being capable of high performance. Students capable of high performance include those with demonstrated achievement or potential ability in any of the following areas:

1. General intellectual ability
2. Specific academic aptitude
3. Creative or productive thinking
4. Leadership ability
5. Visual and performing arts[74]

A more recent definition of giftedness, and one that is widely accepted, suggests that gifted students are those who have demonstrated or shown potential for

1. High ability (including high intelligence)
2. High creativity (the ability to formulate new ideas and apply them to the solution of problems)
3. High task commitment (a high level of motivation and the ability to see a project through to its conclusion)[75]

One of the strengths of using a multiple-criterion definition is that it expands the potential pool of gifted students, and multiple abilities seem to be characteristics of gifted people in practice. Hallaman and Kaufman point out that if one were to use only IQ, the most "gifted" would come from high socioeconomic families, those with fewer siblings, and better educated parents. Furthermore, the notion of intelligence has changed dramatically in recent years. Work by Gardner at Harvard has suggested evidence of multiple intelligences

rather than a single general intelligence. Gardner suggests seven intelligences:[76]

1. Logical-mathematical
2. Linguistic
3. Musical
4. Spatial
5. Bodily-kinesthetic
6. Interpersonal
7. Intrapersonal

Recent work in intelligence, and the reformulation of the definition of giftedness, suggests that giftedness is identifiable only in terms of an interaction between ability, creativity, and task commitment (Figure 11.11).

Two additional aspects of the definition of gifted-ness need to be considered. The first is gifted minority, and the second gifted handicapped.

Gifted minority students have been typically left out of gifted programs, although some strides have been made in recent years.[77] Selecting students for gifted programs needs to take into consideration cultural diversity, language, and parental involvement. Recent suggestions for expanding the talent pool for gifted students include the following practices:

1. Seeking nominations from a variety of persons, professional and nonprofessional, inside and outside school
2. Applying knowledge of the behavioral indicators by which children from different cultures dynamically exhibit giftedness in the development of nomination forms
3. Collecting data from multiple sources, objective and subjective, including performances and products
4. Delaying decision making until all pertinent data can be collected in a case study

[76]Howard Gardner, *Frames of Mind: The Theory of Multiple Intelligences* (New York: Basic Books, 1985).
[77]Mary M. Frasier, "Poor and Minority Students Can be Gifted, Too!" *Educational Leadership*, March, 1989, pp. 16–18.

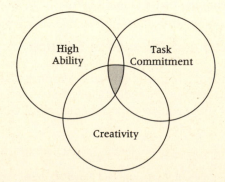

FIGURE 11.11 Giftedness: A Graphic Representation (After Renzulli, Reis, and Smith, *The Revolving Door Identification Model*)

[74]Gearheart, Weishahn, and Gearheart, *The Exceptional Student*, p. 356
[75]J. S. Renzulli, S. M. Reis and L.H. Smith, *The Revolving Door Identification Model*. (Mansfield Center, Conn.: Creative Learning, 1981).

As Hallahan and Kaufman point out, "different cultural and ethnic groups place different values on academic achievement and areas of performance. Stereotypes can easily lead us to overlook intellectual giftedness or to over- or underrate children on any characteristic because they do not conform to our expectations based on their identity or socioeconomic status."[78]

It is estimated that 2 to 3 percent of disabled students are gifted, and it should be a goal to identify these students and make a gifted education available to them. One only needs to look to the field of science to identify a number of eminent scientists who had a severe disability (Charles Steinmetz, Stephen Hawking). Phyllis Stearner has described the lives of scientists who were disabled but pursued careers in science and engineering.[79] These vignettes provide role models and support for the science teacher who encourages the identification of gifted handicapped students.

Characteristics of Gifted Students

There are many characteristics of gifted students, and it is this constellation of characteristics that helps describe gifted young in the broadest sense. Myths such as that gifted persons are physically weak, socially inept, narrow in interests, and prone to emotional instability and early decline need to be offset with more accurate characterizations of gifted students.[80]

Gifted students show the following characteristics:

Academically superior
Apply systems approaches
Apply abstract principles to concrete situations
Careless in handwriting and similar routine tasks
Courteous
Deliberately underachieve under certain
  conditions
Difficulty conversing with age-mates
Emotionally stable
Extraordinarily verbal
Generally curious
Give uncommon responses
Have a high energy level, especially in mental and
  intellectual tasks
Have a high vocabulary level
Imaginative
Intellectually curious
Intellectually superior
Inquisitive

Superior in logical ability
Make individualistic interpretations of new subject
  matter
Obnoxious or rebellious when asked to do repetitive tasks
Original
Outspokenly critical of themselves and others
Perceive and identify significant factors in complex
  situations
Persistent in achieving goals
Have good physical ability
Take pleasure in intellectual pursuits
Have good power of concentration
Recognize and are uncomfortable with unresolved
  ambiguity
Rebel against conformity
Scientifically oriented
Seek older companions
Have sense of humor
Sensitive to problems of others
Skip steps in normal thinking sequence
Socially aware
Have strong sense of responsibility
Superior in ability to remember details
Superior in ability to see relationships
Unhappy with most group-participation projects
  (with normal peers)
Unaccepting of routine classroom rules
Have an unusually good memory
Verbally facile
Verbally flexible

Science Learning Environments for the Gifted

Schools have developed a variety of programs for gifted students. Generally the programs are designed to provide enrichment (providing additional experiences without moving students to higher grades) or acceleration (moving students ahead of their age-mates). The following represent the range of program opportunities for gifted students:

1. Enrichment in the regular classroom
2. Consultation teacher program in which a differentiated program is provided in the regular classroom with the advice of a specialty teacher
3. Resource room-pullout program
4. Community mentor program
5. Independent study program
6. Special class
7. Special school[81]

[78]Daniel P. Hallahan and James M. Kaufman, *Exceptional Children* (Englewood Cliffs, N.J.: Prentice-Hall, 1991), p. 424.
[79]Phyllis Stearner, *Able Scientists—Disabled Persons: Careers in the Sciences,* (Oakbrook, Ill.: John Racila, 1984).
[80]Gearheart, Weishahn, and Gearheart, *The Exceptional Student,* p. 359.
[81]Hallahan and Kaufman, *Exceptional Children,* pp. 426–427.

In this NSTA publication *Gifted Young in Science: Potential through Performance*, Passow also says this about learning environments for the gifted in science:

> Those students who manifest the special interest, motivation, creativity, critical judgement, high intelligence, and other indicators of potential for becoming productive, practicing scientists need opportunities to develop their areas of specialized talent. These opportunities should go beyond the science-technology education experiences provided all gifted students. Students with special talents in science—some latent and some manifest—need differentiated, appropriate experiences in particular contexts or settings.[82]

Passow's suggestion raises the question of special environments for the "gifted in science" versus working with the gifted student in the regular science classroom. For the gifted in science, there are specialty schools within school districts, called "magnets," for science-prone students. Students in these schools are still involved in a general education, but special emphasis is placed on their high ability and interest in science. Some districts and states have created science high schools. For example, a curriculum for students in such a school would involve in-depth study and seminars in mathematics, science, and computer science (Figure 11.12).

There are a number of strategies that the science teacher can use to work with gifted students in science classes. Both individual and cooperative learning strategies are recommended, as well as providing opportunities for gifted students to engage in science inquiry and problem-solving activities.

1. Provide opportunities for self-directed learning—opportunities to assume responsbility for defining and choosing the direction for solving problems and studying topics
2. Engage gifted students in cooperative learning. Although there is controversy surrounding placing gifted students in cooperative teams with "normal" students, Johnson and Johnson provide evidence to support teachers' engaging gifted students in cooperative learning activities in the regular classroom[83]
3. Engage gifted students in high-level problem-solving activities

4. Encourage gifted students to get involved in science fair and olympiad projects
5. Modify the curriculum (curriculum compaction) for gifted students so that they may pursue some topics in more depth, or seek a breadth of topics
6. Use learning contracts (either individually or in pairs), enabling gifted students to break away from the regular content of the course. Students can be encouraged to report back to the entire class

At-Risk Students in Science

In their book *At-Risk, Low Achieving Students in the Classroom*, Lehr and Harris point out that there have been many terms used to characterize students who are labeled "at-risk." Some of these terms include *disadvantaged, underachiever, low ability, slow learner, low socioeconomic status, language-impaired,* and *dropout prone.*[84]

We might think of at-risk students as low-achieving students whose poor performance hinders subsequent success and frequently leads to withdrawal from school.[85] At-risk students typically have low self-esteem, and for school related and academic performance the situation is only exacerbated with continued failure and difficulties in school. Students who are at risk in science class will typically show these characteristics: academic difficulty, lack of structure, inattentive, easily distracted, short attention span, excessive absenteeism, dependence, discipline problem, lack of social skills, inability to face pressure, fear of failure, and apparent lack of motivation.[86] How can student's with these kinds of problems be helped? What can science teachers do to motivate these students to succeed in science?

There is a great deal of controversy surrounding what kinds of interventions are effective with at-risk students. The controversy focuses on remedial instruction versus instruction aimed at high-order thinking. Most educational programs for at-risk students stresses "small mechanical skills and rote memory and regurgitation."[87] The reason for this emphasis is the psychological theory that many educators use to explain the "nature" of at-risk students. The prevailing theory sug-

[82]A. Harry Passow, "The Educating and Schooling of the Community of Artisans in Science," in *Gifted Young in Science: Potential Through Performance*, Paul F. Brandwein et al., eds. (Washington, D.C.: National Science Teachers Association, 1988), p. 31.

[83]Roger T. Johnson and David W. Johnson, "Cooperative Learning and the Gifted Science Student," in *Gifted Young in Science: Potential through Performance*, Paul F. Brandwein and A. Harry Passow, eds. (Washington, D.C.: National Science Teachers Association, 1988). pp. 321–329.

[84]Judy Brown Lehr and Hazel Wiggins Harris, *At-Risk, Low-Achieving Students in the Classroom* (Washington, D.C.: National Education Association, 1988), p. 9.

[85]Definition suggested by Daniel U. Levine, "Teaching Thinking to At-Risk Students: Generalizations and Speculations," in *At-Risk Students and Thinking: Perspectives from Research*, Barbara Z. Presseisen, ed. (Washington, D.C.: National Education Association, 1988), p. 117.

[86]Lehr and Harris, *At-Risk, Low-Achieving Students*, p. 11.

[87]Levine, "Teaching Thinking to At-Risk Students," p. 118.

Year	Mathematics	Science	Seminar	Computer Science
Grade 9	*Course sequence*			
1st semester	Magnet Geometry A&B (1 credit)	Advanced Science 1 Physics (1 credit)	Research and Experimnentation Techniques for Problem Solving 1 including:	Fundamentals of Computer Science A (½ credit)
2nd semester	Magnet Functions A&B (1 credit) or	Advanced Science 2 Chemistry (1 credit)	Probability and Statistics Research methods (½ credit)	Fundamentals of Computer Science B (½ credit)
Grade 10				
1st semester	Magnet Precalculus A,B,C (1½ credits)	Advanced Science 3 Earth Science (½ credit)	Research and Experimentation Techniques for Problem Solving 3 (½ credit)	Algorithms and Data Structures A (½ credit)
2nd semester	Analysis I A&B (1 credit)	Advanced Science 4 Biology (1 credit)	(½ credit)	Algorithms and Data Structures B (½ credit)
Grade 11				
1st semster	Analysis II (½ credit) or	Advanced mini-courses, research, internships, university courses, special topics, cooperatives, etc.	Research and Experimentation Techniques for Problem Solving 3 (1 credit)	Advanced topics in semester and mini-courses, university study, special topic sessions, projects
2nd semester	Linear Algebra or Discrete Mathematics or			
Grade 12	Excursion Topics in Math or	(variable credit)		(variable credit)
1st semester	Guided Research, Internship Cooperatives, University courses, etc.	Examples: Climatology, Tectonics, Metallurgy, Cellular physiology, Biomedical Seminar, Thermodynamics, Optics, Cooperatives	Guided senior project involving research and/or development across discipline lines (1 credit)	Examples: Analysis of Algorithms, Graphics, Survey of Languages, Computer Architecture & Organization, Game Theory
2nd semester				

NOTE: Elective courses for grades 11, 12 include options like: Advanced Placement courses, Game Theory, Topology, Mathematical Programmimg, Abstract Algebra , Cooperative Languages, Robotics, Computer Architecture, Systems Design, Organic Chemistry, Quantitative analysis, Astrophysics, Plant Physiology, Behavior and Brain Chemistry, Calculus in Biology/ Ecology

SOURCE: Montgomery Blair High School, Silver Spring MD, September 1987

FIGURE 11.12 Curriculum of a Mathematics-Science-Computer Magnet Program. Hemisphere Publishing.

gests that at-risk students fail in school because of cognitive deficits, and because by and large, many of these students come from disadvantaged environments. This framework suggests that education should make up these deficits by providing an educational program that remediates cognition and environment. An alternative view suggests that the cognitive deficit theory is biased or misguided, and that these students are capable of learning and thinking at quite high levels, and that the educational programs that have been provided fail to take into account cultural pluralism.[88]

Recent models for working with at-risk students go well beyond the two views just mentioned. New models emphasize "accelerated learning," teaching high-order thinking skills, focusing on student learning styles, and using enrichment programs rather than remedial programs.

Here are some suggestions that science teachers can use to help at-risk students:

1. *Involve at-risk students in hands-on activities using concrete materials*. These students, perhaps more than any group of students, need to be given opportunities to work with real materials. Too many "activities" for at-risk students involve the use of "worksheets," which is counterproductive vis-a-vis student learning styles.

2. *Emphasize high-order thinking skills*. We have emphasized the importance of teaching students scientific

[88]Barbara Z. Presseisen, "Teaching Thinking and At-Risk Students: Defining a Population," in *At-Risk Students and Thinking*, pp. 19–37.

thinking skills such as observing, inferring, hypothesizing, and designing experiments. At-risk students need to be involved in these kinds of activities, just as should other students. One program that lends support to emphasizing high-order thinking for at-risk students is HOTS (High Order Thinking Skills), developed by Pogrow. The program is designed to improve students' cognitive abilities by involving them in computer activities and Socratic teaching using intellectually challenging activities for middle school students. The program focuses on the process of thinking rather than on curriculum objectives, and the results have shown that at-risk students have high levels of intellectual and academic potential, and that achievement can be increased by focusing on thinking skills rather than on the learning of facts.[89]

---

[89]Stanley Pogrow, "Challenging At-Risk Students: Finds from the HOTS Program," *Phi Delta Kappan,* January, 1990, pp. 389–397.

3. *At-risk students benefit from being members of heterogeneous groups.* Cooperative learning is a crucial learning format for at-risk students. Typical remedial programs, which tend to isolate at-risk students and involve them in individual learning activities (usually worksheets), work counter to the students' real needs. These needs include being able to talk with able learners; to work alongside students who are successful; to cooperate with other members of the class; to learn social skills; to learn how to ask questions—in short, to learn how to learn. Cooperative learning strategies have been shown to raise the academic performance of at-risk students in heterogeneous classes.

4. *Teach at-risk students metacognitive strategies—that is, learning-how-to-learn strategies.* In this approach students are taught how to learn by teaching them planning strategies: mind mapping, brainstorming, questioning skills.

5. *Assess the learning style of at-risk students and plan activities accordingly.* Consideration should be made for students who prefer kinesthetic learning or tactile activities, for example.

Think Pieces

- Why should multicultural education be an integral part of science education? Support this position with at least three references from the literature.
- What steps should be taken to ensure that girls are provided equity in terms of science courses and careers in science and engineering?
- What are the major issues regarding multicultural education?
- How can science teachers counteract the effects of racism and sexism in the science classroom?
- What philosophy should guide the actions that science teachers take with regard to exceptional students in the regular science classroom?   ☐

Science Teachers Talk

"How do you accommodate students with exceptional needs, such as the gifted and talented, hearing impaired, visually impaired, mentally retarded, students with learning disabilities, students with behavioral disorders, or potential dropouts? What have you found to be effective working with these students?"

John Ricciardi. On the high school level, I believe the present system of public education is critically limited in its ability to provide effective and meaningful instruction for its "high risk" population of potential dropouts and students with behavioral abnormalities. This is a frightening problem in America's schools today. Rather than trying to understand and

(Continued on p. 422)

Case Study 1. The Experiment

The Case. A university professor has gained permission to plant genetically engineered seeds in a small farming town. The population of this rural town is composed of a small, highly educated white middle class, a large, poorly educated white population of low socioeconomic status, and a large ethnic minority population of low socioeconomic status, most of whom have a high school education.

The Problem. How do you think the different groups that live in the town will respond to the planting of the genetically engineered seeds? Do you think democratic standards and attitudes will be employed when dealing with the ethnic minority and low-income populations?[90]

☐

[90]Based on Mary M. Atwater, "Including Multicultural Education in Science Education: Definitions, Competencies, and Activities," *Journal of Science Teacher Education* 1, no. 1 (Spring 1989): 17–20.

Case Study 2. A New Buzzword?

The Case. Jack Hannapool, having just returned from a summer institute on strategies to enhance the course and career options for women and minorities in middle and high school science programs, prepared to give a report to the rest of the science department at a special meeting for this purpose. The department is composed of 15 faculty members, 5 women and 10 men. During Jack's report, one of the male teachers bursts out and says, "What is this multicultural stuff? Is this just another buzzword?" Jack responds with, "You're joking!"

The teacher says, "Absolutely not!" Two or three other teachers appear to support this buzzword position. Jack flashes back to what the principal said last spring: "Jack, I want a multicultural policy and program implemented in all departments, and since you're the science department chair, I want you to lead the effort."

The Problem. How would you respond to these three or four teachers who don't seem to see much value in multicultural education? How could you get them on your side?   ☐

(Continued from p. 421)

identify the specific educational needs and learning styles within this population group, too many schools are resorting to solutions which essentially segregate, isolate, and intimidate the individual. It's dehumanizing and psychologically counterproductive.

My basic astronomy course (section offered in our high school, four of which I teach) attracts a fair number of "behavioral disordered" students and potential dropouts. I don't treat these kids any differently other than yielding more to their "scatter braininess." I have virtually no discipline problems in my classes, because, I believe, I first show a sense of trust and respect for the individual's integrity and self-worth. When you allow yourself to do this, when you honor the individual, you just keep moving–facilitating, and soon you watch your instructional objectives flow into place.

Ginny Almeder. I have worked primarily with physically handicapped, learning disabled, and behaviorally disordered students. For the physically handicapped students, I have made appropriate room arrangements to accommodate at least one wheelchair. I integrate the student into a regular lab group and make arrangements for one member of the group to make a carbon copy of notes and lab data. On occasion in the past, the Special Education Department has provided an aide for additional assistance during lab activities.

With the learning disabled students, I identify their learning problem with the assistance of the resource teacher and then work with them in developing skills. This approach involves a a combination of individual attention and peer support from their group. I have found that increasing their

participation in class discussions enhances successful learning.

The behaviorally disordered students are the most challenging. Typically, I work with them individually in developing appropriate classroom behavior. A firm but friendly approach usually works. I have found that most students will respond if they believe that I am concerned about their success.

Jerry Pelletier. At the present time I am working mainly with gifted and talented students. I have found that these students need to be challenged with high-quality thinking materials and strategies. Besides accumulating knowledge and comprehending it, these students must learn to analyze and evaluate this knowledge. In order to meet these special needs I use many different types of materials besides the text in our basic science program. These students are asked to do special science projects, research papers and read literature that is associated with various fields of science. It is my feeling that this will enrich their backgrounds and challenge their analytic thinking. It is not my purpose to demand more of these students in the sense of quantity but to demand more in the sense of quality.

Bob Miller. We have been somewhat successful with student involvement in raising animals for classroom use. Students seem to take pride in raising the animal and telling other students about their experience. The approach seems to fit a variety of students with varying disabilities, physical or emotional.

Mary Wilde. I would first like to focus on how to work with the gifted child in the classroom. I am very much opposed to pulling that gifted child out of the classroom for instruction, particularly at the

middle school level. Isolating those students only creates attitude problems, social problems, and morale problems among teachers who do not have those "sparks" in the classroom. I gear my instruction to those students by providing greater course content and problem-solving activities. I then do a lot of small group work so that these students apply what they have learned by helping others learn the same information. Therefore, more students have benefited by an "enrichment" program. It is really neat to see the interaction among gifted students and those who want to learn but need a little more time and a little more instruction. I do use the advanced child to set my instructional goals, however, I allow for learning rate differences among individual students.

The behavioral problem students always create a most difficult situation in the classroom. When a teacher has these students, the reaction is to pull back the reins, take away all freedom, and develop a very structured classroom so that the students have less opportunities to cause havoc. However, this treatment usually makes me miserable, and the learning that takes place in the classroom is minimal. It is very difficult, but I find I have better success when I provide more hands-on activities and a less structured classroom setting. I must spend more time organizing and planning, for it is important that all directions and expectations are very clear. However, the real key to working with the discipline problem student is that you show real interest in that person. You go out of your way to know everything about that child, so that you can comment and ask that child questions that communicate a caring concern for that person. This helps create positive attitudes and

(Continued on p. 423)

(Continued from p. 422)
respect towards the teacher, thus diminishing disrespectful behavior in the classroom. There are times when you don't think your effort is successful, but with a little patience, the rewards will come.

Dale Rosene. I have discovered that open-ended learning assignments allow gifted as well as learning disabled students to work to their abilities. Some activities that I have found very successful and are characterized by their open-endedness include:

- Design a package to protect a raw egg from a three-story fall
- Design a space station, providing modules for all needs
- Design an invention—something that is really needed—and try to sell it to the class

I have also found that using the format of cooperative learning is very powerful in working with students with different needs and abilities. □

Science for All
by Peter J. Fensham[91]

After a decade of stagnation, science education in many countries is seeing renewed interest and nationally supported curriculum efforts. In 1983 new funds in the U.S. initiated a number of projects. In Britain, the first major project since the sixties, the Secondary Science Curriculum Review, was established in 1982. Even earlier, New Zealand set up a Learning in Science Project in 1979 and, in 1982, a corresponding elementary school project. This new activity extends beyond developed or industrialized countries. In 1984 the Asia region of UNESCO, which ranges from Iran to Japan, endorsed a science education program as one of its few priorities for the rest of the eighties.

In both rhetoric and rationale, these programs share in common a strong emphasis on Science for All. This slogan is a compelling and attractive one in societies where applied science and technology are evident in new products and new forms of communications. New jobs emerge and old, familiar ones disappear. There is a cry for new skills and expertise and a chronic toll of unemployed persons who lack technical skills.

This sort of analysis of the curriculum movement of the sixties and seventies and hence of the present state of science in our schools can be used to begin to define characteristics that may be essential if science education is to be effective as Science for All and other characteristics that are at least worth trying.

First, elite or traditional science education must be confined to and contained within an upper level of schooling. It needs to be identified for what it is: a form of vocational preparation. Containment is not achieved by offering alternatives at the levels of schooling where Science for All is to be achieved. It is no good having a proper science for the few and science for the rest.

Second, science must be reexamined and recognized as a variegated source of human knowledge and endeavor. A wider range of appropriate aspects of science needs to be selected for converting into the pedagogical forms of a science curriculum that will have a chance of contributing to effective learning for the great majority of students.

I have argued that present science curriculums are really an induction into science. The ones that might provide Science for All must involve much more learning about and from science.

These curriculum processes are quite fundamentally different. For instance, in the first we use teachers who have themselves been inducted into an acquaintance with some of the basic conceptual knowledge of an area of science to repeat the first steps of this process with their students. Since the teachers usually have little experience of the exciting practical applications of their knowledge or of the process of trying to extend the knowledge of a science, the induction they offer into the corpus of science is through the same abstract route they followed as part of a former elite who could tolerate it and cope with its learning. Few of their students are interested enough to follow.

In the alternative that I envisage for Science for All, students would stay firmly rooted outside the corpus in their society with its myriad examples of technology and its possibilities for science education. Science teachers as persons with some familiarity and confidence with the corpus of science, will need to be helped to be not inductors but couriers between the rich corpus and their students in society.

Third, some clear criteria must be established for selecting the science that is to be the learning of worth....

Such criteria should include, for example, (1) aspects of science that students will very likely use in a relatively short time in their daily lives outside of school; and (2) aspects of natural phenomena that exemplify easily and well to the students the excitement, novelty, and power of scientific knowledge and explanation.

At a recent international curriculum workshop in Cyprus these two criteria were used to spell out a skeletal content for a quite new sort of science curriculum. They were found to be logically applicable in a range of broad topic areas such as senses and measurement; the human body; health, nutrition,

(Continued on p. 424)

(Continued from p. 423)
and sanitation; food; ecology; resources; population; pollution; and use of energy.

A fundamental difference between the sort of science education (and hence curriculums) that we have had hitherto, and what may be needed for a genuine Science for All, is the fact that the "All" must be thought of as existing outside of science. In other words, science is an institutionalized part of all our societies in very definite and varied ways. On the other hand, even in the most highly technical, scientifically advanced societies no more than 20 percent of the population could be even remotely identified with the institutionalized part. The remaining 80 percent are, and for their lives will be, outside of science in this sense.

It is this sense of "outside of science" that I think we must understand and translate into curriculum terms if Science for All is to succeed from our present opportunities. ☐

[91]Peter J. Fensham, "Science for All," *Education leadership*, January, 1987, pp. 18–23. Reprinted with permission of the Association for Supervision and Curriculum Development. Copyright 1987 by ASCD. All rights reserved.

Encouraging Girls in Science Courses and Careers
by Jane Butler Kahle[92]

In the United States women comprise approximately 50 percent of the work force, yet only 9 percent are employed as scientists and engineers. Factors contributing to this situation have been analyzed in research studies. Explanations have ranged from differences in spatial ability related to a sex-linked gene to differences in early childhood toys and games. One study reported a dramatic decline in positive attitudes toward science as girls mature. The authors attribute this

decline to startling inequities in the number of science activities experienced by males and females in elementary and secondary classrooms. In addition, the analysis of the results from the National Assessment of Educational Progress science study indicate that girls continue to score below the national mean on all science achievement items and to express negative attitudes toward science. Although societal, educational, and personal factors are all involved, differences within the science classroom may be a contributing factor to low interest of women in science and scientific careers.

However, some girls like science and continue to study science. In order to determine what motivates these girls to pursue science courses and careers, a group of researchers conducted nationwide surveys to identify teachers who have motivated high school girls to continue in science. In addition to assessing instructional techniques, classroom climate, and teacher-student interactions, a selected sample of students (former and current) responded to questionnaires which assessed attitudes, intellectual and socio-cultural variables.

Two types of research, observational and survey, were used to gather data for this project. The case studies, which were the observational parts of this project, provided information about the student-teacher and student-student interactions. Case studies are limited in the extent to which they may produce generalizations applicable to other situations. Therefore, they were supplemented with survey data, describing the abilities, activities, and aspirations of the involved students and teachers. These research efforts led to the following conclusions:

Teachers who successfully *encourage* girls in science:

- Maintain well-equipped, organized, and perceptually stimulating classrooms.
- Are supported in their teaching activities by the parents of their students and are respected by current and former students.
- Use non-sexist language and examples and include information on women scientists.
- Use laboratories, discussions, and weekly quizzes as their primary modes of instruction and supplement those activities with field trips and guest speakers.
- Stress creativity and basic skills and provide career information.

Factors which discourage girls in science:

- High school counselors who do not encourage further courses in science and mathematics.
- Lack of information about science-related career opportunities and their prerequisites.
- Sex-stereotyped views of science and scientists which are projected by texts, media, and many adults.
- Lack of development of spatial ability skills (which could be fostered in shop and mechanical drawing classes).
- Fewer experiences with science activities and equipment which are stereotyped as masculine (mechanics, electricity, astronomy).

The teachers, both male and female, who were successful in motivating girls to continue to study science practiced "directed intervention." That is, girls were asked to assist with demonstrations; were required to perform, not merely record, in the laboratories; and were encouraged to participate in science-related field trips. In addition, teachers stressed the utility of math and science for future careers.

Both male and female students in the schools identified as "positive toward girls in science" were ques-

(Continued on p. 425)

(Continued from p. 424)
tioned about their attitudes toward science and science careers. When compared with a national sample, the students in these schools had a much more positive outlook. This difference was especially pronounced among girls. When asked how frequently they like to attend science class, 67 percent of the girls responded "often," compared with 32 percent of the girls in the national sample. And when asked if they would like to pursue a science-related job, 65 percent of the girls said "yes," compared with 32 percent of the girls in the national sample.

This research suggests that teaching styles and other school-related factors are important in encouraging girls as well as boys to continue in science courses and careers. The path to a scientific career begins in high school and requires skilled and sensitive teachers. This research identified the following "Do's" and "Don't's" for teachers who want to foster equity in science classrooms.

DO
- Use laboratory and discussion activities
- Provide career information
- Directly involve girls in science activities
- Provide informal academic counseling
- Demonstrate unisex treatment in science classrooms

DON'T
- Use sexist humor
- Use sex-stereotyped examples
- Distribute sexist classroom materials
- Allow boys to dominate discussions or activities
- Allow girls to passively resist ☐

[92]Jane Butler Kahle, "Encouraging Girls in Science Courses and Careers," *Research Matters...To the Science Teacher* Monograph Series, National Association for Research in Science Teaching, Cincinnati, Ohio. Used with permission.

Each Student Is Gifted
by William D. Romey and Mary L. Hibert[93]

Some people clearly are gifted; they do some things better and more easily than other people. Of course no one does everything better than everyone else. Some people, however, manage to do many things very well. The Olympic athletes who compete in the decathlon and pentathlon show exceptional skills in several sports. Yet their scores in any one of those sports generally lag behind those of frontrunners in the individual sports. Even being a jack-of-all-trades is a "special talent."

Giftedness cannot be defined in a blanket way for the population at large. IQ tests, standardized tests, and statistical treatments may be convenient for sorting people into groups and avoiding the appearance of chaos and infinite individuality, but they are false indicators. They accomplish more harm than good both for the individual (especially if he or she is at the lower end of the distribution) and ultimately for society. In general, the only people they really benefit are those in personnel departments and school admissions offices.

Teachers must assume that each student who comes their way, no matter what age or grade level, is gifted or talented in some way. Whether we are teachers in a local Head Start program or in the most elite of graduate schools, we must not ask, "Is this student gifted?" but rather, "In what areas is this student gifted, and how can she or he use the areas of my 'course' to express that giftedness, increase the vitality of the classroom environment, and work toward personal growth and maximization of individual potential?"

Teachers who espouse this view of giftedness can boost students' self-concepts. Rosenthal and Jacobson (*Pygmalion in the Classroom: Teacher Expectation and Pupils' Intellectual Development*, New York: Holt, Rinehart and Winston, 1968) have shown that students tend to behave according to the expectations of the people around them. When teachers treat students as "gifted," the youngsters tend to behave like gifted students. It is equally clear that students treated as stupid will behave that way.

A student may appear to be gifted mainly because she or he has already had practice in that specific area. Another student who initially seems less apt may, after becoming familiar with the task or area, perform better than the first student. It is a mistake to reward the first and take no notice of the latter, because without encouragement the late-bloomer may never exert the energies necessary to develop that particular potential.

The development of any part of a person's potential depends on two basic facts: aptitude and motivation. I believe that motivation is by far the more important. A third factor, the environment, is perhaps the equal of the other two. The environment can help the student by providing freedom to develop in a given direction and by offering appropriate encouragement, assistance, and necessary resources. However, one must be watchful for conditions in the environment that push children further in a direction that they have temporarily or permanently abandoned, because failure to acknowledge changes in children's motivation can have a serious negative impact on their growth. ☐

[93]William D. Romey and Mary L. Hibert, *Teaching the Gifted and Talented in the Science Classroom* (Washington, D.C.: National Education Association, 1988), pp. 11–13. Used with permission.

Problems and Extensions

- Working in a small group, discuss each of the following statements in terms of (a) What are the values and assumptions implicit in the statement? (b) What kinds of power relations are at work here? and (c) What are the implications for practice?
- It is absolutely natural for boys to behave in aggressive, "macho" ways and for girls to be quiet and passive. Men and women are supposed to complement one another and we should not try to change that fact. It is all down to our genes!
- If women teachers have trouble with discipline with boys that is their problem!

- Don't misunderstand me. I'm against sexism like the next person. But I think discussion about sexism in school takes time away from the really important thing—and that's getting kids through the course to take exams, isn't it?[94]
- Find out who the following scientists were, and what their contributions were to science. Design a science lesson using one, a few, or all of these scientists.

Benjamin Banneker
Vilhelm Bjerknes
Elizabeth Blackwell
Sofie Brahe
Charles Richard Drew
Caroline Herschel
Charles Steinmetz

- Write to several organizations serving disabled and exceptional students and report on their services and publications.
- What actions would you take if a student was placed in your class who was wheelchair bound?
- If you were a school board member, what actions would you recommend that your school district take to ensure that all females, minorities, and disabled enrolled in your school district aspire to excellence in science, mathematics and engineering?

[94]Based on Sue Askew and Carol Ross, *Boys Don't Cry* (Milton Keynes, England: Open University Press, 1988), pp. 98–99

RESOURCES

Alic, Margaret. *Hypatia's Heritage: A History of Women in Science from Antiquity through the Nineteenth Century.* Boston: Beacon, 1986.

Hypatia of Alexandria is one of a countless number of women scientists that Margaret Alic rediscovers for us to learn about. This book describes the work of women whose work was left out of history books, whose work was suppressed or stolen, and whose achievements in mathematics, science, and medicine have been denied.

Askew, Sue, and Carol Ross. *Boys Don't Cry: Boys and Sexism in Education.* Milton Keynes, England: Open University Press, 1988.

This explores the authors' work in dealing with the effects that socialization have on boys, and subsequent behavior in classrooms. The book provides an analysis of the ways in which schools may unintentionally reinforce and perpetuate certain aspects of "masculinity" which operate against boys' best interests.

Banks, James A. *Teaching Strategies for Ethnic Studies.* Boston: Allyn & Bacon, 1987.

This book is designed to help present and future teachers obtain the content, strategies, concepts and resources needed to teach comparative ethnic studies, and to integrate ethnic content into the total school curriculum.

Brandwein, Paul F., and A. Harry Passow. *Gifted Young in Science.* Washington, D.C.: National Science Teachers Association, 1988.

This book is a collection of papers by 34 authors probing the philosophy, psychology, and methodologies of gifted and talented science education. A cautionary note is made that we do not know who among the young during their school years will turn to science as their life work, nor do we know enough about learning environments to nourish the initiative to pursue science.

Cole, Michael, and Peg Griffin. *Contextual Factors in Education: Improving Science and Mathematics Education for Minorities and Women.* Madison: Wisconsin Center for Education Research, 1987.

This book describes the implications of research on the development of constructive environments for women and minorities, especially in the content areas of mathematics and science. The authors especially favor cooperative learning and "new technologies" as ways of providing contexts for increasing academic achievement and positive attitudes.

Farnham-Diggory, Sylvia. *Learning Disabilities.* London: Fontana/Open Books, 1981.

This book shows how the fields of cognitive psychology and neurophysiology have helped educators and psychologists understand learning disabilities related to essential processes including reading, writing, spelling, drawing, and calculating.

Gornick, Vivian. *Women in Science.* New York: Simon & Schuster, 1990.

This book is an excellent complement to Alic's *Hypatia's Heritage* because the author writes more than 100 brief vignettes about the lives of contemporary women scientists.

Hofman, Helenmarie H., and Kenneth S. Ricker. *Science Education and the Physically Handicapped.* Washington, D.C.: National Science Teachers Association, 1979.

This sourcebook contains articles and papers giving an overview of the field of science and the handicapped. The authors make suggestions for mainstreaming disabled students, how to work with physically disabled students, and resources.

Humphreys, Sheila A. *Women and Minorities in Science: Strategies for Increasing Participation.* Boulder, Colo.: Westview, 1982.

This book surveys current levels of participation in science by women and minorities and identifies barriers to their participation, and then describes a wide range of intervention programs, including teacher training and career awareness programs.

Kahle, Jane Butler. *Women in Science: A Report from the Field.* Philadelphia: Falmer, 1985.

Jane Butler Kahle, one of the leading researchers on the topic of women in science, has put together the work of other researchers dealing with gender issues in science teaching. The book includes a historical perspective on gender issues, as well as discussion of topics such as minority women (conquering both sexism and racism) and retention of girls in science (case studies of secondary teachers).

Keller, E. C., Jr. et. al. *Teaching the Physically Disabled in the Mainstream Science Class at the Secondary and College Levels.* Morgantown, W.Va.: Printech, 1983.

This resource book is the product of an NSF project aimed at improving the methods of teaching the physically disabled student in the mainstream science class.

The book presents types of mitigative strategies designed to be used with visually, hearing, and motor-orthopedically disabled students.

Kelly, Alison, ed. *Science for Girls?* Milton Keynes: England: Open University Press, 1987.

In this collection of articles, the authors reflect on recent trends in thinking about girls and science education. Topics include why girls don't do science; teachers' views about the importance of science for boys and girls; and increasing the participation and achievement of girls and women in mathematics, science, and engineering.

Lehr, Judy Brown, and Hazel Wiggins Harris. *At-Risk, Low-Achieving Students in the Classroom*. Washington, D.C.: National Education Association, 1988.

This book describes who the at-risk students are in our schools, suggests ways to organize the classroom environment for at-risk students, and how to involve them in learning.

Presseisen, Barbara Z. *At-Risk Students and Thinking: Perspectives from Research*. Washington, D.C.: National Education Association, 1988.

This book raises questions about educational interventions for at-risk students by focusing on thinking strategies for these students. The chapters in the book include topics such as an overview of the at-risk student, thinking success for at-risk students, and teaching strategies to help teachers work with at-risk students.

Ramsey, Patricia G, Edwina Battle Vold, and Leslie R. Williams. *Multicultural Education: A Source Book*. New York: Garland, 1989.

This book contains essays and annotations on a number of issues related to multicultural education, including the evolution of multicultural education, ethnic diversity and children's learning, multicultural programs, curricula, and strategies, and future directions in multicultural education.

Romey, William D., and Mary L. Hibert. *Teaching the Gifted and Talented in the Science Classroom*. Washing-ton, D.C.: National Educational Association, 1988.

Bill Romey, who is a scientist and science educator, presents a view of the gifted that challenges teachers to look for "the gifted in all students." He presents strategies for self-reflection, ways of presenting science to the gifted, and ideas about creating learning environments for gifted students.

Rosser, Sue V. *Female-Friendly Science: Applying Women's Studies Methods and Theories to Attract Students*. New York: Pergamon, 1990.

This book includes chapters on feminist theories and methods, women's ways of knowing, sexism in textbooks, and warming up the classroom climate for women. It addresses the issue of women and science and provides a scholarly analysis of the problem and offers practical solutions.

Skolnick, Joan, Carol Langbort, and Lucille Day. *How to Encourage Girls in Math and Science*. Englewood Cliffs, N.J.: Prentice-Hall, 1982.

This book not only reviews the research on issues related to women in science but provides a series of practical math and science activities designed to help girls increase their confidence in problem solving.

Stearner, Phyllis. *Able Scientists—Disabled Persons: Biographical Sketches Illustrating Careers in the Sciences for Able Disabled Students*. Oakbrook, Ill.: John Ricila Associates, 1984.

This book contains the stories of over thirty disabled scientists who share their life experiences. The accounts serve as examples of how disabled persons, who aspired to become scientists, achieved their goals.

Williams, Robert L. *Cross-Cultural Education: Teaching Toward a Planetary Perspective*. Washington, D.C.: National Education Association, 1977.

One of the earlier book in the field, this book gives not only a multicultural rationale but also includes a global rationale for education.

SCIENCE EDUCATION PROFESSIONAL SOCIETIES

American Association for the Advancement of Science
1333 H Street NW
Washington, DC 20005

American Association of Physics Teachers
Department of Physics and Astronomy
University of Maryland
College Park, MN 20740

American Chemical Society
1155 16th Street NW
Washington, DC 20036

Association for the Education of Teachers in Science
c/o Bill Baird
5040 Haley Center
Auburn University
Auburn, AL 36849

Association for Multicultural Science Education
2020 K Street, NW, Suite 800
Washington, DC 20006

National Association for Research in Science Teaching
c/o Dr. Glenn Markle
401 Teacher College
University of Cincinnati
Cincinnati, Ohio 45221

National Association of Biology Teachers
11250 Roger Beacon Drive #19
Reston, VA 22090

National Council of Teachers of Mathematics
1906 Association Drive
Reston, VA 22091

National Center for Earth Science Education
American Geological Institute
4220 King Street
Alexandria, VA 22302

Native American Science Education Association
1333 H Street, NW
Washington, DC 20005

National Earth Science Teachers Association
c/o Harold Stonehouse
Department of Geological Sciences
Michigan State University
East Lansing, MI 48824

National Science Teachers Association
1742 Connecticut Avenue, NW
Washington, DC 20009

School Science and Math Association
126 Life Science Building
Bowling Green State University
Bowling Green, OH 43403

Science for the Handicapped Association
Science Center
Moorhead State University
Moorhead, MN 56563

B

SOURCES OF FREE AND INEXPENSIVE MATERIALS

Earth Science

American Institute of Aeronautics and Astronautics
1290 Avenue of the Americas
New York, NY 10019

Astrophysical Observatory
60 Garden Street
Cambridge, MA 02138

McDonald Observatory
RLM 15.308
University of Texas
Department of Astronomy
Austin, TX 78712

National Aeronautics and Space Administration
(NASA)
Washington, DC 20546

National Air and Space Museum
Smithsonian Institute
Washington, DC 20560

National Oceanic and Atmospheric Administration
Office of Coastal Zone Management
3300 Whitehaven Street, NW
Washington, DC 20235

U.S. Department of Commerce
National Weather Service
610 Hardesty Street
Kansas City, MO 64124

U.S. Department of the Interior
Geological Survey
Reston, VA 22092

Smithsonian Astrophysical Observatory
Publications Department
60 Garden Street
Cambridge, MA 02138

Environmental Education and S-T-S

American Forest Institute
1619 Massachusetts Avenue, NW
Washington, DC 20036

American Gas Association, Inc.
Educational Services
1515 Witson Boulevard
Arlington, VA 22209

American Museum of Atomic Energy
Oak Ridge Associated Universities
P.O. Box 177
Oak Ridge, TN 37830

American Petroleum Institute
1271 Avenue of the Americas
New York, NY 10020

Chevron Chemical Company
P.O. Box 3744
San Francisco, CA 94119

Council on Family Health
633 Third Street
New York, NY 10017

Johnson & Johnson Products, Inc.
Consumer Services
501 George Street
New Brunswick, NJ 08903

National Coal Association
Education Division-Coal Building
1130 17th Street, N.W.
Washington, DC 20036

Shell Oil Company
Public Relations Department
P.O. Box 2463
Houston, TX 7701

Soil Conservation Society of America
7515 N.W. Ankeny Road
Ankeny, IA 50021

U.S. Department of Agriculture
Forest Service
P.O. Box 1963
Washington, D.C. 20013

U.S. Environmental Protection Agency
401 M Street, S.W.
Washington, DC 20460

Life Science

American Lung Association
1740 Broadway
New York, NY 10019

American Dental Association
211 East Chicago Avenue
Chicago, IL 60611

American Humane Association
P.O. Box 1266
Denver, CO 80201

Animal Protection Institute of America
5894 South Land Park Drive
P.O. Box 22505
Sacramento, CA 95822

Animal Welfare Institute
P.O. Box 3650
Washington, DC 20007

Florida Citrus Commission
Department of Citrus
State of Florida
P.O. Box 148
Lakeland, FL 33802

National Institutes of Health
Department of Health & Human Services
Bethesda, MD 20892

National Wildlife Federation
1413 16th Street, N.W.
Washington, DC 20036

U.S. Department of the Interior
Fish and Wildlife Service
Bureau of Sport Fisheries and Wildlife
Washington, DC 20240

Physical Sciences

Allied Chemical Corporation
P.O. Box 2245R
Morristown, NJ 07960

Aluminum Association, Inc.
Manager, Educational Services
818 Connecticut Ave. N.W.
Washington, DC 20006

American Chemical Society
Career Services
1155 16th Spout. N.W.
Washington, DC 20036

Chemical Manufacturers Association
2501 M Street N.W.
Washington, DC 20036

Edison Electric Institute
1111 19th Street, N.W.
Washington, DC 20036

General Electric Company
Educational Communications
Fairfield, CT 06431

Thomas Alva Edison Foundation
143 Cambridge Office Plaza
18280 West Ten Mile Road
Southfield, MI 48075

U.S. Department of Energy
P.O. Box 62
Oak Ridge, TN 37830

Note: *NSTA Reports!*, a monthly newsletter of the National Science Teachers Association, devotes several pages to free and inexpensive science teaching materials in a column entitled "Spend a Little, Gain a Lot."

C

SCIENCE EQUIPMENT SUPPLIERS

Carolina Biological Supply Co.
2700 York Road
Burlington, NC 27215

Center for Multisensory Learning
Lawrence Hall of Science
University of California
Berkeley, CA 94720

Connecticut Valley Biological Supply Co.
82 Valley Road, P.O. Box 326
Southhampton, MA 01073

Creative Teaching Associates
P.O. Box 7766
Fresno, CA 93747

Delta Education, Inc.
P.O. Box 915
Hudson, NH 03051

Edmund Scientific Co.
101 E. Gloucester Pike
Barrington, NJ 08007

Fisher Scientific Educational Materials Division
4901 W. LeMoyne Street
Chicago, IL 60651

Finn Scientific, Inc.
131 Finn Street
P.O. Box 219
Batavia, IL 65010

LEGO Educational/LEGO Systems, Inc.
555 Taylor Road
Enfield, CT 06082

National Geographic Society Educational Materials
17th & M Streets, N.W.
Washington, DC 20036

Ohaus Corporation
29 Hanover Road
Florham Park, NJ 07932

Science Kit and Boreal Laboratories
777 East Park Drive
Tonawanda, NY 14150

Wards National Science Establishments, Inc.
5100 West Henrietta Rd.
P.O. Box 92912
Rochester, NY 14692

APPENDIX

COMPUTER AND
SOFTWARE SUPPLIERS

American Association of Physics Teachers
5112 Berwyn Road
College Park, MD 20740

Apple Computer, Inc.
20525 Mariani Ave.
Cupertino, CA 95014

Broderbund Software
17 Paul Drive
San Rafael, CA 94903

EduTech
1927 Culver Road
Rochester, NY 14609

Eureka!
Lawrence Hall of Science
University of California
Berkeley, CA 94720

IBM
DEPARTMENT CF1
101 Paragon Drive
Montvale, NJ 07645

MECC
3490 Lexington Ave. North
St. Paul, MN 55126

Scholastic Software
730 Broadway
New York, NY 10003

Sunburst Communications
39 Washington Ave.
Pleasantville, NY 10570

NAME INDEX

SUBJECT INDEX

G

H

I